Lecture Notes in Computational Science and Engineering

107

Editors:

Timothy J. Barth
Michael Griebel
David E. Keyes
Risto M. Nieminen
Dirk Roose
Tamar Schlick

More information about this series at http://www.springer.com/series/3527

Bert Jüttler • Bernd Simeon
Editors

Isogeometric Analysis and Applications 2014

Springer

Editors
Bert Jüttler
Institut für Angewandte Geometrie
Johannes Kepler Universität Linz
Linz, Austria

Bernd Simeon
Fachbereich Mathematik
Technische Universität Kaiserslautern
Kaiserslautern, Germany

ISSN 1439-7358 ISSN 2197-7100 (electronic)
Lecture Notes in Computational Science and Engineering
ISBN 978-3-319-23314-7 ISBN 978-3-319-23315-4 (eBook)
DOI 10.1007/978-3-319-23315-4

Library of Congress Control Number: 2015959791

Springer Cham Heidelberg New York Dordrecht London

Cover figure: Isogeometric simulation using a trivariate multi-patch NURBS model (from Pauley, Nguyen, Mayer, Špeh, Weeger, Jüttler, pp. 1–21)

Printed on acid-free paper

Springer International Publishing AG Switzerland is part of Springer Science+Business Media (www.springer.com)

Foreword for the Proceedings Volume

The very comfortable conference venue in Annweiler am Trifels, Germany, facing the Trifels Castle, perched on the peak of the Sonnenberg, high above the Queich valley, provided a beautiful and inspirational setting for **IGAA 2014, Isogeometric Analysis and Applications**, April 7–10, 2014. Trifels Castle is where Richard the Lionheart, King of England, was imprisoned for three weeks in the year 1193. History was made then and it was made again at **IGAA 2014**, now one of several very important, focused conferences that have been held to assess the state-of-the-art in the emerging field of Isogeometric Analysis. This field began with a single paper published in the 2005 October 1 issue of the international journal *Computer Methods in Applied Mechanics and Engineering*, entitled "Isogeometric analysis: CAD, finite elements, NURBS, exact geometry and mesh refinement." Almost immediately, Isogeometric Analysis attracted considerable attention in the research community, and at the present time it is enjoying explosive growth, as measured by the number of papers published on the topic and the citations to them in the literature. Currently, it represents perhaps the hottest topic within both Computational Mechanics and Computer-Aided Geometrical Modeling, two historically separate disciplines that have joined forces and embraced the vision of Isogeometric Analysis, that is, to reconstitute Computational Mechanics within the framework provided by Computational Geometry, thereby simplifying the analysis model development process and integrating engineering design and analysis into a unified framework. As research activity in Isogeometric Analysis has ramped up, industrial and commercial interest has followed, and we are now seeing considerable growth in these sectors as well. The promise of Isogeometric Analysis, that is, to improve the overall efficiency of the engineering product development process, is the driving force.

An interesting aspect of Isogeometric Analysis is the breadth of activities within it. One can pursue it from a very mathematical perspective, or a very practical engineering perspective, or anywhere in between. There seems to be something in it for everyone! It is a field that suits my broad scientific and engineering interests. From a mathematician's point of view, I may be an engineer. From an engineer's point of view, I may even be a mathematician. So be it. I would claim I am a little bit of both, but I can also claim to be a Mechanical Design Engineer, at least I was 50 years ago! That was my first job at Grumman Aerospace Corporation (now Northrop Grumman) after obtaining my undergraduate degree in Mechanical Engineering. I enjoyed it and learned a lot about engineering in the real world, but it was not long before I gravitated toward engineering analysis. Design and analysis are the cornerstones of Isogeometric Analysis, and I feel I have practical experience with both. **IGAA 2014**, although including a little bit of everything, was somewhat weighted toward the mathematical side of the spectrum, reflecting the interests of the majority of the participants. Nevertheless, my invited talk at the conference was meant to be an overview, and I entitled it "Isogeometric Analysis: Where we are and where we are going." In it, I also tried to identify a few subtopics that I felt were enjoying increased attention, particularly ones that might have substantial impact on the field going forward. I think some of the thoughts presented therein are worth recalling in this Foreword to the book of selected papers emanating from the conference.

Perhaps a main theme that needs to be highlighted is the analysis suitability of design models. Certainly, T-splines and LR-splines are having significant impact due to their watertight and unstructured qualities. An additional theme is performing analysis directly on trimmed NURBS files, which heretofore has been considered unsuitable for analysis. This is important because trimmed NURBS are ubiquitous in contemporary engineering design and give no sign of going away, at least anytime soon. Hierarchically refined NURBS are now reaching maturity and are becoming a standard tool in the design-through-analysis arsenal.

Another theme that has begun to manifest itself is the surprising efficiency of Isogeometric Analysis in practical applications. This is facilitated by the significant gain in accuracy per degree of freedom compared with classical C^0-continuous finite element methods and decreased computational effort engendered by reduced integration and collocation techniques. Collocation techniques, which utilize the strong, rather than weak, form of the governing equations, are perhaps the ultimate reduced integration method. This topic has never had much success in the context of C^0-continuous finite element methods, but is undergoing a renaissance in Isogeometric Analysis due the smoothness of spline basis functions. Efficient solution algorithms that take advantage of the structure of Isogeometric Analysis matrix equations have also contributed to the gains in efficiency, as well as adaptive refinement strategies driven by a posteriori error estimates.

Other areas in which exciting developments are taking place include "thin" shell theory, which has also benefitted from smooth basis functions *without* derivative degrees of freedom. This has given new life to the classical Kirchhoff-Love shell theory, which requires C^1-continuous basis functions in the classical Galerkin weak formulation. These are readily available in Isogeometric Analysis, with the added benefit of the elimination of rotational degrees of freedom. Further developments have even generalized this approach to the Reissner-Mindlin "thick" shell theory and at the same time have eliminated shear locking entirely, independent of basis functions. These indeed are significant developments and are expected to have a profound effect on structural analysis in the coming years.

Smooth, higher-order accurate basis functions and precise geometrical modeling have also created a new beginning in boundary element analysis, whose potential has never before been attained due to inaccurate, faceted geometrical approximations and low-order, C^0-continuous basis functions. Several complete design-through-analysis procedures have been described using the Isogeometric Analysis version of boundary integral formulations. These have included shape optimization directly in the analysis-suitable CAD file, an enormous advantage in streamlining the design-through-analysis process. Shape and topology optimization are likewise general areas of significant Isogeometric Analysis activity.

The breadth of practical applications of Isogeometric Analysis is staggering, and some of applications to complex nonlinear problems and fluid-structure interaction are truly impressive. All this is being buttressed by many important mathematical studies involving complementary functional- and spectral-analysis approaches.

The papers in this volume only represent a small sampling of the many excellent presentations at **IGAA 2014**, but give one a glimpse of the significant recent progress in Isogeometric Analysis, which is rapidly becoming a mainstream analysis technology and a new paradigm for geometric design.

Austin, USA
June 23, 2015

Thomas J.R. Hughes

Preface

This book contains a selection of articles that were presented at the second workshop on *Isogeometric Analysis and Applications – IGAA 2014*, held in Annweiler am Trifels, Germany, from April 7 to 10, 2014. The emerging field of Isogeometric Analysis lives from the interaction of geometric modeling and numerical analysis, and the workshop was a great success in stimulating the exchange of ideas between these two fields. We are glad to see that, over the last years, this interaction has resulted in a series of groundbreaking scientific results, some of which are covered by the articles in this volume. The articles can be structured in four groups that cover the most important branches of recent developments in Isogeometric Analysis.

The first group of four papers deals with various aspects of multi-patch methods in Isogeometric Analysis.

In their paper entitled "Multipatch Discontinuous Galerkin Isogeometric Analysis," U. Langer et al. present Galerkin-type methods that allow for discontinuities across the subdomain boundaries. The required interface conditions are weakly imposed by the terms associated with the boundary of the subdomains. The construction and the corresponding discretization error analysis of the resulting multi-patch schemes are given for heterogeneous diffusion model problems in 2D and 3D domains as well as on open and closed surfaces. The theoretical results are confirmed by extensive numerical experiments.

Next, E. Brivadis and her coauthors study the influence of quadrature errors on isogeometric mortar methods. After reviewing the recent mathematical theory, they investigate the effect of the errors caused by using numerical quadrature for the coupling integrals. Two approaches for designing quadrature rules and their impact on the convergence rate are analyzed.

M. Pauley et al. present a pipeline for the conversion of 3D models into a form suitable for Isogeometric Analysis. Given a boundary represented 3D model, they establish a chain of processes that result in a collection of volumetric NURBS patches. The suitability of the result for Isogeometric Analysis is also exemplified by performing stress simulations.

The fourth paper, entitled "Domain Decomposition Methods and Kirchhoff-Love Shell Multipatch Coupling in Isogeometric Analysis" and contributed by

A. Apostolatos et al., presents and compares several domain decomposition methods, including penalty and Lagrange multiplier methods. The methods, which are extended to account for geometrically nonlinear problems, provide highly accurate results, thus enabling the treatment of Kirchhoff-Love shells on multiple patches.

The second group of papers comprises three articles which are devoted to shell analysis and to T-splines. The last paper of the previous group also contributes to this topic.

C. Adam and coauthors propose a reduced shell element for Reissner-Mindlin geometric nonlinear analysis within the context of T-spline analysis. More precisely, the nonuniform Gauss-Legendre and patchwise reduced integrations for quadratic shape functions are both presented and compared to the standard full Gauss-Legendre scheme. The authors evaluate the effects of mesh distortion and local refinement, using both full and reduced numerical quadratures.

In their paper entitled "Multiresolution shape and topology optimisation with subdivision surfaces," F. Cirak and K. Bandara review the multiresolution shape optimisation technique based on subdivision surfaces and present its application to elastic solids, electrostatic field equations, and thin shells. The geometry of the domain is described with subdivision surfaces, and different resolutions of the same surface are used for optimization and analysis. In their paper, the authors discretize elastic solids with the immersed finite element method, electrostatic field equations with the boundary element method, and thin shells with the subdivision finite element technique. In all situations there is no need to generate and to maintain an analysis-suitable volume discretization.

In the next paper on "Atom Simplification and Quality T-mesh Generation for Multi-resolution Biomolecular Surfaces" by T. Liao et al., the structure of biomolecules is first simplified using an error-bounded atom elimination method. An extended cross field-based parameterization method is then developed to adapt the parametric line spacings to different surface resolutions. From the parameterization results, adaptive and anisotropic T-meshes for the further T-spline surface construction are extracted. A gradient flow-based method is finally used to improve the T-mesh quality, with the anisotropy preserved in the quadrilateral elements.

The two papers in the next group address an important topic that is rarely discussed in detail in the literature, namely, algorithmic and implementation aspects in Isogeometric Analysis.

D. Fußeder and B. Simeon concentrate on these aspects in the context of isogeometric shape optimization. They present a self-contained treatment of gradient-based shape optimization with Isogeometric Analysis, focusing on issues such as the computation of shape gradients in an isogeometric formulation and how to update B-spline and NURBS geometries.

The other paper in this group, entitled "Effective Integration of Sophisticated Operators in Isogeometric Analysis with igatools," has been contributed by N. Cavallini et al. It discusses aspects of igatools, which is a newly released library for operator assembly in Isogeometric Analysis. It is shown that the design of the library makes it an effective tool in assembling isogeometric discretizations of sophisticated differential operators.

Finally, the two remaining papers address advanced topics from the theory of finite element methods and their application in Isogeometric Analysis.

S. Kleiss and S. Tomar present two-sided robust and sharp a posteriori error estimates in the isogeometric discretization of elliptic problems. These estimates, which are fully computable, provide robust, guaranteed, and sharp two-sided bounds of the exact error in the energy norm. The numerical realization and the quality of the computed error distribution are also covered.

The last paper of this volume, entitled "Multilevel Preconditioning for Variational Problems," which has been contributed by A. Kunoth, surveys the main ingredients for multilevel preconditioners in terms of higher-order B-splines for standard second- and fourth-order elliptic partial differential equations in variational form possessing smooth solutions. It addresses three types of multilevel preconditioners for which asymptotic optimality can be shown. Two of them, the so-called additive preconditioners, are specified for Isogeometric Analysis involving linear elliptic partial differential operators in terms of variants of the BPX preconditioner and wavelet preconditioners. The third class is formed by multiplicative preconditioners, specifically by multigrid methods. The performance of the different preconditioners is demonstrated by numerical examples.

Last but not least, the editors acknowledge the support from many people and organizations that has been crucial in organizing the workshop and in compiling this volume. In particular, we are indebted to the reviewers. Their comments have helped us greatly to identify the manuscripts suitable for publication in this book and moreover were useful for improving many of the papers substantially. We also wish to express our thanks to Springer Verlag for the constructive and supportive cooperation during the production of this book. Financial support from the *Fraunhofer Institute for Industrial Mathematics (ITWM)*, Kaiserslautern, from the *Center for Mathematical and Computational Modeling* $(CM)^2$ at Technische Universität Kaiserslautern, and from the program for equal opportunities at Technische Universität Kaiserslautern made it possible to attract participants from all over the world to the workshop. We are now looking forward to the next event in the workshop series on *Isogeometric Analysis and Applications*.

Linz, Austria
Kaiserslautern, Germany
July 2015

Bert Jüttler
Bernd Simeon

Contents

Multipatch Discontinuous Galerkin Isogeometric Analysis

Ulrich Langer, Angelos Mantzaflaris, Stephen E. Moore, and Ioannis Toulopoulos

Abstract Isogeometric Analysis (IgA) uses the same class of basis functions for both representing the geometry of the computational domain and approximating the solution of the boundary value problem under consideration. In practical applications, geometrical patches are used in order to get flexibility in the geometrical representation. This multi-patch representation corresponds to a decomposition of the computational domain into non-overlapping subdomains also called patches in the geometrical framework. We will present discontinuous Galerkin (dG) methods that allow for discontinuities across the subdomain (patch) boundaries. The required interface conditions are weakly imposed by the dG terms associated with the boundary of the subdomains. The construction and the corresponding discretization error analysis of such dG multi-patch IgA schemes is given for heterogeneous diffusion model problems in volumetric 2d and 3d domains as well as on open and closed surfaces. The theoretical results are confirmed by numerous numerical experiments which have been performed in G+SMO. The concept and the main features of the IgA library G+SMO are also described.

1 Introduction

Isogeometric Analysis (IgA), which was introduced by Hughes, Cottrell and Bazilevs in 2005, is a new discretization technology which uses the same class of basis functions for both representing the geometry of the computational domain and approximating the solution of problems modeled by Partial Differential Equations (PDEs), [23]. IgA uses the exact geometry in the class of Computer Aided Design (CAD) geometries, and thus geometrical errors introduced by approximation of the physical domain are eliminated. This feature is especially important in technical applications where the CAD geometry description is directly used in the production

---

U. Langer (✉) • A. Mantzaflaris • S.E. Moore • I. Toulopoulos
RICAM, Altenbergerstr. 69, A-4040 Linz, Austria
e-mail: ulrich.langer@ricam.oeaw.ac.at; ulanger@numa.uni-linz.ac.at; angelos.mantzaflaris@ricam.oeaw.ac.at; stephen.moore@ricam.oeaw.ac.at; ioannis.toulopoulos@ricam.oeaw.ac.at

B. Jüttler, B. Simeon (eds.), *Isogeometric Analysis and Applications 2014*, Lecture Notes in Computational Science and Engineering 107,
DOI 10.1007/978-3-319-23315-4_1

process. Usually, IgA uses basis functions like B-Splines and Non-Uniform Rational B-Splines (NURBS), which are standard in CAD, and have several advantages which make them suitable for efficient and accurate simulation, see [8]. The mathematical analysis of approximation properties, stability and discretization error estimates of NURBS spaces have been well studied in [5]. Furthermore, approximation error estimates due to mesh, polynomial degree and smoothness refinement have been obtained in [6], and likewise a hybrid method that combines a globally C^1-continuous, piecewise polynomial finite element basis with rational NURBS-mappings have also been considered in [27].

In practical applications, the computational domain $\Omega \subset \mathbb{R}^d$ $(d = 2, 3)$ is usually represented by multiple patches leading to non-matching meshes and thus to patch-wise non-conforming approximation spaces. In order to handle non-matching meshes and polynomial degrees across the patch interfaces, the discontinuous Galerkin (dG) technique that is now well established in the Finite Element Analysis (FEA) of different field problems is employed. Indeed, dG methods have been developed and analyzed for many applications including elliptic, parabolic and hyperbolic PDEs. The standard dG finite element methods use approximations that are discontinuous across the boundaries of every finite element of the triangulation. To achieve consistent, stable and accurate schemes, some conditions are prescribed on the inter-element boundaries, see, e.g., Rivière's monograph [35].

In this paper, we present and analyze multipatch dG IgA methods for solving elliptic PDEs in volumetric 2d or 3d computational domains as well as on open and closed surfaces. As model problems, we first consider diffusion problems of the form

$$-\nabla \cdot (\alpha \nabla u) = f \text{ in } \Omega, \text{ and } u = u_D \text{ on } \partial\Omega, \tag{1}$$

where Ω is a bounded Lipschitz domain in $\mathbb{R}^d$, $d = 2, 3$, with the boundary $\partial\Omega, f$ is a given source term, and u_D are given Dirichlet data. Neumann, Robin and mixed boundary conditions can easily be treated in the same dG IgA framework which we are going to analyze in this paper. The same is true for including a reaction term into the PDE (1). The diffusion coefficient α is assumed to be bounded from above and below by strictly positive constants. We allow α to be discontinuous across the inter-patch boundaries with possible large jumps across these interfaces. More precisely, the computational domain Ω is subdivided into a collection $\mathcal{T}_H(\Omega) := \{\Omega_i\}_{i=1}^N$ of non-overlapping subdomains (patches) $\Omega_1, \ldots, \Omega_N$, where the patches Ω_i are obtained by some NURBS mapping Φ_i from the parameter domain $\hat{\Omega} = (0,1)^d \subset \mathbb{R}^d$. In every subdomain, the problem is discretized under the IgA methodology. The dG IgA method is considered in a general case without imposing any matching grid conditions and without any continuity requirements for the discrete solution across the interfaces $F_{ij} = \partial\Omega_i \cap \partial\Omega_j$. Thus, on the interfaces, the dG technique ensures the communication of the solution between the adjacent subdomains. For simplicity, we assume that the coefficients are piecewise constant with respect to the decomposition $\mathcal{T}_H(\Omega_i)_{i=1}^N$. It is well known that solutions of elliptic problems with discontinuous coefficients are generally not smooth [20, 26], Thus, we cannot expect that the dG IgA provides convergence rates like in the smooth case.

Here, the reduced regularity of the solution is of primary concern. Following the techniques developed by Di Pietro and Ern [12, 13] for dG FEA, we present an error analysis for diffusion problems in 2d and 3d computational domains with solutions belonging to $W^{l,p}(\Omega_i)$, $l \geq 2$, $p \in (\max\{1, 2d/(d+2(l-1))\}, 2]$, $i = 1, \ldots, N$. In correspondence with this regularity, we show optimal convergence rates of the discretization error in the classical "broken" dG – norm. Due to the peculiarities of the dG IgA, the proofs are quite technical and can be found in the paper [30] written by two of the co-authors. Solutions of elliptic problems with low regularity can also appear in other cases. For example, geometric singular boundary points (re-entrant corners), points with change of the boundary conditions, and singular source terms can cause low-regularity solutions as well, see, e.g., [20]. Especially, in the IgA, we want to use the potential of the approximation properties of NURBS based on high-order polynomials. In this connection, the error estimates obtained for the low-regularity case can be very useful. Indeed, via mesh grading, that has been used in the FEA for a long time [3, 33], we can recover the full convergence rate corresponding to the underlying polynomial degree. We mention that in the literature other techniques have been proposed for solving two dimensional problems with singularities. For example, in [34] and [24], the original B-Spline finite dimensional space has been enriched by generating singular functions which resemble the types of the singularities of the problem. Also in [7], by studying the anisotropic character of the singularities of the problem, the one dimensional approximation properties of the B-Splines are generalized in order to produce anisotropic refined meshes in the regions of the singular points. Our error analysis is accompanied by a series of numerical tests which fully confirm our convergence rate estimates for the multipatch dG IgA in both the full and the low regularity cases as well as the recovering of the full rate by means of mesh grading techniques. We mention that the detailed proofs of the main theoretical results can be found in [30] and [28], whereas all numerical results are new. It is very clear that our analysis can easily be generalized to more general classes of elliptic boundary value problems like plane strain or stress linear elasticity problems or linear elasticity problems in 3d volumetric computational domains.

We will also consider multipatch dG IgA of diffusion problems of the form (1) on sufficiently smooth, open and closed surfaces Ω in $\mathbb{R}^3$, where the gradient ∇ must now be replaced by the surface gradient ∇_Ω, see, e.g., [18], and the patches Ω_i, into which Ω is decomposed, are now images of the parameter domain $\hat{\Omega} = (0,1)^2$ by the mapping $\Phi_i : \hat{\Omega} \subset \mathbb{R}^2 \rightarrow \Omega_i \subset \mathbb{R}^3$. The case of matching meshes was considered and analyzed in [29] by two of the co-authors, see also [36] for a similar work. It is clear that the results for non-matching spaces and for mesh grading presented here for the volumetric (2d and 3d) case can be generalized to diffusion problems on open and closed surfaces. Until now, the most popular numerical method for solving PDEs on surfaces is the surface Finite Element Method (FEM). The surface FEM was first applied to compute approximate solutions of the Laplace-Beltrami problem, where the finite element solution is constructed from the variational formulation of the surface PDEs in the finite element space that is living on the triangulated surface approximating the real surface [15]. This method has been extended to the parabolic equations fixed surfaces by Dziuk and Elliot [17]. To treat conservation laws on moving surfaces, Dziuk and Elliot proposed the evolving surface finite element methods, see [16]. Recently, Dziuk and Elliot have published the survey paper [18] which provides a comprehensive presentation of different finite element approaches to the solution of PDEs on surfaces with several applications. The first dG finite element scheme, which extends Dziuk's approach, was proposed by Dedner et al. [11] and has very recently been extended to adaptive dG surface FEM [10] and high-order dG finite element approximations on surfaces [2]. However, since all these approaches rely on the triangulated surface, they have an inherent geometrical error which becomes more complicated when approximating problems with complicated geometries. This drawback of the surface FEM can be overcome by the IgA technology, at least, in the class of CAD surfaces which can be represented exactly by splines or NURBS. Dede and Quateroni have introduced the surface IgA for fixed surfaces which can be represented by one patch [9]. They presented convincing numerical results for several PDE problems on open and closed surfaces. However, in many practical applications, it is not possible to represent the surface Ω with one patch. The surface multipatch dG IgA that allows us to use patch-wise different approximations spaces on non-matching meshes is then the natural choice.

The new paradigm of IgA brings challenges regarding the implementation. Even though the computational domain is partitioned into subdomains (patches) and elements (parts of the domain delimited by images of knot-lines or knot-planes), the information that is accessible by the data structure (parametric B-spline patches) is quite different than that of a classical finite element mesh (collection of triangles or simplices providing the triangulation of the domain). In this realm, existing finite element software libraries cannot easily be adapted to the isogeometric setting. Apart from the data structures used, another issue is the fact that FEA codes are focused on treating nodal shape function spaces, contrary to isogeometric function spaces. To provide a unified solution to the above (and many other) issues, we present the G+SMO[1] C++ library, which provides a unified, object-oriented development framework suitable to implement advanced isogeometric techniques, such as dG methods see [25].

The rest of the paper is organized as follows. In Sect. 2, we present and analyze the multipatch dG IgA for diffusion problems in volumetric 2d and 3d computational domains including the case of low-regularity solutions. We also study the mesh grading technology which allows us to recover the full convergence rates defined by the degree k of the underlying polynomials. The numerical analysis is accompanied by numerical experiments fully confirming the theoretical results. Section 3 is devoted to multipatch dG IgA of diffusion problems on open and closed surfaces. We present and discuss new numerical results including the case of jumping coefficients and non-matching meshes. The concept and the main features of the G+SMO library is presented in Sect. 4. All numerical experiments presented in this paper have been performed in G+SMO.

[1] *Geometry plus Simulation Modules*, http://www.gs.jku.at/gismo.

2 Multipatch dG IgA for PDEs in 3d Computational Domains

2.1 Multipatch dG IgA Discretization

The weak formulation of (1) reads as follows: find a function u from the Sobolev space $W^{1,2}(\Omega)$ such that $u = u_D$ on $\partial\Omega$ and satisfies the variational formulation

$$a(u, v) = l(v), \ \forall v \in W_0^{1,2}(\Omega) := \{v \in W^{1,2}(\Omega) : v = 0 \text{ on } \partial\Omega\}, \tag{2}$$

where the bilinear form $a(\cdot,\cdot)$ and the linear form $l(\cdot)$ are defined by the relations

$$a(u, v) = \int_\Omega \alpha \nabla u \cdot \nabla v \, dx \quad \text{and} \quad l(v) = \int_\Omega f v \, dx,$$

respectively. Beside Sobolev's Hilbert spaces $H^l(\Omega) = W^{l,2}(\Omega)$, we later also need Sobolev's Banach spaces $W^{l,p}(\Omega)$, $p \in [1,\infty)$, see, e.g., [1]. The existence and uniqueness of the solution u of problem (2) can be derived by Lax-Milgram's Lemma [19]. In order to apply the dG IgA methodology to problem (1), the domain Ω is subdivided into a union of subdomains $\mathscr{T}_H(\Omega) := \{\Omega_i\}_{i=1}^N$ such that

$$\bar{\Omega} = \bigcup_{i=1}^{N} \bar{\Omega}_i, \quad \text{with} \quad \Omega_i \cap \Omega_j = \emptyset, \text{ if } j \neq i. \tag{3}$$

As we mentioned in the introduction, the subdivision of Ω is assumed to be compatible with the discontinuities of α, i.e. they are constant in the interior of Ω_i, that is $\alpha|_{\Omega_i} := \alpha^{(i)}$, and their discontinuities appear only across the interfaces $F_{ij} = \partial\Omega_i \cap \partial\Omega_j$, cf., e.g., [12–14]. Throughout the paper, we will use the notation $a \sim b$ meaning that there are positive constants c and C such that $ca \leq b \leq Ca$.

As it is common in IgA, we assume a parametric domain $\hat{\Omega}$ of unit length, e.g., $\hat{\Omega} = (0,1)^d$. For any Ω_i, we associate d knot vectors $\Xi_n^{(i)}$, $n = 1, \ldots, d$, on $\hat{\Omega}$, which create a mesh $T_{h_i,\hat{\Omega}}^{(i)} = \{\hat{E}_m\}_{m=1}^{M_i}$, where $\hat{E}_m$ are the micro-elements, see details in [23]. We refer to $T_{h_i,\hat{\Omega}}^{(i)}$ as the *parametric mesh of* Ω_i. For every $\hat{E}_m \in T_{h_i,\hat{\Omega}}^{(i)}$, we denote by $h_{\hat{E}_m}$ *its diameter* and by $h_i = \max\{h_{\hat{E}_m}\}$ *the mesh size* of $T_{h_i,\hat{\Omega}}^{(i)}$. We assume the following properties for every $T_{h_i,\hat{\Omega}}^{(i)}$:

- quasi-uniformity: for every $\hat{E}_m \in T_{h_i,\hat{\Omega}}^{(i)}$ holds $h_i \sim h_{\hat{E}_m}$,
- for the micro-element edges $e_{\hat{E}_m} \subset \partial\hat{E}_m$ holds $h_{\hat{E}_m} \sim e_{\hat{E}_m}$.

On every $T^{(i)}_{h_i,\hat{\Omega}}$, we construct the finite dimensional space $\hat{\mathbb{B}}^{(i)}_{h_i}$ spanned by B-spline basis functions of degree k, see [23],

$$\hat{\mathbb{B}}^{(i)}_{h_i} = span\{\hat{B}^{(i)}_j(\hat{x})\}_{j=0}^{dim(\hat{\mathbb{B}}^{(i)}_{h_i})}, \tag{4}$$

where every basis function $\hat{B}^{(i)}_j(\hat{x})$ in (4) is derived by means of tensor products of one-dimensional B-spline basis functions, e.g.,

$$\hat{B}^{(i)}_j(\hat{x}) = \hat{B}^{(i)}_{j_1}(\hat{x}_1)\cdots\hat{B}^{(i)}_{j_d}(\hat{x}_d). \tag{5}$$

For simplicity, we assume that the basis functions of every $\hat{\mathbb{B}}^{(i)}_{h_i}, i = 1,\ldots,N$, are of the same degree k.

Every subdomain $\Omega_i \in \mathscr{T}_H(\Omega)$, $i = 1,\ldots,N$, is exactly represented through a parametrization (one-to-one mapping), cf. [23], having the form

$$\Phi_i : \hat{\Omega} \to \Omega_i, \quad \Phi_i(\hat{x}) = \sum_j C^{(i)}_j \hat{B}^{(i)}_j(\hat{x}) := x \in \Omega_i, \tag{6}$$

where $C^{(i)}_j$ are the control points and $\hat{x} = \Psi_i(x) := \Phi_i^{-1}(x)$. Using Φ_i, we construct a mesh $T^{(i)}_{h_i,\Omega_i} = \{E_m\}_{m=1}^{M_i}$ for every Ω_i, whose vertices are the images of the vertices of the corresponding mesh $T^{(i)}_{h_i,\hat{\Omega}}$ through Φ_i. If $h_{\Omega_i} = \max\{h_{E_m} : E_m \in T^{(i)}_{h_i,\Omega_i}\}$ is the subdomain Ω_i mesh size, then, based on definition (6) of Φ_i, we have the equivalence relation $h_i \sim h_{\Omega_i}$.

The mesh of Ω is considered to be $T_h(\Omega) = \bigcup_{i=1}^N T^{(i)}_{h_i,\Omega_i}$, where we note that there are no matching mesh requirements on the interior interfaces $F_{ij} = \partial\Omega_i \cap \partial\Omega_j, i \neq j$. For the sake of brevity in our notations, the interior faces of the boundary of the subdomains are denoted by $\mathscr{F}_I$ and the collection of the faces that belong to $\partial\Omega$ by $\mathscr{F}_B$, e.g. $F \in \mathscr{F}_B$ if there is a Ω_i such that $F = \partial\Omega_i \cap \partial\Omega$. We denote the set of all subdomain faces by $\mathscr{F}$.

Lastly, we define the B-spline space $\mathbb{B}_h(\mathscr{T}_H(\Omega)) = \mathbb{B}^{(1)}_{h_1} \times \ldots \times \mathbb{B}^{(N)}_{h_N}$ on Ω, where every $\mathbb{B}^{(i)}_{h_i}$ is defined on $T^{(i)}_{h_i,\Omega_i}$ as follows

$$\mathbb{B}^{(i)}_{h_i} := \{B^{(i)}_{h_i}|_{\Omega_i} : B^{(i)}_h(x) = \hat{B}^{(i)}_h \circ \Psi_i(x),\ \forall \hat{B}^{(i)}_h \in \hat{\mathbb{B}}^{(i)}_{h_i}\}. \tag{7}$$

We assume that the mappings Φ_i are regular in the sense that there exist positive constants c_m and c_M such that

$$c_m \leq |det(J_i(\hat{x}))| \leq c_M, \text{ for } i = 1,\ldots,N, \text{ for all } \hat{x} \in \hat{\Omega}, \tag{8}$$

where $J_i(\hat{x})$ denotes the Jacobian $\partial\Phi_i(\hat{x})/\partial(\hat{x})$ of the mapping Φ_i.

Now, for any $\hat{u} \in W^{m,p}(\hat{\Omega}), m \geq 0, p > 1$, we define the function

$$\mathscr{U}(x) = \hat{u}(\Psi_i(x)), \ x \in \Omega_i, \tag{9}$$

and the following relation holds true, see [30],

$$C_m \|\hat{u}\|_{W^{m,p}(\hat{\Omega})} \leq \|\mathscr{U}\|_{W^{m,p}(\Omega_i)} \leq C_M \|\hat{u}\|_{W^{m,p}(\hat{\Omega})}, \tag{10}$$

where the constants C_m and C_M depending on $C_m := C_m(\max_{m_0 \leq m}(\|D^{m_0}\Phi_i\|_\infty), \|det(\Psi_i^{'})\|_\infty)$ and $C_M := C_M(\max_{m_0 \leq m}(\|D^{m_0}\Psi_i\|_\infty), \|det(\Phi_i^{'})\|_\infty)$. The usefulness of inequalities (10) in the analysis is the following: every required relation can be proved in the parametric domain and then using (10) we can directly have the expression on the physical subdomain.

We use the B-spline spaces $\mathbb{B}_h^{(i)}$ defined in (7) for approximating the solution of (2) in every subdomain Ω_i. Continuity requirements for $\mathbb{B}_h(\mathscr{T}_H(\Omega))$ are not imposed on the interfaces F_{ij} of the subdomains, clearly $\mathbb{B}_h(\mathscr{T}_H(\Omega)) \subset L^2(\Omega)$ but $\mathbb{B}_h(\mathscr{T}_H(\Omega)) \nsubseteq W^{1,2}(\Omega)$. Thus, problem (2) is discretized by discontinuous Galerkin techniques on F_{ij}, see, e.g., [14]. Using the notation $v_h^{(i)} := v_h|_{\Omega_i}$, we define the average and the jump of v_h on $F_{ij} \in \mathscr{F}_I$, respectively, by

$$\{v_h\} := \frac{1}{2}(v_h^{(i)} + v_h^{(j)}), \quad \text{and} \quad [\![v_h]\!] := v_h^{(i)} - v_h^{(j)}, \tag{11}$$

and for $F_i \in \mathscr{F}_B$

$$\{v_h\} := v_h^{(i)}, \quad \text{and} \quad [\![v_h]\!] := v_h^{(i)}. \tag{12}$$

The dG-IgA scheme reads as follows: find $u_h \in \mathbb{B}_h(\mathscr{T}_H(\Omega))$ such that

$$a_h(u_h, v_h) = l(v_h) + p_D(u_D, v_h), \ \forall v_h \in \mathbb{B}_h(\mathscr{T}_H(\Omega)), \tag{13}$$

where

$$a_h(u_h, v_h) = \sum_{i=1}^{N} a_i(u_h, v_h) - \sum_{i=1}^{N} \Big(\frac{1}{2} s_i(u_h, v_h) - p_i(u_h, v_h)\Big), \tag{14}$$

with the bilinear forms

$$a_i(u_h, v_h) = \int_{\Omega_i} \alpha \nabla u_h \nabla v_h \, dx,$$

$$s_i(u_h, v_h) = \sum_{F_{ij}\subset\partial\Omega_i} \int_{F_{ij}} \{\alpha\nabla u_h\}\cdot \mathbf{n}_{F_{ij}} [\![v_h]\!] + \{\alpha\nabla v_h\}\cdot \mathbf{n}_{F_{ij}} [\![u_h]\!]\, ds,$$

$$p_i(u_h, v_h) = \begin{cases} \sum_{F_{ij}\subset\partial\Omega_i} \int_{F_{ij}} \mu\Big(\frac{\alpha^{(j)}}{h_j} + \frac{\alpha^{(i)}}{h_i}\Big)[\![u_h]\!][\![v_h]\!]\, ds, & \text{if } F_{ij} \in \mathscr{F}_I \\ \sum_{F_i\subset\partial\Omega_i} \int_{F_i} \mu\frac{\alpha^{(i)}}{h_i}[\![u_h]\!][\![v_h]\!]\, ds, & \text{if } F_i \in \mathscr{F}_B \end{cases}$$

$$p_D(u_D, v_h) = \sum_{F_i\subset\partial\Omega_i} \int_{F_i} \mu\frac{\alpha^{(i)}}{h_i} u_D v_h\, ds, \quad F_i \in \mathscr{F}_B$$

where $\alpha^{(i)} := \alpha|_{\Omega_i}$ and the unit normal vector $\mathbf{n}_{F_{ij}}$ is oriented from Ω_i towards the interior of Ω_j and the parameter $\mu > 0$ will be specified later in the error analysis, cf. [14].

For notation convenience in what follows, we will use the following expression

$$\int_{F_{ij}} \mu(\frac{\alpha^{(j)}}{h_j} + \frac{\alpha^{(i)}}{h_i})[\![u_h]\!][\![v_h]\!]\, ds,$$

for both cases, $F_{ij} \in \mathscr{F}_I$ and $F_i \in \mathscr{F}_B$. In the latter case, we will assume that $\alpha^{(j)} = 0$.

2.2 *Auxiliary Results*

We will use the following auxiliary results which have been shown in [30].

Lemma 1 *Let $u \in W^{l,p}(\mathscr{T}_H(\Omega))$ with $l \geq 2$ and $p > 1$. Then there is a constant $C \geq 0$ determined according to the C_m and C_M of (10), such that for $F_{ij} \subset \partial\Omega_i$ holds*

$$\int_{F_{ij}} |u|^p\, ds \leq C\Big(h_i^{-1}\int_{\Omega_i} |u|^p\, dx + h_i^{p-1}\int_{\Omega_i} |\nabla u|^p\, dx\Big). \tag{15}$$

Lemma 2 *For all $v_h \in \mathbb{B}_{h_i}^{(i)}$ defined on $T_{h_i,\Omega_i}^{(i)}$, there is a positive constant C, depending on the mesh quasi-uniformity parameters and C_m and C_M of (10) but not on h_i, such that*

$$\|\nabla v_h\|_{L^p(\Omega_i)}^p \leq C h_i^{-p} \|v_h\|_{L^p(\Omega_i)}^p. \tag{16}$$

Lemma 3 *For all $v_h \in \mathbb{B}_{h_i}^{(i)}$ defined on $T_{h_i,\Omega_i}^{(i)}$ and for all $F_{ij} \subset \partial\Omega_i$, there is a positive constant C, which depends on the mesh quasi-uniformity parameters C_m and C_M of (10) but not on h_i, such that*

$$\|v_h\|_{L^p(F_{ij})}^p \leq C h_i^{-1} \|v_h\|_{L^p(\Omega_i)}^p. \tag{17}$$

Lemma 4 *Let* $v_h \in \mathbb{B}_{h_i}^{(i)}$ *such that* $v_h \in W^{l,p}(E) \cap W^{m,q}(E)$, $E \in T_{h_i,\Omega_i}^{(i)}$, *and* $0 \leq m \leq l$, $1 \leq p, q \leq \infty$. *Then there is a positive constant C, depended on the mesh quasi-uniformity parameters* C_m *and* C_M *of (10) but not on* h_i, *such that*

$$|v_h|_{W^{l,p}(E)} \leq C h_i^{m-l-\frac{d}{q}+\frac{d}{p}} |v_h|_{W^{m,q}(E)}. \tag{18}$$

2.3 Analysis of the dG IgA Discretization

Next, we study the convergence estimates of the method (13) under the following regularity assumption for the weak solution $u \in W^{1,2}(\Omega) \cap W^{l,p}(\mathscr{T}_H(\Omega))$ with $l \geq 2$ and $p \in (\max\{1, \frac{2d}{d+2(l-1)}\}, 2]$. For simplicity of the presentation, we assume that $l \leq k + 1$. Nevertheless, for the case of highly smooth solutions, the estimates given below, see Lemma 5, can be expressed in terms of the underlying polynomial degree k. More precisely, the estimate $\delta(l, p, d) = l + (d/2 - d/p - 1)$, must be replaced by $\delta(l, p, d) = \min\{l + (d/2 - d/p - 1), k\}$. We use the enlarged space $W_h^{l,p} := W^{1,2}(\Omega) \cap W^{l,p}(\mathscr{T}_H(\Omega)) + \mathbb{B}_h(\mathscr{T}_H(\Omega))$, and will show that the dG IgA method converges in optimal rate with respect to $\|.\|_{dG}$ norm

$$\|u\|_{dG}^2 = \sum_{i=1}^{N} \left(\alpha^{(i)} \|\nabla u^{(i)}\|_{L^2(\Omega_i)}^2 + p_i(u^{(i)}, u^{(i)})\right), \; u \in W_h^{l,2}. \tag{19}$$

For the error analysis, it is necessary to show the continuity and coercivity properties of the bilinear form $a_h(.,.)$ of (14) and interpolation estimates in $\|.\|_{dG}$ norm. We start by providing estimates on how well the quasi-interpolant $\Pi_h u$ approximates $u \in W^{l,p}(\Omega_i)$, see proof in [30].

Lemma 5 *Let* $u \in W^{l,p}(\Omega_i)$ *with* $l \geq 2$ *and* $p \in (\max\{1, \frac{2d}{d+2(l-1)}\}, 2]$ *and let* $E = \Phi_i(E), E \in T_{h_i,\Omega}^{(i)}$. *Then for* $0 \leq m \leq l \leq k + 1$, *there exist constants* $C_i := C_i\big(\max_{l_0 \leq l}(\|D^{l_0}\Phi_i\|_{L^\infty(\Omega_i)}), \|u\|_{W^{l,p}(\Omega_i)}\big)$, *such that*

$$\sum_{E \in T_{h_i,\Omega_i}^{(i)}} |u - \Pi_h u|_{W^{m,p}(E)}^p \leq C_i h_i^{p(l-m)}. \tag{20}$$

Moreover, we have the following estimates for $F_{ij} = \partial\Omega_i \cap \partial\Omega_j$:

$$(i)\ h_i^{\beta}\|\nabla u^{(i)} - \nabla \Pi_h u^{(i)}\|^p_{L^p(F_{ij})} \leq C_i C_{d,p} h_i^{p(l-1)-1+\beta},$$

$$(ii)\ \Big(\frac{\alpha^{(j)}}{h_j} + \frac{\alpha^{(i)}}{h_i}\Big)\|[\![u - \Pi_h u]\!]\|^2_{L^2(F_{ij})} \leq$$

$$C_i\alpha^{(j)}\frac{h_i}{h_j}\Big(h_i^{\delta(l,p,d)}\|u\|^p_{W^{l,p}(\Omega_i)}\Big)^2 + C_j\alpha^{(i)}\frac{h_j}{h_i}\Big(h_j^{\delta(l,p,d)}\|u\|_{W^{l,p}(\Omega_j)}\Big)^2 +$$

$$C_j\Big(h_j^{\delta(l,p,d)}\|u\|_{W^{l,p}(\Omega_j)}\Big)^2 + C_i\Big(h_i^{\delta(l,p,d)}\|u\|_{W^{l,p}(\Omega_i)}\Big)^2,$$

$$(iii)\ \|u - \Pi_h u\|^2_{dG} \leq \sum_{i=1}^{N} C_i\Big(h_i^{\delta(l,p,d)}\|u\|_{W^{l,p}(\Omega_i)}\Big)^2 +$$

$$\sum_{i=1}^{N}\sum_{F_{ij}\subset\partial\Omega_i} C_i\alpha^{(j)}\frac{h_i}{h_j}\Big(h_i^{\delta(l,p,d)}\|u\|_{W^{l,p}(\Omega_i)}\Big)^2,$$

where $\delta(l,p,d) = l + (d/2 - d/p - 1)$.

We mention that the proof of estimate *(iii)* in Lemma 5 can be derived by using the estimates *(i)*, *(ii)* and Lemma 1.

Lemma 6 *Suppose $u_h \in \mathbb{B}_h(\mathscr{S}(\Omega))$. There exist a positive constant C, independent of α and h_i, such that*

$$a_h(u_h, u_h) \geq C\,\|u_h\|^2_{dG}, \quad \forall\, u_h \in \mathbb{B}_h(\mathscr{S}(\Omega)). \tag{21}$$

Proof By (13), we have that

$$a_h(u_h,u_h) = \sum_{i=1}^{N} a_i(u_h,u_h) - \frac{1}{2}\sum_{i=1}^{N} s_i(u_h,u_h) + p_i(u_h,u_h) =$$

$$\sum_{i=1}^{N}\alpha_i\|\nabla u_h\|^2_{L^2(\Omega_i)} - 2\sum_{F_{ij}\in\mathscr{F}}\int_{F_{ij}}\{\alpha\nabla u_h\}\cdot\mathbf{n}_{F_{ij}}[\![u_h]\!]\,ds$$

$$+\sum_{F_{ij}\in\mathscr{F}}\mu\Big(\frac{\alpha^{(i)}}{h_i} + \frac{\alpha^{(j)}}{h_j}\Big)\|[\![u_h]\!]\|^2_{L^2(F_{ij})}. \tag{22}$$

For the second term on the right hand side, Lemma 2 and the trace inequality (17) expressed on $F_{ij} \in \mathscr{F}$ yield the bound

$$
-\sum_{F_{ij}\in\mathscr{F}}\int_{F_{ij}}\{\alpha\nabla u_h\}\cdot\mathbf{n}_{F_{ij}}[\![u_h]\!]\,ds \geq
$$

$$
-C_{1,\varepsilon}\sum_{i=1}^{N}\alpha_i\|\nabla u_h\|^2_{L^2(\Omega_i)}-\sum_{F_{ij}\in\mathscr{F}}\frac{1}{C_{2,\varepsilon}}\Big(\frac{\alpha^{(i)}}{h_i}+\frac{\alpha^{(j)}}{h_j}\Big)\|[\![u_h]\!]\|^2_{L^2(F_{ij})}. \tag{23}
$$

Inserting (23) into (22) and choosing $C_{1,\varepsilon} < 1/2$ and $\mu > 2/C_{2,\varepsilon}$, we obtain (21). □

Note that similar discrete coercivity result has been shown in [30] for numerical fluxes without symmetrized terms. Due to the assumed regularity of the solution, the normal interface fluxes $(\alpha\nabla u)|_{\Omega_i}\cdot\mathbf{n}_{F_{ij}}$ belongs (in general) to $L^p(F_{ij})$. The following bound for the interface fluxes in $\|.\|_{L^p}$ setting has been shown in [30].

Lemma 7 *There is a constant C such that the following inequality for $(u, v_h) \in W_h^{l,p}\times\mathbb{B}_h(\mathscr{S}(\Omega))$ holds true*

$$
\sum_{i=1}^{N}\sum_{F_{ij}\subset\partial\Omega_i}\int_{F_{ij}}\{\alpha\nabla u\}\cdot\mathbf{n}_{F_{ij}}[\![v_h]\!]\,ds \leq \tag{24}
$$

$$
C\Big(\sum_{F_{ij}\in\mathscr{F}}\alpha^{(i)}h_i^{1+\gamma_{p,d}}\|\nabla u^{(i)}\|^p_{L^p(F_{ij})}+\alpha^{(j)}h_j^{1+\gamma_{p,d}}\|\nabla u^{(j)}\|^p_{L^p(F_{ij})}\Big)^{\frac{1}{p}}\|v_h\|_{dG},
$$

where $\gamma_{p,d} = d(p-2)/2$.

Proof We use Hölder's inequality and then the results of Lemmas 3 and 4. □

Applying similar procedure as this in Lemma 7, we can show for the symmetrizing terms that there is a positive constant independent of grid size such that

$$
\sum_{i=1}^{N}\sum_{F_{ij}\subset\partial\Omega_i}\int_{F_{ij}}\{\alpha\nabla v_h\}\cdot\mathbf{n}_{F_{ij}}[\![u]\!]\,ds \leq C_1\|v_h\|_{dG}\|u\|_{dG}. \tag{25}
$$

Using the results (24) and (25), we can show the boundedness of the bilinear form, see details in [30].

Lemma 8 *There is a C independent of h_i such that for $(u, v_h) \in W_h^{l,p}\times\mathbb{B}_h(\mathscr{S}(\Omega))$*

$$
a_h(u, v_h) \leq C(\|u\|^p_{dG}+\sum_{F_{ij}\in\mathscr{F}}h_i^{1+\gamma_{p,d}}\alpha^{(i)}\|\nabla u^{(i)}\|^p_{L^p(F_{ij})}+ \tag{26}
$$

$$
h_j^{1+\gamma_{p,d}}\alpha^{(j)}\|\nabla u^{(j)}\|^p_{L^p(F_{ij})}\Big)^{\frac{1}{p}}\|v_h\|_{dG},\ \textit{where}\ \gamma_{p,d} = d(p-2)/2.
$$

Next, we give the main error estimate for the dG IgA method.

Theorem 1 *Let $u \in W^{1,2}(\Omega) \cap W^{l,p}(\mathcal{T}_H(\Omega))$, with $l \geq 2$ and $p \in (p(d,l), 2]$ be the solution of (2). Let $u_h \in \mathbb{B}_h(\mathcal{S}(\Omega))$ be the dG IgA solution of (13). Then there are $C_i := C_i\Big(\max_{l_0 \leq l}\Big(\|D^{l_0}\Phi_i\|_{L^\infty(E)}\Big), \|u\|_{W^{l,p}(\Omega_i)}\Big)$, such that*

$$\|u - u_h\|_{dG} \leq \sum_{i=1}^{N}\Big(C_i\Big(h_i^{\delta(l,p,d)} + \sum_{F_{ij}\subset\partial\Omega_i} \alpha^{(j)}\frac{h_i}{h_j}h_i^{\delta(l,p,d)}\Big)\|u\|_{W^{l,p}(\Omega_i)}\Big), \tag{27}$$

where $\delta(l,p,d) = l + (d/2 - d/p - 1)$ and $p(d,l) = \max\{1, \frac{2d}{d+2(l-1)}\}$.

Proof First we need to prove the consistency of u, i.e. u satisfies (13). Then, we use a variation of Cea's Lemma (expressed in the dG framework), the results of Lemmas 6 and 8, as well the quasi-interpolation estimates of Lemma 5. A complete proof can be found in [30]. □

2.4 Numerical Examples

In this section, we present a series of numerical examples to validate the theoretical results, which have been presented.

2.4.1 Smooth and Low-Regularity Solutions

We restrict ourselves to a model problem in $\Omega = (-1,1)^3$, with $\Gamma_D = \partial\Omega$. The domain Ω is subdivided into four equal subdomains Ω_i, $i = 1,\dots,4$, where for simplicity every Ω_i is initially partitioned into a mesh $T^{(i)}_{h_i,\Omega_i}$, with $h := h_i = h_j, i \neq j$, and $i,j = 1,\dots,4$. Successive uniform refinements are performed on every $T^{(i)}_{h_i,\Omega_i}$ in order to compute numerically the convergence rates. We set the diffusion coefficient equal to one.

In the first test, the data u_D and f in (1) are determined such that the exact solution is $u(\mathbf{x}) = \sin(5\pi x_1)\sin(5\pi x_2)\sin(5\pi x_3)$ (smooth test case). The first three columns of Table 1 display the convergence rates. As it was expected, the convergence rates are optimal. In the second case, the exact solution is $u(x) = |x|^\lambda$. The parameter λ is chosen such that $u \in W^{l,p=1.67}(\Omega)$. Specifically, for $l = 2$, we get $\lambda = 0.2$ and, for $l = 3$, $\lambda = 1.2$. In the last columns of Table 1, we display the convergence rates for degree $k = 2, k = 3$ and $k = 4$ in case of having $l = 2$ and $l = 3$. We observe that, for each of the two different tests, the error in the dG-norm behaves according to the main error estimate given by (27).

Table 1 The numerical convergence rates of the dG IgA method

	Highly smooth			$k = 2$		$k = 3$		$k = 4$	
$\frac{h}{2^s}$	$k = 2$	$k = 3$	$k = 4$	$l = 2$	$l = 3$	$l = 2$	$l = 3$	$l = 2$	$l = 3$
Convergence rates									
$s = 0$	–	–	–	–	–	–	–	–	–
$s = 1$	0.445	2.916	0.083	0.510	1.425	0.406	1.124	0.369	0.926
$s = 2$	2.342	2.463	3.576	0.670	1.645	0.615	1.582	0.562	1.465
$s = 3$	2.100	3.141	4.089	0.699	1.669	0.694	1.692	0.683	1.674
$s = 4$	2.031	3.041	3.989	0.700	1.680	0.700	1.700	0.700	1.700

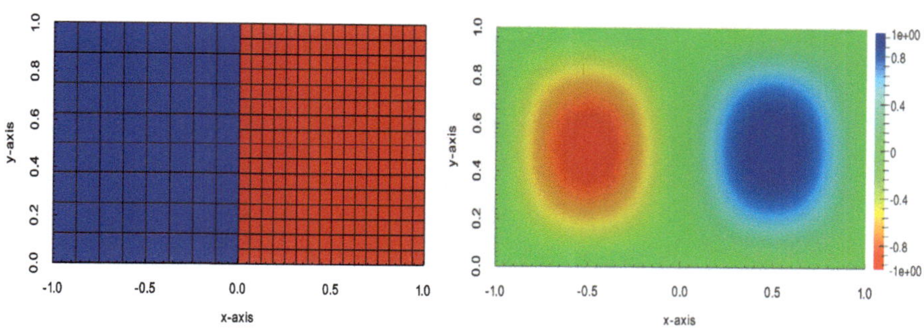

Fig. 1 Non-matching meshes: decomposition into 2 patches with underlying grid of ratio $R = 2$ (*left*), contours of u_h (*right*)

2.4.2 Non-matching Meshes

We consider the boundary value problem (1) with exact solution $u(x, y) = \sin(\pi x) \sin(\pi y)$. The computational domain consist of two unit square patches, $\Omega_1 = (-1, 0) \times (0, 1)$ and $\Omega_2 = (0, 1) \times (0, 1)$, see Fig. 1 (left). The knot vectors representing the geometry are given by $\Xi_1^{(1,2)} = \Xi_2^{(1,2)} = \{0, 0, 1, 1\}$. We refine the mesh of the patches to a ratio R, i.e., the ratio of the grid sizes is $R = h_1/h_2$, see Fig. 1 (left). We solved the problem using equal B-spline degree k on all patches. We plot in Fig. 1 (right) the dG IgA solution u_h computed with $k = 2$. In Fig. 2, we present the decay of the L_2 and dG errors for $k = 1, 2, 3$ and ratios $R = 2^r$ with r from 1 up to 5. In Table 2, we display the convergence rates for large ratio $R = 40$. In both cases, we observe that the rates are the expected ones and are not affected by the different grid sizes of the meshes.

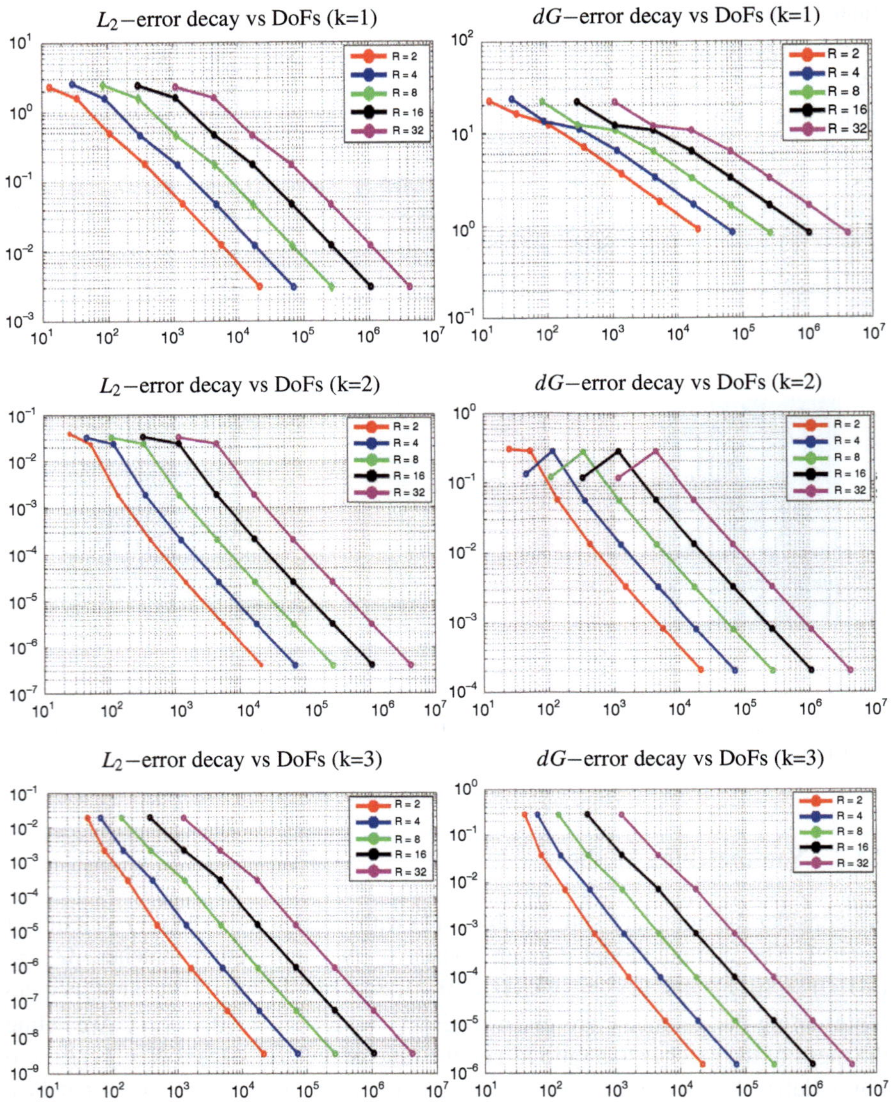

Fig. 2 Error decay in the L_2 (*left*) and dG (*right*) norms for polynomial degree $k = 1, 2, 3$ (*top* to *bottom*) and ratio $R = 2^r$ with r running from 1 to 5

Table 2 Non-matching meshes: error estimates for degrees $k = 1, 2, 3$ and $R = 40$

Dofs	L_2 error	Conv. rate	DG error	Conv. rate
Degree $k = 1$				
1685	0.31667	0	2.1596	0
6570	0.0867936	1.86732	0.97857	1.14202
25,946	0.0249492	1.79859	0.49992	0.968979
103,122	0.00638919	1.96529	0.251392	0.991758
411,170	0.00160452	1.99349	0.125887	0.997804
1.64205×10^6	0.000401493	1.9987	0.0629683	0.999435
6.56295×10^6	0.000100393	1.99972	0.0314873	0.999857
Degree $k = 2$				
1773	0.0330803	0	0.117511	0
6740	0.0237587	0.477516	0.278058	−1.2426
26,280	0.00186298	3.67278	0.0555191	2.32433
103,784	0.000205886	3.17769	0.0130802	2.0856
412,488	2.53126×10^{-5}	3.02392	0.00321704	2.02358
1.64468×10^6	3.17795×10^{-6}	2.99369	0.000800278	2.00716
6.5682×10^6	3.99389×10^{-7}	2.99223	0.000199727	2.00247
Degree $k = 3$				
1865	0.0196108	0	0.281761	0
6914	0.00216825	3.17704	0.037403	2.91324
26,618	0.00030589	2.82545	0.00706104	2.4052
104,450	1.60210×10^{-5}	4.25498	0.000803958	3.13469
413,810	9.49748×10^{-7}	4.07627	9.7687×10^{-5}	3.04088
1.64731×10^6	5.85542×10^{-8}	4.0197	1.21191×10^{-5}	3.01089
6.57346×10^6	3.64710×10^{-9}	4.00495	1.51195×10^{-6}	3.00280

2.5 Graded Mesh Partitions for the dG IgA Methods

We saw in the previous numerical tests that the presence of singular points reduces the convergence rates. In this section, we will study this subject in a more general form. We will focus on solving the model problem in domains with re-entrant corners on the boundary. Due to these singular corner points, the regularity of the solution (at least in a small vicinity) is reduced in comparison with the solutions in smooth domains [20]. As a result, the numerical methods applied on quasi uniform meshes for solving these problems do not yield the optimal convergence rate and thus a particular treatment must be applied. We will devise the popularly known graded mesh techniques which have widely been applied so far for finite element methods [3, 4].

2.5.1 Regularity Properties of the Solution Around the Boundary Singular Points

Let us assume a domain $\Omega \subset \mathbb{R}^2$ and let $P_s \in \partial\Omega$ be a boundary point with internal angle $\omega \in (\pi, 2\pi)$. We consider the local cylindrical coordinates (r, θ) with pole P_s, and define the cone, see Fig. 3 (left),

$$\mathscr{C} = \{(x, y) \in \Omega : x = r\cos(\theta), y = r\sin(\theta), 0 < r < R, 0 < \theta < \omega\}. \quad (28)$$

Then the solution in $\mathscr{C}$ can be written, [20],

$$u = u_r + u_s, \quad (29)$$

where $u_r \in W^{l\geq 2,2}(\Omega)$ and

$$u_s = \xi(r)\gamma r^\lambda \sin(\lambda\theta), \quad (30)$$

where γ is the *stress intensity factor* (is a real number depending only on f), and $\lambda = \pi/\omega \in (0, 1)$ is an exponent which determines the strength of the singularity. Since $\lambda < 1$, by an easy computation, we can show that the singular function u_s does not belong to $W^{2,2}(\Omega)$ but $u \in W^{2,p}(\Omega)$ with $p = 2/2 - \lambda$. The representation (30) of u_s helps us to reduce our study to the examination of the behavior of u in the vicinity of the singular point, since the regularity properties of u are determined by the regularity of u_s. The main idea is the following: based on the a priori knowledge of the analytical form of u_s in $\mathscr{C}$, we carefully construct a locally adapted mesh in $\mathscr{C}$ by introducing a grading control parameter $\mu := \mu(\lambda, k)$, such as allows us to prove that the approximation order of the method applied on this adapted mesh for u_s is similar with the order of the method applied on the rest of the mesh (maybe quasi uniform) for u_r.

2.5.2 The Graded Mesh for $\mathscr{T}_H(\Omega)$ and Global Approximation Estimates

The area $U_s := \{x \in \Omega : |P_s - x| \leq R, R \geq N_Z h,\ N_Z \geq 2\}$ is further subdivided into ring zones Z_ζ, $\zeta = 0, .., \zeta_M < N_Z$, with distance from P_s equal to $D_{(Z_\zeta, P_s)} := C(n_\zeta h)^{\frac{1}{\mu}}$, where $1 \leq n_\zeta < N_Z$ and $\frac{1}{2} \leq C \leq 1$. The radius of every zone is defined to be $R_{Z_\zeta} := D_{(Z_{\zeta+1}, P_s)} - D_{(Z_\zeta, P_s)} = C(n_{\zeta+1}h)^{\frac{1}{\mu}} - C(n_\zeta h)^{\frac{1}{\mu}}$.

For convenience, we assume that the initial subdivision $\mathscr{T}_H$ fulfill the following conditions, for an illustration see Fig. 3 (right) with $\zeta_M = 2$:

- The subdomains can be grouped into those which belong (entirely) to the area U_s and those that belong (entirely) to $\Omega \setminus U_s$. This means that there is no Ω_i, $i = 1, \ldots, N$ such that $U_s \cap \Omega_i \neq \emptyset$ and $(\Omega \setminus U_s) \cap \Omega_i \neq \emptyset$.

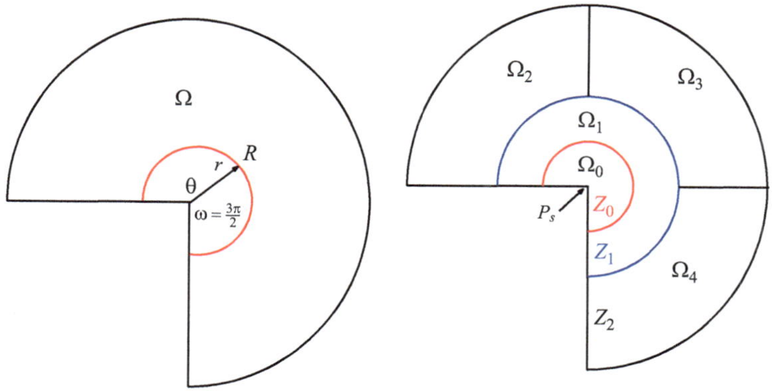

Fig. 3 *Left*: 2d domain with corner singularity. *Right*: zone partition and the subdomians Ω_i of $\mathcal{T}_H(\Omega)$

- Every ring zone Z_ζ is partitioned into "circular" subdomains Ω_{i_ζ} which have radius Ω_{i_ζ} equal to the radius of the zone, that is $R_{\Omega_{i_\zeta}} = R_{Z_\zeta}$. For computational efficiency reasons, we prefer -if its possible- every zone to be only represented by one subdomain.
- The zone Z_0 is represented by one subdomain, say Ω_0, and the mesh $T^{(0)}_{h_0}(\Omega_0)$ includes all $E \in T_h(\Omega)$ such that $\partial E \cap P_s \neq \emptyset$.

The graded meshes $T^{(i_\zeta)}_{h_{i_\zeta}}(\Omega_{i_\zeta})$ are mainly determined by the grading parameter $\mu(\lambda, k) \in (0, 1]$ and the mesh sizes h_{i_ζ} are chosen to satisfy the following properties: for Ω_{i_ζ} with distance $R_{\Omega_{i_\zeta}}$ from P_s, the mesh size h_{i_ζ} is defined to be $h_{i_\zeta} = \mathcal{O}(hR^{1-\mu}_{\Omega_{i_\zeta}})$ and for $T^{(i_0)}_{h_0}(\Omega_0)$ the mesh size is of order $h_{i_0} = \mathcal{O}(h^{\frac{1}{\mu}})$, more details are given in [28]. Based on previous properties of the $T^{(i_\zeta)}_{h_{i_\zeta}}(\Omega_{i_\zeta})$ meshes, we can conclude the relations

$$C_m h^{\frac{1}{\mu}} \leq h_{i_\zeta} \leq C_M h^{\frac{1}{\mu}}, \qquad \text{if } \overline{\Omega}_{i_\zeta} \cap P_s \neq \emptyset, \tag{31}$$

$$C_m h R^{1-\mu}_{\Omega_{i_\zeta}} \leq h_{i_\zeta} \leq C_M h D^{1-\mu}_{(Z_\zeta, P_s)}, \qquad \text{if } \overline{\Omega}_{i_\zeta} \cap P_s = \emptyset. \tag{32}$$

Using the local interpolation estimate of Lemma 5 in every $\Omega_{i_\zeta} \subset U_s$ and the characteristics of the meshes $T^{(i_\zeta)}_{h_{i_\zeta}}(\Omega_{i_\zeta})$, we can easily obtain the estimate

$$\|u_s - \Pi_h u_s\|_{dG(U_s)} \leq \sum_{i_\zeta} C_{i_\zeta} h^{\lambda}_{i_\zeta}, \tag{33}$$

since $u_s \in W^{l=1+\lambda,2}(\Omega)$ (and also $u \in W^{2,p=\frac{2}{2-\lambda}}(\Omega)$) and we do not consider non-matching grid interfaces. Using the mesh properties (31) and (32), estimate (33) and

Lemma 5, we can prove the following global error estimate of the proposed dG IgA method applied to problems with boundary singularities, see [28].

Theorem 2 *Let Z_ξ be a partition of $\Omega \subset \mathbb{R}^2$ into ring zones and let $\mathscr{T}_H$ to be a subdivision to Ω with the properties as listed in the previous paragraphs. Let $T^{(i)}_{h_i}(\Omega_i)$ be the meshes of subdomains as described above. Then for the solution u of (2) and the dG IgA solution u_h, we have the approximation result*

$$\|u - u_h\|_{dG} \leq Ch^r, \text{ with } r = \min(k, \lambda/\mu), \tag{34}$$

where the constant $C > 0$ depends on the characteristics of the mesh and on the mappings Φ_i (see (6)) but not on h_i.

2.5.3 Numerical Examples

In this section, we present a series of numerical examples in order to validate the theoretical analysis on the graded mesh in Sect. 2.5. The first example concerns a two-dimensional problem with a boundary point singularity (L-shape domain). The second example is the interior point singularity problem of the Sect. 2.4.

1. **Boundary Singular Point**
   One of the classical test cases is the singularity due to a re-entrant corner. The L-shape domain given by $(-1,1)^2 \setminus (-1,0)^2$. In Fig. 4 (left), the subdivision of Ω into two subdomains is presented. The exact solution is $u = r^{\frac{\pi}{\omega}} \sin(\theta\pi/\omega)$, where $\omega = 3\pi/2$. We set $\Gamma_D = \partial\Omega$ and the data f, u_D of (2) are specified by the given exact solution. The problem has been solved using B-splines of degree $k = 1$ and $k = 2$ and the grading parameter is $\mu = 0.6$ and $\mu = 0.3$ respectively. In Fig. 4 (middle), the graded mesh for $\mu = 0.6$ is presented and in Fig. 4 (right) the contours of the numerical solution are plotted. In Table 3, we present the convergence rates of the method without grading (left columns). As we can see, the convergence rates are determined by the regularity of the solution around the singular boundary point. In the right columns, we present the convergence rates corresponding to the graded meshes. We can see that the rates tend to be optimal with respect the B-spline degree.
2. **Interior Point Singularity**
   The domain is $\Omega = (-1,1)^2$. We consider a solution u of (1) with a point singularity at the origin given by $u(x) = |x|^\lambda$. We set $\lambda = 0.6$ and is easy to show that $u \in W^{l=1.6,2}(\Omega)$. We set $\alpha = 1$ in Ω. In the left columns of Table 3, we display the convergence rates for degrees $k = 1$ and $k = 2$ without mesh grading. The convergence rates are suboptimal and follow the approximation estimate (27). The problem has been solved again on graded meshes with $\mu = 0.6$ for $k = 1$ and $\mu = 0.3$ for $k = 2$, see Fig. 5. We display the convergence rates in the right columns of Table 3. The rates tend to be optimal as it was expected.

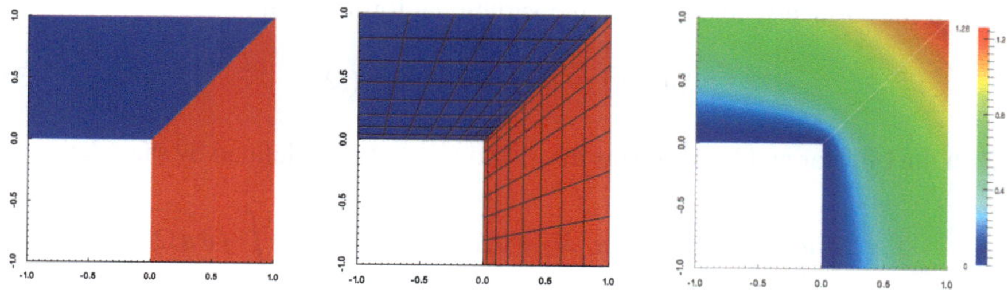

Fig. 4 L-shape test. *Left*: subdomains, *middle*: graded mesh with $\mu = 0.6$, *right*: contours of u_h

Table 3 The convergence rates for the L-shape test (*left*) and for the internal singularity test (*right*)

	No grading		With grading	
			$k=1$	$k=2$
$\frac{h}{2^s}$	$k=1$	$k=2$	$\mu=0.6$	$\mu=0.3$
Convergence rates				
$s=0$	–	–	–	–
$s=1$	0.636	0.650	0.915	1.672
$s=2$	0.641	0.657	0.933	1.919
$s=3$	0.647	0.661	0.946	1.987
$s=4$	0.652	0.666	0.957	2.071

	No grading		With grading	
			$k=1$	$k=2$
$\frac{h}{2^s}$	$k=1$	$k=2$	$\mu=0.6$	$\mu=0.3$
Convergence rates				
$s=0$	–	–	–	–
$s=1$	0.508	0.573	0.855	1.661
$s=2$	0.547	0.580	0.930	1.828
$s=3$	0.570	0.586	0.969	1.928
$s=4$	0.583	0.591	0.990	1.951

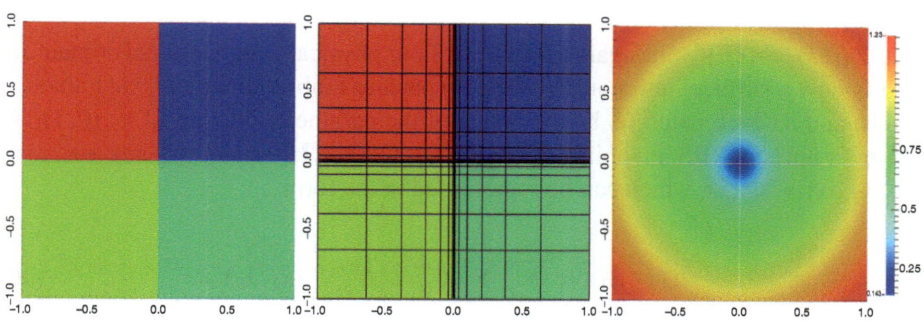

Fig. 5 Interior singularity test. *Left*: the subdomains Ω_i, $i = 1, \ldots, 4$, *middle*: the graded mesh with $\mu = 0.3$, *right*: the contours of u_h

3 Multipatch dG IgA for PDEs on Surfaces

3.1 *Diffusion Problems on Open and Closed Surfaces*

Let us now consider a diffusion problem of the form (1) on a sufficiently smooth, open surface Ω, the weak formulation of which can formally be written in the same form as (2) in Sect. 2: find a function $u \in W^{1,2}(\Omega)$ such that $u = u_D$ on the boundary

$\partial\Omega$ of the surface Ω and satisfies the variational formulation

$$a(u, v) = l(v), \ \forall v \in W_0^{1,2}(\Omega), \tag{35}$$

with the bilinear and linear forms $a(\cdot,\cdot)$ and $l(\cdot)$, but now defined by the relations

$$a(u, v) = \int_{\Omega} \alpha \, \nabla_{\Omega} u \cdot \nabla_{\Omega} v \, d\Omega \quad \text{and} \quad l(v) = \int_{\Omega} f v \, d\Omega,$$

respectively, where ∇_{Ω} denotes the surface gradient, see, e.g., Definition 2.3 in [18] for its precise description. For simplicity of the presentation, we here assume Dirichlet boundary condition. It is clear that other boundary conditions can be treated in the same framework as it was done in [29] for mixed boundary conditions. In the case of open surfaces with pure Neumann boundary condition and closed surfaces, we look for a solution $u \in W^{1,2}(\Omega)$ satisfying the uniqueness condition $\int_{\Omega} u(x)dx = 0$ and the variational equation (2) under the solvability condition $l(1) = 0$. In Sect. 3.4, we present and discuss the numerical results obtained for different diffusion problems on an open (Car) and on two closed (Sphere, Torus) surfaces.

3.2 Multipatch dG IgA Discretization

Let $\mathcal{T}_H(\Omega) = \{\Omega_i\}_{i=1}^N$ be again a partition of our physical computational domain Ω, that is now a surface, into non-overlapping patches (sub-domains) Ω_i such that (3) holds, and let each patch Ω_i be the image of the parameter domain $\hat{\Omega} = (0,1)^2 \subset \mathbb{R}^2$ by some NURBS mapping $\Phi_i : \hat{\Omega} \to \Omega_i \subset \mathbb{R}^3$, $\hat{x} = (\hat{x}_1, \hat{x}_2) \mapsto x = (x_1, x_2, x_3) = \Phi_i(\hat{x})$, which can be represented in the form

$$\Phi_i(\hat{x}_1, \hat{x}_2) = \sum_{k_1=1}^{n_1} \sum_{k_2=1}^{n_2} C^{(i)}_{(k_1,k_2)} \hat{B}^{(i)}_{(k_1,k_2)}(\hat{x}_1, \hat{x}_2) \tag{36}$$

where $\{\hat{B}^{(i)}_{(k_1,k_2)}\}$ are the bivariate NURBS basis functions, and $\{C^{(i)}_{(k_1,k_2)}\}$ are the control points, see [8] for a detailed description. We always assume that the mapping Φ_i is regular. Therefore, the inverse mapping $\hat{x} = \Psi_i(x) := \Phi_i^{-1}(x)$ is well defined for all patches Ω_i, $i = 1, \ldots, N$.

Now the dG IgA scheme for solving our surface diffusion problem (35) can formally be written in the form (13) as in Sect. 2: find $u_h \in \mathbb{B}_h(\mathcal{T}_H(\Omega))$ such that

$$a_h(u_h, v_h) = l(v_h) + p_D(u_D, v_h), \ \forall v_h \in \mathbb{B}_h(\mathcal{T}_H(\Omega)), \tag{37}$$

where $a_h(\cdot,\cdot)$ and $p_D(\cdot,\cdot)$ are defined by (14) provided that we replace the gradient ∇ by the surface gradient ∇_{Ω}. Since the dG bilinear form $a_h(\cdot,\cdot)$ is positive on

$\mathbb{B}_h(\mathcal{T}_H(\Omega)) \setminus \{0\}$ for sufficiently large μ, cf. Lemma 6, there exist a unique dG solution $u_h \in \mathbb{B}_h(\mathcal{T}_H(\Omega))$. The dG IgA scheme (37) is equivalent to a system of algebraic equations of the form

$$K_h \underline{u}_h = \underline{f}_h, \tag{38}$$

the solution $\underline{u}_h$ of which gives us the coefficients (control points) of u_h. In order to generate the entries of the system matrix K_h and the right-hand side $\underline{f}_h$, we map the patches Ω_i composing the physical domain, i.e., our surface Ω, into the parameter domain $\hat{\Omega} = (0,1)^2$. For instance, for the broken part $a_i(u_h, v_h)$ of the bilinear form $a_h(u_h, v_h)$, we obtain

$$\begin{aligned} a_i(u_h, v_h) &= \int_{\Omega_i} \alpha^{(i)}\, \nabla_\Omega u_h(x) \cdot \nabla_\Omega u_h(x)\, d\Omega \\ &= \int_{\hat{\Omega}_i} \alpha^{(i)}\, [J_i(\hat{x}) F_i^{-1}(\hat{x}) \hat{\nabla} \hat{u}_i(\hat{x})]^\top [J_i(\hat{x}) F_i^{-1}(\hat{x}) \hat{\nabla} \hat{v}_i(\hat{x})] g_i(\hat{x})\, d\hat{x} \\ &= \int_{\hat{\Omega}_i} \alpha^{(i)}\, (\hat{\nabla} \hat{u}_i(\hat{x}))^\top\, F_i^{-1}(\hat{x})\, \hat{\nabla} \hat{v}_i(\hat{x})\, g_i(\hat{x}) d\hat{x}, \end{aligned}$$

where $J_i(\hat{x}) = \partial \Phi_i(\hat{x})/\partial \hat{x}$, $F_i(\hat{x}) = (J_i(\hat{x}))^\top (J_i(\hat{x}))$ and $g_i(\hat{x}) = (\det F_i(\hat{x}))^{1/2}$ denote the Jacobian, the first fundamental form and the square root of the associated determinant, respectively. These terms, coming from the parameterization of the domain, can be exploited for deriving efficient matrix assembly methods, cf. [31, 32]. Furthermore, we use the notations $\hat{u}_i(\hat{x}) = u_h(\Phi_i(\hat{x}))$ and $\hat{\nabla} = (\frac{\partial}{\partial \hat{x}_1}, \frac{\partial}{\partial \hat{x}_2})^\top$.

3.3 *Discretization Error Estimates*

In [29], we derived discretization error estimates of the form

$$\|u - u_h\|_{dG}^2 \le C \sum_{i=1}^{N} \alpha^{(i)} h_i^{2t} \|u\|_{H^{1+t}(\Omega_i)}^2, \tag{39}$$

with $t := \min\{s, k\}$, provided that the solution u of our surface diffusion problem (35) belongs to $H^{1+s}(\mathcal{T}_H(\Omega)) = W^{1+s,2}(\mathcal{T}_H(\Omega))$ with some $s > 1/2$. In the case $t = k$, estimate (39) yields the convergence rate $\mathcal{O}(h^k)$ with respect to the dG norm, whereas the Aubin-Nitsche trick provides the faster rate $\mathcal{O}(h^{k+1})$ in the L_2 norm. Here k always denotes the underlying polynomial degree of the NURBS. This convergence behavior is nicely confirmed by our numerical experiments presented in [29] and in the next subsection.

In [29], we assumed matching meshes and some regularity of the solution of (35), namely $u \in H^{1+s}(\mathcal{T}_H(\Omega))$. It is clear that the results of Theorem 1, which includes no-matching meshes and low-regularity solutions, can easily be carried over to diffusion problems on open and closed surfaces. The same is true for mesh grading techniques presented in Sect. 2.5.

3.4 Numerical Examples

3.4.1 Sphere

Let us start with a diffusion problem on a closed surface Ω that is given by the sphere $\Omega = \{(x, y, z) \in (-1, 1)^3 : x^2 + y^2 + z^2 = 1\}$ with unitary radius. The computational domain Ω is decomposed into 6 patches, see left-hand side of Fig. 6. The knot vectors representing the geometry of each patch are given as $\Xi_{1,2} = (0, 0, 0, 0, 0, 1, 1, 1, 1, 1)$ in both directions. Since the surface is closed, we impose the uniqueness constraint $\int_\Omega u \, d\Omega = 0$ on the solution. The right-hand side $f(r, \phi, \theta) = 12u(r, \phi, \theta)$, where the solution $u(r, \phi, \theta) = 12 \sin(3\phi) \sin^3(\theta)$ is an eigenfunction of the Laplace-Beltrami operator $(-\Delta_\Omega)$ satisfying the compatibility condition $\int_\Omega f \, d\Omega = 0$. The example can also be found in [21]. The diagrams displayed in Fig. 7 show the error decay with respect to the L_2 (left) and dG (right) norms for polynomial degrees $k = 1, 2, 3$, and 4. As expected by our theoretical results, we observe the full convergence rates $\mathcal{O}(h^{k+1})$ and $\mathcal{O}(h^k)$ for the L_2 norm and the dG norm, respectively. In Table 4, we compare the L_2 errors of dG IgA solutions with those produced by the corresponding continuous (standard) Galerkin (cG) IgA scheme for the polynomial degree $k = 5$. In the case of smooth solutions, the dG IgA is as good as the cG counterpart. The same is true for the errors with respect to the dG norm.

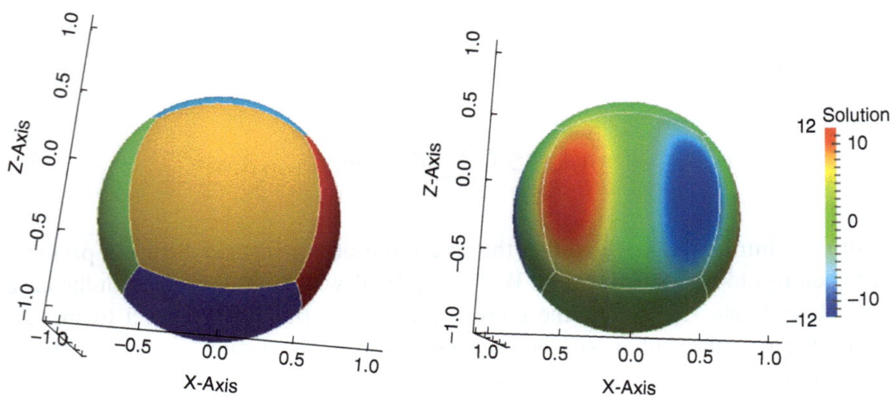

Fig. 6 Sphere: geometry and decomposition into 6 patches (*left*) and solution for the Laplace-Beltrami problem (*right*)

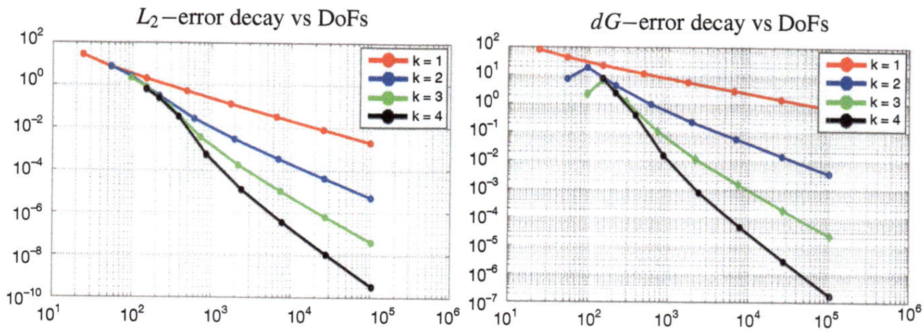

Fig. 7 Sphere: error decay in the L_2 (*left*) and dG (*right*) norms for polynomial degrees 1 to 4

Table 4 Sphere: comparison of the cG and dG IgA error decay in the L_2 norm for $k = 5$

$k = 5$	cG-IgA		dG-IgA	
Dofs	L_2 error	Conv. rate	L_2 error	Conv. rate
216	0.168803	0	0.166582	0
294	0.0602254	1.48689	0.0599854	1.16908
486	0.00900833	2.74104	0.00898498	2.55496
1014	8.90909×10^{-5}	6.65984	8.90774×10^{-5}	6.537
2646	8.64021×10^{-7}	6.68807	8.63906×10^{-7}	6.85905
8214	1.16592×10^{-8}	6.21152	1.16582×10^{-8}	6.27626
28,566	1.75119×10^{-10}	6.05699	1.75127×10^{-10}	6.06894

3.4.2 Torus

We now consider the closed surface

$$\Omega = \{(x,y) \in (-3,3)^2, z \in (-1,1) : \; r^2 = z^2 + (\sqrt{x^2+y^2} - R^2)\}$$

that is nothing but a torus that is decomposed into 4 patches, see Fig. 8 left. The knot vectors $\Xi_1 = \{0,0,0,0.25,0.25,0.50,0.50,0.75,0.75,1,1,1\}$ and $\Xi_2 = \{0,0,0,1,1,1\}$ describe the NURBS used for the geometrical representation of the patches. We first consider the surface Poisson equation, also called Laplace-Beltrami equation, with the right-hand side

$$\begin{aligned} f(\phi,\theta) &= r^{-2}\left[9\sin(3\phi)\cos(3\theta+\phi)\right] \\ &\quad - \left[(R+r\cos(\theta))^{-2}(-10\sin(3\phi)\cos(3\theta+\phi) - 6\cos(3\phi)\sin(3\theta+\phi))\right] \\ &\quad - \left[(r(R+r\cos(\theta))^{-1})(3\sin(\theta)\sin(3\phi)\sin(3\theta+\phi))\right], \end{aligned}$$

where $\phi = \arctan(y/x)$, $\theta = \arctan(z/(\sqrt{x^2+y^2} - R))$, $R = 2$, and $r = 1$. The exact solution is given by $u = \sin(3\phi)\cos(3\theta+\phi)$, cf. also [21]. We mention

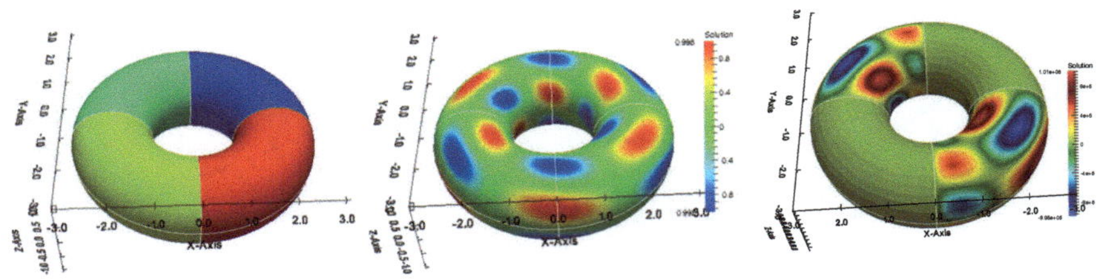

Fig. 8 Torus: geometry and decomposition into 4 patches (*left*), solution for the Laplace-Beltrami problem (*middle*), solution for jumping coefficients (*right*)

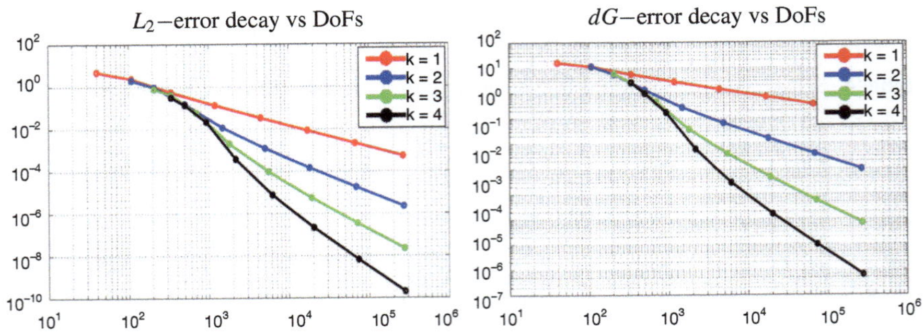

Fig. 9 Torus: error decay in the L_2 (*left*) and dG (*right*) norms for polynomial degrees 1 to 4

Table 5 Torus: comparison of the cG and dG IgA error decay in the L_2 norm for $k = 5$

$k = 5$	cG-IgA		dG-IgA	
Dofs	L_2 error	Conv. rate	L_2 error	Conv. rate
504	0.0974255	0	0.0973029	0
700	0.0433491	1.1683	0.043271	1.16908
1188	0.00736935	2.55639	0.00736339	2.55496
2548	7.9296×10^{-5}	6.53814	7.92948×10^{-5}	6.537
6804	6.83083×10^{-7}	6.85904	6.83068×10^{-7}	6.85905
21,460	8.8131×10^{-9}	6.27627	8.81294×10^{-9}	6.27626
75,348	1.3131×10^{-10}	6.0686	1.31276×10^{-10}	6.06894

that the functions u and f are chosen such that the zero mean compatibility condition holds. The IgA approximation to this solution is depicted in middle picture of Fig. 8. The diagrams displayed in Fig. 9 show the error decay with respect to the L_2 (left) and dG (right) norms for polynomial degrees $k = 1, 2, 3$, and 4. As expected by our theoretical results, we observe the full convergence rates $\mathcal{O}(h^{k+1})$ and $\mathcal{O}(h^k)$ for the L_2 norm and the dG norm, respectively. In Table 5, we compare the L_2 errors of dG IgA solutions with those produced by the corresponding cG IgA scheme for the polynomial degree $k = 5$. In the case of smooth solutions, the dG IgA is as good as the cG counterpart. The same is true for the errors with respect to the dG norm.

Table 6 Torus: L_2 and energy norm errors with degree $k = 2$

Dofs	L_2 error	Conv. rate	dG error	Conv. rate
108	2.04841×10^6	0	8.56088×10^6	0
208	1.03579×10^6	0.98377	4.89856×10^6	0.805403
504	128,215	3.0141	1.09244×10^6	2.1648
1480	15,431.9	3.05458	258,162	2.08121
4968	1808.17	3.09331	64,084.3	2.01023
18,088	213.252	3.0839	173,61.8	1.88406
268,840	Reference	Solution	Reference	Solution

Now we consider the case when the diffusion coefficients α have large jumps across the boundaries of the patches in which the torus was decomposed. More precisely, we assume that the diffusion coefficient $\alpha = \alpha^{(i)} > 0$ in the patch Ω_i, $i = 1, 2, 3, 4$, where $\alpha^{(2)} = \alpha^{(4)} = 1$ and $\alpha^{(1)} = \alpha^{(3)} = 10^{-6}$. The patches are arranged from blue Ω_1 to red Ω_4. The right-hand side f is the same as given above for the case of the Laplace-Beltrami problem, i.e. for $\alpha = 1$. Now the solution is not known, but we know that the solution is smooth in the patches $\overline{\Omega_i}$ and has steep gradients towards $\partial\Omega_2$ and $\partial\Omega_4$, see also of Fig. 8 right. Due to our theory, we can expect full convergence rates since we have an exact representation of the geometry. Indeed, we realize the full convergence rate by choosing a fine grid as the reference solution and comparing the solutions at each refinement step against this reference solution as shown in Table 6.

3.4.3 Car

We now consider the diffusion problem on an open, free-form surface. In order to demonstrate the fact that our results are general and not limited to academic examples, we apply our methods to a CAD model representing a car shell.

The surface is composed of eight quadratic B-spline patches, shown in Fig. 10 (left). The model exhibits several small bumps in the interior of the surface and sharp corners on the boundary. In addition, the patches have varying areas and meet along curved one-dimensional (quadratic) B-spline interfaces.

We choose a constant diffusion coefficient on the whole domain and we prescribe homogeneous Dirichlet conditions along the boundary. For the right-hand side, we used a (globally defined in $\mathbb{R}^3$) linear function which we restricted on the surface, in order to obtain a smooth solution to our problem.

As suggested by the isogeometric paradigm, we used quadratic B-spline basis functions, forwarded on the surface, as discretization basis. We started with a coarse grid, which was $h-$refined uniformly several times, in both parametric directions.

An analytic formula for the exact solution is not available. Therefore, we have chosen a fine grid of approximately one million degrees of freedom as the reference solution. Comparing against this solution (obtained by the corresponding cG IgA

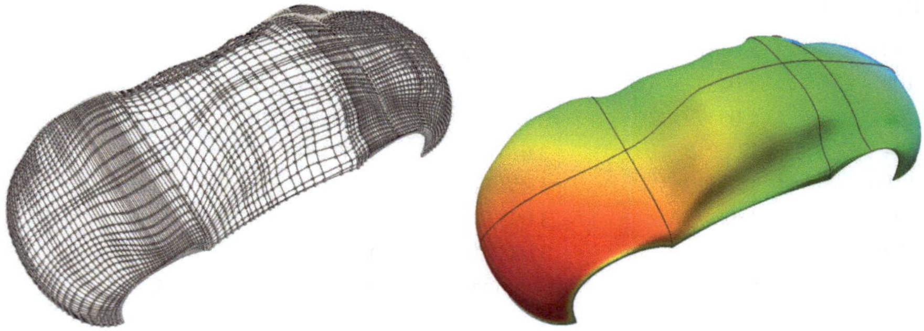

Fig. 10 Car: control points of the geometry (*left*), IgA solution of the Dirichlet surface diffusion problem (*right*). *Black lines* indicate patch interfaces

Table 7 Diffusion on a car-shell model. Numerical results for continuous Galerkin IgA with strong imposition of Dirichlet boundary (cG IgA), weak imposition (Nitsche-type) and finally patch-discontinuous Galerkin (dG IgA)

$k = 2$	cG IgA		Nitsche-type BCs		dG IgA	
DoFs	L_2 error	Rate	L_2 error	Rate	L_2 error	Rate
96	1.87598		1.77144		1.77029	
192	1.23006	0.608922	1.19201	0.571527	1.19188	0.570749
480	0.633648	0.956971	0.623767	0.934312	0.623764	0.934161
1440	0.241842	1.38962	0.239835	1.37897	0.239835	1.37896
4896	0.0659275	1.87511	0.0655902	1.87049	0.0655902	1.87049
17,952	0.0115491	2.5131	0.011502	2.5116	0.011502	2.5116
68,640	0.00155178	2.89578	0.00154625	2.89504	0.00154624	2.89504
∼268 K	0.00017567	3.14298	0.000175096	3.14256	0.000175094	3.14257
∼1 M	Reference	Reference	Reference	Reference	Reference	Reference

method), the expected convergence rates have been observed. Table 7 contains the numeric results for the L^2 norm, obtained using either the continuous (with strong or weak imposition of Dirichlet boundaries) or dG IgA method. Apart from the observed order of convergence, the magnitude of the error agrees in all cases as well.

4 The G+Smo C++ Library

Isogeometric analysis requires seamless integration of Finite Element Analysis (FEA) and Computer-aided design (CAD) software. The existing software libraries, however, cannot be adapted easily to the rising new challenges since they have been designed and developed for different purposes. In particular, FEA codes are traditionally implemented by means of functional programming, and are focused on treating nodal shape function spaces. In CAD packages, on the other hand, the central objects are free-form curves and surfaces, defined by control points, which are realized in an object-oriented programming environment.

G+SMO is an object-oriented, template C++ library, that implements a generic concept for IGA, based on abstract classes for geometry map, discretization basis, assemblers, solvers and so on [25]. It makes use of object polymorphism and inheritance techniques in order to support a variety of different discretization bases, namely B-spline, Bernstein, NURBS bases, hierarchical and truncated hierarchical B-spline bases of arbitrary polynomial order, and so on.

The library is open-source and is licenced under the Mozilla Public License v2.0. The source code, together with a reference manual and wiki pages can be reached at http://www.gs.jku.at/gismo.

Our design allows the treatment of geometric entities such as surfaces or volumes through dimension independent code, realized by means of template meta-programming. Available features include simulations using continuous and discontinuous Galerkin approximation of PDEs, over conforming and non-conforming multi-patch computational domains. PDEs on surfaces as well as integral equations arising from elliptic boundary value problems. Boundary conditions may be imposed both strongly and weakly. In addition to advanced discretization and generation techniques, efficient solvers like multigrid iteration schemes are available. Methods for solving non-linear problems are under development. Finally, we aim to employ existing high-end libraries for large-scale parallelization.

In the following paragraphs we shall provide more details on the design and features of the library.

4.1 Description of the Main Modules

The library is partitioned into smaller entities, called modules. Currently, there are six (6) modules in G+SMO namely Core, Matrix, NURBS, Modeling, Input/Output and Solver modules.

The **Core module** is the backbone of the library. Here an abstract interface is defined for a *basis*, that is, a set of real-valued functions living on a parameter domain. At this level, we do not specify how these functions (or its derivatives) should be evaluated. However, a number of virtual member functions define an interface that should be implemented by derived classes of this type. Another abstract class is the *geometry* class. This object consists of a (still abstract) basis and a coefficient vector, and represents a patch. Note that parameter or physical dimension are not specified at this point. There are four classes directly derived from the geometry class; these are curve, surface, volume and bulk. These are parametric objects with known parameter space dimension 1, 2, 3 and 4 respectively.

Another abstract class is a *function* class. The interface for this class includes evaluation, derivations and other related operations. The *geometry* abstract class is actually deriving from the function class, demonstrating the fact that parametric geometries can be simply viewed as (vector) functions. Another interesting object is the *multipatch* object. CAD models are composed of many patches. Therefore, a multipatch structure is of great importance. It contains two types of information; first, geometric information, essentially a list of geometry patches. Second, topological information between the patches, that is, the adjacency graph between patch boundaries, degenerate points, and so on. Let us also mention the *field* class, which is the object that typically represents the solution of a PDE. A *field* is a mathematical scalar or vector field which is defined on a parametric patch, or multipatch object. It may be evaluated either on the parameter or physical space, as the isogeometric paradigm suggests.

The **Matrix module** contains all the linear algebra related infrastructure. It is based on the third party library Eigen [22]. The main objects are dense and sparse matrices and vectors. Typical matrix decompositions such as LU, QR, SVD, and so on, are available. Furthermore, the user has also access to iterative solvers like conjugate gradient methods with different preconditioners. Finally, one can use popular high-end linear solver packages like SuperLU[2] through a common interface.

The **NURBS module** consists of B-Splines, NURBS, Bézier of arbitrary degrees and knot-vectors, tensor-product B-splines of arbitrary spatial dimension.

The **Modeling module** provides data structures and geometric operations that are needed in order to prepare CAD data for analysis. It contains trimmed surfaces, boundary represented (B-rep) solids and triangle meshes. Regarding modeling operations, B-Spline fitting, smoothing of point clouds, Coon's patches, and volume segmentation methods are available.

The **Input/Output module** is responsible for visualization as well as file reading and writing. For visualization, we employ Paraview or Axel (http://axel.inria.fr) and VTK at INRIA, France. An important issue is file formats. Even if NURBS is an industrial standard, a variety of different formats are used in the CAD industry to exchange NURBS data. In G+SMO, we have established I/O with popular CAD formats, which include the 3DM file format of Rhinoceros 3D modeler, the `X_T` format of Siemens' NX platform as well as an (exported) format used by the LS-DYNA general-purpose finite element program.

The **Assembler module** can already treat a number of PDE problems like convection-diffusion problems, linear elasticity, Stokes equations as well as diffusion problems on surfaces by means of continuous or discontinuous Galerkin methods, including divergence preserving discretizations for Stokes equations. Strong or weak imposition Dirichlet boundary conditions and Neumann-type conditions are provided. Boundary element IgA collocation techniques are also available.

Apart from the modules described above, there are several more which are under development. These include a hierarchical bases module, an optimization module and a triangular Bézier module, see Fig. 11.

---

[2]http://acts.nersc.gov/superlu

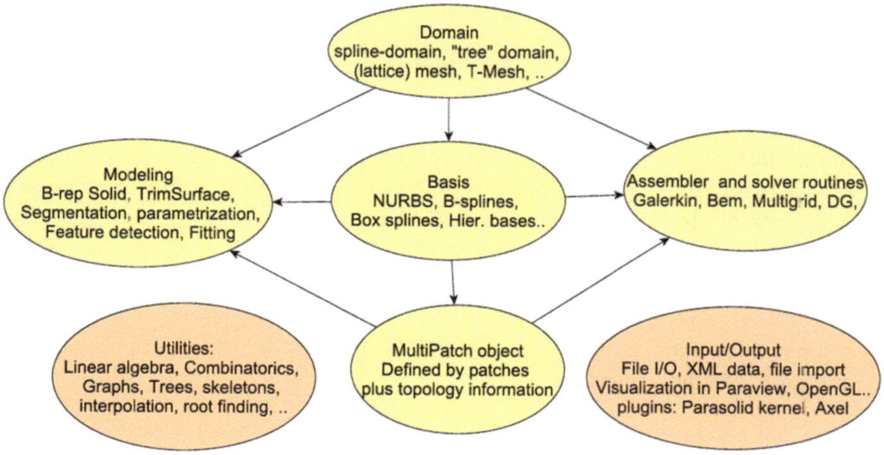

Fig. 11 A diagram of the different parts in the library and their interactions

4.2 Development Framework

The library uses a set of standard C++ tools in order to achieve cross-platform functionality. Some compilers that have been tested includes Microsoft's Visual C++, Mingw32, GCC, Clang and the Intel compiler.

The main development tools used are:

- CMake cross-platform build system for configuration of the code. A small set of options are available in order to enable optional parts of the library that require external packages. Furthermore, the user is able to enable debug mode build, that disables code optimization and facilitates development, by attaching debug information in the code.
- The code is available in an SVN repository. We have chosen the continuous integration as development policy. That is, all developers commit their contributions in a mainline repository.
- Trac software management system is used for bug reporting. Additionally, we use the integrated wiki for documentation and user guidance.
- A CDASH software testing server is employed for regular compilation and testing of the mainline code. Nightly builds are executed on different platforms, as well as continuous builds after user commits. This allows to easily correct errors and ensure the quality of the library. Coverage analysis and memory checks are also performed regularly.
- The in-source documentation system Doxygen is used heavily in the code.
- A mailing list is available for communication and user support.

Regarding tools that are employed in the library:

- The C++ Standard Library is employed. This is available by default on C++ modern installations.
- The Eigen C++ Linear Algebra library is used for linear algebra operations. This library features templated coefficient type, dense and sparse matrices and vectors, typical matrix decompositions (LU, QR, SVD,...) as well as iterative solvers and wrappers for high-end packages (SuperLU, PaStiX, SparseSuite,...).
- XML reader/writer for input/output of XML files, and other CAD formats (OFF, STL, OBJ, GeoPDEs, 3DM, Parasolid).
- Mathematical Expression Toolkit Library (ExprTk). This is an expression-tree evaluator for mathematical function expressions that allows input of functions similar to Matlab's interface.

4.3 Additional Features and Extensions

Apart from the basic set of tools provided in the library, the user has the possibility to enable features that require third-party software. The following connections to external tools is provided:

- OpenNurbs library used to support Rhino's 3DM CAD file format. This allows file exchange with standard CAD software.
- Connection to Parasolid geometric kernel. Enabling this feature allows input and output of the `x_t` file format. Also, the user can employ advanced modeling operations like intersection, trimming, boolean operations, and so on.
- Connection to LS-DYNA. The user can output simulation data that can be used by LS-DYNA's Generalized element module system, for performing for instance simulations on shells.
- Connection to IPOPT nonlinear optimization library. With this feature one can use powerful interior point constrained optimization algorithms in order to perform, for instance, isogeometric shape optimization.
- MPFR library for multi-precision floating point arithmetic. With this feature critical geometric operations can be performed with arbitrary precision, therefore guaranteeing a verified result.
- Graphical interface, interaction and display using Axel modeler.

4.4 Plugins for Third-Party Platforms

Finally, there are currently two plugins under development. The plugins allow third-party software to employ and exchange data with G+SMO.

- Axel Modeler is an open-source spline modeling package based on Qt and VTK. Our plugin allows to use G+SMO within this graphical interface, and provides user interaction (e.g. control point editing) and display.
- A Matlab interface is under development. This allows to use operations available in G+SMO within Matlab.

Acknowledgements This research was supported by the Research Network "Geometry + Simulation" (NFN S117) funded by the Austrian Science Fund (FWF) under the grand S117-03.

References

1. R.A. Adams, J.J.F. Fournier, *Sobolev Spaces* Pure and Applied Mathematics, vol. 140, 2nd edn. (Elsevier/Academic Press, Amsterdam, 2003)
2. P. Antonietti, A. Dedner, P. Madhavan, S. Stangalino, B. Stinner, M. Verani, High order discontinuous Galerkin methods on surfaces. SIAM J. Numer. Anal. **53**(2), 1145–1171 (2015)
3. T. Apel, F. Milde, Comparison of several mesh refinement strategies near edges. Commun. Numer. Methods Eng. **12**, 373–381 (1996)
4. T. Apel, A.-M. Sändig, J.R. Whiteman, Graded mesh refinement and error estimates for finite element solutions of elliptic boundary value problems in non-smooth domains. Math. Methods Appl. Sci. **19**(30), 63–85 (1996)
5. Y. Bazilevs, L. Beirão da Veiga, J. Cottrell, T. Hughes, G. Sangalli, Isogeometric analysis: approximation, stability and error estimates for *h*-refined meshes. Comput. Methods Appl. Mech. Eng. **194**, 4135–4195 (2006)
6. L. Beirão da Veiga, A. Buffa, J. Rivas, G. Sangalli, Some estimates for *hpk*-refinement in isogeometric analysis. Numerische Mathematik **118**(2), 271–305 (2011)
7. L. Beirão da Veiga, D. Cho, G. Sangalli, Anisotropic NURBS approximation in isogeometric analysis. Comput. Methods Appl. Mech. Eng. **209–212**(0):1–11 (2012)
8. J. Cottrell, T. Hughes, Y. Bazilevs, *Isogeometric Analysis: Toward Integration of CAD and FEA* (Wiley, Chichester, 2009)
9. L. Dede, A. Quarteroni, Isogeometric analysis for second order partial differential equations on surfaces. Comput. Methods Appl. Mech. Eng. **284**, 807–834 (2015)
10. A. Dedner, P. Madhavan, Adaptive discontinuous Galerkin methods on surfaces. Numerische Mathematik (2015). doi: 10.1007/s00211-015-0719-4
11. A. Dedner, P. Madhavan, B. Stinner, Analysis of the discontinuous Galerkin method for elliptic problems on surfaces. IMA J. Numer. Anal. **33**(3), 952–973 (2013)
12. D.A. Di Pietro, A. Ern, Analysis of a discontinuous Galerkin method for heterogeneous diffusion problems with low-regularity solutions. Numer. Methods Partial Differ. Equ. **28**(4), 1161–1177 (2012)
13. D.A. Di Pietro, A. Ern, *Mathematical Aspects of Discontinuous Galerkin Methods*. Mathématiques et Applications, vol. 69 (Springer, Berlin/Heidelberg, 2012)
14. M. Dryja, On discontinuous Galerkin methods for elliptic problems with discontinuous coefficients. Comput. Methods Appl. Math. **3**, 76–85 (2003)
15. G. Dziuk, Finite elements for the Beltrami operator on arbitrary surfaces, in *Partial Differential Equations and Calculus of Variations*, ed. by S. Hildebrandt, R. Leis. Lecture Notes in Mathematics, vol. 1357 (Springer, Berlin/Heidelberg, 1988), pp. 142–155
16. G. Dziuk, C. Elliott, Finite elements on evolving surfaces. IMA J. Num. Anal. **27**(2), 262–292 (2007)
17. G. Dziuk, C. Elliott, Surface finite elements for parabolic equations. J. Comput. Math. **25**(4), 385–407 (2007)

18. G. Dziuk, C. Elliott, Finite element methods for surface PDEs. Acta Numer. **22**, 289–396 (2013)
19. L.C. Evans, *Partial Differential Equestions*. Graduate Studies in Mathematics, vol. 19, 1st edn. (American Mathematical Society, Rhode Island, 1998)
20. P. Grisvard, *Elliptic Problems in Nonsmooth Domains*. Classics in Applied Mathematics, vol. 69 (SIAM, Philadelphia, 2011)
21. S. Gross, A. Reusken, *Numerical Methods for Two-phase Incompressible Flows*. Springer Series in Computational Mathematics, vol. 40 (Springer, Berlin/Heidelberg, 2011)
22. G. Guennebaud, B. Jacob et al., Eigen v3. http://eigen.tuxfamily.org (2010)
23. T. Hughes, J. Cottrell, Y. Bazilevs, Isogeometric analysis: CAD, finite elements, NURBS, exact geometry and mesh refinement. Comput. Methods Appl. Mech. Eng. **194**, 4135–4195 (2005)
24. J.W. Jeong, H.-S. Oh, S. Kang, H. Kim, Mapping techniques for isogeometric analysis of elliptic boundary value problems containing singularities. Comput. Methods Appl. Mech. Eng. **254**, 334–352 (2013)
25. B. Jüttler, U. Langer, A. Mantzaflaris, S.E. Moore, W. Zulehner, Geometry + Simulation Modules: Implementing Isogeometric Analysis. PAMM **14**(1), 961–962 (2014)
26. R.B. Kellogg, On the Poisson equation with intersecting interfaces. Appl. Anal. **4**, 101–129 (1975)
27. S.K. Kleiss, B. Jüttler, W. Zulehner, Enhancing isogeometric analysis by a finite element-based local refinement strategy. Comput. Methods Appl. Mech. Eng. **213–216**, 168–182 (2012)
28. U. Langer, A. Mantzaflaris, S.E. Moore, I. Toulopoulos, Mesh Grading in Isogeometric Analysis. Comput. Math. Appl. (2015). http://dx.doi.org/10.1016/j.camwa.2015.03.011
29. U. Langer, S. Moore, Discontinuous Galerkin Isogeometric Analysis of elliptic PDEs on surfaces. NFN Technical Report 12, Johannes Kepler University Linz, NFN Geometry and Simulation, Linz, 2014. http://arxiv.org/abs/1402.1185 and accepted for publication in the DD22 proceedings
30. U. Langer, I. Toulopoulos, Analysis of multipatch discontinuous Galerkin IgA approximations to elliptic boundary value problems. RICAM Reports 2014-08, Johann Radon Institute for Computational and Applied Mathematics, Austrian Academy of Sciences, Linz, 2014. http://arxiv.org/abs/1408.0182
31. A. Mantzaflaris, B. Jüttler, Exploring matrix generation strategies in isogeometric analysis, in *Mathematical Methods for Curves and Surfaces*, ed. by M. Floater, T. Lyche, M.-L. Mazure, K. Mørken, L. Schumaker. Lecture Notes in Computer Science, vol. 8177 (Springer, Berlin/Heidelberg, 2014), pp. 364–382
32. A. Mantzaflaris, B. Jüttler, Integration by interpolation and look-up for Galerkin-based isogeometric analysis. Comput. Methods Appl. Mech. Eng. **284**, 373–400 (2015)
33. L. Oganesjan, L. Ruchovetz, *Variational Difference Methods for the Solution of Elliptic Equations* (Isdatelstvo Akademi Nank Armjanskoj SSR, Erevan, 1979) (in Russian)
34. H.-S. Oh, H. Kim, J.W. Jeong, Enriched isogeometric analysis of elliptic boundary value problems in domains with cracks and/or corners. Int. J. Numer. Methods Eng **97**(3), 149–180 (2014)
35. B. Rivière, *Discontinuous Galerkin Methods for Solving Elliptic and Parabolic Equations: Theory and Implementation* (SIAM, Philadelphia, 2008)
36. F. Zhang, Y. Xu, F. Chen, Discontinuous Galerkin methods for isogeometric analysis for elliptic equations on surfaces. http://staff.ustc.edu.cn/~yxu/paper/IGA-DG.pdf (2014)

The Influence of Quadrature Errors on Isogeometric Mortar Methods

Ericka Brivadis, Annalisa Buffa, Barbara Wohlmuth, and Linus Wunderlich

Abstract Mortar methods have recently been shown to be well suited for isogeometric analysis. We review the recent mathematical analysis and then investigate the variational crime introduced by quadrature formulas for the coupling integrals. Motivated by finite element observations, we consider on the interface a quadrature rule purely based on the slave mesh as well as a method using on the interface quadrature rules based on the slave mesh and on the master mesh, resulting in a non-symmetric saddle point problem. While in the first case reduced convergence rates can be observed, in the second case the influence of the variational crime is less significant.

1 Introduction

Isogeometric analysis, introduced in 2005 by Hughes et al. in [17], is a family of methods that use B-splines and non-uniform rational B-splines (NURBS) as basis functions to construct numerical approximations of partial differential equations (PDEs). With isogeometric methods, the computational domain is generally split into patches. Within this framework, techniques to couple the numerical solution on

E. Brivadis (✉)
IUSS (Istituto Universitario di Studi Superiori di Pavia), Palazzo del Broletto, Piazza della Vittoria 15, 27100 Pavia, Italy

IMATI (Istituto di Matematica Applicata e Tecnologie Informatiche del CNR), Via Ferrata 1, 27100 Pavia, Italy
e-mail: ericka.brivadis@iusspavia.it

A. Buffa
IMATI (Istituto di Matematica Applicata e Tecnologie Informatiche del CNR), Via Ferrata 1, 27100 Pavia, Italy
e-mail: annalisa@imati.cnr.it

B. Wohlmuth • L. Wunderlich
M2 – Zentrum Mathematik, Technische Universität München, Boltzmannstraße 3, 85748 Garching, Germany
e-mail: wohlmuth@ma.tum.de; linus.wunderlich@ma.tum.de

B. Jüttler, B. Simeon (eds.), *Isogeometric Analysis and Applications 2014*, Lecture Notes in Computational Science and Engineering 107,
DOI 10.1007/978-3-319-23315-4_2

different patches are required. To retain the flexibility of the meshes at the interfaces, mortar methods are very attractive.

Mortar methods are a popular tool for the coupling of non-matching meshes, originally introduced for spectral and finite element methods [4–6]. They were successfully applied in the context of isogeometric analysis [1, 13, 16]. A mathematical analysis enlightening the use of different dual spaces was recently presented in [8]. In this paper, starting from these latter results, we focus on one particular challenge in the realization of a mortar method, namely, the evaluation of the interface integrals which contain a product of functions, each of which being defined on a different mesh.

This article is structured as follows. In Sect. 2, we recall the basics of isogeometric mortar methods. In Sect. 3, we consider a review of numerical quadrature for mortar integrals such as additional aspects in the case of isogeometric analysis, illustrated by numerical results in Sect. 4.

2 Isogeometric Mortar Methods

In the following, we briefly present isogeometric mortar methods, for more details we refer to [8]. After stating the problem setting, we review isogeometric parametrizations, describe the domain decomposition into several NURBS patches and finally discuss suitable coupling spaces.

Let $\Omega \subset \mathbb{R}^d$, d the dimension being 2 or 3, be a bounded domain, $\alpha,\ \beta \in L^\infty(\Omega)$, $\alpha > \alpha_0 > 0$, and $\beta \geq 0$. We consider the following second order elliptic boundary value problem with homogeneous Dirichlet conditions

$$-\operatorname{div}(\alpha\nabla u) + \beta u = f \quad \text{in } \Omega, \tag{1a}$$

$$u = 0 \quad \text{on } \partial\Omega_D = \partial\Omega. \tag{1b}$$

We assume α and β to be piecewise sufficiently smooth.

2.1 Isogeometric Parametrization

Here, we present isogeometric concepts and notations used throughout the paper and refer to the classical literature [2, 10, 20, 23] for more details.

Let us denote by p the degree of the univariate B-splines and by Ξ an open univariate knot vector, where the first and last entries are repeated $(p + 1)$-times, i.e.,

$$\Xi = \{0 = \xi_1 = \ldots = \xi_{p+1} < \xi_{p+2} \leq \ldots \leq \xi_n < \xi_{n+1} = \ldots = \xi_{n+p+1} = 1\}.$$

Let us define $Z = \{\zeta_1, \zeta_2, \ldots, \zeta_E\}$ as the knot vector without any repetition, also called breakpoint vector. For each breakpoint ζ_j of Z, we define its multiplicity m_j as its number of repetitions in Ξ. The Cox-de Boor algorithm, see [10], defines n univariate B-splines $\hat{B}_i^p(\zeta)$, $i = 1, \ldots, n$, based on the univariate knot vector Ξ and the degree p. We denote by $S^p(\Xi) = \text{span}\{\hat{B}_i^p(\zeta),\ i = 1, \ldots, n\}$ the corresponding spline space. The smoothness of B-splines is defined by the breakpoint multiplicity. More precisely, each basis function is C^{p-m_j} at each $\zeta_j \in Z$.

To define multivariate spline spaces, we introduce the multivariate knot vector $\boldsymbol{\Xi} = (\Xi_1 \times \Xi_2 \times \ldots \times \Xi_d)$, and for simplicity of notations assume in the following that the degree is the same in all parametric directions and denote it by p. Multivariate B-splines $\hat{B}_{\mathbf{i}}^p(\boldsymbol{\zeta})$ are defined by tensor product of univariate B-splines for each multi-index $\mathbf{i} \in \mathbf{I} = \{(i_1, \ldots, i_d) : 1 \le i_\delta \le n_\delta\}$. We denote by $S^p(\boldsymbol{\Xi})$ the corresponding spline space in the parametric domain.

Given a set of positive weights $\omega_{\mathbf{i}}$, we define NURBS functions $\hat{N}_{\mathbf{i}}^p(\boldsymbol{\zeta})$ as rational functions of B-splines and the weight function $\hat{W} = \sum_{\mathbf{i}\in\mathbf{I}} \omega_{\mathbf{i}} \hat{B}_{\mathbf{i}}^p(\boldsymbol{\zeta})$. We set $N^p(\boldsymbol{\Xi})$ as the multivariate NURBS space in the parametric domain.

For a set of control points $\mathbf{C}_{\mathbf{i}} \in \mathbb{R}^d$, $\mathbf{i} \in \mathbf{I}$, we define a parametrization of a NURBS surface ($d = 2$) or solid ($d = 3$) as a linear combination of NURBS and control points

$$\mathbf{F}(\boldsymbol{\zeta}) = \sum_{\mathbf{i}\in\mathbf{I}} \mathbf{C}_{\mathbf{i}} \hat{N}_{\mathbf{i}}^p(\boldsymbol{\zeta}),$$

and assume the regularity stated in [3, Assumption 3.1].

The knot vector $\boldsymbol{\Xi}$ forms a mesh in the parametric space $\hat{\Omega}$. We define the physical mesh $\mathcal{M}$ as the image of this parametric mesh through $\mathbf{F}$, and denote by $\mathbf{O}$ its elements. The h-refinement procedure, see [3, Section 2.1.3], yields a family of meshes denoted $\mathcal{M}_h$, each mesh being a refinement of the initial one. We assume quasi-uniformity for each mesh.

2.2 Description of the Computational Domain

Let the domain Ω be decomposed into K non-overlapping domains Ω_k, i.e.,

$$\overline{\Omega} = \bigcup_{k=1}^{K} \overline{\Omega}_k, \text{ and } \Omega_i \cap \Omega_j = \emptyset, i \neq j.$$

Each subdomain is a NURBS geometry, i.e., there exists a NURBS parametrization $\mathbf{F}_k$ based on a knot vector $\boldsymbol{\Xi}_k$ and a degree p_k, see Sect. 2.1, such that Ω_k is the image of the parametric space $\hat{\Omega} = (0, 1)^d$ by $\mathbf{F}_k$.

For $1 \le k_1 < k_2 \le K$, we define the interface as the interior of the intersection of the boundaries, i.e., $\overline{\gamma}_{k_1k_2} = \partial\Omega_{k_1} \cap \partial\Omega_{k_2}$, where $\gamma_{k_1k_2}$ is open. Let the non-empty

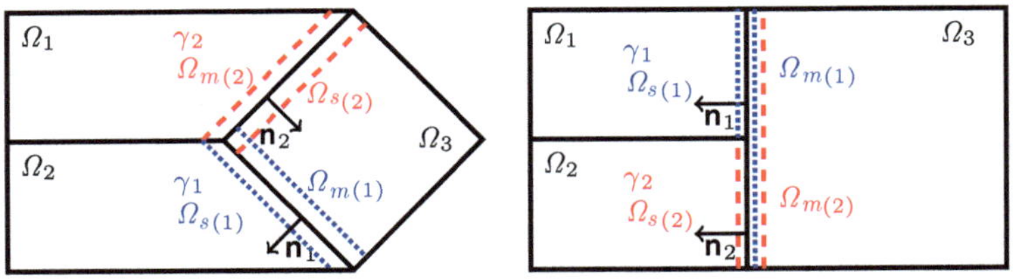

Fig. 1 Geometrical conforming case (*left*) and slave conforming case (*right*)

interfaces be enumerated by γ_l, $l = 1, \ldots, L$, and let us define the skeleton $\Gamma = \bigcup_{l=1}^{L} \gamma_l$ as the union of all interfaces. For each interface, one of the adjacent subdomains is chosen as the master side and one as the slave side. We denote the index of the former by $m(l)$, the index of the latter one by $s(l)$, and thus $\overline{\gamma}_l = \partial\Omega_{m(l)} \cap \partial\Omega_{s(l)}$. On the interface γ_l, we define the outward normal $\mathbf{n}_l$ of the master side $\partial\Omega_{m(l)}$.

We assume that for each interface the pull-back with respect to the slave domain is a whole face of the unit d-cube in the parametric space, which we call a *slave conforming* situation, see the right setting in Fig. 1. If we also assume that the pull-back with respect to the master domain is a whole face of the unit d-cube, we are in a fully geometrically conforming situation, see the left picture of Fig. 1. We note that in this latter case, the master-slave choice is arbitrary.

For each Ω_k, we introduce the space $H^1_*(\Omega_k) = \{v_k \in H^1(\Omega_k), v_{k|\partial\Omega\cap\partial\Omega_k} = 0\}$, where we use standard Sobolev spaces, as defined in [15], endowed with their usual norms. In order to set a global functional framework on Ω, we consider the broken Sobolev space $V = \prod_{k=1}^{K} H^1_*(\Omega_k)$, endowed with the broken norm $\|v\|_V^2 = \sum_{k=1}^{K} \|v\|^2_{H^1(\Omega_k)}$. For any interface $\gamma_l \subset \partial\Omega_{s(l)}$, we define by $H^{-1/2}(\gamma_l)$ the dual space of $H^{1/2}_{00}(\gamma_l)$, which is the space of all functions that can be trivially extended on $\partial\Omega_{s(l)} \setminus \gamma_l$ by zero to an element of $H^{1/2}(\partial\Omega_{s(l)})$.

The mortar method is based on a weak coupling between different subdomains. Each subdomain is discretized independently and a weak coupling is performed on each interface.

2.3 *Isogeometric Mortar Discretization*

In the following, we set our non-conforming approximation framework. On each subdomain Ω_k, based on the NURBS parametrization, we introduce the approximation space $V_{k,h} = \{v_k = \hat{v}_k \circ \mathbf{F}_k^{-1}, \hat{v}_k \in N^{p_k}(\boldsymbol{\Xi}_k)\}$. We recall that under the assumptions on the mesh $\mathcal{M}_{k,h}$ and on the parametrization $\mathbf{F}_k$, this NURBS space has optimal approximation properties, see, e.g., [2]. On Ω, we define the discrete product space $V_h = \prod_{k=1}^{K} V_{k,h} \subset V$, which forms a $H^1(\Omega)$ non-conforming space discontinuous over the interfaces. We denote in the following the maximal mesh size $h = \max_k h_k$ as the mesh parameter.

On the skeleton Γ, we define the discrete Lagrange multiplier product space M_h as $M_h = \prod_{l=1}^{L} M_{l,h}$, where $M_{l,h}$ has to be selected carefully.

From now on, we assume that jumps of α and β are solely located at the skeleton, and we define the bilinear and linear forms $a: V \times V \to \mathbb{R}$ and $f: V \to \mathbb{R}$, such that

$$a(u, v) = \sum_{k=1}^{K} \int_{\Omega_k} \alpha \nabla u \cdot \nabla v + \beta\, u\, v \, \mathrm{d}\mathbf{x}, \quad f(v) = \sum_{k=1}^{K} \int_{\Omega_k} f v \, \mathrm{d}\mathbf{x}$$

as well as

$$b(\lambda, v) = \sum_{l=1}^{L} \int_{\gamma_l} \lambda [v]_l \, \mathrm{d}\sigma,$$

where $[\cdot]_l$ denotes the jump from the master to the slave side over γ_l. The saddle point formulation of the isogeometric mortar method reads as follows: Find $(u_h, \lambda_h) \in V_h \times M_h$, such that

$$a(u_h, v_h) + b(\lambda_h, v_h) = f(v_h), \quad v_h \in V_h, \tag{2a}$$

$$b(\mu_h, u_h) = 0, \quad \mu_h \in M_h. \tag{2b}$$

We note that the Lagrange multiplier λ_h gives an approximation of the normal flux across the skeleton.

It is well known from the theory of mixed and mortar methods that two abstract requirements on each interface guarantee the method to be well-posed and of optimal order, see [4]. Namely, an appropriate approximation order of the dual space and a uniform inf-sup stability between the primal space and the dual space.

The first choice $M^0_{l,h}$ is a spline space of degree $p_{s(l)}$, defined on the interface γ_l based on the interface knot vector of the slave body $\Omega_{s(l)}$. Note that in the presence of any cross point, a suitable modification, e.g., a local degree reduction as presented in [8, Section 4.3], has to be applied. The alternative choice $M^2_{l,h}$ is an order $(p_{s(l)}-2)$ spline space defined on the interface γ_l based on the interface knot vector of the slave body $\Omega_{s(l)}$ for which the definition requires the trace space of $V_{s(l),h}$ to be a subset of $C^1(\gamma_l)$. We point out that although this combination is uniformly inf-sup stable with respect to the mesh-size, it results in an order reduction of $1/2$. The alternative candidate, a spline space of degree $(p_{s(l)} - 1)$ is not uniformly inf-sup stable, so we do not consider it any further.

Note that for simplicity of notations, we assume the same type of dual space to be used for all interfaces. The following theorem is shown in [8] and guarantees a-priori bounds.

Theorem 1 *Let* $\theta = 0$ *if* $M_{l,h} = M^0_{l,h}$ *and* $\theta = 1/2$ *if* $M_{l,h} = M^2_{l,h}$. *For* $u \in H^{\sigma+1}(\Omega)$, $1/2 < \sigma \leq \min_k(p_k - \theta)$, *the weak solution of* (1) *and* (u_h, λ_h) *the* *non-conforming approximation, see* (2)*, it holds*

$$\frac{1}{h^2}\|u-u_h\|^2_{L^2(\Omega)}+\|u-u_h\|^2_V\leq C\sum_{k=1}^{K}h_k^{2\sigma}\|u\|^2_{H^{\sigma+1}(\Omega_k)}$$

for the primal solution u and

$$\sum_{l=1}^{L}\|\,\alpha\nabla u\cdot\mathbf{n}_l|_{\Omega_{s(l)}}-\lambda_h\|^2_{H^{-1/2}(\gamma_l)}\leq C\sum_{k=1}^{K}h_k^{2\sigma}\|u\|^2_{H^{\sigma+1}(\Omega_k)}$$

for the dual solution $\lambda=\alpha\nabla u\cdot\mathbf{n}_l|_{\Omega_{s(l)}}$*. With* $0<C<\infty$ *a generic constant that is independent of the mesh sizes but possibly depends on* p_k*.*

We highlight that while for both pairings the inf-sup stability is satisfied, the approximation order of the lower order dual space $M^2_{l,h}$ is sub-optimal. Indeed, in this case an order of one half is lost in the convergence order.

3 Mortar Integrals

To evaluate the bilinear form $b(\lambda,v)$ (and analogously $b(\mu,u)$), we need to evaluate for each interface γ_l the mortar integrals $\int_{\gamma_l}\lambda\ v^+\mathrm{d}\sigma$ and $\int_{\gamma_l}\lambda\ v^-\mathrm{d}\sigma$, where v^+ denotes the trace of v from the master domain $\Omega_{m(l)}$, and v^- the trace of v from the slave domain $\Omega_{s(l)}$. The Lagrange multiplier λ is based on the mesh of the slave domain $\Omega_{s(l)}$. To simplify the notation, let us restrict ourselves to the case of one single interface and drop the index l in the following.

One particular challenge in the realization of a mortar method is the evaluation of the first interface integral, referred as the master-slave mortar integral, due to the product $\lambda\ v^+$ of functions each of which being defined on a different mesh, see [7] for a method to bypass it in a finite element/wavelet context. Any quadrature rule based on the slave mesh does not respect the mesh lines of the master mesh and vice versa for a quadrature based on the master mesh.

It is obvious that the use of a suitable quadrature rule based on a merged mesh, i.e., a mesh which respects the reduced smoothness of the master and slave functions at their respective mesh lines, leads to an exact evaluation of the integral. However, the construction of this auxiliary mesh commonly named segmentation process is challenging, especially in the three dimensional case since the shape of the elements varies and is difficult to determine, see, e.g., [12, 16, 19, 21, 22]. Note that in an isogeometric context the merged mesh needs to be constructed in the physical space and then pulled back to the parametric space for each subdomain. The complexity of constructing such a mesh becomes even more severe in the case of non-linear and time-dependent problems, where the relative position of the meshes changes in every time or load step which implies to recompute the merged mesh at every step.

Due to this computational complexity, it has been seen very appealing to use a higher order quadrature rule either based on the slave mesh or on the master mesh to approximate the master-slave mortar integral, see [11, 14, 24] for some applications in the finite element and the isogeometric analysis context. However in the finite element case, early results in [9, 18] showed that this strategy does not necessarily yield optimal methods. More precisely, in the case that only the master mesh is chosen, the best approximation error is affected, while for the case only the slave mesh is chosen it is the consistency error. Numerical results confirmed the lack of optimality with the master integration approach, while with the slave integration approach reasonable results were obtained although not optimal in terms of the Lagrange multiplier norm.

Due to the global smoothness of splines, one could expect the sensitivity with respect to the quadrature rules for isogeometric methods to be less than for finite element methods. In the mortar context, according to the finite element results, it seems interesting to consider a slave integration rule. And, in case of maximal regularity, i.e., $V_{k,h} \subset C^{p_k-1}(\Omega_k)$ one also might expect the quadrature error on non-matching meshes to be significantly smaller than in the finite element case. These preliminary observations motivate us to study the different cases numerically.

Let us denote the quadrature rule based on the boundary mesh of the slave domain as $\sum_-$, i.e., $\int_\gamma \lambda\ v^+ \mathrm{d}\sigma \approx \sum_- \lambda\ v^+$. We precise that in the examples a Gaussian quadrature rule is used, and we vary the number of Gauss nodes. In all cases, we choose sufficiently many nodes, such that the integration on a merged mesh would have been exact. The mortar method with pure slave integration is obtained by evaluating all interface integrals in (2) using this quadrature rule, i.e., the discrete system reads as follows: Find $(\tilde{u}_h, \tilde{\lambda}_h) \in V_h \times M_h$, such that

$$a(\tilde{u}_h, v_h) + \sum\nolimits_- (v_h^+ - v_h^-)\ \tilde{\lambda}_h = f(v_h), \quad v_h \in V_h,$$
$$\sum\nolimits_- (\tilde{u}_h^+ - \tilde{u}_h^-)\ \mu_h = 0, \quad \mu_h \in M_h.$$

The notation $\tilde{\cdot}$ is used to stress the difference to the discrete solution with exact integration.

In the next section, we present numerical examples which show severe deviations even in the isogeometric case. Hence, even though the global smoothness of the integrated function is increased compared to the finite element case, an integration approach based only on one mesh reduces the convergence order drastically.

Moreover, we consider an alternative approach which was proposed in [9, 18] using both integration rules. Additionally denoting $\sum_+$ a quadrature rule based on the boundary mesh of the master domain Ω_m, this approach, resulting in a non-symmetric saddle point problem, reads as follows: Find $(\tilde{u}_h, \tilde{\lambda}_h) \in V_h \times M_h$, such that

$$a(\tilde{u}_h, v_h) + \sum\nolimits_+ v_h^+\ \tilde{\lambda}_h - \sum\nolimits_- v_h^-\ \tilde{\lambda}_h = f(v_h), \quad v_h \in V_h,$$
$$\sum\nolimits_- (\tilde{u}_h^+ - \tilde{u}_h^-)\ \mu_h = 0, \quad \mu_h \in M_h.$$

The non-symmetric saddle point problem, which corresponds to a Petrov–Galerkin approach in the primal formulation, was motivated by different requirements for the integration of the primal and dual test functions. Numerical examples showed error values very close to the case of exact integration, but we note that from the theoretical side even the well-posedness of the non-symmetric saddle point problem remains open. In the next section, we present numerical examples which show that also in an isogeometric context, the results are generally close to those from the exact integration case.

4 Numerical Results

In this section, we consider two-dimensional and three-dimensional settings in order to observe the effects of inexact quadrature rules, as presented in Sect. 3, on the optimality of the mortar method. We first set the problem settings and then give the results of several studies.

4.1 Two-Dimensional Example

As a first example, let us consider the Poisson problem $-\Delta u = f$ solved on the domain $\Omega = (0,1)\times(-1,1)$ which is decomposed into two patches by the interface $\gamma = \{(x,y) \in \Omega,\ y = 0\}$. The upper domain is set as the slave domain. The internal load and the boundary conditions are manufactured to have the analytical solution

$$u(x,y) = \cos(\pi x)\,(\cos\left(\frac{\pi}{2}y\right) + \sin(2\pi y)).$$

The normal derivative on the interface is given by $\partial u/\partial \mathbf{n}(x) = 2\pi\cos(\pi x)$, see Fig. 2. Neumann conditions are applied on the left and right boundary parts, such that no cross point modification is necessary.

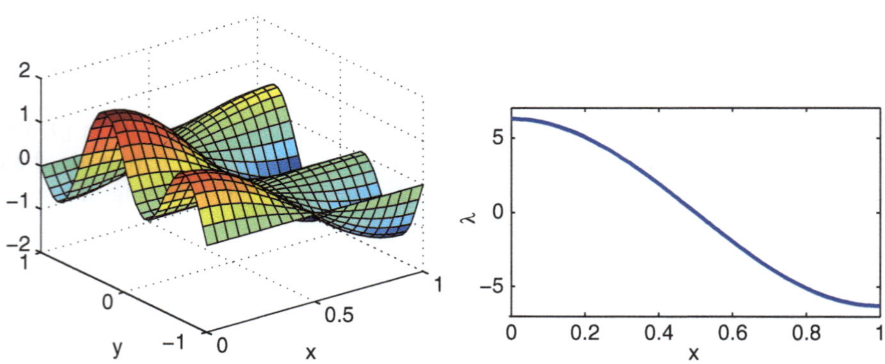

Fig. 2 *Left*: Primal solution on Ω. *Right*: Lagrange multiplier along the interface

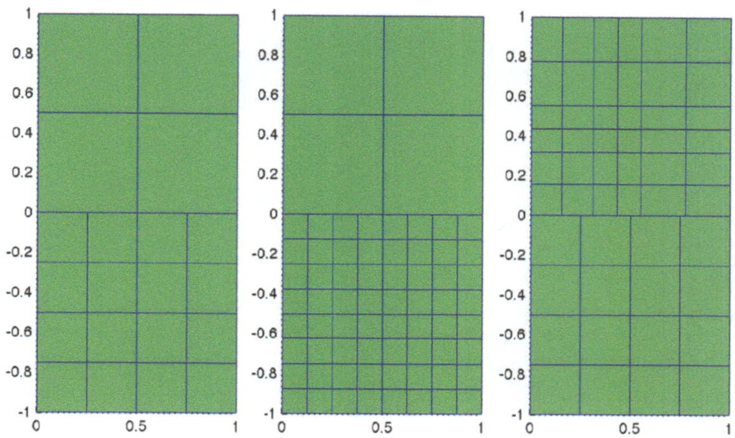

Fig. 3 Different meshes at mesh refinement level 1. From the left to the right: $M1$ to $M3$

Regarding the meshes, we consider three different cases, presented in Fig. 3. In the first two cases, the initial master mesh is a refinement of the initial slave mesh. The initial slave mesh consists of just one element. In the case $M1$, one uniform refinement step is applied to build the master mesh, in the case $M2$ two uniform refinement steps. Case $M3$ was chosen such that at no refinement level parts of the slave and master boundary meshes do coincide. The initial interior knots of the slave domain were chosen as $\{\pi/10, 1-\pi/7\}$ in both parametric directions, yielding nine elements. The initial master mesh consists of four uniform elements.

The cases $M1$ and $M2$ serve as a simple test to investigate the influence of the quadrature error. We note that inverting the role of the master and slave domains is not interesting in these cases as both integration approaches are able to exactly evaluate the master-slave mortar integral.

In the following, we provide different numerical error studies. Starting from the initial mesh, denoted by refinement level 0, we perform uniform refinements for the slave and the master domains. We note that the inter-element smoothness of the dual functions can influence the accuracy of the quadrature based on the master mesh, but not the one based on the slave mesh. Therefore for the slave integration approach, the equal order pairing with maximal smoothness is considered, i.e., $M_h = M_h^0 \subset C^{p-1}(\gamma)$, while for the non-symmetric approach we vary the dual degree. In all cases, the primal $L^2(\Omega)$ and the dual $L^2(\gamma)$ errors are computed by a comparison with the analytical solution stated above.

4.1.1 Slave Integration Approach

Firstly, we consider the case $M3$, see Fig. 3, to measure the impact of the integration error in a general situation. A numerical error study is provided in Fig. 4 for a

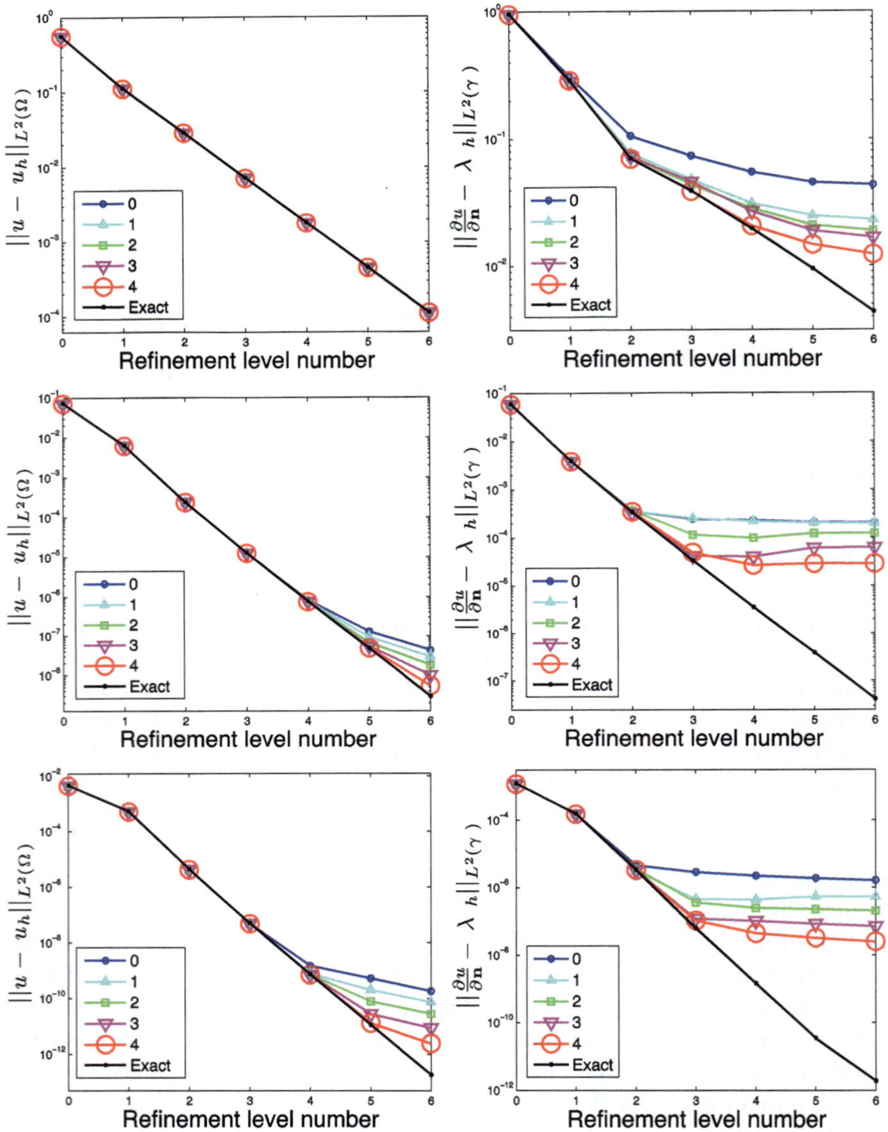

Fig. 4 2D results – L^2 primal (*left*) and dual (*right*) error curves for the case $M3$: equal order pairings with $p = 1, 3, 5$ (from top to bottom) for the slave integration approach and a different number of additional quadrature points

different number of additional Gauss points and different spline degrees. For a spline of degree p, we start with $p + 1$ Gauss points and investigate the effect of using a higher number of Gauss nodes. It can clearly be seen that the primal and dual solutions are both affected by the inexact quadrature, leading to non-optimal

Table 1 2D results – Last estimated order of convergence of the primal and dual L^2 errors for the cases $M1$ and $M2$: pairing $P5 - P5$ for the slave integration approach and a different number of additional quadrature points

Add.	Primal error		Dual error	
q.p.	Case $M1$	Case $M2$	Case $M1$	Case $M2$
0	1.63	1.74	0.50	0.50
1	1.63	1.54	0.50	0.50
2	1.63	1.55	0.50	0.50
3	1.63	1.58	0.50	0.50
4	1.63	1.56	0.50	0.50
5	1.63	1.50	0.50	0.50

methods. In all cases, the same characteristic behavior can be seen. Up to a certain refinement level, the results with inexact quadrature rules coincide with the ones with no quadrature error. Then, at a certain refinement level, the convergence order is reduced and the error is significantly larger than the exact integration one. The level, where this effect starts depends on the considered error norm, the order p and the number of additional Gauss nodes. Moreover, in this situation the higher order splines are more sensitive to the numerical quadrature approximation than the lower order splines.

In almost all cases of Fig. 4, we observe poor approximation results and a reduced convergence order which is numerically independent of the spline degree. Especially, the rate of the $L^2(\gamma)$ dual error is very low and in some cases no convergence can be observed anymore.

Secondly, we consider a simpler situation to show that even then the impact of the slave integration is noticeable. Let us focus on the cases $M1$ and $M2$, see Fig. 3, for which the master mesh is a refinement of the slave mesh. See Fig. 5 for a comparison of results between the cases $M1$ and $M2$ for a spline degree $p = 3$. We note that the low convergence orders of the primal and dual solutions, as remarked above, already appear in this simple context. Moreover, for a fixed number of slave elements, the error is increasing with the number of master elements. This is expected as there are more points of reduced smoothness which are not taken into account by the quadrature rule. Let us now consider the final numerical convergence rate in more details. In Table 1, estimated convergence orders for degree $p = 5$ are given. We notice that the dual $L^2(\gamma)$ rate breaks down to an order of $1/2$, while the $L^2(\Omega)$ primal rate lies about $3/2$.

Thirdly, we have additionally compared the case $M3$ with a similar situation in which the master and slave roles are reverted. The results also show that the integration error is increasing with the increase of the master element number. Thus, in accordance to the practical applications, in a slave integration context it seems worthwhile to choose the slave domain as the finest one.

Moreover, it can be observed that on coarse meshes using the slave integration method it is possible to recover the accuracy of the optimal mortar method simply by increasing the number of quadrature points, see Fig. 6. However, it has also been shown that the number of necessary quadrature points is drastically increasing with the refinement level. It can easily be seen that the number of Gauss points gets soon

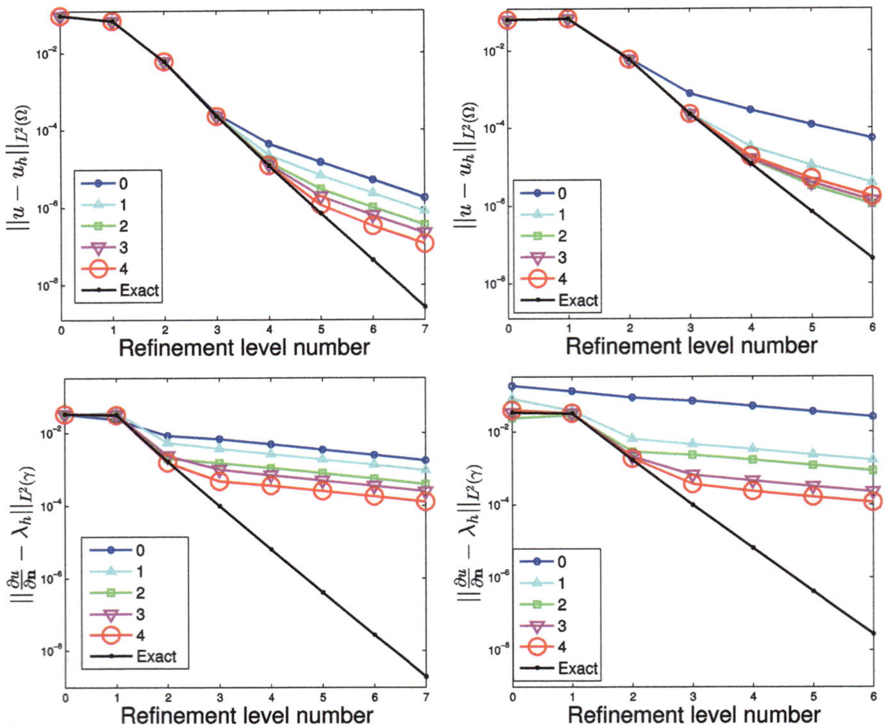

Fig. 5 2D results – L^2 primal (*top*) and dual (*bottom*) error curves for the cases $M1$ (*left*) and $M2$ (*right*): pairing $P3 - P3$ for the slave integration approach and a different number of additional quadrature points

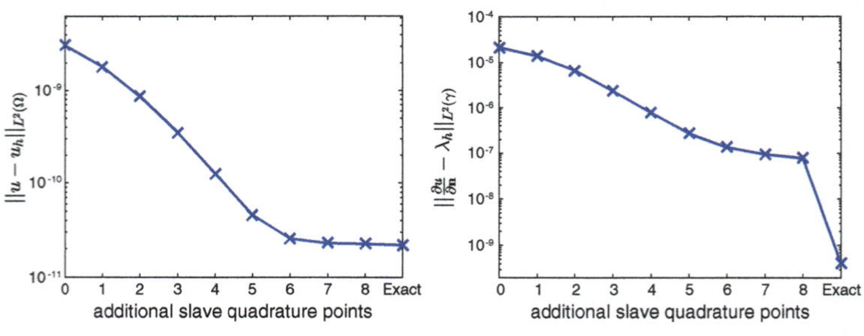

Fig. 6 2D results – L^2 primal (*left*) and dual (*right*) errors at refinement level number 6 as a function of the number of additional quadrature points for the case $M3$: pairing $P3 - P3$ for the slave integration approach

impracticably large, see the right picture of Fig. 6. Furthermore, in several cases, the deviation to the mortar method has been observed to be more severe for higher order functions.

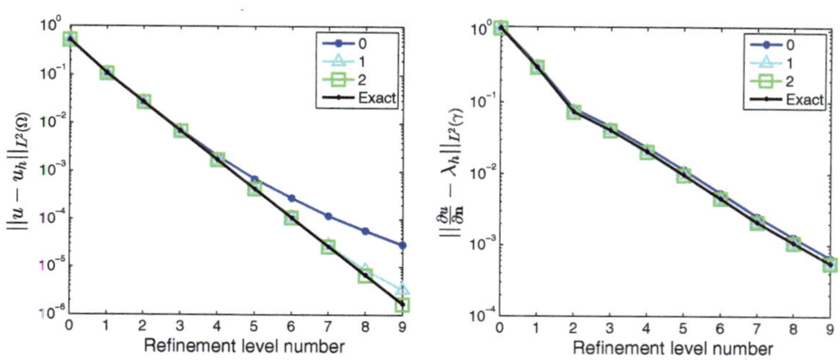

Fig. 7 2D results – L^2 primal (*left*) and dual (*right*) error curves for the case $M3$: equal order pairing $p = 1$ for the non-symmetric approach and a different number of additional quadrature points

4.1.2 Non-symmetric Approach

The non-symmetric saddle point problem based on the two different quadrature rules, see Sect. 3, was introduced to overcome the non-optimality of the pure slave integration approach in a finite element context. Due to the suboptimal results seen in the previous section, it is also interesting to consider it in an isogeometric context.

Firstly, we consider same degree pairings. In almost all tested cases, the results of the non-symmetric approach are comparable to the results of the exact integration case. However, we note that differences could still be seen in some cases. For example, for a degree $p = 1$ in the case $M3$, we obtained a non-optimal method, see in Fig. 7 the corresponding primal and dual error curves. Note that we do not show any curves in the cases where no disturbance is observed. For example for degree $p = 5$, we observed convergence almost up to machine precision without any remarkable difference compared to the exact integration case.

Secondly, we consider dual spaces with lower degrees than the primal ones. Note that in [8] stability for these pairings was only observed if the primal and the dual degrees have the same parity. Similar to the previously considered equal order case, the dual error did not show a significant deviation by the non-symmetric approach. In Fig. 8, primal error curves are shown for all stable different degree pairings up to a primal degree $p = 4$. We note that theoretically, we expect sub-optimal primal error rates even in the exact integration case, although often improved convergence rates were observed. For a dual degree $p - 2k$, $k \in \mathbb{N}$, we can expect a convergence of order $\mathcal{O}(h^{p-2k+5/2})$ in the $L^2(\Omega)$ norm, see the dashed lines in Fig. 8. For the $P4 - P2$ and $P3 - P1$ pairings, we observe small differences compared to the exact integration results, but note that the convergence rate is not significantly different than the theoretical expectation. The situation is different for the $P4 - P0$ and $P2 - P0$ pairings, for which the rate is more disturbed and even below the theoretical expectation. This can be explained by the discontinuity of the dual basis functions which introduces large errors in the integration approximation done with a rule based on the master mesh, which does not respect these discontinuities.

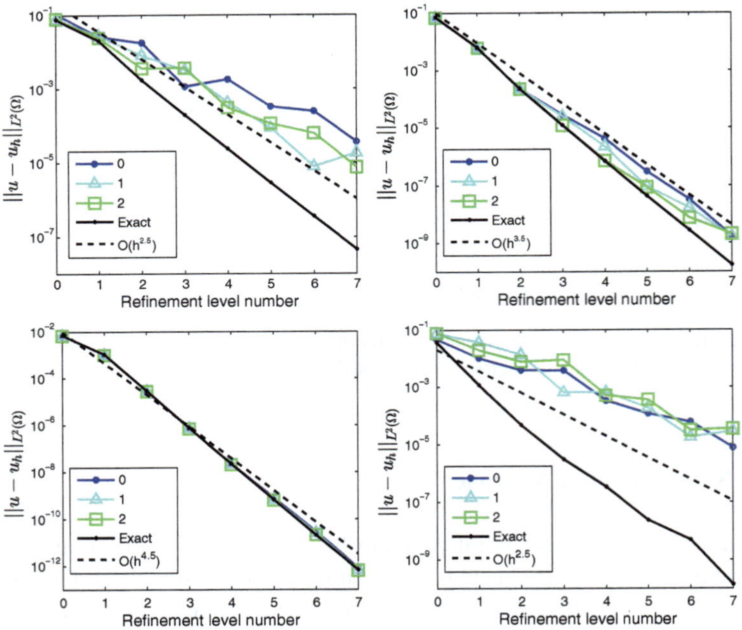

Fig. 8 2D results – L^2 primal error curves for the case $M3$: different order pairings for the non-symmetric approach and a different number of additional quadrature points. *Top left*: $P2 - P0$. *Top right*: $P3 - P1$. *Bottom left*: $P4 - P2$. *Bottom right*: $P4 - P0$

4.2 Three-Dimensional Example

As a second example, we consider a three-dimensional problem with a curved interface. Precisely, we consider the Poisson problem $-\Delta u = f$ on the domain $\Omega = (0, 1)^3$, which is divided into two patches by the interface $\gamma = \{(x, y, \rho(x)), (x, y) \in (0, 1)^2\}$, with $\rho(x, y) = 1/8\,(1 + x)(1 + y^2) + 1/5$, see Fig. 9. The bottom domain is set as the slave domain. The internal load and the boundary conditions are manufactured to have the analytical solution

$$u(x, y, z) = \cos(2\pi x)\cos(2\pi y)\sin(2\pi z).$$

Note that due to the curved interface, the normal derivative has a complex form, but is still explicitly computable. Neumann conditions are applied such that no cross point modification is necessary. The initial master mesh has 8 uniform elements, while the initial slave mesh has 8 elements given by the breakpoint vector $\{0, \pi/5, 1\}$ in each direction. In the following, we provide some numerical error studies, considering the slave integration approach as well as the non-symmetric approach.

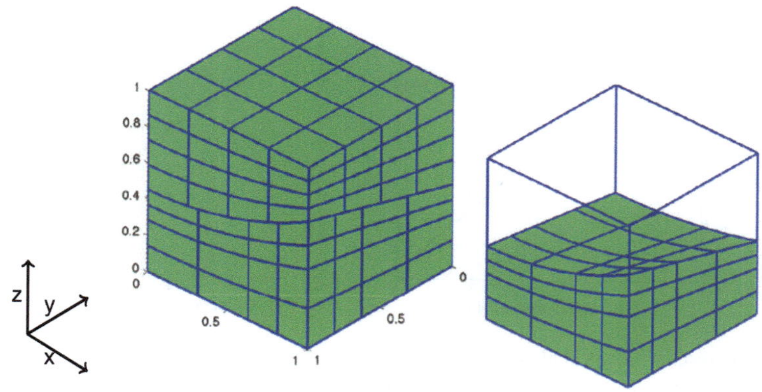

Fig. 9 Meshes at refinement level 1 (*left*) and the slave domain (*right*) illustrating the curved interface

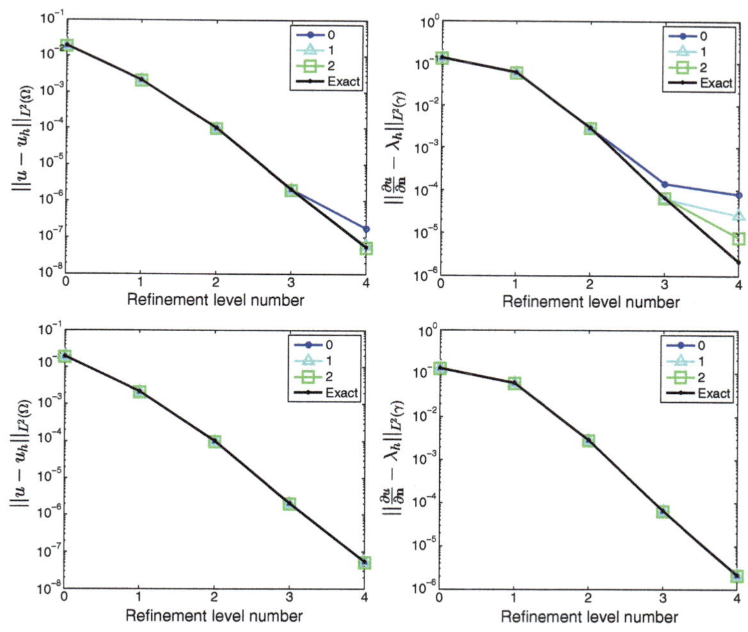

Fig. 10 3D results – L^2 primal (*left*) and dual (*right*) error curves for the pairing $P4 - P4$, for the slave integration approach (*top*) and the non-symmetric approach (*bottom*). Each of the curves being obtained with a different number of additional quadrature points

The obtained results are in accordance with the two-dimensional case results for both approaches. In Fig. 10, the deviation for the slave integration approach is shown for the $P4 - P4$ pairing. Although not shown here, we note that the results for the $P2 - P2$ and $P3 - P3$ pairing have a similar behavior. The non-symmetric approach does not lead to reduced rates considering equal order pairings, i.e., $M_h = M_h^0$, on

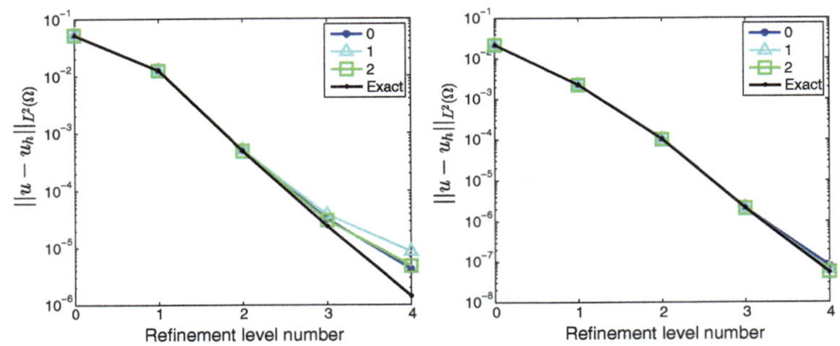

Fig. 11 3D results – L^2 primal error curves for the pairings $P3 - P1$ (*left*) and $P4 - P2$ (*right*), for the non-symmetric approach and a different number of additional quadrature points

the refinement levels we considered. As previously, using a lower order dual space, a difference to the exact integration case can be seen. See Fig. 11 for the disturbance in the primal variable of the $P3 - P1$ and $P4 - P2$ pairings.

5 Conclusion

After briefly reviewing optimal isogeometric mortar methods, a study on the possibility to approximate the mortar integrals by efficient numerical quadrature rules was performed.

The precise evaluation of mortar integrals is of a high computational complexity, since it requires the construction of a merged mesh, combining the non-matching meshes at the interface. While it would be desirable to use a quadrature rule based on the slave mesh, numerical examples show that this induces large deviations to the mortar method. Especially the convergence rate of the Lagrange multiplier is reduced to at most $1/2$, independently of the spline degree. While the method improves by increasing the number of quadrature points, the amount of points necessary to obtain nearly optimal results is not predictable.

To overcome these difficulties, we have considered a non-symmetric saddle point problem based on both master and slave integration rules, which was previously introduced in the finite element context. Numerical examples demonstrate that for most cases the reached accuracy is close to the optimal one.

Acknowledgements The first author has been supported by Michelin under the contract A10-4087. The third and the fourth authors have been supported by the International Research Training Group IGDK 1754, funded by the German Research Foundation (DFG) and the Austrian Science Fund (FWF), and by the German Research Foundation (DFG) in Project WO 671/15-1 and the Priority Programme "Reliable Simulation Techniques in Solid Mechanics. Development of Non-standard Discretisation Methods, Mechanical and Mathematical Analysis" (SPP 1748). The last author was supported by the Elite Network of Bavaria through its graduate program TopMath and the TUM Graduate School through its Thematic Graduate Center TopMath. All supports are gratefully acknowledged.

References

1. A. Apostolatos, R. Schmidt, R. Wüchner, K.U. Bletzinger, A Nitsche-type formulation and comparison of the most common domain decomposition methods in isogeometric analysis. Int. J. Numer. Methods Eng. **97**, 473–504 (2014)
2. Y. Bazilevs, L. Beirão da Veiga, J.A. Cottrell, T.J.R. Hughes, G. Sangalli, Isogeometric analysis: approximation, stability and error estimates for h-refined meshes. Math. Models Methods Appl. Sci. **16**(7), 1031–1090 (2006)
3. L. Beirão Da Veiga, A. Buffa, G. Sangalli, R. Vásquez, Mathematical analysis of variational isogeometric methods. Acta Numer. **23**, 157–287 (2014)
4. F. Ben Belgacem, The mortar finite element method with Lagrange multipliers. Numer. Math. **84**, 173–197 (1999)
5. F. Ben Belgacem, Y. Maday, The mortar finite element method for three dimensional finite elements. Math. Model. Numer. Anal. **31**(2), 289–302 (1997)
6. C. Bernardi, Y. Maday, A.T. Patera, A new nonconforming approach to domain decomposition: the mortar element method, in *Nonlinear Partial Differential Equations and Their Applications*, vol. XI, ed. by H. Brezis, J.L. Lions (Collège de France, Paris, 1994), pp. 13–51
7. S. Bertoluzza, S. Falletta, An object-oriented implementation of the mortar method with approximate constraint. Technical report, Istituto di Matematica Applicata e Tecnologie Informatiche, Consiglio Nazionale delle Ricerche, Pavia (2004)
8. E. Brivadis, A. Buffa, B. Wohlmuth, L. Wunderlich, Isogeometric mortar methods. Comput. Methods Appl. Mech. Eng. **284**, 292–319 (2015)
9. L. Cazabeau, C. Lacour, Y. Maday, Numerical quadrature and mortar methods, in *Computational Science for the 21st Century* (Wiley, Chichester, 1997), pp. 119–128
10. J.A. Cottrell, T.J.R. Hughes, Y. Bazilevs, *Isogeometric Analysis. Towards Integration of CAD and FEA* (Wiley, Chichester, 2009)
11. L. De Lorenzis, I. Temizer, P. Wriggers, G. Zavarise, A large deformation frictional contact formulation using NURBS-based isogeometric analysis. Int. J. Numer. Methods Eng. **87**, 1278–1300 (2011)
12. M. Dittmann, M. Franke, I. Temizer, C. Hesch, Isogeometric analysis and thermomechanical mortar contact problems. Comput. Methods Appl. Mech. Eng. **274**, 192–212 (2014)
13. W. Dornisch, G. Vitucci, S. Klinkel, The weak substitution method – an application of the mortar method for patch coupling in NURBS-based isogeometric analysis. Int. J. Numer. Methods Eng. (2015). doi:10.1002/nme.4918
14. K. Fischer, P. Wriggers, Frictional 2D contact formulation for finite deformations based on the mortar method. Comput. Mech. **36**, 226–244 (2005)
15. P. Grisvard, *Elliptic Problems in Nonsmooth Domains* (SIAM, Philadelphia, 2011)
16. C. Hesch, P. Betsch, Isogeometric analysis and domain decomposition methods. Comput. Methods Appl. Mech. Eng. **213–216**, 104–112 (2012)
17. T.J.R. Hughes, J.A. Cottrell, Y. Bazilevs, Isogeometric analysis: CAD, finite elements, NURBS, exact geometry and mesh refinement. Comput. Methods. Appl. Mech. Eng. **194**, 4135–4195 (2005)
18. Y. Maday, F. Rapetti, B.I. Wohlmuth, The influence of quadrature formulas in 2D and 3D mortar element methods, in *Recent Developments in Domain Decomposition Methods. Some Papers of the Workshop on Domain Decomposition*, ETH Zürich, 7–8 June 2001 (Springer, 2002), pp. 203–221
19. T. McDevitt, T. Laursen, A mortar-finite element formulation for frictional contact problems. Int. J. Numer. Methods Eng. **48**, 1525–1547 (2000)

20. L. Piegl, W. Tiller, *The NURBS Book* (Springer, Berlin/New York, 1997)
21. M. Puso, A 3D mortar method for solid mechanics. Int. J. Numer. Methods Eng. **59**, 315–336 (2004)
22. M. Puso, T. Laursen, A mortar segment-to-segment contact method for large deformation solid mechanics. Comput. Methods Appl. Mech. Eng. **193**, 601–629 (2004)
23. L. Schumaker, *Spline Functions: Basic Theory*, 3rd edn. (Cambridge University Press, Cambridge, 2007)
24. M. Tur, F. Fuenmayor, P. Wriggers, A mortar-based frictional contact formulation for large deformations using Lagrange multipliers. Comput. Methods Appl. Mech. Eng. **198**, 2860–2873 (2009)

The Isogeometric Segmentation Pipeline

Michael Pauley, Dang-Manh Nguyen, David Mayer, Jaka Špeh, Oliver Weeger, and Bert Jüttler

Abstract We present a pipeline for the conversion of 3D models into a form suitable for isogeometric analysis (IGA). The input into our pipeline is a boundary represented 3D model, either as a triangulation or as a collection of trimmed non-uniform rational B-spline (NURBS) surfaces. The pipeline consists of three stages: computer aided design (CAD) model reconstruction from a triangulation (if necessary); segmentation of the boundary-represented solid into topological hexahedra; and volume parameterization. The result is a collection of volumetric NURBS patches. In this paper we discuss our methods for the three stages, and demonstrate the suitability of the result for IGA by performing stress simulations with examples of the output.

1 Introduction

In isogeometric analysis (IGA), the space of functions used to approximate the solution of a differential equation is the same space of functions used to represent the geometry. As such, IGA methods work directly with geometry represented by splines and have begun to close the gap between the geometry representations used for analysis and those used for design [1]. An additional benefit of IGA is that it can achieve the same accuracy as the finite element method (FEM) with fewer degrees of freedom (see [4]).

In order to realize the benefits of IGA when dealing with complex geometries, it is necessary to segment them into IGA-suitable pieces. A rich theory of FEM-suitable mesh generation exists (see e.g., [5, 8, 14, 15, 17, 21, 24]) but these methods

M. Pauley (✉) • D.-M. Nguyen • J. Špeh • B. Jüttler • D. Mayer
Institute of Applied Geometry, Johannes Kepler University, Altenberger Straße 69, 4040 Linz, Austria
e-mail: Michael.Pauley@jku.at; dmnguyen@cise.ufl.edu; Jaka.Speh@jku.at; Bert.Juettler@jku.at; david_mayer@aon.at

O. Weeger
Faculty of Mathematics, TU Kaiserslautern, P.O. Box 3049, 67653 Kaiserslautern, Germany
e-mail: oliver_weeger@sutd.edu.sg

B. Jüttler, B. Simeon (eds.), *Isogeometric Analysis and Applications 2014*, Lecture Notes in Computational Science and Engineering 107,
DOI 10.1007/978-3-319-23315-4_3

aim at a segmentation of a solid into a large number of (approximately uniform) small tetrahedra and hexahedra. The goal of the *isogeometric segmentation pipeline* is to convert a boundary represented model (e.g. something designed in computer aided design (CAD) software) into an IGA-suitable volumetric model. For us this means segmenting the geometry into a small number of *topological hexahedra*, that is, deformed cubes.

Methods for parameterization of solids by a single B-spline or T-spline cube are developed in [11, 19, 25, 26]. A method for multiple patches is given in [22] which requires a predefined segmentation as input. Several papers [9, 10, 20, 23] provide methods for segmentation and parameterization of a solid by B-spline or T-spline cubes, allowing for singularities in the output.

The input into the pipeline can be either a triangulated boundary representation, or a trimmed non-uniform rational B-spline (NURBS) surface boundary representation. A trimmed surface consists of a *master spline* mapping a square into $\mathbb{R}^3$, together with a collection of *trimming curves* defining a domain of the surface as a subset of the square. The boundary representation for the pipeline consists of a collection of trimmed surfaces, together with adjacency information. The output of the pipeline is a representation of the solid as a collection of volumetric NURBS. The pipeline consists of the three following stages.

CAD model reconstruction. A triangulated model is converted into a trimmed NURBS surface boundary representation, for input into the remaining stages. This step is not necessary for models that are already represented with trimmed surfaces. It is necessary when only a triangulated solid is available, or when CAD data is available but uses features not supported by the remainder of the pipeline. In the latter case, the CAD model can be triangulated and then reconstructed to produce a solid in a format compatible with the pipeline.

Isogeometric segmentation. A trimmed NURBS surface boundary represented solid is segmented into a collection of boundary represented topological hexahedra. Our segmentation approach, which began development in [6, 12], is based on the edge graph of the solid – the graph defined by the vertices and edges between the trimmed surfaces.

In its present state, our isogeometric segmentation method requires manual interaction in two scenarios. Firstly, it is applicable to the class of contractible solids with 3-vertex-connected edge graphs. This means that a manual preprocessing step is required to ensure that the input satisfies these assumptions. In our ongoing work we are developing methods to automatically ensure these assumptions. Secondly, a post-processing step may be necessary if the segmentation is required to be free of T-joints. So far, we only have a way to eliminate them in special cases. Our belief is that T-joints will not be a barrier to IGA, as contemporary research is developing methods for handling non-matching interfaces [7].

Volume parameterization. A collection of boundary represented topological hexahedra is converted to a collection of volumetric splines.

Sections 2, 3,and 4 of the paper provide details and examples for the above stages of the pipeline. In Sect. 5, a segmentation produced by our pipeline is used to conduct mechanical simulations, showing the suitability of our results for isogeometric analysis. Section 6 summarizes the state of the pipeline and outlines our ongoing and future work.

2 Segmentation of Triangulated Surfaces and CAD Model Reconstruction

Consider a complex model for which IGA-based simulation is required. If a collection of trimmed NURBS surfaces is provided describing its boundary, we can segment it into topological hexahedra using the methods in Sects. 3 and 4. However, often we only have access to a triangular mesh describing the boundary. This would happen, for example, if a triangulation of a model is available but the original model is not. Furthermore, if a CAD model is of low quality or uses features that are not supported by the segmentation software, it is useful to convert it into a triangle mesh from which a more suitable trimmed surface model can be created. Finally, a more complex CAD model is often obtained by combining several models produced by several users, possibly with different CAD systems. In this situation, using a triangulated CAD model is often the only feasible option.

Sophisticated methods exist for the reverse engineering of CAD data (see [18] and its references). The method we use can afford to be relatively straightforward because (i) we are primarily interested in engineering objects, which tend to have simple faces and sharp edges, compared to general, more free-form designs; (ii) we assume a high quality triangulation is provided as input, not noisy point cloud data; (iii) while more sophisticated methods can reproduce blend surfaces, we do not want them as the later stages of our pipeline are not prepared to deal with them.

- The triangulated boundary is segmented into regions using a *region growing algorithm*. Two triangles are part of the same region if there exists a path between them which passes only between triangles that meet at an angle close to π. Post-processing can be applied to further segment the large regions with a more strict angle tolerance, or to merge small regions with a less strict angle tolerance. An example is shown in Fig. 1.
- For each region, a triangulated domain is constructed whose triangles are in correspondence with the triangles of the region. The points in the interior of the

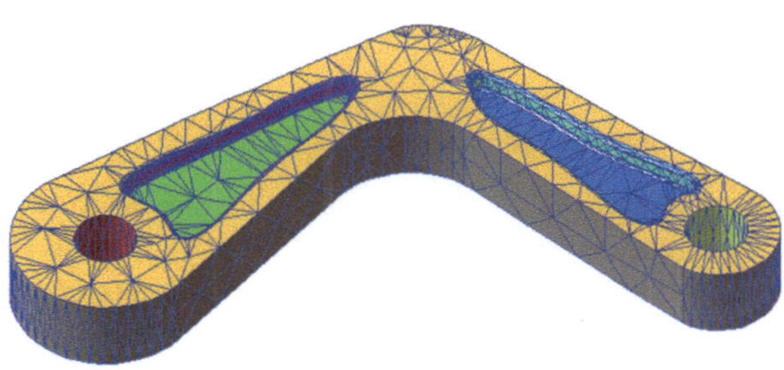

Fig. 1 The region growing algorithm segments the triangulation of the TERRIFIC demonstrator part into nine regions. Different regions are shaded different colors

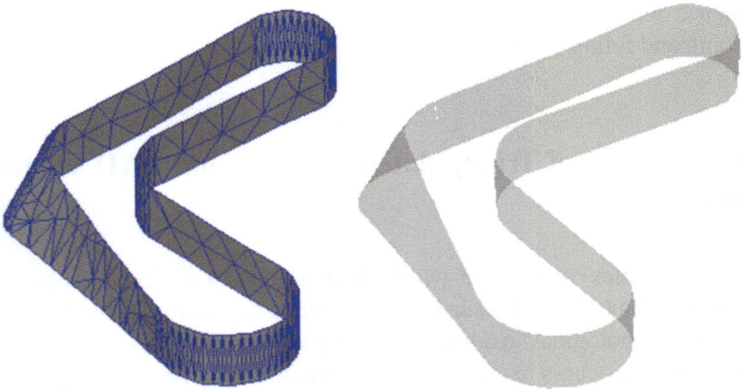

Fig. 2 Example of a NURBS trimmed surface fit to triangulated data

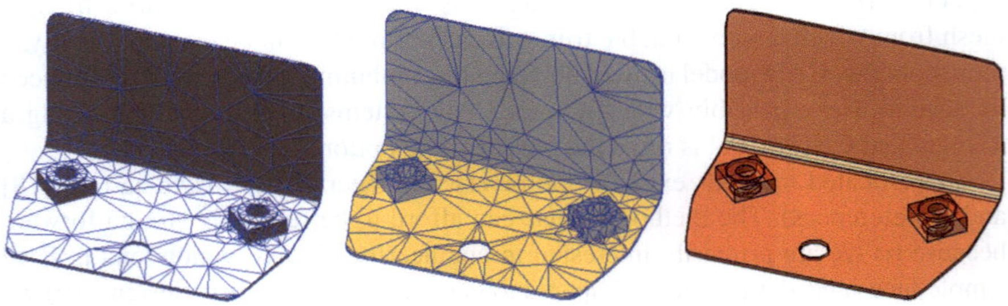

Fig. 3 (**a**) The triangulated car part. (**b**) The model is separated into regions. We highlight one here. (**c**) The CAD model reconstruction (translucent)

domain, and the positions of holes in the domain, are chosen by a method based on [2, 3].

- A tensor product NURBS master surface is constructed, by fitting the triangulation data using a least-squares approach. Finally, sharp corners in the domains are smoothed out where appropriate. An example of a surface fit to a detected region is shown in Fig. 2.

2.1 Examples

We demonstrate the procedure with the following examples.

Car part example. The procedure is applied to a car part consisting of 1,220 triangles. Angle tolerance is set to 40° and a secondary region growing phase is applied, with a tolerance of 20°, to those regions with area more than 20 % of the average. The result is shown in Fig. 3a–c.

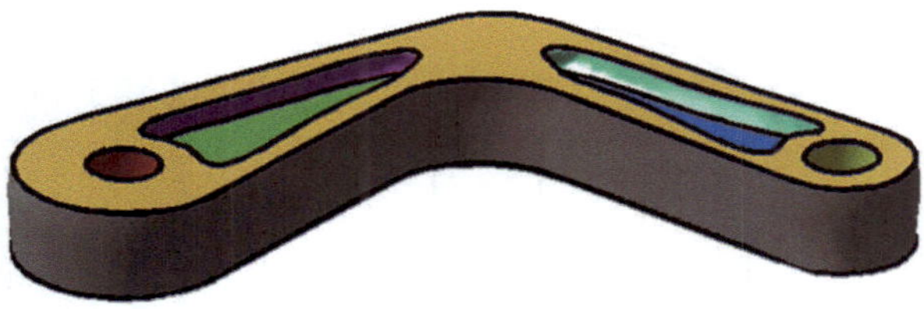

Fig. 4 The TERRIFIC Demonstrator: reconstructed CAD model

TERRIFIC Demonstrator. The TERRIFIC demonstrator model was created as part of the TERRIFIC European Commission project for testing and demonstrating the project's research and development. The demonstrator in its original format is a boundary represented CAD model. However, feeding a triangulated version of the model through the CAD model reconstruction step allows us to clean the model up, eliminating the blends which the later steps of our pipeline do not support.

The model has 3,672 triangles. The region growing algorithm is used with a cutoff of 12° and regions are merged together if they have area less than 200 % of the average. The result is shown in Fig. 4.

3 Isogeometric Segmentation of Boundary Represented Solids

Here we describe our approach to the problem of segmenting a boundary-represented solid into topological hexahedra. The solid has sharp features, and its faces are described by NURBS-based trimmed surfaces.

The faces, edges and vertices of the solid form a structure called the *edge graph* of the solid (see Fig. 5). A graph is 3-vertex-connected if it is connected, it has at least four vertices and any two vertices can be deleted without disconnecting the graph. We make the following assumptions about the solid:

- the solid is contractible;
- the edge graph is simple and 3-vertex-connected.

Once these assumptions are satisfied, we have an automatic process to obtain a segmentation into topological hexahedra. The process is as follows.

1. Ensure that the solid satisfies the assumptions.
2. For each existing non-convex edge, segment the solid using a cutting loop (Sects. 3.1 and 3.2) or an extending loop (Sect. 3.4) containing the edge.
3. If the solid has no non-convex edges, segment it further using cutting loops until we arrive at pre-defined *base solids*.
4. Segment the base solids in a predefined way into topological hexahedra.

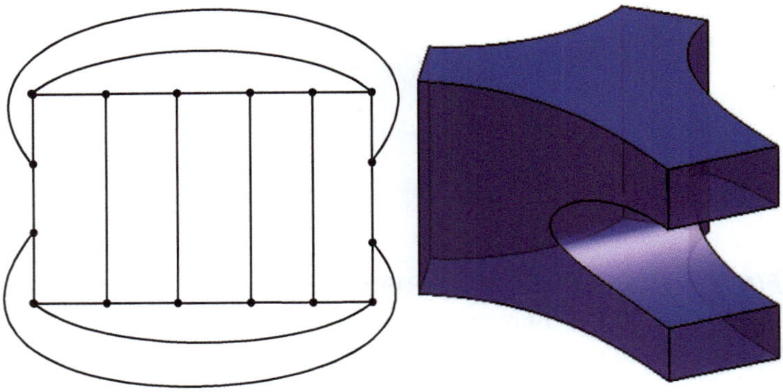

Fig. 5 An example of a contractible solid and a planar representation of its edge graph

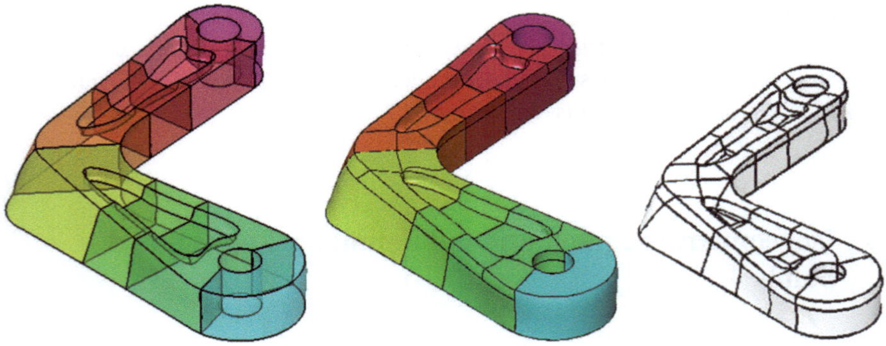

Fig. 6 (**a**) The TERRIFIC Demonstrator, preprocessed to cut it into contractible pieces with 3-vertex-connected edge graphs. (**b**) Segmentation into topological hexahedra. (**c**) The solid is further segmented to eliminate T-joints

Item 1 above is ongoing work: methods based on the concept of a Reeb graph from Morse theory, which has previously been used to good effect in *flow volume decomposition* [16], show promise for dealing with non-contractible solids. Items 2–4 are explained below in Sects. 3.1, 3.2, 3.3 and 3.4.

Figure 6 shows an example: this stage of the pipeline is applied to the TERRIFIC demonstrator. The demonstrator is first preprocessed to cut it into contractible pieces and artificial edges are added to make its graph 3-vertex-connected. Then these pieces are segmented into hexahedra using the methods of cutting loops and extending loops described in the following sections.

3.1 Theory for Contractible Solids with Only Convex Edges

An edge **e** of the solid is *convex* at a point **p** if the two incident faces at **e** meet at a convex interior angle at **p**. The edge **e** is called *convex* if it is convex at every point, and non-convex otherwise. Non-convex edges have a significant impact on the options for segmentation. In this section we restrict to solids with no non-convex edges; we extend our approach to the more general case in Sect. 3.2.

Our approach is based on the concept of a *cutting loop*; a cycle in the edge graph consisting of existing edges as well as *auxiliary edges* which can be created between any two vertices on the same face. A cutting loop is *valid* if it can be used as the boundary of a surface with the following properties:

- the surface is contained in the solid;
- the surface cuts the solid into exactly two pieces;
- both pieces satisfy our original assumptions, i.e., they are contractible and the edge graphs are simple and 3-vertex-connected.

For solids with only convex edges, validity is essentially a combinatorial condition, i.e., the structure of the graph matters but the geometry of the trimmed surfaces does not. This allows us to guarantee the existence of a valid cutting loop.

Theorem 1 ([6]) *Assume that all edges are convex. If the edge graph of the solid is not that of a tetrahedron, then a valid cutting loop exists.*

3.2 Theory for Contractible Solids with Non-convex Edges

Non-convex edges create additional complications. They must be eliminated during the segmentation process, because they prevent smooth parameterization (see Fig. 7). Furthermore, they create new validity criteria for cutting loops, which can sometimes require *auxiliary vertices* (new vertices created along the existing edges of the edge graph) to guarantee the existence (See [12, Section 3]) (Fig. 8).

Theorem 2 ([12]) *Given a non-convex edge of the solid, there exists a valid cutting loop containing that edge and possibly containing auxiliary vertices, such that cutting the solid with this loop eliminates the non-convex edge.*

This enables an extension of the algorithm outlined in Sect. 3.1. Cutting loops are used to cut the solid to remove all the non-convex edges. Once the solid has been segmented into pieces with only convex edges, the method of Sect. 3.1 can be applied.

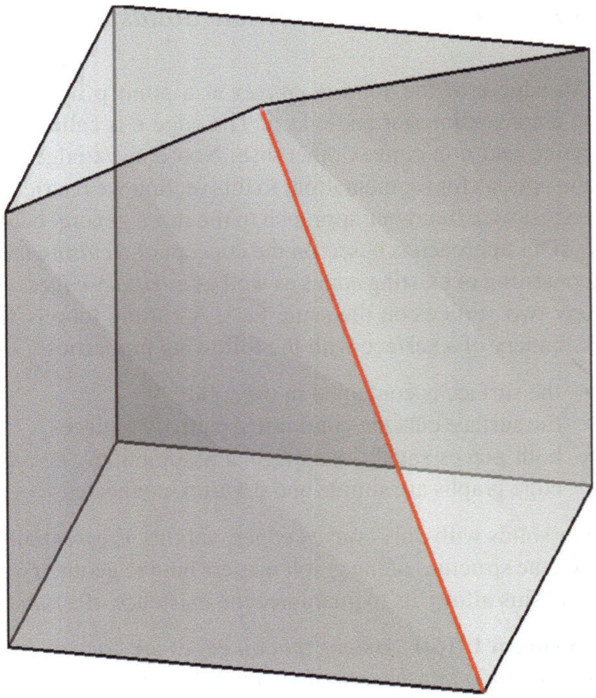

Fig. 7 A solid with a non-convex edge. The edge graph is isomorphic to that of a cube. However, the non-convex edge prevents C^1 volume parameterization by a cube. The segmentation algorithm eliminates non-convex edges as its first priority so that the resulting topological hexahedra have only convex edges

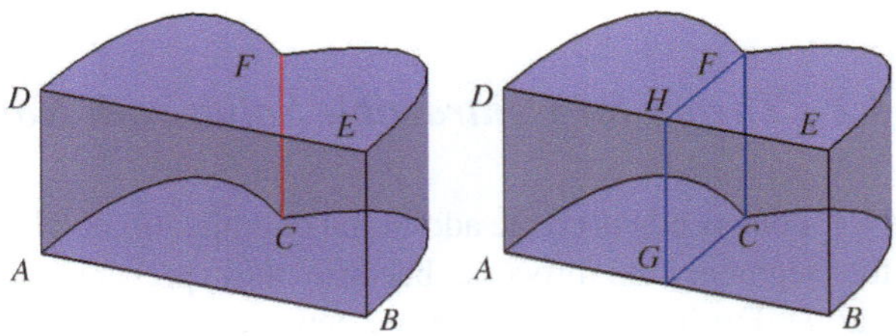

Fig. 8 *Left*: a solid with a non-convex edge CF. Without using auxiliary vertices, there is no valid cutting loop to cut CF, but it becomes possible with the addition of the vertices G and H

3.3 Description of the Algorithm

Theorems 1 and 2 mean that the following algorithm successfully segments the solid into topological hexahedra, given its edge graph structure:

- search for a valid cutting loop;
- use the cutting loop to construct a surface which cuts the solid into two pieces. The geometric aspects of this step are difficult. Firstly, any auxiliary edges of the loop must be turned into real edges by constructing a curve cutting the face. This problem is dealt with in the paper [13]. Secondly, a trimmed surface must be constructed having the cutting loop as its boundary;

- repeat the procedure for the two pieces until we arrive at *base solids*: a class of solids which have predefined segmentations into topological hexahedra. Our base solids include tetrahedra, hexahedra and prisms.

Each of the base solids is further split into topological hexahedra using pre-defined templates. More details of the algorithm are provided in [6, 12].

Typically, many valid cutting loops exist so a decision procedure is needed to pick one. We decide based on a cost which is a combination of the following components:

Combinatorial component. Aiming for a small number of hexes, our cost includes a component based on the graph theoretic structure:

- the number of edges of the cutting loop (e.g. preferring loops of length 4 since they more quickly lead to topological hexahedra).
- the number of edges of the newly created faces when a face is split by an auxiliary edge (e.g. preferring to split a 6-sided face into two 4-sided faces vs. a 5- and a 3-sided face).

Geometric component. Aiming for hexes with good shape, we include measurements of the planarity of the cutting loop.

A more detailed discussion, with examples, is provided in [12, Sections 5–6].

3.4 Surface Extension

The cutting loop approach can sometimes miss elegant segmentations that could be created by extending existing surfaces. For example, [12] provided an example where using the cutting loop method alone results in a segmentation with 31 topological hexahedra. Furthermore, they do not have ideal shape. Extending existing faces would allow a segmentation with only four hexes and higher quality shape (see Fig. 9). Surface extension can be done by modifying cutting loops into *extending loops* (see Fig. 10). In contrast to cutting loops, extending loops result in the construction of two different faces for the two new solids.

Extending loops can be defined as cutting loops with additional structure: each edge of the loop carries data of whether it extends one of the two incident faces, and if so, which one. The validity of an extending loop can be determined from the way faces meet at vertices and edges. Specifically,

- *it is only possible to extend a surface through a given edge* $\mathbf{e}$ *if* $\mathbf{e}$ *is everywhere concave*. This is because, if $\mathbf{e}$ is convex at any point, then the surface would be extended to outside the solid;
- at a vertex $\mathbf{v}$, consider the set of unit tangent vectors pointing into the solid. This set is a region of the unit sphere, whose boundary is a spherical polygon. See Fig. 11 for an example. Three faces $\mathbf{A}, \mathbf{B}, \mathbf{C}$ of a solid meet at a vertex $\mathbf{v}$; the resulting spherical polygon has three corresponding great arcs $\tilde{\mathbf{A}}, \tilde{\mathbf{B}}, \tilde{\mathbf{C}}$. *The geometry of this polygon determines the options available for extending a surface*

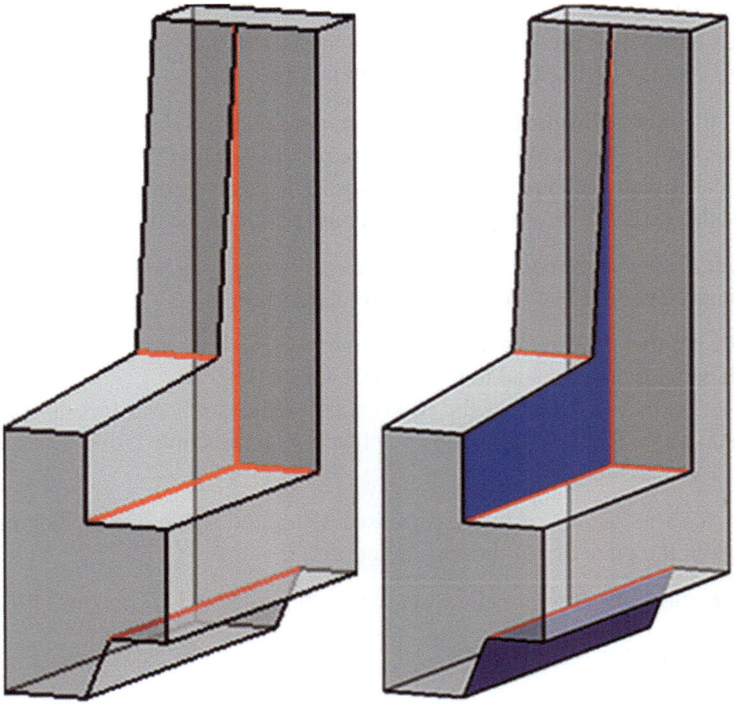

Fig. 9 Half-chair example. The solid in (**a**) was segmented into 31 topological hexahedra in [6]. However, a segmentation into just four topological hexahedra can be produced by extending the faces marked in (**b**) and merging them together as the first step. This motivates our ongoing work on automatic surface extension as a step in the segmentation procedure

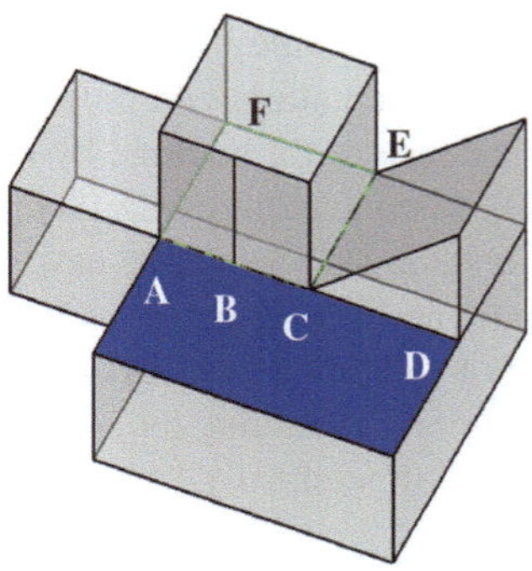

Fig. 10 Example of a solid with some surfaces that can be extended. Loop **A-B-C-E-F** is an extending loop which can be used to extend the highlighted face and cut the solid. An extending loop cannot contain the path **A-B-C-D** due to the geometry at vertex **C**

through **v**. In the example of Fig. 11, face **B** can be extended through **v**, resulting in face **A** being cut. This is because extending arc $\tilde{\mathbf{B}}$ into the interior of the

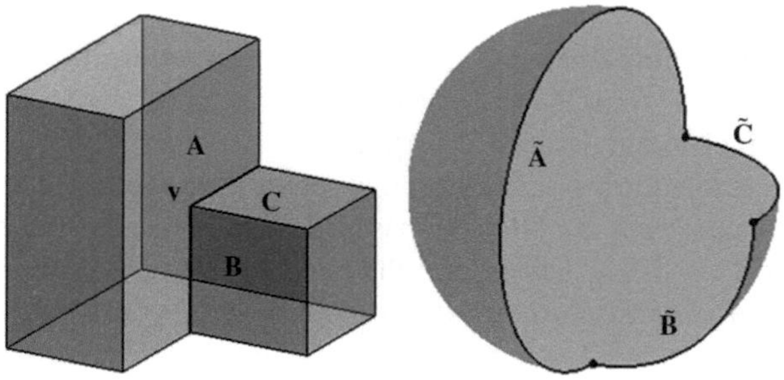

Fig. 11 In the solid (*left*), three faces **A**, **B**, **C** are incident to a vertex **v**. The unit vectors at **v** pointing into the solid define a filled spherical polygon with arcs $\tilde{\mathbf{A}}, \tilde{\mathbf{B}}, \tilde{\mathbf{C}}$ containing the tangent vectors pointing along the faces **A**, **B**, **C**. The geometry of the spherical polygon determines the available options for extending these faces through **v**

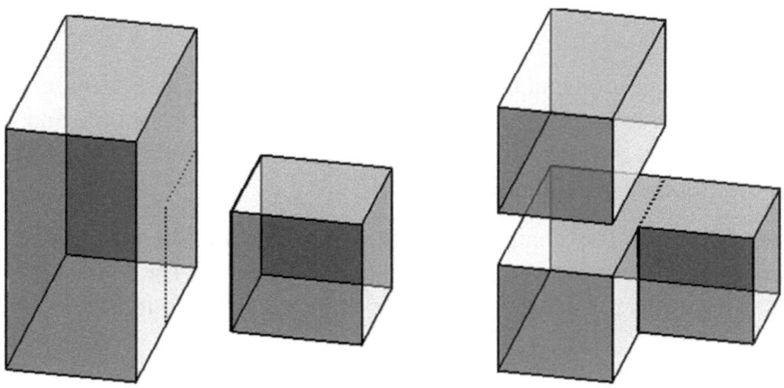

Fig. 12 Two possible results of cutting the solid of Fig. 11 (*left*) by extending surfaces

spherical polygon, it eventually intersects arc $\tilde{\mathbf{A}}$. A similar situation arises when extending face **C**.

A slightly different situation occurs for extending face **A** through **v**. If arc $\tilde{\mathbf{A}}$ is extended into the interior of the polygon, it does not intersect the interior of any of the other arcs, but instead closes up into a great circle. As a consequence, face **A** can be extended through **v** in such a way that the vertex **v** can be deleted from one of the two solids resulting from the cut. The question of whether two faces, meeting at a vertex, can be extended simultaneously into a single face, can also be answered using the geometry of the spherical polygon.

Compare these facts with [12, Proposition 2], where it is shown that the *spherical visibility problem* for the spherical polygon constructed above plays an important role in determining the validity of a cutting loop.

Two possible results of cutting the solid of Fig. 11 (left) are shown in Fig. 12.

The selection of an extending loop is similar to one for a cutting loop. A cost is assigned based on combinatorial and geometric properties. Minor complications arise from the fact that vertices can be deleted, and the new faces in the two resulting solids can have a different number of vertices.

There remains the question of the geometric realization of the extended surface, i.e., how do we construct the NURBS-based trimmed surface structure for a face which includes the existing face(s) and the newly constructed part? This question is still open; our current attitude is that it is better to only construct the new part, represent the extended face as a union of trimmed surfaces, and merge them together only at the time of volume fitting. This avoids creating an extra source of approximation.

Examples

We demonstrate the segmentation process for several models. For these models, a preprocessing step is applied to ensure that the model is contractible and the edge graph is simple and 3-vertex connected. This pre-processing step is ad-hoc as we have not developed a general method.

Chair stand piece. The chair stand piece (Fig. 13) was created to test various preprocessing strategies and demonstrate the algorithm for solids with convex edges. The result for one preprocessing strategy is shown in Fig. 14. A second

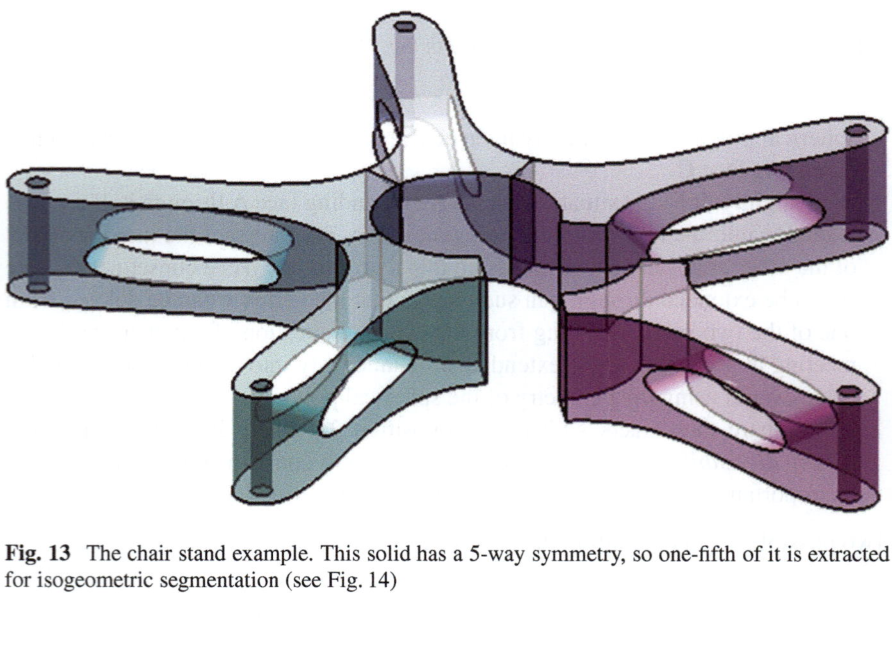

Fig. 13 The chair stand example. This solid has a 5-way symmetry, so one-fifth of it is extracted for isogeometric segmentation (see Fig. 14)

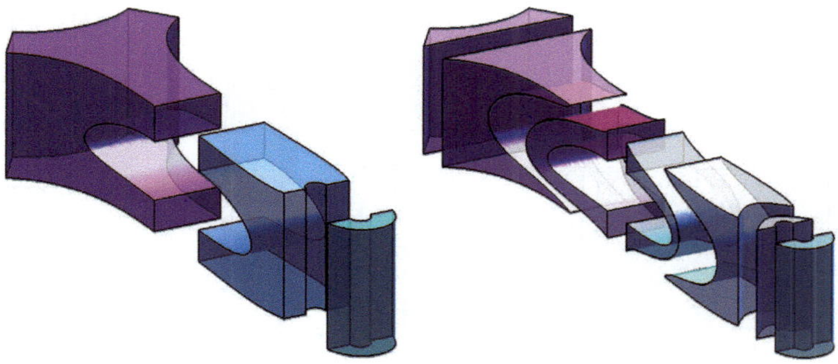

Fig. 14 *Left*: the chair stand piece, preprocessed to cut it into contractible pieces. *Right*: the result of the segmentation algorithm

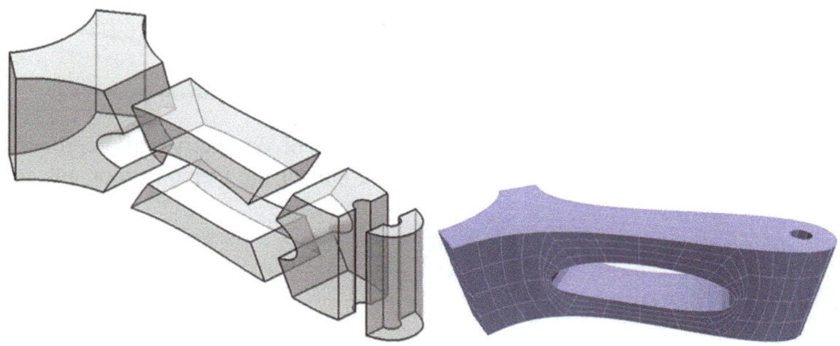

Fig. 15 (**a**) The chair stand piece, preprocessed to cut it into contractible pieces, using a different strategy from Fig. 14. (**b**) Representation of the chair stand as a watertight collection of tensor-product NURBS volumes

preprocessing strategy is shown in Fig. 15 where the resulting topological hexahedra are volume parameterized using the techniques in Sect. 4.

TERRIFIC Demonstrator The preprocessed TERRIFIC Demonstrator is shown in Fig. 6a. The solid is segmented into topological hexahedra using the surface extension and cutting loop methods described in the previous sections. The result is shown in Fig. 6b. The result of the removal of T-joints is shown in Fig. 6c.

Car part The segmentation procedure is applied to the CAD model reconstruction of the car part (See Sect. 2 and Fig. 3a–c). The preprocessed solid (Fig. 16, left) has a plane of symmetry. For one half, the resulting contractible pieces are segmented into topological hexahedra (Fig. 16, right). Data courtesy of Engineering Center Steyr.

Fig. 16 *Left*: the car part, divided into two along its plane of symmetry and preprocessed to cut it into contractible pieces. *Right*: segmentation of one half of the Demonstrator into topological hexahedra

3.5 T-Joint Removal

For some simulation methods, it may be necessary to post-process the model to eliminate T-joints. We do not have a general method for this. A special-purpose method was developed which suffices for our segmentation of the TERRIFIC demonstrator. During the segmentation process, we keep track of adjacency information for the solids. When the segmentation of a solid results in the creation of a T-joint, new edges must be added to the adjacent solid(s) which must themselves be segmented. This segmentation may create new T-joints, resulting in a propagation of segmentations through the solid. The result for the TERRIFIC Demonstrator is shown in Fig. 6c. Ongoing research on advanced IGA techniques aim to handle meshes with T-joints directly (see Sect. 1).

4 Hexahedron Parameterization

The aim is to convert a boundary represented topological hexahedron into a volume parameterization. Besides the possibility of quite distorted hexes, difficulty arises from the fact that the boundary faces are trimmed surfaces, since the domain of such a surface is not a rectangle in general.

A single topological hexahedron is parameterized by the following procedure.

1. Sample each of the four edges of a boundary face at parameters such that the distance between consecutive points is approximately constant. This is necessary because the curves from adjacent boundaries do not generally have the same parameterization;
2. Apply 2-dimensional Coon's patches to the sampled points, to produce a grid of points in the domain;
3. Evaluate the grid points, to produce a grid on the 3D surface;
4. Applying the above steps for each of the six faces of the topological hexahedron, we now have a grid of sampled boundary points. Grid points agree along the edges because of the way the edges are sampled in Step 1. Apply 3D Coon's patches to the boundary points to produce a grid of interior points. (It may be possible to skip this step if the topological hexahedron is not too heavily distorted);
5. Fit a tensor product B-spline to the sampled points using a least squares method.

The fitting method could be easily modified to include more complex quality measures.

Applying the procedure to a collection of topological hexahedra produces a multi-patch domain (i.e., a domain composed of multiple volume-parameterized topological hexahedra) suitable for isogeometric analysis (see Sect. 5). If a segmentation of a solid into topological hexahedra has no T-joints, then we can also ensure watertightness of the collection of patches by the following procedure. (The collection is not automatically watertight because the construction of cutting faces during the segmentation, as well as the fitting of the volume patch, can cause errors in the corresponding control points of incident patches.)

(i) Find all the incident faces between patches (the easiest way to do this is to hold on to the adjacency information from the segmentation procedure);
(ii) Wherever two or more patches meet (more than two patches can meet at vertices or edges), the corresponding control points of all such patches are moved to the average.

This simple adjustment is made possible by the following fact: If two topological hexahedra meet face-to-face in a quadrilateral patch, then this patch is represented by identical trimmed NURBS surfaces for both hexahedra. This is guaranteed by our construction of the splitting surface, cf. Sect. 3.3. The steps 1–3 of our parameterization procedure therefore give identical parameterization of the matching boundary faces in both patches. Thus, the only possible deviations between the control points are due to the approximation of the inner points in step 5. They can safely be expected to be relatively minor and can be dealt with by the averaging step (ii).

5 Simulation

As applications of the segmentation presented in previous sections, we use the TERRIFIC Demonstrator and the chair stand in mechanical simulations, solving the linear elasticity partial differential equation using an isogeometric finite element discretization. In Sect. 5.1 we briefly present the linear elasticity problem. In Sect. 5.2 we describe the method for solution on a multi-patch domain. Results are presented in Sects. 5.3 and 5.4.

5.1 The Linear Elasticity Problem

The linear elasticity problem on a domain $\Omega \subset \mathbb{R}^3$ means finding a solution $\mathbf{u} : \Omega \to \mathbb{R}^3$ of the following boundary value problem:

$$\begin{aligned} -\operatorname{div} \boldsymbol{\sigma}(\mathbf{u}) &= \mathbf{b} \quad && \text{in } \Omega, \\ \mathbf{u} &= \mathbf{g} && \text{on } \Gamma_u, \\ \boldsymbol{\sigma}(\mathbf{u}) \cdot \mathbf{n} &= \mathbf{t} && \text{on } \Gamma_n, \end{aligned} \tag{1}$$

where $\Gamma_u \subset \delta\Omega$ is the Dirichlet boundary, $\Gamma_n \subset \delta\Omega$ the Neumann boundary of the domain, the Cauchy stress $\boldsymbol{\sigma}$ is expressed by the linear St. Venant-Kirchhoff constitutive law,

$$\boldsymbol{\sigma}(\mathbf{u}) = 2\mu\, \boldsymbol{\varepsilon}(\mathbf{u}) + \lambda \ \operatorname{trace} \boldsymbol{\varepsilon}(\mathbf{u})\, \mathbf{I}$$

using Lamé constants $\mu, \lambda > 0$ and the linear strain tensor

$$\boldsymbol{\varepsilon}(\mathbf{u}) = \frac{1}{2}\left(\nabla \mathbf{u}^T + \nabla \mathbf{u}\right).$$

For a finite element discretization the weak form of (1) is required, i.e. seeking $\mathbf{u} \in H^1(\Omega)$, $\mathbf{u}|_{\Gamma_u} = \mathbf{g}$:

$$\int_{\Omega} \boldsymbol{\varepsilon}(\mathbf{v}) : \boldsymbol{\sigma}(\mathbf{u})\, d\mathbf{x} = \int_{\Omega} \mathbf{v}^T \mathbf{b}\, d\mathbf{x} + \int_{\Gamma_n} \mathbf{v}^T\, \mathbf{t}\, d\Gamma \quad \forall \mathbf{v} \in H_0^1(\Omega).$$

Using the 4th-order elasticity tensor $\mathbb{C}$,

$$\mathbb{C}_{ijkl} = \lambda\, \delta_{ij}\delta_{kl} + \mu\left(\delta_{ik}\delta_{jl} + \delta_{il}\delta_{kj}\right), \quad i, j, k, l = 1, 2, 3,$$

the equation can be rewritten as:

$$\int_{\Omega} \nabla \mathbf{v} : \mathbb{C} : \nabla \mathbf{u} \, d\mathbf{x} = \int_{\Omega} \mathbf{v}^T \mathbf{b} \, d\mathbf{x} + \int_{\Gamma_n} \mathbf{v}^T \, \mathbf{t} \, d\Gamma \quad \forall \mathbf{v} \in H_0^1(\Omega). \tag{2}$$

5.2 *Isogeometric Treatment with Multi-patch Domain*

The basis of an isogeometric method is a spline representation of the domain Ω and also the numerical solution field $\mathbf{u}^h$ [1]. This means that there exists a global mapping

$$\mathbf{g} : \Omega_0 \to \Omega, \quad \mathbf{x} = \mathbf{g}(\boldsymbol{\xi}) = \sum_{i=1}^{n} N_i^p(\boldsymbol{\xi}) \, \mathbf{c}_i, \tag{3}$$

which parameterizes the whole domain as a B-Spline or NURBS volume, using tensor product basis functions N_i^p of degree p, which are defined on the parameter domain Ω_0 using knot vectors $\Xi \in \mathbb{R}^{n+p+1}$, and control points $\mathbf{c}_i \in \mathbb{R}^3$. Then $\mathbf{u}^h$ is discretized as the push-forward of these spline functions onto the physical domain Ω and a set of displacement control points $\mathbf{d}_i \in \mathbb{R}^3$:

$$\mathbf{u}^h(\mathbf{x}) = \sum_{i=1}^{n} N_i^p \left(\mathbf{g}^{-1}(\mathbf{x}) \right) \, \mathbf{d}_i. \tag{4}$$

Finally substitution of (4) into (2) and numerical evaluation of the integrals leads to the well-known matrix-vector form of the finite element discretization of the elasticity problem:

$$\mathbf{K}\mathbf{d} = \mathbf{f}, \tag{5}$$

where $N = 3n$ is the dimension of the system, $\mathbf{d}$ is the vector of unknown control point displacements, $\mathbf{K}$ the stiffness matrix and $\mathbf{f}$ the external force vector.

We have to deal with multi-patch partitionings of the initial domain Ω into b subdomains (patches):

$$\overline{\Omega} = \bigcup_{i=1}^{b} \overline{\Omega^i}, \quad \Omega^i \cap \Omega^j = \emptyset.$$

If two patches Ω_i and Ω_j are adjacent, i.e.

$$\overline{\Omega^i} \cap \overline{\Omega^j} = \Gamma^{ij} \neq \emptyset,$$

we call Γ^{ij} the interface between Ω^i and Ω^j. Thus we can discretize the solutions $\mathbf{u}^i := \mathbf{u}|_{\Omega^i}$ independently, but we have to enforce the conditions $\mathbf{u}^i = \mathbf{u}^j$ on Γ^{ij}. As we deal with conforming parameterizations, i.e. $\mathbf{g}_i|_{\Gamma_{ij}} \equiv \mathbf{g}_j|_{\Gamma_{ij}} \;\forall\; \Gamma_{ij}$, this continuity constraint simply means corresponding displacement control points have to be equal, i.e.

$$\mathbf{d}^i_k = \mathbf{d}^j_l \quad \forall\; \mathbf{d}^i_k \in \mathbf{u}^{h^i}|_{\Gamma_{ij}},\; \mathbf{d}^j_l \in \mathbf{u}^{h\,j}|_{\Gamma_{ij}} \text{ with } \mathbf{c}^i_k = \mathbf{c}^j_l. \tag{6}$$

For the enforcement of these constraints basically two possibilities exist, either adding each constraint as additional equations in a Lagrangian multiplier approach, or eliminating degrees of freedom from the system.

5.3 TERRIFIC Demonstrator

The isogeometric volume parameterization consists of 70 patches of cubic B-Spline volumes, with a total of 24,010 control points and 72,030 degrees of freedom (DOFs). Coupling of displacement control points on 132 interfaces has to be enforced. When Lagrangian multipliers are used for coupling, the isogeometric finite element discretization of the model has 90,048 DOFs, otherwise when elimination is used, it has only 53,925 DOFs.

As material parameters of linear elasticity we chose

$$\lambda = 54.0\text{ GPa}, \qquad \mu = 27.8\text{ GPa}.$$

As boundary conditions we take a clamping of the right bore hole (see Fig. 17) by a zero Dirichlet condition at Γ_u and a surface traction on the left bore hole as a Neumann boundary condition on Γ_n:

$$\begin{aligned} \mathbf{u} &= \mathbf{0}\text{ m} \quad \text{on } \Gamma_u,\\ \mathbf{t} &= (60.0,\ -42.0,\ 0.0)^T\text{ MPa} \quad \text{on } \Gamma_n. \end{aligned}$$

Running the simulation, we get a maximal norm of displacement of 7.06 mm and internal energy of 70.1 J. The displaced object can be seen in Fig. 17. Comparison with a commercial finite element software shows a very good correspondence of results. Using ANSYS® with a mesh of 20,855 quadratic tetrahedral elements (type SOLID187) and 36,181 nodes resp. 108,543 DOFs, we have a maximum displacement of 7.00 mm and energy 70.2 J. It is also worth remarking that the stress distribution obtained from IGA in Fig. 17b looks very smooth.

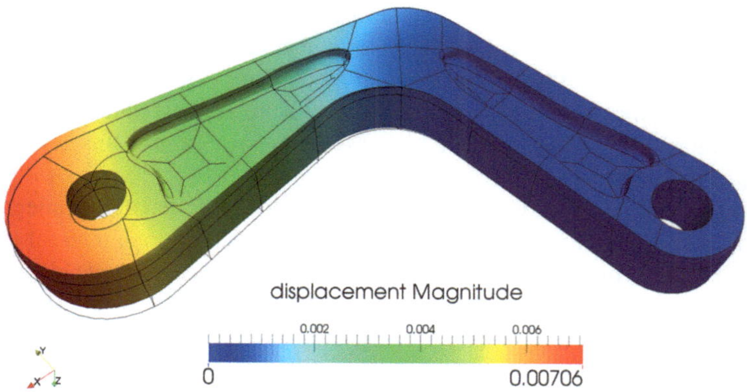

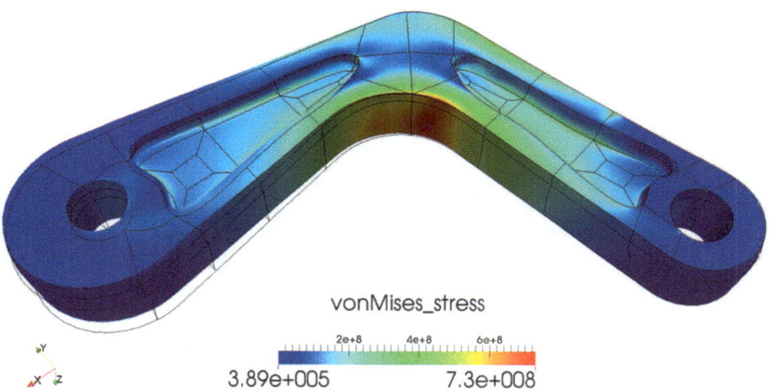

Fig. 17 Linear elasticity simulation of TERRIFIC Demonstrator. Displaced part, colored by (**a**) norm of displacement in [m] and (**b**) von Mises stress in [Pa]

5.4 Chair Stand

The chair stand model consists of 12 cubic B-Spline volumes with a total of 3,969 control points and 11,907 DOFs. Using coupling of patches with the Lagrangian multiplier approach we have a total of 14,231 DOFs.

For a realistic simulation of the chair stand we have added an inner cylinder connecting the stand with the seating and a small cylinder connecting to the wheels made from aluminum, and assumed the main parts of the stand are made from plastic (Polyamid P6). Thus the material parameters are:

$$\lambda_{Al} = 54.0\text{ GPa}, \quad \mu_{Al} = 27.8\text{ GPa},$$
$$\lambda_{P6} = 3.19\text{ GPa}, \quad \mu_{P6} = 0.899\text{ GPa}.$$

Assuming a person weighing 81.5 kg sitting on the chair, this would result in a vertical force of 0.10 MPa acting as Neumann boundary condition on the inner aluminum cylinder. Furthermore symmetry boundary conditions are applied on displacements, since we only model one fifth of the total chair stand. As the smaller cylinder connects to the wheel, which we assume as a frictionless link to the ground, no vertical displacement is possible there.

Running the simulation, we get a small vertical displacement of the inner cylinder of 0.82 mm and a maximal von Mises stress of 1.6 MPa. The total internal energy is 64.2 mJ. The deformed stand is visualized in Fig. 18. We also validate our results by a comparison to commercial finite element software ANSYS®. Automated mesh generation provides a discretization using a mix of 13,437 quadratic hexahedral and tetrahedral elements (types SOLID186 and SOLID187) with 22,518 nodes and 57,554 DOFs. The simulation results in an evaluated z-displacement of 0.81 mm, a von Mises stress of 1.6 MPa and a total energy of 62.8 mJ.

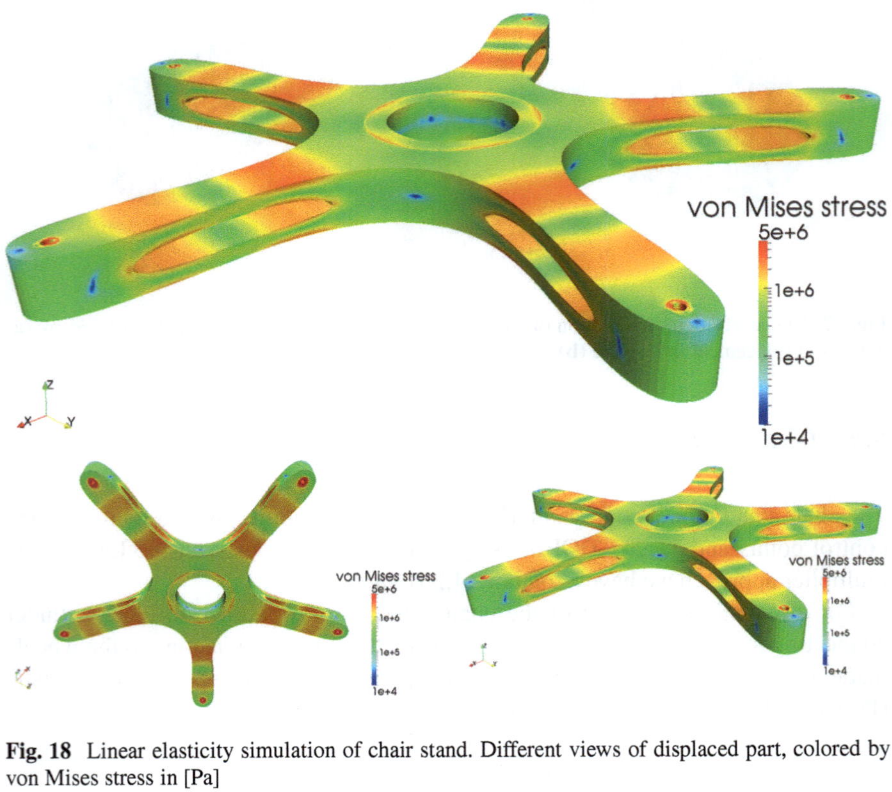

Fig. 18 Linear elasticity simulation of chair stand. Different views of displaced part, colored by von Mises stress in [Pa]

6 Conclusions and Further Work

We created a pipeline, including algorithms and software together with a rigorous mathematical treatment, for the the conversion of 3D models into an IGA-suitable form. The first step of the pipeline is the reconstruction, if necessary, of a CAD-like boundary representation from triangulated data. Then the boundary represented solid is segmented into a collection of topological hexahedra. Finally, volumetric NURBS patches are constructed for the hexahedra.

We used outputs of the pipeline to conduct simulations based on the linear elasticity problem. Results had good agreement with commercial finite element software, demonstrating the suitability of the segmentation and parameterization for simulation.

While the theory and software developed forms the core of the pipeline, there are several improvements already under development:

- improved geometric construction of the cutting curves and surfaces, with theoretical guarantees;
- addressing solids which are non-contractible or do not have 3-vertex-connected edge graphs without human interaction;
- improving the quality of isogeometric segmentation, in terms of the number and shape of the resulting topological hexahedra, using the surface extension methods described in Sect. 3.4.

Potential future developments include the use of splines allowing local refinement, such as LR- or THB-splines, and improving the reconstruction of cylindrical surfaces using periodic splines.

Acknowledgements This work has been funded by the European Commission projects TERRIFIC (Grant Agreement 284981), EXAMPLE (Grant Agreement 324340) and ITN INSIST (Grant Agreement 289361), and the Austrian Science Fund (FWF) project Geometry and Simulation (Project Number S 117). We thank Martin Schifko at Engineering Center Steyr and Stefan Boschert at Siemens Corporate Technology for some of the input data used in our work.

References

1. J.A. Cottrell, T.J.R. Hughes, Y. Bazilevs, *Isogeometric Analysis: Toward Integration of CAD and FEA* (Wiley, Chichester/Hoboken, 2009)
2. M.S. Floater, Parametrization and smooth approximation of surface triangulations. Comput. Aided Geom. Des. **14**(3), 231–250 (1997)
3. M.S. Floater, Mean value coordinates. Comput. Aided Geom. Des. **20**(1), 19–27 (2003)
4. D. Großmann, B. Jüttler, H. Schlusnus, J. Barner, A.-V. Vuong, Isogeometric simulation of turbine blades for aircraft engines. Comput. Aided Geom. Des. **29**(7), 519–531 (2012). Geometric Modeling and Processing 2012
5. S. Han, J. Xia, Y. He, Constructing hexahedral shell meshes via volumetric polycube maps. Comput.-Aided Des. **43**(10), 1222–1233 (2011)

6. B. Jüttler, M. Kapl, D.-M. Nguyen, Q. Pan, M. Pauley, Isogeometric segmentation: the case of contractible solids without non-convex edges. Comput.-Aided Des. **57**, 74–90 (2014)
7. U. Langer, I. Toulopoulos, Analysis of multipatch discontinuous Galerkin IgA approximations to elliptic boundary value problems. Technical Report 1408.0182, arxiv.org, 2014
8. Y. Lu, R. Gadh, T.J. Tautges, Feature based hex meshing methodology: feature recognition and volume decomposition. Comput.-Aided Des. **33**(3), 221–232 (2001)
9. T. Martin, E. Cohen, Volumetric parameterization of complex objects by respecting multiple materials. Comput. Graph. **34**(3), 187–197 (2010). Shape Modelling International (SMI) Conference 2010
10. T. Martin, E. Cohen, R.M. Kirby, Volumetric parameterization and trivariate B-spline fitting using harmonic functions. Comput. Aided Geom. Des. **26**(6), 648–664 (2009)
11. T. Nguyen, B. Jüttler, Parameterization of contractible domains using sequences of harmonic maps, in *Curves and surfaces*. Lecture Notes in Computer Science, vol. 6920 (Springer, Heidelberg, 2012), pp. 501–514
12. D.-M. Nguyen, M. Pauley, B. Jüttler, Isogeometric segmentation. Part II: on the segmentability of contractible solids with non-convex edges. Graph. Models **76**(5), 426–439 (2014). Geometric Modeling and Processing 2014
13. D.-M. Nguyen, M. Pauley, B. Jüttler, Isogeometric segmentation: construction of auxiliary curves. Comput.-Aided Des. **70**, 89–99 (2016). Proceedings of GDSPM 2015, conditionally accepted
14. M. Nieser, U. Reitebuch, K. Polthier, CubeCover – parameterization of 3D volumes. Comput. Graph. Forum **30**(5), 1397–1406 (2011)
15. A. Sheffer, M. Etzion, A. Rappoport, M. Bercovier, Hexahedral mesh generation using the embedded Voronoi graph, in *Proceedings of the 7th International Meshing Roundtable*, Sandia National Laboratories (1999), pp. 347–364
16. B. Strodthoff, M. Schifko, B. Jüttler, Horizontal decomposition of triangulated solids for the simulation of dip-coating processes. Comput.-Aided Des. **43**(12), 1891–1901 (2011)
17. T.J. Tautges, T. Blacker, S.A. Mitchell, The whisker weaving algorithm: a connectivity-based method for constructing all-hexahedral finite element meshes. Int. J. Numer. Methods Eng. **39**(19), 3327–3349 (1996)
18. T. Varady, R. Martin, Reverse engineering, in *Handbook of Computer Aided Geometric Design*, ed. by G. Farin, J. Hoschek, M.-S. Kim (North-Holland, Amsterdam, 2002), pp. 651–681
19. X. Wang, X. Qian, An optimization approach for constructing trivariate B-spline solids. Comput.-Aided Des. **46**, 179–191 (2014)
20. W. Wang, Y. Zhang, L. Liu, T.J.R. Hughes, Trivariate solid T-spline construction from boundary triangulations with arbitrary genus topology. Comput.-Aided Des. **45**(2), 351–360 (2013)
21. D. White, L. Mingwu, S.E. Benzley, G.D. Sjaardema, Automated hexahedral mesh generation by virtual decomposition, in *Proceedings of the 4th International Meshing Roundtable*, Sandia National Laboratories (1995), pp. 165–176
22. G. Xu, B. Mourrain, R. Duvigneau, A. Galligo, Analysis-suitable volume parameterization of multi-block computational domain in isogeometric applications. Comput.-Aided Des. **45**(2), 395–404 (2013)
23. W. Yu, K. Zhang, S. Wan, X. Li, Optimizing polycube domain construction for hexahedral remeshing. Comput.-Aided Des. **46**, 58–68 (2014)
24. Y. Zhang, C. Bajaj, Adaptive and quality quadrilateral/hexahedral meshing from volumetric data. Comput. Methods Appl. Mech. Eng. **195**, 942–960 (2006)
25. Y. Zhang, W. Wang, T.J.R. Hughes, Solid T-spline construction from boundary representations for genus-zero geometry. Comput. Methods Appl. Mech. Eng. **249/252**, 185–197 (2012)
26. Y. Zhang, W. Wang, T.J.R. Hughes, Conformal solid T-spline construction from boundary T-spline representations. Comput. Mech. **51**(6), 1051–1059 (2013)

Domain Decomposition Methods and Kirchhoff-Love Shell Multipatch Coupling in Isogeometric Analysis

Andreas Apostolatos, Michael Breitenberger, Roland Wüchner, and Kai-Uwe Bletzinger

Abstract The necessity for solving the isogeometric Kirchhoff-Love shell problem into multiple domains has been exemplified especially in cases where the geometry comprises multipatches. In fact, geometries taken from *Computer Aided Geometric Design* involve in principle trimmed multipatches. Herein, the application and comparison of the most common *Domain Decomposition Methods* for the coupling of Kirchhoff-Love shell multipatches in isogeometric analysis is presented. The investigated methods comprise *Penalty* and *Lagrange Multipliers* methods. All methods are extended to account for geometrically nonlinear problems. The aforementioned methods provided highly accurate results, thus extending the Kirchhoff-Love shell analysis from a single to multiple patches which is a prerequisite for solving practical engineering problems using isogeometric analysis.

1 Introduction

In the past decades light-weight shell structures have been playing an ever-increasing role in various engineering applications as the cost can be significantly reduced by keeping vital mechanical properties such as the load carrying capacity. The Kirchhoff-Love or the Reissner-Mindlin shell models [1] were used among plenty of studies.

Isogeometric Analysis [2] (IGA) is a generalization of the well-established classical *Finite Element Analysis* (FEA) [3, 4] using a geometric basis, as used in *Computer-Aided Geometric Design* (CAGD), within the analysis process. The *Non-Uniform Rational B-Splines* (NURBS) basis functions [5] form the most common geometric basis and therefore are also used within this study, see [6]. The successful application of isogeometric analysis for Kirchhoff-Love shell problems was demonstrated in [7], which was initially confined into single-patch

A. Apostolatos (✉) • M. Breitenberger • R. Wüchner • K.-U. Bletzinger
Technische Universität München, Arcisstraße 21, D-80333 München, Germany
e-mail: andreas.apostolatos@tum.de; michael.breitenberger@tum.de; wuechner@tum.de; kub@tum.de

B. Jüttler, B. Simeon (eds.), *Isogeometric Analysis and Applications 2014*,
Lecture Notes in Computational Science and Engineering 107,
DOI 10.1007/978-3-319-23315-4_4

geometries. However, geometries taken from CAGD comprise typically trimmed multipatches [8]. Another issue is the tensor product structure that multidimensional NURBS basis has, making local refinement over the NURBS basis in its original form impossible. However, studies on local refinement and adaptivity over a modified NURBS basis can be found among other studies in [9, 10]. It is worth mentioning that since the variational index of the Kirchhoff-Love shell problem is equal to two, the basis functions must be at least one time continuously differentiable and therefore geometries involving kinks are ab initio excluded in standard NURBS-based IGA. Finally it is demonstrated in [11] that isogeometric analysis for shell structures can be extended to account for trimmed multipatch geometries which are not necessarily watertight, can join with an arbitrary angle along their common interfaces and which can have an independent from each other discretization level.

Within this study, three different *Domain Decomposition Methods* (DDMs) are introduced and compared for the multipatch coupling of the Kirchhoff-Love shell problem in both linear and nonlinear isogeometric analysis: a *Penalty*, a *Lagrange Multipliers* and an *augmented Lagrange Multipliers* method.

The Penalty method, firstly introduced in [12], aimed at enlarging the test and the solution spaces by not restricting the corresponding functions to comply with the Dirichlet boundary conditions. Since then it has been widely employed for the coupling of non-conforming finite element meshes, see for example [13–15]. Variationally consistent extensions of the Penalty method for boundary value problems have been proposed among other studies in [16, 17] and specifically for the coupling of linear multipatch thin and blended shell structures in [18] within IGA. The *Bending Strip Method* [19] (BSM) utilized a penalization over the bending stiffness of Kirchhoff-Love shell multipatches along their common interfaces, thus extending isogeometric shell analysis to multipatches. However, this method is restricted into matching interface parametrizations. In the present contribution, a Penalty method involving two penalty terms, for the interface displacement and rotations Dirichlet compatibility conditions, is presented, which releases this substantial restriction of matching parametrizations along the patch interfaces. As common for the Penalty methods, the underlying formulation is positive definite [12] but the accuracy of the numerical scheme depends on the choice of the penalty parameters. A rule for the choice of the penalty parameters involving scaling with the material matrices norms and the element size is herein proposed. The method is applicable for multipatches which join arbitrarily along their common interfaces.

The Lagrange Multipliers method is one of the most common methods for enforcing constraints within a finite element equation system, see for instance [14, 20, 21]. Having at hand two interface Dirichlet conditions, one for the displacement and one for the rotation field, two independent Lagrange Multipliers fields are chosen which act as the interface traction force and traction moment fields from a physical viewpoint, respectively. As being typical for the Lagrange Multipliers methods, the resulting linear equation system belongs to the category of saddle point problems, which necessitates that the discrete spaces satisfy the *Ladyzenskaja-Babuska-Brezzi* [22] condition so that a unique solution to the problem can be guaranteed. The numerical investigations showed satisfactory results for the decomposed Kirchhoff-Love shell problem, provided that a suitable discretization for the Lagrange Multipliers fields is found.

At last, the augmented Lagrange Multipliers method for the Kirchhoff-Love shell problem is elaborated and discussed. This method is a combination of the Lagrange Multipliers and the Penalty method [15, 23]. This method belongs also to the class of saddle point problems and hence, the same restrictions as for the Lagrange Multipliers method apply in this case.

The application and comparison of the aforementioned methods is demonstrated using two linear and two nonlinear problems. Thus, isogeometric Kirchhoff-Love shell analysis is extended into multiple domains, towards the integration of CAGD and FEA into a large scale computational environment as in [11].

2 Theory

2.1 The Strong Form of the Problem

The model problem is an elastic body represented by a surface embedded in 3D space $\overline{\Omega} \subset \mathbb{R}^3$, where $\overline{\Omega}$ stands for the closure of Ω in the Euclidean space. The surface is described using a convective covariant space $\overline{\Omega_0} \subset \mathbb{R}^2$ through the map $\mathbf{S} : \overline{\Omega_0} \rightarrow \overline{\Omega}$. The mechanical behaviour of the elastic body is described by the *Kirchhoff-Love shell* theory. The process is assumed to be static, i.e. no inertia forces are to be considered. For each material particle $\mathbf{X} \in \overline{\Omega}$, the unknown displacement field $\mathbf{d} = \mathbf{d}(\mathbf{X})$ is defined in a Lagrangian description of the motion, that is, the current configuration is defined by the collection of material particles $\mathbf{x} = \mathbf{X} + \mathbf{d}$. In this way, the displacement field can have a representation in both the physical and the covariant space, that is [1],

$$\mathbf{d} = d_i^0 \mathbf{e}_i = \mathbf{d}_{\|} + d_3 \mathbf{A}_3 = d^{\alpha} \mathbf{A}_{\alpha} + d_3 \mathbf{A}_3 \quad \forall \mathbf{X} \in \overline{\Omega} \,, \tag{1}$$

where d_i^0, $\mathbf{d}_{\|}$ and d_3 stands for the components of the displacement field on the Cartesian space, the displacement field on the tangent space of the shell's surface and the displacement component normal to the shell's surface, respectively. According to standard notations, Greek and Latin indices range from 1 to 2 and from 1 to 3, respectively, if not stated otherwise. Moreover, $\mathbf{e}_i$ and $\{\mathbf{A}_{\alpha}, \mathbf{A}_3\}$ stand for the base vectors of the global Cartesian and the covariant space in the reference configuration, respectively. In particular, the covariant base vectors are given by [1],

$$\mathbf{A}_{\alpha} := \mathbf{X}_{,\alpha} \,, \tag{2a}$$

$$\mathbf{A}_3 := \overline{j}^{-1} \mathbf{A}_1 \times \mathbf{A}_2 \,, \tag{2b}$$

in $\overline{\Omega}$ with $\bar{j} = \|\mathbf{A}_1 \times \mathbf{A}_2\|_2$ and the commas standing for the partial derivatives with respect to the convective coordinates $(\theta^1, \theta^2) \in \overline{\Omega_0}$. The latter implies that $\mathbf{A}^3 = \mathbf{A}_3$ in what concerns the contravariant basis, meaning that the contravariant and the covariant components of any first order tensor defined over the convective space lateral to the shell's surface are equal. Hence notation d_3 is employed for both the covariant and the contravariant component of the displacement field in $\mathbf{A}_3$ direction, see [1]. In addition, the map $\mathbf{S}$ is assumed to be regular almost everywhere in Ω_0 meaning that $\bar{j}$ is non zero almost everywhere in Ω_0, see [24]. The boundary of the surface $\Gamma = \partial\Omega$ is assumed to be a piecewise C^1-continuous curve, so that the tangent, the normal and the binormal vectors can be uniquely defined almost everywhere in $\overline{\Omega}$. Let also $\mathfrak{S}^m$ be the space of all m-th order symmetric tensors. Then, the curvature tensor $\mathbf{B} \in \mathfrak{S}^2$ is given by [24],

$$\mathbf{B} = B_{\alpha\beta}\ \mathbf{A}^\alpha \otimes \mathbf{A}^\beta := -\mathbf{A}_\alpha \cdot \mathbf{A}_{3,\beta}\ \mathbf{A}^\alpha \otimes \mathbf{A}^\beta = \mathbf{A}_{\alpha,\beta} \cdot \mathbf{A}_3\ \mathbf{A}^\alpha \otimes \mathbf{A}^\beta \quad \forall \mathbf{X} \in \overline{\Omega}\ , \tag{3}$$

where $\{\mathbf{A}^\alpha, \mathbf{A}_3\}$ stands for the contravariant base vectors in the reference configuration. The small rotations assumption results in a linear relation between the rotations and the displacements. In this way, the total rotation field $\boldsymbol{\omega} \in \mathfrak{S}^1$ can be defined at each point $\mathbf{X} \in \overline{\Omega}$ by [1],

$$\boldsymbol{\omega}(\mathbf{d}) = -\boldsymbol{\epsilon} \cdot \left(\nabla d_3 + \mathbf{B} \cdot \mathbf{d}_{\|}\right) = \epsilon^{\alpha\beta} \left(d_{3,\beta} + B_{\beta\gamma}\, d^\gamma\right) \mathbf{A}_\alpha\ , \tag{4}$$

where $\boldsymbol{\epsilon} = \epsilon^{\alpha\beta}\ \mathbf{A}_\alpha \otimes \mathbf{A}_\beta$ stands for the permutation second order tensor with [1],

$$\epsilon^{\alpha\beta} = \frac{1}{\bar{j}} \begin{cases} 1, \text{ if } (\alpha, \beta) = (1, 2)\ , \\ -1, \text{ if } (\alpha, \beta) = (2, 1)\ , \\ 0, \text{ if } \alpha = \beta\ . \end{cases} \tag{5}$$

In addition, ∇ stands for the gradient operator over the convective space. The linear Hooke's law is employed for the description of the constitutive behaviour of the elastic material. Thus, the force and moment tensors corresponding to the second Piola-Kirchhoff stress measure $\mathbf{n}, \mathbf{m} \in \mathfrak{S}^2$ are linearly dependent on the membrane and bending strain tensors $\boldsymbol{\varepsilon}, \boldsymbol{\kappa} \in \mathfrak{S}^2$ through [25],

$$\mathbf{n}(\mathbf{d}) = \frac{Eh}{1-\nu^2} \mathbf{C} : \boldsymbol{\varepsilon}(\mathbf{d}) = \mathbf{C}_{\mathrm{m}} : \boldsymbol{\varepsilon}(\mathbf{d})\ , \tag{6a}$$

$$\mathbf{m}(\mathbf{d}) = \frac{Eh^3}{12(1-\nu^2)} \mathbf{C} : \boldsymbol{\kappa}(\mathbf{d}) = \mathbf{C}_{\mathrm{b}} : \boldsymbol{\kappa}(\mathbf{d})\ , \tag{6b}$$

in $\overline{\Omega}$, where $\mathbf{C} \in \mathfrak{S}^4$, E, ν and h is the elasticity tensor, the Young's modulus, the Poisson's ratio and the shell's thickness, respectively. Tensors $\mathbf{C}_{\mathrm{m}}, \mathbf{C}_{\mathrm{b}} \in \mathfrak{S}^4$ stand for the particular elasticity tensors for the membrane and the bending action of the Kirchhoff-Love shell, respectively. In what concerns the kinematics of the Kirchhoff-Love shell, the nonlinear membrane and bending strain tensors corresponding to the Green-Lagrange strain measure are respectively defined by [7],

$$\boldsymbol{\varepsilon} = \varepsilon_{\alpha\beta}\ \mathbf{A}^{\alpha} \otimes \mathbf{A}^{\beta} = \frac{1}{2}\left(\mathbf{a}_{\alpha} \cdot \mathbf{a}_{\beta} - \mathbf{A}_{\alpha} \cdot \mathbf{A}_{\beta}\right)\ \mathbf{A}^{\alpha} \otimes \mathbf{A}^{\beta}\ , \tag{7a}$$

$$\boldsymbol{\kappa} = \kappa_{\alpha\beta}\ \mathbf{A}^{\alpha} \otimes \mathbf{A}^{\beta} = \left(B_{\alpha\beta} - b_{\alpha\beta}\right)\ \mathbf{A}^{\alpha} \otimes \mathbf{A}^{\beta}\ , \tag{7b}$$

where $\mathbf{a}_{\alpha}$ and $b_{\alpha\beta}$ stand for the base vectors of the covariant space and covariant components of the curvature tensor (3) over the current configuration. The linearised version of the strain components (7a) and (7b) writes [25],

$$\varepsilon_{\alpha\beta} := \frac{1}{2}\left(\mathbf{A}_{\beta} \cdot \mathbf{d}_{,\alpha} + \mathbf{A}_{\alpha} \cdot \mathbf{d}_{,\beta}\right)\ , \tag{8a}$$

$$\begin{aligned}\kappa_{\alpha\beta} := -\mathbf{A}_3 \cdot \mathbf{d}_{,\alpha\beta} + \mathbf{A}_{\alpha,\beta} \cdot \mathbf{A}_3 \frac{1}{\bar{j}}\left((\mathbf{A}_2 \times \mathbf{A}_3) \cdot \mathbf{d}_{,1} - (\mathbf{A}_1 \times \mathbf{A}_3) \cdot \mathbf{d}_{,2}\right) \\ + \frac{1}{\bar{j}}\left(\left(\mathbf{A}_{\alpha,\beta} \times \mathbf{A}_2\right) \cdot \mathbf{d}_{,1} - \left(\mathbf{A}_{\alpha,\beta} \times \mathbf{A}_1\right) \cdot \mathbf{d}_{,2}\right)\ .\end{aligned} \tag{8b}$$

Displacement and/or rotation Dirichlet boundary conditions are prescribed along a portion of the body's boundary $\Gamma_{\mathrm{d}} \subset \Gamma$, confining ourselves into the homogeneous Dirichlet boundary conditions without loss of generality. Forces $\mathbf{p}$ and/or moments $\mathbf{r}$ are applied over another portion of the shell's boundary $\Gamma_{\mathrm{n}} \subset \Gamma$ forming the so-called Neumann boundary conditions. Finally, it is also assumed that the elastic body is subject to body forces designated by $\mathbf{b} = b^{\alpha}\mathbf{A}_{\alpha} + b_3\mathbf{A}_3$. Subsequently, domain Ω is decomposed into a finite set of subdomains $\Omega^{(i)}$, $i = 1, \ldots, n \in \mathbb{N}$ for which the following relations hold,

$$\bigcup_{i=1}^{n} \overline{\Omega^{(i)}} = \overline{\Omega}\ , \tag{9a}$$

$$\overline{\Omega^{(i)}} \cap \overline{\Omega^{(j)}} = \overline{\Gamma_{\mathrm{c}}^{(i,j)}} \quad \forall\, (i,j) \in \mathcal{I}\ , \tag{9b}$$

$$\bigcup_{(i,j)\in\mathcal{I}} \overline{\Gamma_{\mathrm{c}}^{(i,j)}} = \overline{\Gamma_{\mathrm{c}}}\ , \tag{9c}$$

$\mathcal{I}$ being the set of all $n_{\mathrm{I}} \in \mathbb{N}$ intersections between the subdomains excluding pairs (i, i) and $\Gamma_{\mathrm{c}}^{(i,j)}$ being a set of piecewise C^1-continuous curves on the Euclidean space in the context of the non-overlapping DDMs. The Dirichlet and the Neumann boundaries, Γ_{d} and Γ_{n} respectively, are decomposed in the same fashion as Ω in (9a)–(9b). The aforementioned problem is depicted in Fig. 1. Subsequently, it is also assumed that the displacement, the rotation, the force and the moment fields are defined piecewisely in $\overline{\Omega}$ at each subdomain $\overline{\Omega^{(i)}}$. Whenever the displacement, the rotation, the force or the moment field is restricted into one subdomain, a superscript $i = 1, \ldots, n$ is used to indicate the domain of restriction, that is, $\mathbf{d}_{|\overline{\Omega^{(i)}}} = \mathbf{d}^{(i)}$,

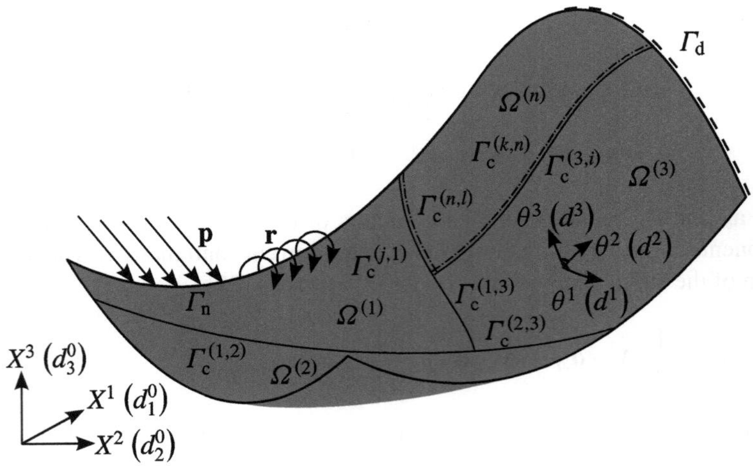

Fig. 1 Problem placement: multipatch Kirchhoff-Love shell structure

$\boldsymbol{\omega}\left(\mathbf{d}^{(i)}\right) = \boldsymbol{\omega}^{(i)}$, $\mathbf{n}\left(\mathbf{d}^{(i)}\right) = \mathbf{n}^{(i)}$ and $\mathbf{m}\left(\mathbf{d}^{(i)}\right) = \mathbf{m}^{(i)}$, respectively. Then, the Kirchhoff-Love of n-domain decomposed *Boundary Value Problem* (BVP) writes:

$$n^{\alpha\beta}\big|_{\alpha} - q^{\alpha} B_{\alpha}^{\beta} + b^{\beta} = 0\,, \quad \text{in } \Omega_0 \setminus \Gamma_{\mathrm{c}}\,, \tag{10a}$$

$$n^{\alpha\beta} B_{\alpha\beta} + q^{\alpha}\big|_{\alpha} + b_3 = 0\,, \quad \text{in } \Omega_0 \setminus \Gamma_{\mathrm{c}}\,, \tag{10b}$$

$$m^{\alpha\beta}\big|_{\alpha} - q^{\beta} = 0\,, \quad \text{in } \Omega_0 \setminus \Gamma_{\mathrm{c}}\,, \tag{10c}$$

$$\mathbf{d} = \mathbf{0}\,, \quad \text{on } \subset \Gamma_{\mathrm{d}}\,, \tag{10d}$$

$$\boldsymbol{\omega} = \mathbf{0}\,, \quad \text{on } \subset \Gamma_{\mathrm{d}}\,, \tag{10e}$$

$$\mathbf{n}^{\alpha} u_{\alpha} = \mathbf{p}\,, \quad \text{on } \subset \Gamma_{\mathrm{n}}\,, \tag{10f}$$

$$\mathbf{m}^{\alpha} u_{\alpha} = \mathbf{r}\,, \quad \text{on } \subset \Gamma_{\mathrm{n}}\,, \tag{10g}$$

$$\mathbf{d}^{(i)} - \mathbf{d}^{(j)} = \mathbf{0}\,, \quad \text{on } \Gamma_{\mathrm{c}}^{(i,j)}\ \forall i,j = 1,\ldots,n\,, \tag{10h}$$

$$\boldsymbol{\omega}^{(i)} + \boldsymbol{\omega}^{(j)} = \mathbf{0}\,, \quad \text{on } \Gamma_{\mathrm{c}}^{(i,j)}\ \forall i,j = 1,\ldots,n\,, \tag{10i}$$

where $\mathbf{u}$, $\mathbf{n}^{\alpha} = n^{\alpha\beta}\mathbf{A}_{\beta} + q^{\alpha}\mathbf{A}_3$ and $\mathbf{m}^{\alpha} = m^{\alpha\beta}\mathbf{A}_{\beta}$ stand for the normal to the Neumann boundary Γ_{n}, the traction force and the traction moment vector, respectively. The force and moment tensor components $n^{\alpha\beta}$ and $m^{\alpha\beta}$, respectively, that appear in the above mentioned BVP correspond to the first Piola-Kirchhoff stress measure for the geometrically nonlinear case. Those are connected to the second Piola-Kirchhoff stress measure through the deformation gradient tensor. However, in the geometrically linear case both stress measures coincide, for more details see [1]. Additionally, q^{α} stand for the contravariant components of the shear force vector. The compatibility of the displacement and the rotation field across the interface Γ_c of the subdomains is ensured through the interface continuity conditions (10h)–(10i).

It is important to note that rotation continuity equation (10i) implies that the interface tangent to surfaces $\Omega^{(i)}$ and $\Omega^{(j)}$ vectors around which rotation vectors $\boldsymbol{\omega}^{(i)}$ and $\boldsymbol{\omega}^{(j)}$ are defined, have opposite direction. In case those interface vectors have the same direction, continuity condition (10i) should be reformulated such that the rotation vectors from each neighbouring patch pair $\Omega^{(i)}$ and $\Omega^{(j)}$ are equal and not opposite.

2.2 The Weak Form of the Decoupled Problem

The variational formulation of the BVP (10a), (10b), (10c), (10d), (10e), (10f) and (10g) without the consideration of the interface conditions writes: Find $\mathbf{d} \in \mathcal{V}$ such that:

$$a(\mathbf{h}, \mathbf{d}) = L(\mathbf{h}) \quad \forall \mathbf{h} \in \mathcal{V}, \tag{11}$$

where $\mathcal{V} \subset \left(H^2(\Omega \setminus \Gamma_c)\right)^3$ is defined by,

$$\mathcal{V} = \left\{ \mathbf{h} \in \left(H^2(\Omega \setminus \Gamma_c)\right)^3 \middle| \, \mathbf{h} = \mathbf{0} \text{ on } \subset \Gamma_d \text{ and } \boldsymbol{\omega}(\mathbf{h}) = \mathbf{0} \text{ on } \subset \Gamma_d \right\} . \tag{12}$$

On the other hand, form $a : \mathcal{V} \times \mathcal{V} \to \mathbb{R}$ and linear functional $L : \mathcal{V} \to \mathbb{R}$ in (11) are defined as follows,

$$a(\mathbf{h}, \mathbf{d}) := \int_{\Omega \setminus \Gamma_c} \boldsymbol{\varepsilon}(\mathbf{h}) : \mathbf{n}(\mathbf{d}) \, \mathrm{d}\Omega + \int_{\Omega \setminus \Gamma_c} \boldsymbol{\kappa}(\mathbf{h}) : \mathbf{m}(\mathbf{d}) \, \mathrm{d}\Omega , \tag{13a}$$

$$L(\mathbf{h}) := \int_{\Omega \setminus \Gamma_c} \mathbf{h} \cdot \mathbf{b} \, \mathrm{d}\Omega + \int_{\Gamma_n} \mathbf{h} \cdot \mathbf{p} + \boldsymbol{\omega}(\mathbf{h}) \cdot \mathbf{r} \, \mathrm{d}\Gamma . \tag{13b}$$

Space $\mathcal{V}$ is naturally equipped with the energy norm [26],

$$\|\mathbf{h}\|_{\mathcal{V}} = \sqrt{a(\mathbf{h}, \mathbf{h})} \qquad \forall \mathbf{h} \in \mathcal{V} , \tag{14}$$

In [26] it was shown that for the linear kinematics case, see Eqs. (8a) and (8b), the bilinear form (13a) is bounded and elliptic in $\mathcal{V}$ under the energy norm (14), meaning that variational problem (11) has a unique solution in $\mathcal{V}$. However, in the present case this solution is discontinuous across the subdomains, i.e. it does not represent the solution to boundary value problem (10a), (10b), (10c), (10d), (10e), (10f), (10g), (10h) and (10i), as it does not account for the interface continuity conditions (10h) and (10i).

2.3 The Weak Form of the Penalty Approach

Let us introduce space $\mathfrak{L} \subset \mathcal{V}$,

$$\mathfrak{L} := \left\{ \mathbf{h} \in \mathcal{V} \,\middle|\, \exists \text{ c.e. } \mathbf{h}^{(i)}, \mathbf{h}^{(j)} \text{ on } \Gamma_{\mathrm{c}}^{(i,j)} : \mathbf{h}^{(i)} - \mathbf{h}^{(j)} \in \left(L^2\left(\Gamma_{\mathrm{c}}^{(i,j)}\right)\right)^3 \right.$$
$$\left. \text{and } \boldsymbol{\omega}\left(\mathbf{h}^{(i)}\right) + \boldsymbol{\omega}\left(\mathbf{h}^{(j)}\right) \in \left(L^2\left(\Gamma_{\mathrm{c}}^{(i,j)}\right)\right)^3 \right\} , \tag{15}$$

which becomes an inner product space when equipped with the inner product $\langle \bullet, \bullet \rangle_{\mathfrak{L}} : \mathfrak{L} \times \mathfrak{L} \to \mathbb{R}$,

$$\langle \mathbf{h}, \mathbf{d} \rangle_{\mathfrak{L}} = a\left(\mathbf{h}, \mathbf{d}\right) + \langle \boldsymbol{\chi}_{\mathrm{d}}\left(\mathbf{h}\right), \boldsymbol{\chi}_{\mathrm{d}}\left(\mathbf{d}\right) \rangle_{0,\Gamma_{\mathrm{c}}} + \langle \boldsymbol{\chi}_{\omega}\left(\mathbf{h}\right), \boldsymbol{\chi}_{\omega}\left(\mathbf{d}\right) \rangle_{0,\Gamma_{\mathrm{c}}} , \tag{16}$$

for all $\mathbf{h}, \mathbf{d} \in \mathfrak{L}$, where $\boldsymbol{\chi}_{\mathrm{d}}$ and $\boldsymbol{\chi}_{\omega}$ stand for the jump operators for the displacement and the rotation fields across the interface Γ_{c}, respectively. Notation $\langle \bullet, \bullet \rangle_{0,\Gamma_{\mathrm{c}}}$ stands for the inner product in the $L^2\left(\Gamma_{\mathrm{c}}\right)$ space, meaning that the second term in (16) writes,

$$\langle \boldsymbol{\chi}_{\mathrm{d}}\left(\mathbf{h}\right), \boldsymbol{\chi}_{\mathrm{d}}\left(\mathbf{d}\right) \rangle_{0,\Gamma_{\mathrm{c}}} = \int_{\Gamma_{\mathrm{c}}} \boldsymbol{\chi}_{\mathrm{d}}\left(\mathbf{h}\right) \cdot \boldsymbol{\chi}_{\mathrm{d}}\left(\mathbf{d}\right) \, \mathrm{d}\Gamma . \tag{17}$$

Moreover, space $\mathfrak{L}$ is equipped naturally with the norm $\|\bullet\|_{\mathfrak{L}} : \mathfrak{L} \to \mathbb{R}$ induced by the inner product (16),

$$\|\mathbf{h}\|_{\mathfrak{L}} = \sqrt{\langle \mathbf{h}, \mathbf{h} \rangle_{\mathfrak{L}}} \qquad \forall \mathbf{h} \in \mathfrak{L} . \tag{18}$$

Following [15] the variational formulation of the coupled problem using the Penalty approach writes: Find $\mathbf{d} \in \mathfrak{L}$ with $\mathbf{d}^{(i)} - \mathbf{d}^{(j)} = \boldsymbol{\chi}_{\mathrm{d}}^{(i,j)}\left(\mathbf{d}\right) = \boldsymbol{\chi}_{\mathrm{d}}\left(\mathbf{d}\right)_{|_{\Gamma_{\mathrm{c}}^{(i,j)}}}$ and $\boldsymbol{\omega}\left(\mathbf{d}^{(i)}\right) + \boldsymbol{\omega}\left(\mathbf{d}^{(j)}\right) = \boldsymbol{\chi}_{\omega}^{(i,j)}\left(\mathbf{d}\right) = \boldsymbol{\chi}_{\omega}\left(\mathbf{d}\right)_{|_{\Gamma_{\mathrm{c}}^{(i,j)}}}$ on $\Gamma_{\mathrm{c}}^{(i,j)}$, $i, j = 1, \ldots, n$ such that,

$$a\left(\mathbf{h}, \mathbf{d}\right) + \langle \boldsymbol{\chi}_{\mathrm{d}}\left(\mathbf{h}\right), \overline{\alpha}_{\mathrm{d}} \boldsymbol{\chi}_{\mathrm{d}}\left(\mathbf{d}\right) \rangle_{0,\Gamma_{\mathrm{c}}} + \langle \boldsymbol{\chi}_{\omega}\left(\mathbf{h}\right), \overline{\alpha}_{\omega} \boldsymbol{\chi}_{\omega}\left(\mathbf{d}\right) \rangle_{0,\Gamma_{\mathrm{c}}} = L\left(\mathbf{h}\right) , \tag{19}$$

for all $\mathbf{h} \in \mathfrak{L}$ with $\mathbf{h}^{(i)} - \mathbf{h}^{(j)} = \boldsymbol{\chi}_{\mathrm{d}}^{(i,j)}\left(\mathbf{h}\right) = \boldsymbol{\chi}_{\mathrm{d}}\left(\mathbf{h}\right)_{|_{\Gamma_{\mathrm{c}}^{(i,j)}}}$ and $\boldsymbol{\omega}\left(\mathbf{h}^{(i)}\right) + \boldsymbol{\omega}\left(\mathbf{h}^{(j)}\right) = \boldsymbol{\chi}_{\omega}^{(i,j)}\left(\mathbf{h}\right) = \boldsymbol{\chi}_{\omega}\left(\mathbf{h}\right)_{|_{\Gamma_{\mathrm{c}}^{(i,j)}}}$ on $\Gamma_{\mathrm{c}}^{(i,j)}$, $i, j = 1, \ldots, n$. Additionally, parameters $\overline{\alpha}_{\mathrm{d}}$ and $\overline{\alpha}_{\omega}$ which appear in Eq. (19) stand for the penalty factors for the displacement and the rotation field over the interface Γ_{c}, respectively, and they are strictly positive. Their distribution depends on the interface discretizations and a rule for their choice is proposed in Sect. 4. The penalty form $a_{\mathrm{p}} : \mathfrak{L} \times \mathfrak{L} \to \mathbb{R}$ governing variational formulation (19) is given by,

$$a_{\mathrm{p}}\left(\mathbf{h}, \mathbf{d}\right) := a\left(\mathbf{h}, \mathbf{d}\right) + \langle \boldsymbol{\chi}_{\mathrm{d}}\left(\mathbf{h}\right), \overline{\alpha}_{\mathrm{d}} \boldsymbol{\chi}_{\mathrm{d}}\left(\mathbf{d}\right) \rangle_{0,\Gamma_{\mathrm{c}}} + \langle \boldsymbol{\chi}_{\omega}\left(\mathbf{h}\right), \overline{\alpha}_{\omega} \boldsymbol{\chi}_{\omega}\left(\mathbf{d}\right) \rangle_{0,\Gamma_{\mathrm{c}}} , \tag{20}$$

for all functions $\mathbf{h}, \mathbf{d} \in \mathfrak{L}$. For the linear kinematics case, see Eqs. (8a) and (8b), form (20) is bilinear, by construction symmetric, bounded and coercive in $\mathfrak{L}$ which renders the formulation in general stable, meaning that one unique solution to problem (19) exists. However, this solution is not the same as for the strong form of the problem (10a), (10b), (10c), (10d), (10e), (10f), (10g), (10h) and (10i) for finite penalty parameters $\overline{\alpha}_{\mathrm{d}}$ and $\overline{\alpha}_{\omega}$, see [12]. This is the reason why the Penalty method is often called *variationally inconsistent*. The symmetry of the bilinear form (20) is obvious. Boundedness of a_{p} is easy to be shown using the *Cauchy-Schwarz* inequality in $L^2(\Gamma_{\mathrm{c}})$ space for the interface inner products. Coercivity of a_{p} follows from the positivity of the penalty parameters along the interface Γ_{c}. Numerical issues occurring due to the size of the penalty factors are discussed in Sect. 3.

2.4 The Weak Form of the Lagrange Multipliers Approach

The establishment of a weak formulation for the coupled system using the *Lagrange Multipliers* method demands two independent Lagrange Multipliers fields $\boldsymbol{\lambda}$ and $\boldsymbol{\zeta}$, i.e. one for each interface condition (10h) and (10i) respectively. In this way, the weak form of the coupled system using the Lagrange Multipliers writes [15]: Find $\mathbf{d} \in \mathfrak{L}$ with $\mathbf{d}^{(i)} - \mathbf{d}^{(j)} = \boldsymbol{\chi}_{\mathrm{d}}^{(i,j)}(\mathbf{d}) = \boldsymbol{\chi}_{\mathrm{d}}(\mathbf{d})_{|_{\Gamma_{\mathrm{c}}^{(i,j)}}}$, $\boldsymbol{\omega}\left(\mathbf{d}^{(i)}\right) + \boldsymbol{\omega}\left(\mathbf{d}^{(j)}\right) = \boldsymbol{\chi}_{\omega}^{(i,j)}(\mathbf{d}) = \boldsymbol{\chi}_{\omega}(\mathbf{d})_{|_{\Gamma_{\mathrm{c}}^{(i,j)}}}$ on $\Gamma_{\mathrm{c}}^{(i,j)}$, $i,j = 1,\ldots,n$ and $\boldsymbol{\lambda}, \boldsymbol{\zeta} \in \left(L^2(\Gamma_{\mathrm{c}})\right)^3$ such that,

$$
\begin{aligned}
a(\mathbf{h},\mathbf{d}) + \langle \boldsymbol{\chi}_{\mathrm{d}}(\mathbf{h}), \boldsymbol{\lambda}\rangle_{0,\Gamma_{\mathrm{c}}} + \langle \boldsymbol{\chi}_{\omega}(\mathbf{h}), \boldsymbol{\zeta}\rangle_{0,\Gamma_{\mathrm{c}}} + \langle \boldsymbol{\mu}, \boldsymbol{\chi}_{\mathrm{d}}(\mathbf{d})\rangle_{0,\Gamma_{\mathrm{c}}} +& \\
\langle \boldsymbol{\xi}, \boldsymbol{\chi}_{\omega}(\mathbf{d})\rangle_{0,\Gamma_{\mathrm{c}}} = L(\mathbf{h})& \ ,
\end{aligned}
\tag{21}
$$

for all $\mathbf{h} \in \mathfrak{L}$ with $\mathbf{h}^{(i)} - \mathbf{h}^{(j)} = \boldsymbol{\chi}_{\mathrm{d}}^{(i,j)}(\mathbf{h}) = \boldsymbol{\chi}_{\mathrm{d}}(\mathbf{h})_{|_{\Gamma_{\mathrm{c}}^{(i,j)}}}$ and $\boldsymbol{\omega}\left(\mathbf{h}^{(i)}\right) + \boldsymbol{\omega}\left(\mathbf{h}^{(j)}\right) = \boldsymbol{\chi}_{\omega}^{(i,j)}(\mathbf{h}) = \boldsymbol{\chi}_{\omega}(\mathbf{h})_{|_{\Gamma_{\mathrm{c}}^{(i,j)}}}$ on $\Gamma_{\mathrm{c}}^{(i,j)}$, $i,j = 1,\ldots,n$, for all $\boldsymbol{\mu}$ and for all $\boldsymbol{\xi}$ in $\left(L^2(\Gamma_{\mathrm{c}})\right)^3$, $\boldsymbol{\mu}$ and $\boldsymbol{\xi}$ being the Lagrange Multipliers test functions. From a physical viewpoint, the Lagrange Multipliers fields $\boldsymbol{\lambda}$ and $\boldsymbol{\zeta}$ represent the independent force traction and moment traction fields on the interface Γ_{c} which are necessary for the satisfaction of the interface force and moment equilibrium, respectively. Variational problem (21) is a saddle point problem since its weak solution is defined in $\mathcal{V}_{\mathrm{L}} = \mathcal{V} \times \left(L^2(\Gamma_{\mathrm{c}})\right)^3 \times \left(L^2(\Gamma_{\mathrm{c}})\right)^3$, thus having not to be necessarily existent or unique.

2.5 The Weak Form of the Augmented Lagrange Multipliers Approach

As already discussed in [15], the augmented *Lagrange Multipliers* method is a combination of the Lagrange Multipliers with the Penalty method, namely: Find

$\mathbf{d} \in \mathfrak{L}$ with $\mathbf{d}^{(i)} - \mathbf{d}^{(j)} = \boldsymbol{\chi}_{\mathrm{d}}^{(i,j)}(\mathbf{d}) = \boldsymbol{\chi}_{\mathrm{d}}(\mathbf{d})|_{\Gamma_{\mathrm{c}}^{(i,j)}}$ and $\boldsymbol{\omega}\left(\mathbf{d}^{(i)}\right) + \boldsymbol{\omega}\left(\mathbf{d}^{(j)}\right) = \boldsymbol{\chi}_{\omega}^{(i,j)}(\mathbf{d}) = \boldsymbol{\chi}_{\omega}(\mathbf{d})|_{\Gamma_{\mathrm{c}}^{(i,j)}}$ on $\Gamma_{\mathrm{c}}^{(i,j)}$, $i,j = 1,\ldots,n$ and $\boldsymbol{\lambda}, \boldsymbol{\zeta} \in \left(L^2(\Gamma_{\mathrm{c}})\right)^3$ such that,

$$\begin{aligned} &a(\mathbf{h},\mathbf{d}) + \langle \boldsymbol{\chi}_{\mathrm{d}}(\mathbf{h}), \boldsymbol{\lambda}\rangle_{0,\Gamma_{\mathrm{c}}} + \langle \boldsymbol{\chi}_{\mathrm{d}}(\mathbf{h}), \overline{\alpha}_{\mathrm{d}}\boldsymbol{\chi}_{\mathrm{d}}(\mathbf{d})\rangle_{0,\Gamma_{\mathrm{c}}} + \langle \boldsymbol{\chi}_{\omega}(\mathbf{h}), \boldsymbol{\zeta}\rangle_{0,\Gamma_{\mathrm{c}}} + \\ &\langle \boldsymbol{\chi}_{\omega}(\mathbf{h}), \overline{\alpha}_{\omega}\boldsymbol{\chi}_{\omega}(\mathbf{d})\rangle_{0,\Gamma_{\mathrm{c}}} + \langle \boldsymbol{\mu}, \boldsymbol{\chi}_{\mathrm{d}}(\mathbf{d})\rangle_{0,\Gamma_{\mathrm{c}}} + \langle \boldsymbol{\xi}, \boldsymbol{\chi}_{\omega}(\mathbf{d})\rangle_{0,\Gamma_{\mathrm{c}}} = L(\mathbf{h}) \,, \end{aligned} \tag{22}$$

for all $\mathbf{h} \in \mathfrak{L}$ with $\mathbf{h}^{(i)} - \mathbf{h}^{(j)} = \boldsymbol{\chi}_{\mathrm{d}}^{(i,j)}(\mathbf{h}) = \boldsymbol{\chi}_{\mathrm{d}}(\mathbf{h})|_{\Gamma_{\mathrm{c}}^{(i,j)}}$ and $\boldsymbol{\omega}\left(\mathbf{h}^{(i)}\right) + \boldsymbol{\omega}\left(\mathbf{h}^{(j)}\right) = \boldsymbol{\chi}_{\omega}^{(i,j)}(\mathbf{h}) = \boldsymbol{\chi}_{\omega}(\mathbf{h})|_{\Gamma_{\mathrm{c}}^{(i,j)}}$ on $\Gamma_{\mathrm{c}}^{(i,j)}$, $i,j = 1,\ldots,n$, for all $\boldsymbol{\mu}$ and for all $\boldsymbol{\xi}$ in $\left(L^2(\Gamma_{\mathrm{c}})\right)^3$. The weak solution of (22) is also defined in $\mathcal{V}_{\mathrm{L}}$, thus the augmented Lagrange Multipliers formulation belongs also to the family of saddle point problems. However, the penalty factors in this case do not have to be as large as in the Penalty method to achieve the same accuracy level in the interface conditions.

3 Discretization

3.1 Discrete Spaces

The numerical method used for the approximation of the given weak forms is IGA, as mentioned in Sect. 1. Within IGA the discrete spaces for the field approximation are taken from the geometry parametrization which is herein assumed to be the NURBS basis functions since they are dominant in CAGD. The convective coordinates $\left(\theta^1, \theta^2\right)$ introduced in Sect. 1 are thereafter designated by (ξ, η), complying with standard NURBS notations, see [2]. For each subdomain $\Omega^{(i)}$, $i = 1,\ldots,n$ and each interface $\Gamma_{\mathrm{c}}^{(k,l)}$, $(k,l) \in \mathcal{I}$, subdivisions $\mathcal{T}_{\mathrm{h}}^{(i)}$ and $\mathfrak{T}_{\mathrm{h}}^{(k,l)}$ are respectively chosen, which subdivide each $\Omega^{(i)}$ and $\Gamma_{\mathrm{c}}^{(k,l)}$ into isogeometric finite elements $\mathcal{T}_j^{(i)}$, $j = 1,\ldots,m^{(i)} \in \mathbb{N}$ and $\mathfrak{T}_j^{(k,l)}$, $j = 1,\ldots,m_{\lambda}^{(k,l)} \in \mathbb{N}$, respectively, for which the following relations hold,

$$\mathcal{T}_j^{(i)} \cap \mathcal{T}_k^{(i)} = \emptyset \quad \forall \mathcal{T}_j^{(i)}, \mathcal{T}_k^{(i)} \in \mathcal{T}_{\mathrm{h}}^{(i)} \text{ with } j \neq k \,, \tag{23a}$$

$$\mathfrak{T}_j^{(k,l)} \cap \mathfrak{T}_r^{(k,l)} = \emptyset \quad \forall \mathfrak{T}_j^{(k,l)}, \mathfrak{T}_r^{(k,l)} \in \mathfrak{T}_{\mathrm{h}}^{(k,l)} \text{ with } j \neq r \,, \tag{23b}$$

$$\bigcup_{j=1}^{m^{(i)}} \overline{\mathcal{T}_j^{(i)}} = \overline{\Omega^{(i)}} \,, \tag{23c}$$

$$\bigcup_{j=1}^{m_{\lambda}^{(k,l)}} \overline{\mathfrak{T}_j^{(k,l)}} = \overline{\Gamma_{\mathrm{c}}^{(k,l)}} \,, \tag{23d}$$

In the context of IGA the finite dimensional subspaces $\mathfrak{L}_h \subset \mathfrak{L}$ and $\mathcal{L}_h \subset \left(L^2\left(\Gamma_c\right)\right)^3$ are chosen as,

$$\mathfrak{L}_h = \left\{ \mathbf{h}_h \in \mathfrak{L} \middle| \mathbf{h}_h \in \left(\mathcal{R}^{p,q}\left(\mathcal{T}_j^{(i)}\right)\right)^3 \ \forall \mathcal{T}_j^{(i)} \in \mathcal{T}_h^{(i)} \right\} , \tag{24a}$$

$$\mathcal{L}_h = \left\{ \boldsymbol{\lambda}_h \in \left(L^2\left(\Gamma_c\right)\right)^3 \middle| \boldsymbol{\lambda}_h \in \left(\mathcal{P}^0\left(\mathfrak{T}_j^{(k,l)}\right)\right)^3 \ \forall \mathfrak{T}_j^{(k,l)} \in \mathfrak{T}_h^{(k,l)} \right\} , \tag{24b}$$

$\mathcal{R}^{p,q}\left(\mathcal{T}_j^{(i)}\right)$ being the space of all rational polynomials of polynomial orders p and q in the parametric image of $\mathcal{T}_j^{(i)}$. This parametric image is a rectangular domain $(\xi, \eta) \in \Xi^{(i)} \times H^{(i)} = \overline{\Omega_0^{(i)}}$, where $\overline{\Omega_0^{(i)}}$ is the image of $\overline{\Omega^{(i)}}$ under the inverse map $\mathbf{S}^{-1}$, within classical NURBS based isogeometric analysis, and $\Xi^{(i)}, H^{(i)}$ stand for the underlying knot vectors of patch $\Omega^{(i)}$, see [2] for more details. Additionally, the aforementioned rational polynomials are C^{p-m_k}-continuous across the elements within each patch $\Omega^{(i)}$, where p stands for the polynomial order of the basis crossing the element boundary and m_k the repetition of the knot at the element boundary, see [27] for more details. Equivalently, $\mathcal{P}^0\left(\mathfrak{T}_j^{(k,l)}\right)$ stands for the space of all constant functions in the parametric image of $\mathfrak{T}_j^{(k,l)}$. Assume that $\Phi^{(i)} = \left(\boldsymbol{\phi}_j^{(i)}\right)_j$, $j = 1, \ldots, \dim\left(\bigcup_{k=1}^{m^{(i)}} \mathcal{R}^{p,q}\left(\mathcal{T}_k^{(i)}\right)\right)^3$ and $\Psi^{(k,l)} = \left(\boldsymbol{\psi}_j^{(k,l)}\right)_j$, $j = 1, \ldots, \dim\left(\bigcup_{r=1}^{m_\lambda^{(k,l)}} \mathcal{P}^0\left(\mathfrak{T}_r^{(k,l)}\right)\right)^3 = 3m_\lambda^{(k,l)}$ are the set of basis functions at each subdomain $\Omega^{(i)}$ and each interface $\Gamma_c^{(k,l)}$, respectively. Notation $\bigcup_{k=1}^{m^{(i)}} \mathcal{R}^{p,q}\left(\mathcal{T}_k^{(i)}\right)$ is herein abused for the space that the NURBS basis functions of arbitrary inter-element continuity span at each patch $\Omega^{(i)}$. In this way, a basis can be formed for the discrete spaces $\mathfrak{L}_h$ and $\mathcal{L}_h$ as $\Phi = \bigcup_{i=1}^{n} \Phi^{(i)} = \left(\boldsymbol{\phi}_j\right)_j$, $j = 1, \ldots, \dim \Phi$ and $\Psi = \bigcup_{(k,l)\in\mathcal{I}} \Psi^{(k,l)} = \left(\boldsymbol{\psi}_j\right)_j$, $j = 1, \ldots, \dim \Psi$, respectively. These bases are in turn discontinuous across each patch $\partial\Omega^{(i)}$ and interface $\partial\Gamma_c^{(k,l)}$ boundary. Then, for any $\mathbf{h}_h \in \mathfrak{L}_h$ and any $\boldsymbol{\lambda}_h \in \mathcal{L}_h$ there exist reals $\hat{h}_i$ and $\hat{\lambda}_i$ such that,

$$\mathbf{h}_h = \sum_{i=1}^{\dim \Phi} \hat{h}_i \boldsymbol{\phi}_i \quad \text{and} \quad \boldsymbol{\lambda}_h = \sum_{i=1}^{\dim \Psi} \hat{\lambda}_i \boldsymbol{\psi}_i . \tag{25}$$

Important is to note that since $\mathfrak{L}_h$ represents the discrete admissible space for the displacement field, the multiplicity of the knots within each patch $\Omega^{(i)}$ must be chosen smaller than the corresponding polynomial order of the basis such that the basis functions are at least one time continuously differentiable and hence the bending strain is well defined, see Eqs. (7b) and (8b). The same restriction does not apply for the discrete admissible space $\mathcal{L}_h$ corresponding to the Lagrange Multipliers fields, and therefore discontinuous across the elements $\mathfrak{T}_j^{(k,l)}$ functions are chosen for the Lagrange Multipliers discretization. For all the forthcoming numerical examples in Sect. 4, the multiplicity of the interior to the patches knots is chosen equal to one.

3.2 Discrete Equation System of the Decoupled Problem

Subsequently, the *Bubnov-Galerkin* approximation is employed, that is both the test and the trial spaces are projected onto the same finite dimensional subspace, as those are defined in (24a) and (24b). Then, projection of the weak form corresponding to the decoupled system (11) onto $\mathfrak{L}_h \times \mathfrak{L}_h$ yields the following discrete equation system for the geometrically linear setting,

$$\begin{bmatrix} \mathbf{K}^{(1)} & \cdots & \mathbf{0} \\ \vdots & \ddots & \vdots \\ \mathbf{0} & \cdots & \mathbf{K}^{(n)} \end{bmatrix} \begin{bmatrix} \hat{\mathbf{d}}^{(1)} \\ \vdots \\ \hat{\mathbf{d}}^{(n)} \end{bmatrix} = \begin{bmatrix} \hat{\mathbf{F}}^{(1)} \\ \vdots \\ \hat{\mathbf{F}}^{(n)} \end{bmatrix}, \tag{26}$$

where $\mathbf{K}^{(i)}$, $\hat{\mathbf{F}}^{(i)}$ and $\hat{\mathbf{d}}^{(i)}$ stand for the stiffness matrix with entries $K_{jk}^{(i)} = a\left(\boldsymbol{\phi}_j^{(i)}, \boldsymbol{\phi}_k^{(i)}\right)$, the load vector with entries $F_j^{(i)} = L\left(\boldsymbol{\phi}_j^{(i)}\right)$ and the discrete Control Point solution vector at each patch $\Omega^{(i)}$, respectively. For the geometrically nonlinear setting, the $\hat{i}$-th Newton-Rapshon iteration consists in the solution of the following linear system,

$$\begin{bmatrix} \mathbf{K}_{\mathrm{T}}^{(1)}\left(\hat{\mathbf{d}}_{\hat{i}}^{(1)}\right) & \cdots & \mathbf{0} \\ \vdots & \ddots & \vdots \\ \mathbf{0} & \cdots & \mathbf{K}_{\mathrm{T}}^{(n)}\left(\hat{\mathbf{d}}_{\hat{i}}^{(n)}\right) \end{bmatrix} \begin{bmatrix} \delta\hat{\mathbf{d}}_{\hat{i}}^{(1)} \\ \vdots \\ \delta\hat{\mathbf{d}}_{\hat{i}}^{(n)} \end{bmatrix} = - \begin{bmatrix} \mathbf{R}^{(1)}\left(\hat{\mathbf{d}}_{\hat{i}}^{(1)}\right) \\ \vdots \\ \mathbf{R}^{(n)}\left(\hat{\mathbf{d}}_{\hat{i}}^{(i)}\right) \end{bmatrix}, \tag{27}$$

where $\mathbf{K}_{\mathrm{T}}^{(i)}$, $\delta\hat{\mathbf{d}}_{\hat{i}}^{(i)} = \hat{\mathbf{d}}_{\hat{i}+1}^{(i)} - \hat{\mathbf{d}}_{\hat{i}}^{(i)}$ and $\mathbf{R}^{(i)}\left(\hat{\mathbf{d}}_{\hat{i}}^{(i)}\right) = \mathbf{K}\left(\hat{\mathbf{d}}_{\hat{i}}^{(i)}\right)\hat{\mathbf{d}}_{\hat{i}}^{(i)} - \hat{\mathbf{F}}^{(i)}$ denote the tangent stiffness matrix, the Control Point displacement increment and the residual in patch $\Omega^{(i)}$, respectively.

3.3 Discrete Equation System of the Penalty Approach

Projection of the weak form corresponding to the Penalty approach (19) onto $\mathfrak{L}_h \times \mathfrak{L}_h$ results into the following discrete equation system for the linear setting,

$$\begin{bmatrix} \mathbf{K}^{(1)} + \mathbf{K}_{\mathrm{p}}^{(1)} & \cdots & \mathbf{C}_{\mathrm{p}}^{(1,n)} \\ \vdots & \ddots & \vdots \\ \mathbf{C}_{\mathrm{p}}^{(n,1)} & \cdots & \mathbf{K}^{(n)} + \mathbf{K}_{\mathrm{p}}^{(n)} \end{bmatrix} \begin{bmatrix} \hat{\mathbf{d}}^{(1)} \\ \vdots \\ \hat{\mathbf{d}}^{(n)} \end{bmatrix} = \begin{bmatrix} \hat{\mathbf{F}}^{(1)} \\ \vdots \\ \hat{\mathbf{F}}^{(n)} \end{bmatrix}, \tag{28}$$

where the entries of the additional stiffness and Penalty coupling matrices $\mathbf{K}_{\mathrm{p}}^{(i)}$ and $\mathbf{C}_{\mathrm{p}}^{(i,j)}$ are respectively defined as,

$$K_{\mathrm{p},k,l}^{(i)} := \left\langle \boldsymbol{\phi}_k^{(i)}, \overline{\alpha}_{\mathrm{d}} \boldsymbol{\phi}_l^{(i)} \right\rangle_{0,\Gamma_\mathrm{c}} + \left\langle \boldsymbol{\omega}\left(\boldsymbol{\phi}_k^{(i)}\right), \overline{\alpha}_{\omega} \boldsymbol{\omega}\left(\boldsymbol{\phi}_l^{(i)}\right) \right\rangle_{0,\Gamma_\mathrm{c}}, \tag{29a}$$

$$C_{\mathrm{p},k,l}^{(i,j)} := -\left\langle \boldsymbol{\phi}_k^{(i)}, \overline{\alpha}_{\mathrm{d}} \boldsymbol{\phi}_l^{(j)} \right\rangle_{0,\Gamma_\mathrm{c}} + \left\langle \boldsymbol{\omega}\left(\boldsymbol{\phi}_k^{(i)}\right), \overline{\alpha}_{\omega} \boldsymbol{\omega}\left(\boldsymbol{\phi}_l^{(j)}\right) \right\rangle_{0,\Gamma_\mathrm{c}}. \tag{29b}$$

For the geometrically nonlinear problem, the Newton-Raphson subproblem at the $\hat{i}$-th iteration, defined in (27), is written as,

$$\begin{bmatrix} \mathbf{K}_{\mathrm{T}}^{(1)}\left(\hat{\mathbf{d}}_{\hat{i}}^{(1)}\right) + \mathbf{K}_{\mathrm{p}}^{(1)} & \cdots & \mathbf{C}_{\mathrm{p}}^{(1,n)} \\ \vdots & \ddots & \vdots \\ \mathbf{C}_{\mathrm{p}}^{(n,1)} & \cdots & \mathbf{K}_{\mathrm{T}}^{(n)}\left(\hat{\mathbf{d}}_{\hat{i}}^{(n)}\right) + \mathbf{K}_{\mathrm{p}}^{(n)} \end{bmatrix} \begin{bmatrix} \delta\hat{\mathbf{d}}_{\hat{i}}^{(1)} \\ \vdots \\ \delta\hat{\mathbf{d}}_{\hat{i}}^{(n)} \end{bmatrix} = \\ -\begin{bmatrix} \mathbf{R}^{(1)}\left(\hat{\mathbf{d}}_{\hat{i}}^{(1)}\right) \\ \vdots \\ \mathbf{R}^{(n)}\left(\hat{\mathbf{d}}_{\hat{i}}^{(n)}\right) \end{bmatrix} - \begin{bmatrix} \mathbf{K}_{\mathrm{p}}^{(1)} & \cdots & \mathbf{C}_{\mathrm{p}}^{(1,n)} \\ \vdots & \ddots & \vdots \\ \mathbf{C}_{\mathrm{p}}^{(n,1)} & \cdots & \mathbf{K}_{\mathrm{p}}^{(n)} \end{bmatrix} \begin{bmatrix} \hat{\mathbf{d}}_{\hat{i}}^{(1)} \\ \vdots \\ \hat{\mathbf{d}}_{\hat{i}}^{(n)} \end{bmatrix}. \tag{30}$$

Note also that the additional interface terms which show up in the Penalty formulation (19) are linear with respect to the displacement field, thus making the extension from the geometrically linear to the geometrically nonlinear problem straightforward. It must be emphasized that the resulting system is positive definite given that the penalty parameters do not destroy its conditioning, and therefore must be carefully chosen. A rule for their choice is given in Sect. 4.1 where an assessment of this choice with respect to the domain and interface errors in the L^2-norm is in addition provided.

3.4 Discrete Equation System of the Lagrange Multipliers Approach

As discussed in Sect. 2.4 two independent Lagrange Multipliers fields are used for the corresponding variational form (21). Therefore, also two independent discrete Lagrange Multipliers spaces, one for the Lagrange Multipliers field representing the interface traction forces and one for the Lagrange Multipliers field representing the interface traction moments are chosen as $\mathcal{L}_\mathrm{h}^\lambda$ and $\mathcal{L}_\mathrm{h}^\zeta$, respectively, following (24b). Let also the set of basis functions for each space $\mathcal{L}_\mathrm{h}^\lambda$ and $\mathcal{L}_\mathrm{h}^\zeta$ be $\Psi^\lambda = \left(\boldsymbol{\psi}_j^\lambda\right)_j$, $j = 1, \dots, \dim \Psi^\lambda$ and $\Psi^\zeta = \left(\boldsymbol{\psi}_j^\zeta\right)_j$, $j = 1, \dots, \dim \Psi^\zeta$, respectively. Then, projection of the Lagrange Multipliers weak form (21) onto $\left(\mathfrak{L}_\mathrm{h} \times \mathcal{L}_\mathrm{h}^\lambda \times \mathcal{L}_\mathrm{h}^\zeta\right) \times \left(\mathfrak{L}_\mathrm{h} \times \mathcal{L}_\mathrm{h}^\lambda \times \mathcal{L}_\mathrm{h}^\zeta\right)$

provides us with the following equation system for the geometrically linear setting,

$$\begin{bmatrix} \mathbf{K}^{(1)} & \cdots & \mathbf{0} & \boldsymbol{\Lambda}^{(1)} & \mathbf{Z}^{(1)} \\ \vdots & \ddots & \vdots & \vdots & \vdots \\ \mathbf{0} & \cdots & \mathbf{K}^{(n)} & \boldsymbol{\Lambda}^{(n)} & \mathbf{Z}^{(n)} \\ \left(\boldsymbol{\Lambda}^{(1)}\right)^{\mathrm{T}} & \cdots & \left(\boldsymbol{\Lambda}^{(n)}\right)^{\mathrm{T}} & \mathbf{0} & \mathbf{0} \\ \left(\mathbf{Z}^{(1)}\right)^{\mathrm{T}} & \cdots & \left(\mathbf{Z}^{(n)}\right)^{\mathrm{T}} & \mathbf{0} & \mathbf{0} \end{bmatrix} \begin{bmatrix} \hat{\mathbf{d}}^{(1)} \\ \vdots \\ \hat{\mathbf{d}}^{(n)} \\ \hat{\boldsymbol{\lambda}} \\ \hat{\boldsymbol{\xi}} \end{bmatrix} = \begin{bmatrix} \hat{\mathbf{F}}^{(1)} \\ \vdots \\ \hat{\mathbf{F}}^{(n)} \\ \mathbf{0} \\ \mathbf{0} \end{bmatrix}, \tag{31}$$

where the entries of matrices $\boldsymbol{\Lambda}^{(i)}$ and $\mathbf{Z}^{(i)}$ are given by,

$$\Lambda_{k,l}^{(i)} := \sum_{j=1}^{n} (-1)^{\alpha^{(i,j)}} \left\langle \boldsymbol{\phi}_k^{(i)}, \boldsymbol{\psi}_l^{\lambda} \right\rangle_{0,\Gamma_{\mathrm{c}}^{(i,j)}}, \tag{32a}$$

$$Z_{k,l}^{(i)} := \sum_{j=1}^{n} \left\langle \boldsymbol{\omega}\left(\boldsymbol{\phi}_k^{(i)}\right), \boldsymbol{\psi}_l^{\xi} \right\rangle_{0,\Gamma_{\mathrm{c}}^{(i,j)}}, \tag{32b}$$

with $\alpha^{(i,j)} = 0$ if $i < j$ or $\alpha^{(i,j)} = 1$ if $j < i$. Additionally, $\hat{\boldsymbol{\lambda}}$ and $\hat{\boldsymbol{\xi}}$ denote the discrete solution vectors for both Lagrange Multiplier fields. Due to the linearity of the weak form on the Lagrange Multipliers fields, the extension to the nonlinear setting is also in this case straightforward. Thus, the $\hat{i}$-th Newton-Raphson iteration in this case writes,

$$\begin{aligned} &\begin{bmatrix} \mathbf{K}_{\mathrm{T}}^{(1)}\left(\hat{\mathbf{d}}_{\hat{i}}^{(1)}\right) & \cdots & \mathbf{0} & \boldsymbol{\Lambda}^{(1)} & \mathbf{Z}^{(1)} \\ \vdots & \ddots & \vdots & \vdots & \vdots \\ \mathbf{0} & \cdots & \mathbf{K}_{\mathrm{T}}^{(n)}\left(\hat{\mathbf{d}}_{\hat{i}}^{(n)}\right) & \boldsymbol{\Lambda}^{(n)} & \mathbf{Z}^{(n)} \\ \left(\boldsymbol{\Lambda}^{(1)}\right)^{\mathrm{T}} & \cdots & \left(\boldsymbol{\Lambda}^{(n)}\right)^{\mathrm{T}} & \mathbf{0} & \mathbf{0} \\ \left(\mathbf{Z}^{(1)}\right)^{\mathrm{T}} & \cdots & \left(\mathbf{Z}^{(n)}\right)^{\mathrm{T}} & \mathbf{0} & \mathbf{0} \end{bmatrix} \begin{bmatrix} \delta\hat{\mathbf{d}}_{\hat{i}}^{(1)} \\ \vdots \\ \delta\hat{\mathbf{d}}_{\hat{i}}^{(n)} \\ \delta\hat{\boldsymbol{\lambda}}_{\hat{i}} \\ \delta\hat{\boldsymbol{\xi}}_{\hat{i}} \end{bmatrix} = \\ &- \begin{bmatrix} \mathbf{R}^{(1)}\left(\hat{\mathbf{d}}_{\hat{i}}^{(1)}\right) \\ \vdots \\ \mathbf{R}^{(n)}\left(\hat{\mathbf{d}}_{\hat{i}}^{(n)}\right) \\ \mathbf{0} \\ \mathbf{0} \end{bmatrix} - \begin{bmatrix} \mathbf{0} & \cdots & \mathbf{0} & \boldsymbol{\Lambda}^{(1)} & \mathbf{Z}^{(1)} \\ \vdots & \ddots & \vdots & \vdots & \vdots \\ \mathbf{0} & \cdots & \mathbf{0} & \boldsymbol{\Lambda}^{(n)} & \mathbf{Z}^{(n)} \\ \left(\boldsymbol{\Lambda}^{(1)}\right)^{\mathrm{T}} & \cdots & \left(\boldsymbol{\Lambda}^{(n)}\right)^{\mathrm{T}} & \mathbf{0} & \mathbf{0} \\ \left(\mathbf{Z}^{(1)}\right)^{\mathrm{T}} & \cdots & \left(\mathbf{Z}^{(n)}\right)^{\mathrm{T}} & \mathbf{0} & \mathbf{0} \end{bmatrix} \begin{bmatrix} \hat{\mathbf{d}}_{\hat{i}}^{(1)} \\ \vdots \\ \hat{\mathbf{d}}_{\hat{i}}^{(n)} \\ \hat{\boldsymbol{\lambda}}_{\hat{i}} \\ \hat{\boldsymbol{\xi}}_{\hat{i}} \end{bmatrix}. \end{aligned} \tag{33}$$

It must be remarked that the discrete spaces chosen for the displacement and the Lagrange Multipliers fields must fulfil the so-called Ladyzenskaja-Babuska-Brezzi condition, alternatively known as *inf-sup* condition, so that one unique solution to the finite dimensional problem can be guaranteed. This is not a trivial problem and an improper choice of the discrete spaces renders the discrete equation system ill-conditioned or even indefinite. The numerical investigations suggest the following way of choosing the discretization for the Lagrange Multipliers fields: As mentioned in Sect. 3.1 both Lagrange Multipliers fields are discretized with constant, discontinuous across the elements functions. In addition, field $\boldsymbol{\lambda}|_{\Gamma_c^{(k,l)}}$ is discretized using as many elements $n_\lambda^{(k,l)}$ as the coarsest of the subdomains $\Omega^{(k)}$ and $\Omega^{(l)}$ has and field $\boldsymbol{\zeta}|_{\Gamma_c^{(k,l)}}$ is discretized with $n_\zeta^{(k,l)} = \left\lceil \frac{n_\lambda^{(k,l)}}{2} \right\rceil$. These rules have only numerically proven to produce in most of the cases stable and not ill-conditioned discrete equation systems, thus further mathematical investigations are urged for this case.

3.5 Discrete Equation System of the Augmented Lagrange Multipliers Approach

As mentioned in Sect. 2.5, the augmented Lagrange Multipliers method is a combination of the Penalty and the Lagrange Multipliers method. Therefore, the discrete equation systems for both the geometrically linear and geometrically nonlinear problems write,

$$\begin{bmatrix} \mathbf{K}^{(1)} + \mathbf{K}_\mathrm{p}^{(1)} & \cdots & \mathbf{C}_\mathrm{p}^{(1,n)} & \boldsymbol{\Lambda}^{(1)} & \mathbf{Z}^{(1)} \\ \vdots & \ddots & \vdots & \vdots & \vdots \\ \mathbf{C}_\mathrm{p}^{(n,1)} & \cdots & \mathbf{K}^{(n)} + \mathbf{K}_\mathrm{p}^{(n)} & \boldsymbol{\Lambda}^{(n)} & \mathbf{Z}^{(n)} \\ \left(\boldsymbol{\Lambda}^{(1)}\right)^\mathrm{T} & \cdots & \left(\boldsymbol{\Lambda}^{(n)}\right)^\mathrm{T} & \mathbf{0} & \mathbf{0} \\ \left(\mathbf{Z}^{(1)}\right)^\mathrm{T} & \cdots & \left(\mathbf{Z}^{(n)}\right)^\mathrm{T} & \mathbf{0} & \mathbf{0} \end{bmatrix} \begin{bmatrix} \hat{\mathbf{d}}^{(1)} \\ \vdots \\ \hat{\mathbf{d}}^{(n)} \\ \hat{\boldsymbol{\lambda}} \\ \hat{\boldsymbol{\zeta}} \end{bmatrix} = \begin{bmatrix} \hat{\mathbf{F}}^{(1)} \\ \vdots \\ \hat{\mathbf{F}}^{(n)} \\ \mathbf{0} \\ \mathbf{0} \end{bmatrix}, \tag{34}$$

$$\begin{aligned} &\begin{bmatrix} \mathbf{K}_\mathrm{T}^{(1)}\left(\hat{\mathbf{d}}_{\hat{i}}^{(1)}\right) + \mathbf{K}_\mathrm{p}^{(1)} & \cdots & \mathbf{C}_\mathrm{p}^{(1,n)} & \boldsymbol{\Lambda}^{(1)} & \mathbf{Z}^{(1)} \\ \vdots & \ddots & \vdots & \vdots & \vdots \\ \mathbf{C}_\mathrm{p}^{(n,1)} & \cdots & \mathbf{K}_\mathrm{T}^{(n)}\left(\hat{\mathbf{d}}_{\hat{i}}^{(n)}\right) + \mathbf{K}_\mathrm{p}^{(n)} & \boldsymbol{\Lambda}^{(n)} & \mathbf{Z}^{(n)} \\ \left(\boldsymbol{\Lambda}^{(1)}\right)^\mathrm{T} & \cdots & \left(\boldsymbol{\Lambda}^{(n)}\right)^\mathrm{T} & \mathbf{0} & \mathbf{0} \\ \left(\mathbf{Z}^{(1)}\right)^\mathrm{T} & \cdots & \left(\mathbf{Z}^{(n)}\right)^\mathrm{T} & \mathbf{0} & \mathbf{0} \end{bmatrix} \begin{bmatrix} \delta\hat{\mathbf{d}}_{\hat{i}}^{(1)} \\ \vdots \\ \delta\hat{\mathbf{d}}_{\hat{i}}^{(n)} \\ \delta\hat{\boldsymbol{\lambda}}_{\hat{i}} \\ \delta\hat{\boldsymbol{\zeta}}_{\hat{i}} \end{bmatrix} = \\ &- \begin{bmatrix} \mathbf{R}^{(1)}\left(\hat{\mathbf{d}}_{\hat{i}}^{(1)}\right) \\ \vdots \\ \mathbf{R}^{(n)}\left(\hat{\mathbf{d}}_{\hat{i}}^{(n)}\right) \\ \mathbf{0} \\ \mathbf{0} \end{bmatrix} - \begin{bmatrix} \mathbf{K}_\mathrm{p}^{(1)} & \cdots & \mathbf{C}_\mathrm{p}^{(1,n)} & \boldsymbol{\Lambda}^{(1)} & \mathbf{Z}^{(1)} \\ \vdots & \ddots & \vdots & \vdots & \vdots \\ \mathbf{C}_\mathrm{p}^{(n,1)} & \cdots & \mathbf{K}_\mathrm{p}^{(n)} & \boldsymbol{\Lambda}^{(n)} & \mathbf{Z}^{(n)} \\ \left(\boldsymbol{\Lambda}^{(1)}\right)^\mathrm{T} & \cdots & \left(\boldsymbol{\Lambda}^{(n)}\right)^\mathrm{T} & \mathbf{0} & \mathbf{0} \\ \left(\mathbf{Z}^{(1)}\right)^\mathrm{T} & \cdots & \left(\mathbf{Z}^{(n)}\right)^\mathrm{T} & \mathbf{0} & \mathbf{0} \end{bmatrix} \begin{bmatrix} \hat{\mathbf{d}}_{\hat{i}}^{(1)} \\ \vdots \\ \hat{\mathbf{d}}_{\hat{i}}^{(n)} \\ \hat{\boldsymbol{\lambda}}_{\hat{i}} \\ \hat{\boldsymbol{\zeta}}_{\hat{i}} \end{bmatrix}. \end{aligned} \tag{35}$$

As in the case of the Lagrange Multipliers method, the linear equation system in this case might also be indefinite, depending on whether the chosen discrete spaces fulfil the inf-sup condition or not. For the discretization of the Lagrange Multipliers fields, the same empirical rules as in Sect. 3.4 are employed.

4 Numerical Results

In this section, the DDMs are compared and systematically evaluated over two linear and two nonlinear problems. For the linear problems convergence studies over domain and interface L^2-norms are performed. The L^2-norm of a tensor $\boldsymbol{\tau} \in \mathfrak{S}^m$ over a set Ω is given by,

$$\|\boldsymbol{\tau}\|_{0,\Omega} := \left(\int_{\Omega} \sum_{i_1=1}^{m} \cdots \sum_{i_m=1}^{m} \tau_{i_1 \ldots i_m} \tau_{i_1 \ldots i_m} \, \mathrm{d}\Omega \right)^{\frac{1}{2}} . \tag{36}$$

On the other hand in what concerns the nonlinear problems, the corresponding load-displacement curves are given and compared against the corresponding results in [28]. The various quantities depicted in the following graphs are measured in *Systéme International d'Unités* (SI), if not otherwise stated.

4.1 Circular Plate Subject to Uniform Pressure Load

The first geometrically linear problem over which a convergence study for all the DDMs is performed is that of a circular plate with radius $\overline{R} = 50$ cm subject to uniform distributed load $\overline{p} = 0.01\ \mathrm{KN/m^2}$ at its surface. The problem is decomposed into five patches, see Fig. 2. An analytical solution in terms of

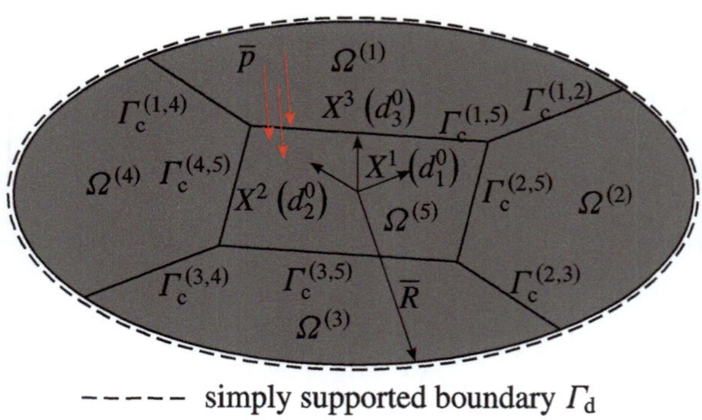

Fig. 2 Circular plate subject to uniform pressure load: problem placement

displacements for this problem exists [29],

$$\mathbf{d}\left(X^1,X^2\right) = -\frac{3\overline{p}\left(\overline{R}^2 - r^2\right)\left(1-\nu^2\right)}{16Eh^3}\left(\frac{5+\nu}{1+\nu}\overline{R}^2 - r^2\right)\mathbf{e}_3\,, \tag{37}$$

where $r^2 = \left(X^1\right)^2 + \left(X^2\right)^2 \leq \overline{R}^2$. The Young's modulus, Poisson's ratio and plate's thickness are chosen as $E = 1.8 \times 10^3\,\mathrm{KN/m^2}$, $\nu = 0.3$ and $h = 1\,\mathrm{cm}$, respectively. A convergence study is performed by refining patches $\Omega^{(1)}$, $\Omega^{(2)}$, $\Omega^{(3)}$, $\Omega^{(4)}$ and $\Omega^{(5)}$ with $\left\lceil\frac{5}{9}i\right\rceil^2$, $\left\lceil\frac{6}{9}i\right\rceil^2$, $\left\lceil\frac{4}{9}i\right\rceil^2$, $\left\lceil\frac{7}{9}i\right\rceil^2$ and i^2 biquadratic and bicubic elements, respectively, where $i = 4, \ldots, 40$ stands for the refinement level index. The isogeometric discretization of each patch is chosen equal in both surface parametric directions ξ and η. The resulting scaled deformation and the 2-norm of the displacement field $\|\mathbf{d}\|_2$ using the Penalty method, corresponding to refinement level $i = 8$ for the bicubic elements are shown in the set of Fig. 3. Within this study, the penalty parameters $\overline{\alpha}_{\mathrm{d}}$ and $\overline{\alpha}_{\omega}$ are chosen constant throughout the whole interface Γ_{c}, as a scaling of the matrix norms $\|\mathbf{D}_{\mathrm{m}}\|_{\mathrm{F}}$ and $\|\mathbf{D}_{\mathrm{b}}\|_{\mathrm{F}}$ corresponding to the material matrices for the membrane and the bending stiffness of the Kirchhoff-Love shell, respectively. In addition, this scaling is chosen as the inverse of the minimum element area size within the multipatch geometry, that is $h_{-1} = \left(\min_{\mathcal{T}_j \in \bigcup_{i=1}^{5} \mathcal{T}_{\mathrm{h}}^{(i)}} |\mathcal{T}_j|\right)^{-1}$. For the case depicted in Fig. 3, $h_{-1} = 6.1169 \times 10^2\,\mathrm{m^{-2}}$ resulting in choosing the penalty parameters as $\overline{\alpha}_{\mathrm{d}} = 1.5729 \times 10^4\,\mathrm{KN/m^3}$ and $\overline{\alpha}_{\omega} = 0.1310\,\mathrm{KN/m}$. This choice is found to be optimal with respect to the domain error and interface displacement jump, for both the Penalty and the augmented Lagrange Multipliers methods, see Fig. 4. Then, in what concerns the refinement studies, the scaling related to the penalty parameters is chosen as 10^4 and 10^2 for the Penalty and the augmented Lagrange Multipliers method, respectively. The refinement studies consist in biquadratic as well as bicubic elements, see set of Fig. 5. The Penalty method is reaching its threshold in the order of 10^{-4} and 10^{-8} for the relative domain error and the interface jump in the L^2-norm, respectively.

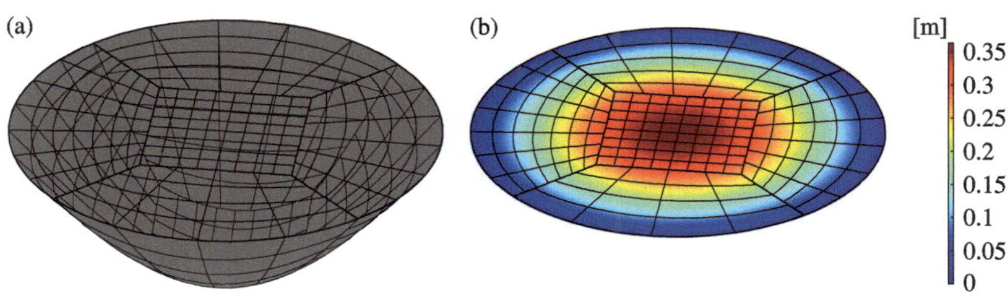

Fig. 3 Circular plate subject to uniform pressure load: postprocessing (Penalty). Setting corresponding to refinement level i=8. Patches $\Omega^{(1)}$, $\Omega^{(2)}$, $\Omega^{(3)}$, $\Omega^{(4)}$ and $\Omega^{(5)}$ modelled using 25, 36, 16, 49 and 64 bicubic elements, respectively. (**a**) Plate's deformation scaled by $\frac{3}{2}$. (**b**) The displacement field in 2-norm $\|\mathbf{d}\|_2$

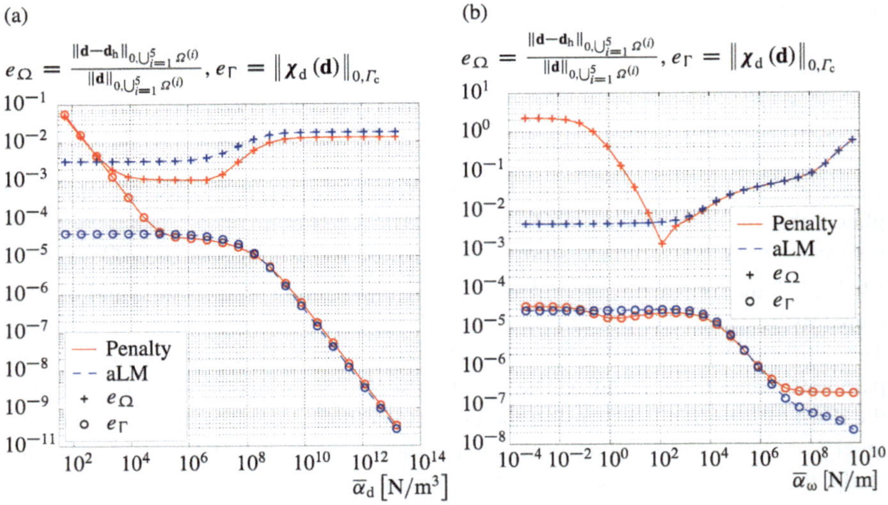

Fig. 4 Circular plate subject to uniform pressure load: penalty parameters. Relative domain error and interface displacement jump in the L^2-norm against the penalty parameters corresponding to refinement level i=8, see Fig. 3. (**a**) Error depending on $\overline{\alpha}_d$ maintaining $\overline{\alpha}_\omega = 0.1310\,\text{KN/m}$. (**b**) Error depending on $\overline{\alpha}_\omega$ maintaining $\overline{\alpha}_d = 1.5729 \times 10^4\,\text{KN/m}^3$

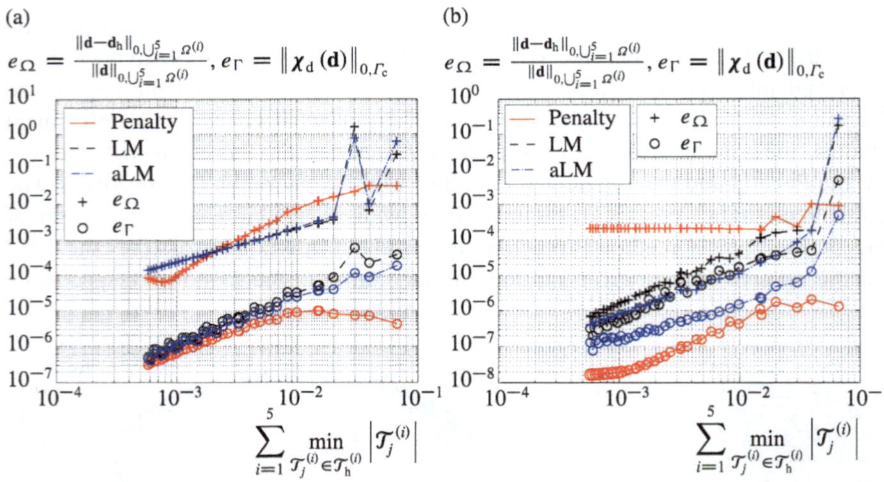

Fig. 5 Circular plate subject to uniform pressure load: convergence study. Relative domain error and interface displacement jump in the L^2-norm against the sum of the minimum element areas in the multipatch geometry. (**a**) Biquadratic basis. (**b**) Bicubic basis

Both Lagrange Multipliers methods are demonstrating uniform convergence with the augmented Lagrange Multipliers showing a slightly improved convergence compared to the Lagrange Multipliers method. That is due to the penalty factors whose magnitude is controlling the level of satisfaction to the interface constraints

(10h) and (10i). Important for the augmented Lagrange Multipliers method is not to use penalization parameters much larger than for the Penalty method, cause in this case the method would be dominated by the penalty terms destroying the uniform convergence expected from a Lagrange Multipliers method. The latter can be observed in Fig. 4 where although for small penalty parameters the error levels obtained by the Lagrange Multipliers are lower than the corresponding Penalty ones, when the penalization parameters are increased both methods yield essentially the same results, meaning that the Lagrange Multipliers play no more role.

4.2 Scordelis-Lo Roof

The second geometrically linear example which is employed for the comparison of the DDMs is the Scordelis-Lo roof shell [30]. Its Young's modulus, Poisson's ratio and thickness are chosen equal to $E = 4.32 \times 10^5\,\mathrm{KN/m^2}$, $\nu = 0$ and $h = 25\,\mathrm{cm}$, respectively. Its curvature in the X^1-direction is zero with length equal to $L = 50\,\mathrm{m}$, whereas in the X^2-X^3 plane it is representing a part of a cylinder with radius and arc length size equal to $\overline{R} = 25\,\mathrm{m}$ and $\overline{\phi}_0 = \frac{4\pi}{9}$, respectively. At $X^1 = -\frac{L}{2}$ and $X^2 = \frac{L}{2}$ the shell is supported with diaphragms. The latter means that along the Dirichlet boundary Γ_d the motion of the shell in X^1 and X^2 directions is confined. The problem is confronted in its original geometric setting without reducing it into one quarter by apply the necessary symmetry conditions. Therefore, the corner $\left(X^1, X^2, X^3\right) \equiv \left(-\frac{L}{2}, -\overline{R}\sin\frac{\overline{\phi}_0}{2}, \overline{R}\cos\frac{\overline{\phi}_0}{2}\right)$ has to be fixed so that the indeterminacy of the problem is eliminated. In addition, a uniform constant distributed load $\overline{p} = 9 \times 10^{-2}\,\mathrm{KN/m^2}$ is applied on the shell's surface in the X^3-direction. Then, the problem is decomposed symmetrically into four patches, see Fig. 6.

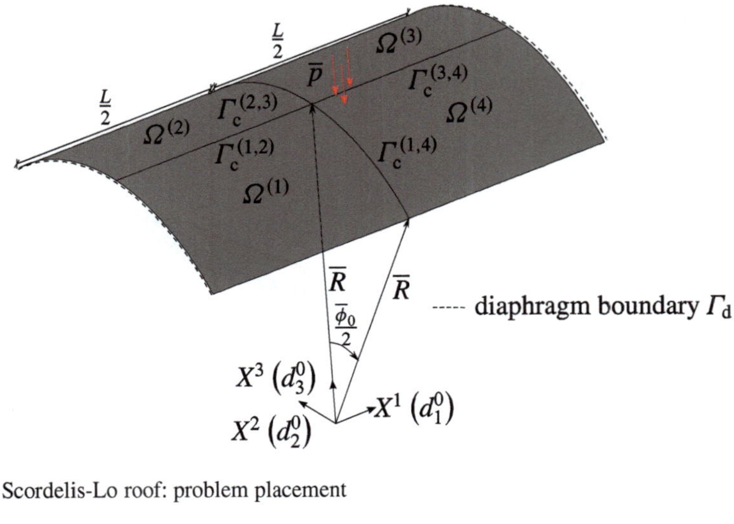

Fig. 6 Scordelis-Lo roof: problem placement

A convergence study is performed for low and high polynomial bases of the patches. As reference solution for the displacement vector $\mathbf{d}$, force $\mathbf{n}(\mathbf{d})$ and moment $\mathbf{m}(\mathbf{d})$ tensors fields, it is considered the solution of the same problem modelled with a single patch attaining a bidecic basis, sixty and forty elements in the X^1 and $X^2 \sin\overline{\phi} + X^3 \cos\overline{\phi}$ direction, respectively. The low polynomial order case consists in patch pairs $\Omega^{(1)}$, $\Omega^{(3)}$ and $\Omega^{(2)}$, $\Omega^{(4)}$ being modelled with bicubic and biquadratic basis, respectively. On the other hand, the high polynomial order case represents the patch pair $\Omega^{(1)}$, $\Omega^{(3)}$ and $\Omega^{(2)}$, $\Omega^{(4)}$ modelling with biquartic and bicubic basis respectively. The refinement study is performed by refining patches $\Omega^{(1)}$, $\Omega^{(2)}$, $\Omega^{(3)}$ and $\Omega^{(4)}$ with $\lceil \frac{4}{3}i \rceil \times \lceil \frac{4}{3}i \rceil$, $(\lceil \frac{7}{4}i \rceil + 2) \times (\lceil \frac{7}{4}i \rceil + 3)$, $\lceil \frac{6}{4}i \rceil \times (\lceil \frac{6}{4}i \rceil - 2)$ and $(\lceil \frac{8}{5}i \rceil + 4) \times (\lceil \frac{8}{5}i \rceil + 1)$ elements in ξ and η parametric directions, respectively, with $i = 1, \ldots, 26$. The parametric coordinates ξ and η, are chosen for all the patches to coincide with the longitudinal $X^1 \in \left[-\frac{L}{2}, \frac{L}{2}\right]$ and $\left(X^2 \sin\overline{\phi}, X^3 \cos\overline{\phi}\right)$ with $\overline{\phi} \in \left[-\frac{\overline{\phi}_0}{2}, \frac{\overline{\phi}_0}{2}\right]$ circumferential direction, respectively. Moreover, the refinement is chosen such that the resulting elements have a good aspect ratio as well as the interface parametrizations are kept highly non-conforming at each refinement level. The scaling of the penalty factors for the displacement and rotation penalization corresponding to the Penalty method are chosen as $10^3\,\mathrm{m}^{-2}$ and $0.25 \times 10^3\,\mathrm{m}^{-2}$, respectively. Accordingly, the scaling of the penalty factors for the displacement and rotation penalization corresponding to the augmented Lagrange Multipliers method are chosen as 1 and $0.25\,\mathrm{m}^{-2}$, respectively. Then, the seven times scaled deformation and the 2-norm of the rotation field obtained using the Lagrange Multipliers method and corresponding to refinement level $i = 5$ for the high polynomial order case can be seen in set of Fig. 7.

The convergence of the displacement field at each subdomain using the relative L^2-norm for the low and high polynomial order of the underlying bases can be seen in set of Fig. 8. More specifically, the Penalty method is clearly reaching its threshold at the relative displacement error order of 10^{-3} and 10^{-4} for patch pair $\Omega^{(1)}$, $\Omega^{(3)}$ and $\Omega^{(2)}$, $\Omega^{(4)}$, respectively.

In what concerns the Lagrange Multipliers methods, the resulting numerical displacement fields converge uniformly as the element area size goes to zero with

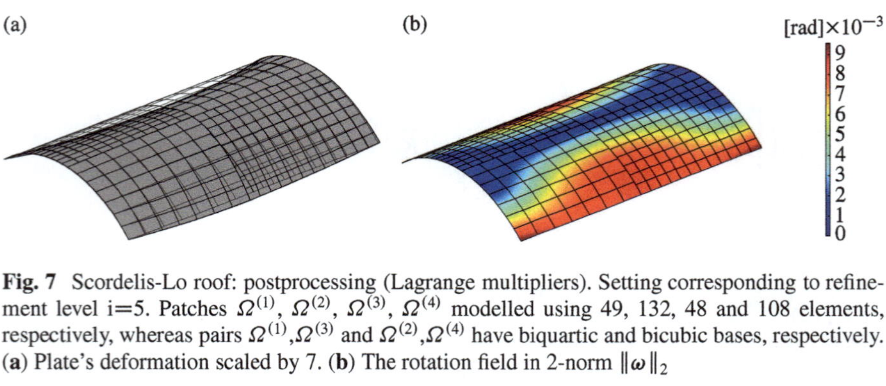

Fig. 7 Scordelis-Lo roof: postprocessing (Lagrange multipliers). Setting corresponding to refinement level i=5. Patches $\Omega^{(1)}$, $\Omega^{(2)}$, $\Omega^{(3)}$, $\Omega^{(4)}$ modelled using 49, 132, 48 and 108 elements, respectively, whereas pairs $\Omega^{(1)}$,$\Omega^{(3)}$ and $\Omega^{(2)}$,$\Omega^{(4)}$ have biquartic and bicubic bases, respectively. (**a**) Plate's deformation scaled by 7. (**b**) The rotation field in 2-norm $\|\boldsymbol{\omega}\|_2$

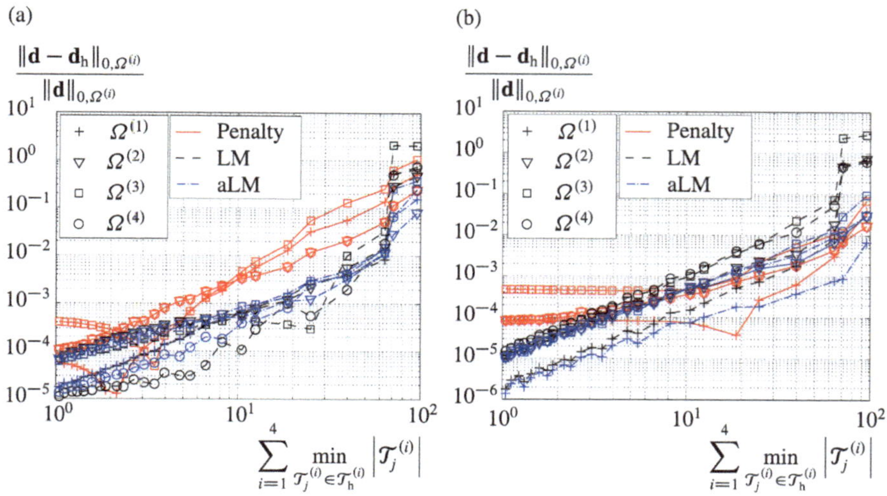

Fig. 8 Scordelis-Lo roof: convergence study. Relative displacement error in the L^2-norm at each subdomain against the sum of the minimum element areas for the multipatch geometry. (**a**) $\Omega^{(1)}$, $\Omega^{(3)}$ bicubic and $\Omega^{(2)}$, $\Omega^{(4)}$ biquadratic bases. (**b**) $\Omega^{(1)}$, $\Omega^{(3)}$ biquartic and $\Omega^{(2)}$, $\Omega^{(4)}$ bicubic bases

the augmented Lagrange Multipliers method demonstrating lower error levels in the coarse region of the isogeometric field discretization. On the other hand, both Lagrange Multipliers methods produce numerical fields whose relative errors are almost identical in the fine discretization region. This is explained by the fact that the penalty factors are kept constant with the refinement level while the Lagrange multipliers fields are successively refined and therefore becoming dominant as the element area size goes to zero.

As last, the convergence of the force and moment tensors, which are computed using Eqs. (6a) and (6b), is demonstrated for both polynomial order cases in the set of Figs. 9 and 10, respectively. Uniform convergence is observed for all but the Penalty method, which is reaching its threshold in the force tensor at the relative error of order 10^{-3} and 10^{-4} for the patch pairs $\Omega^{(2)}$, $\Omega^{(3)}$ and $\Omega^{(1)}$, $\Omega^{(4)}$, respectively, see Fig. 9b. Equivalently, the aforementioned method is reaching its threshold in the moment tensor at the relative error of order 10^{-2}, 10^{-3} and 10^{-4} for patches $\Omega^{(2)}$, $\Omega^{(4)}$ and pair $\Omega^{(1)}$, $\Omega^{(3)}$, respectively, see Fig. 10b. Important is to note that although the Penalty method is converging slower for the low order polynomial case in the displacement field before reaching its threshold, see Fig. 8a, the same does not necessarily hold for the resultant fields, that is the force and the moment tensors, see Fig. 9a and 10a, respectively. The Lagrange Multipliers methods in contrast to the Penalty method belong to the family of mixed formulations which result into saddle point problems. As mentioned in Sect. 2, a unique solution to the formulations resulting by using those methods is not guaranteed to exist or to be unique, a fact which is closely linked to the choice of the discrete Lagrange Multipliers fields. Moreover, it has been observed [14]

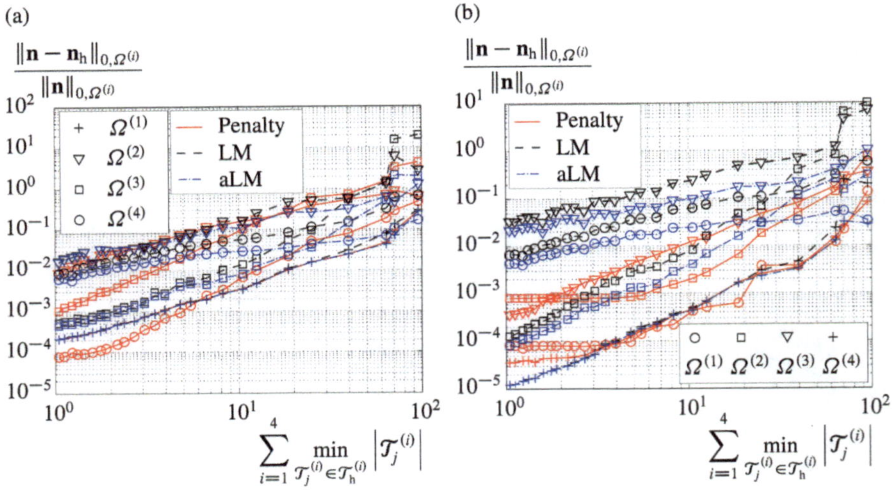

Fig. 9 Scordelis-Lo roof: convergence study. Relative force tensor error in the L^2-norm at each subdomain against the sum of the minimum element areas for the multipatch geometry. (**a**) $\Omega^{(1)}$, $\Omega^{(3)}$ bicubic and $\Omega^{(2)}$, $\Omega^{(4)}$ biquadratic bases. (**b**) $\Omega^{(1)}$, $\Omega^{(3)}$ biquartic and $\Omega^{(2)}$, $\Omega^{(4)}$ bicubic bases

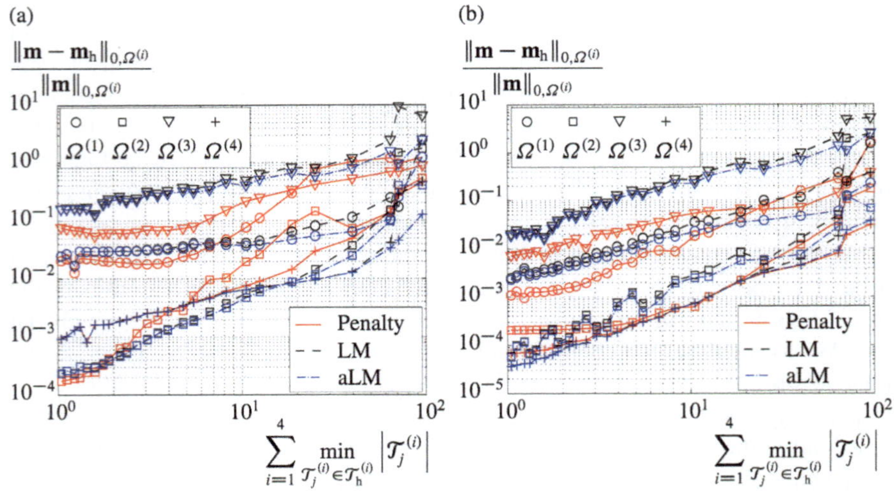

Fig. 10 Scordelis-Lo roof: convergence study. Relative moment tensor error in the L^2-norm at each subdomain against the sum of the minimum element areas for the multipatch geometry. (**a**) $\Omega^{(1)}$, $\Omega^{(3)}$ bicubic and $\Omega^{(2)}$, $\Omega^{(4)}$ biquadratic bases. (**b**) $\Omega^{(1)}$, $\Omega^{(3)}$ biquartic and $\Omega^{(2)}$, $\Omega^{(4)}$ bicubic bases

that mixed formulations may perform well for the approximation of the primary variables of a problem, in this case the displacement field, but not necessarily for the their derivatives, which in this case are the force and the moment traction fields.

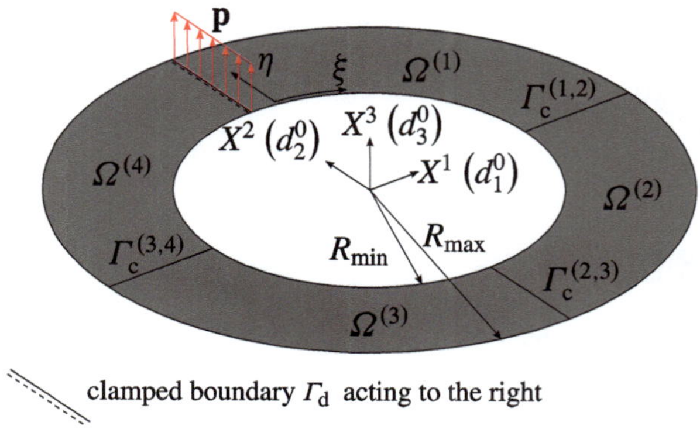

Fig. 11 Slit annular plate subject to tip lifting line force: problem placement

4.3 Slit Annular Plate Subject to Tip Lifting Line Force

The first nonlinear problem to be studied is that of a slit annular plate subject to tip lifting load. As reference, the results provided in [28] are used. The inner radius, the outer radius, the thickness of the plate and the load magnitude are $R_{\min} = 6\,\mathrm{m}$, $R_{\max} = 10\,\mathrm{m}$, $t = 3\,\mathrm{cm}$ and $\mathbf{p} = p\,\mathbf{e}_3$ with $p = 8 \times 10^{-4}\,\mathrm{KN/m}$, respectively. The Young's modulus and the Poisson's ratio are $E = 21 \times 10^3\ \mathrm{KN/m^2}$ and $\nu = 0$, respectively. The problem is decomposed into four subdomains, see Fig. 11. Two settings for solving the aforementioned problem are used; the *coarse setting* for which patch $\Omega^{(1)}$, patch $\Omega^{(2)}$, patch $\Omega^{(3)}$ and patch $\Omega^{(4)}$ are discretized with thirty-six, eighteen, twenty-eight and ten elements, respectively, all having C^1-continuous biquadratic bases. On the other hand, the *fine setting* consists in discretizing patch $\Omega^{(1)}$, patch $\Omega^{(2)}$, patch $\Omega^{(3)}$ and patch $\Omega^{(4)}$ with two hundred nine, one hundred five, two hundred seventy-three and one hundred fifty-three elements, respectively, where patches $\Omega^{(1)}$, $\Omega^{(3)}$ have a C^1-continuous biquadratic basis and patches $\Omega^{(2)}$, $\Omega^{(4)}$ a C^2-continuous bicubic basis. The scaling of the penalty factors for the displacement and rotation penalization corresponding to the Penalty method are chosen as 10^2 and $10^3\,\mathrm{m}^{-2}$, respectively. In what concerns the augmented Lagrange multipliers method, those factors are chosen as 10^{-1} and $10^0\,\mathrm{m}^{-2}$, respectively.

For the solution of the nonlinear problem, sixty-one load steps are used with Newton-Raphson tolerance equal to 10^{-6}. In the set of Fig. 12, it is shown the ultimate deformation of the plate and the distribution of the physical moment component $\overline{m}^{22}$ along the plate for the fine setting corresponding to the solution provided by the Penalty method. The smoothness of the results across the patches which have strongly non-conforming bases exemplifies the applicability of the DDMs for the solution of the nonlinear Kirchhoff-Love shell problem in multiple domains. As last, the load-displacement curves corresponding to the d_3^0 component of the displacement field at points $\mathbf{X}_\mathrm{A}$ and $\mathbf{X}_\mathrm{B}$, with displaced coordinates $\mathbf{x}_\mathrm{A} = \mathbf{X}_\mathrm{A} + \mathbf{d}(\mathbf{X}_\mathrm{A})$ and $\mathbf{x}_\mathrm{B} = \mathbf{X}_\mathrm{B} + \mathbf{d}(\mathbf{X}_\mathrm{B})$, see Fig. 12a, for both the coarse and the fine settings and subsequently are given in Fig. 13 and compared to the results provided in [28]. Note that for both settings the DDMs deliver results very close

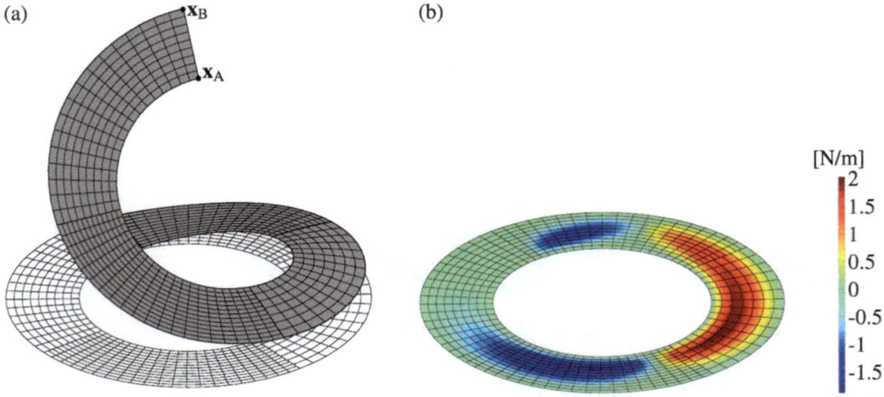

Fig. 12 Slit annular plate subject to tip lifting line force: postprocessing (Penalty). Results corresponding to the fine setting. (**a**) Ultimate deformation. (**b**) Ultimate moment field $\overline{m}^{22}$

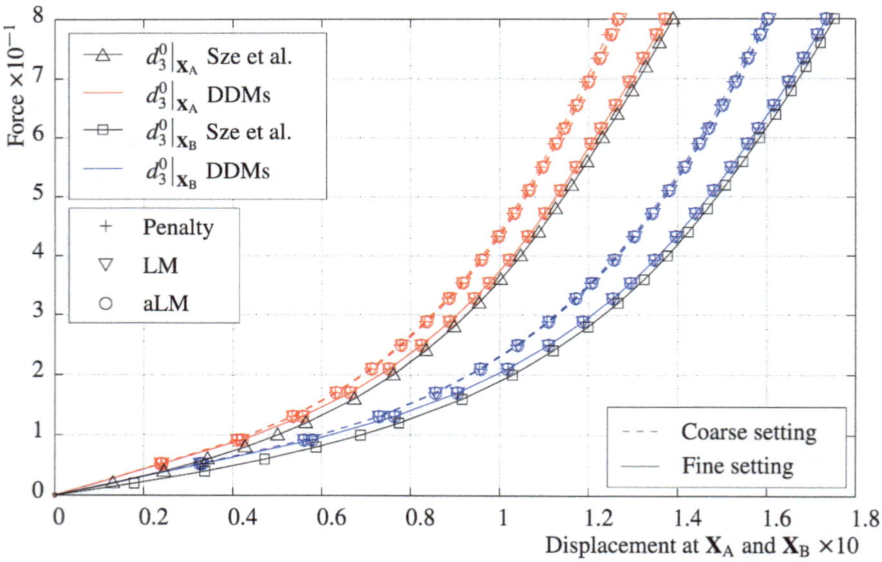

Fig. 13 Slit annular plate subject to tip lifting line force: load-displacement curves

to each other, and they are almost identical in the fine setting. The latter shows that all variational methods produce almost identical results as the mesh size goes to zero, given that the coupling parameters and Lagrange Multipliers discretizations are carefully chosen. It is worth mentioning that the reference results correspond to a Reissner-Mindlin shell formulation and therefore the reference load-displacement curves cannot be identical with the ones provided by the DDMs for a Kirchhoff-Love shell. However the problem considers a relatively thin structure and thus the results from both theories are very close to each other, as expected.

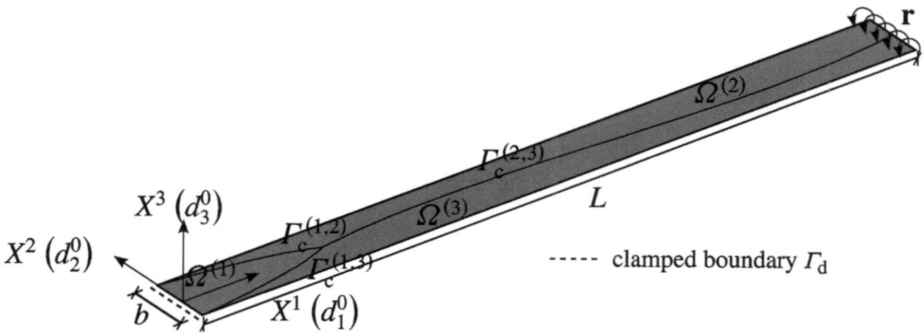

Fig. 14 Cantilever subjected to end moment: problem placement

4.4 Cantilever Subjected to End Moment

The second nonlinear problem to be studied is that of a cantilever subject to end moment. The material characteristics of the plate, that is, Young's modulus and Poisson's ratio are equal to $E = 1.2 \times 10^3\,\mathrm{KN/m^2}$ and $\nu = 0$, respectively. On the other hand, the geometrical characteristics of the plate, that is, its length, its width and its thickness are equal to $L = 12\,\mathrm{m}$, $b = 1\,\mathrm{m}$ and $h = 10\,\mathrm{cm}$, respectively. The cantilever is subject to end bending moment $\mathbf{r}$, with amplitude $\|\mathbf{r}\|_2 = 2\pi \frac{EI}{L} = 52.3333\,\mathrm{Nm}$, where $I = \frac{bh^3}{12} = 8.3333\,\mathrm{m}^4$ stands for the plate's moment of inertia. It is known that under this loading, the plate must bend into a circle of radius $\overline{R} = \frac{EI}{\|\mathbf{r}\|_2} = 1.9099\,\mathrm{m}$, see [28]. The computational domain Ω is decomposed into three subdomains, see Fig. 14. Those subdomains are neighbouring through three interface boundaries which are chosen curved. Interface boundary $\overline{\Gamma_c^{(1,2)}}$ is represented by a quadratic Bézier curve between the control points $\hat{\mathbf{X}}_1^{(1,2)} \equiv (0, -0.5, 0)$, $\hat{\mathbf{X}}_2^{(1,2)} \equiv (0.15, -0.3, 0)$ and $\hat{\mathbf{X}}_3^{(1,2)} \equiv (0.2, 0, 0)$. Similarly, interface boundary $\overline{\Gamma_c^{(1,3)}}$ is also represented by a quadratic Bézier curve between the control points $\hat{\mathbf{X}}_1^{(1,3)} \equiv (0, 0.5, 0)$, $\hat{\mathbf{X}}_2^{(1,3)} \equiv (0.15, 0.3, 0)$ and $\hat{\mathbf{X}}_3^{(1,3)} \equiv (0.2, 0, 0)$. Finally, interface boundary $\overline{\Gamma_c^{(2,3)}}$ is represented by a cubic Bézier curve between the control points $\hat{\mathbf{X}}_1^{(2,3)} \equiv (0.2, 0, 0)$, $\hat{\mathbf{X}}_2^{(2,3)} \equiv (0.4667, 0.5, 0)$, $\hat{\mathbf{X}}_3^{(2,3)} \equiv (0.7333, -0.5, 0)$ and $\hat{\mathbf{X}}_4^{(2,3)} \equiv (1, 0, 0)$. Note that within this decomposition, patch $\Omega^{(1)}$ attains a parametric singularity on its boundary point $\overline{\Gamma_c^{(1,2)}} \cap \overline{\Gamma_c^{(1,3)}}$ as its rectangular NURBS parametric space is degenerated into a triangle in the Cartesian space.

The tolerance for the Newton-Rapshon iterations is chosen 10^{-5} for twenty load steps in relation to the employed nonlinear scheme. In addition, the scaling of the both the displacement and the rotation penalty factors is chosen as 10^2 and 10^{-2} for the Penalty and the augmented Lagrange Multipliers method, respectively. Two settings, one *coarse* and one *fine*, are chosen for the current problem as in Sect. 4.3. The coarse setting consists in parametrizing patch $\Omega^{(1)}$, $\Omega^{(2)}$ and $\Omega^{(3)}$

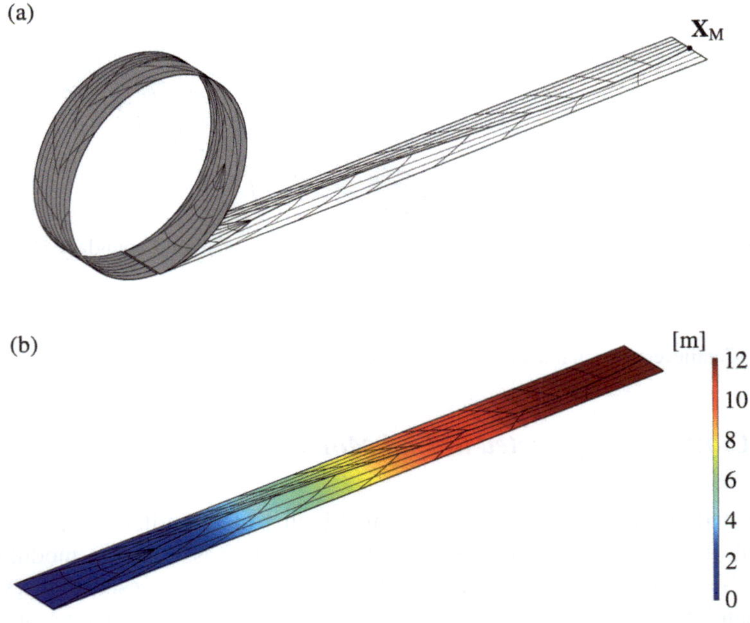

Fig. 15 Cantilever subjected to end moment: postprocessing (augmented Lagrange multipliers). Results corresponding to the fine setting. (**a**) Ultimate plate's deformation. (**b**) The ultimate displacement field in $\|\mathbf{d}\|_2$

with four, ten and twenty-one elements, respectively, all attaining C^2-continuous bicubic polynomial bases. In what concerns the fine setting, patches $\Omega^{(1)}$, $\Omega^{(2)}$ and $\Omega^{(3)}$ are parametrized with sixteen, twenty-four and fifty-five elements, respectively, all attaining C^3-continuous biquartic polynomial bases. The ultimate plate's deformation and the ultimate displacement field in the 2-norm using the augmented Lagrange Multipliers method and corresponding to the fine configuration is shown in set of Fig. 15. As expected, the plate is bent into a circle of the expected radius and the DDMs were proven to handle pointwise singularities even when those appear on the interface Γ_c. It must be however noted that the deformed configuration does not represent an exact circle if the control point weights of the NURBS parametrization are not chosen appropriately, see [7]. However, the finer the geometry space is chosen the better the approximation of the circle becomes.

Subsequently, load-displacement curves for the d_3^0 and d_1^0 components of the displacement field at point $\mathbf{X}_M$ are given in Fig. 16, which is lying in the middle of the plate's tip, see Fig. 15a. The DDMs delivered highly accurate results in the fine setting and for the given choice of parameters as it can be verified by the almost overlapping curves with the analytical ones provided in [28]. On the other hand, for the coarse setting the results are as expected poor, as the discretization of the displacement field is not rich enough, with the Lagrange Multipliers methods

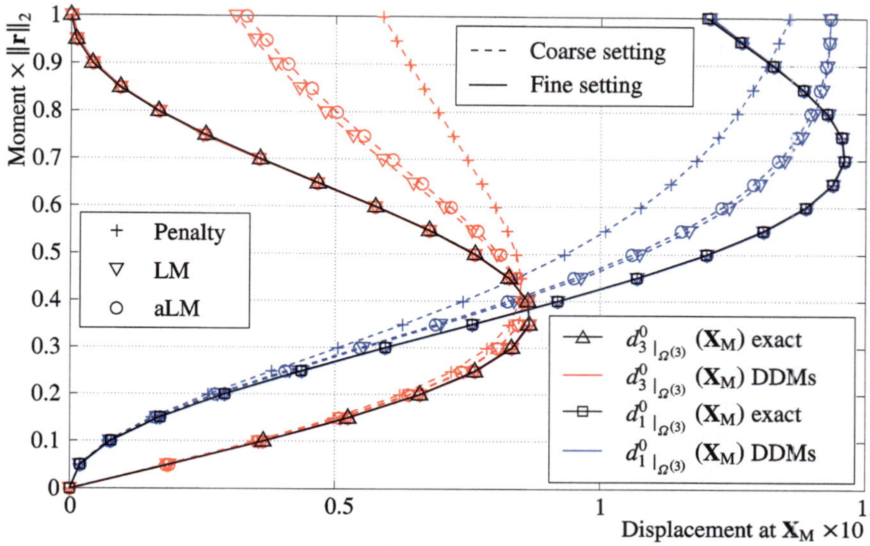

Fig. 16 Cantilever subjected to end moment: load-displacement curves

demonstrating better behaviour than the Penalty method. Additionally, it can be observed that the produced by the augmented Lagrange Multipliers method results are lying between those of the Penalty and the Lagrange Multipliers methods. The latter verifies that the augmented Lagrange Multipliers method is essentially a compensation between the other two methods, with the magnitude of the penalty parameters being decisive on its behaviour. The larger the penalty parameters are chosen, the better the interface continuity conditions (10h) and (10i) are satisfied but on the same time the overall error is increased and vice versa, see set of Fig. 4.

5 Conclusions and Outlook

In this study, three different DDMs are presented and compared for the decomposed Kirchhoff-Love shell problem in both linear and nonlinear isogeometric analysis. Investigated are a Penalty, the Lagrange Multipliers and the augmented Lagrange Multipliers method.

The Penalty method results in a positive definite discrete equation system given that the penalty parameters do not affect the conditioning of the equation system. The method performs robustly and its implementation is relatively simple. One drawback of the method is that its accuracy depends on the size of the penalty parameters which is typical for the Penalty methods. However, a discretization-dependent rule is proposed and numerically tested.

The Lagrange Multipliers method results in a saddle point discrete equation system. The latter means that the finite dimensional spaces must fulfil the Ladyzenskaja-Babuska-Brezzi condition so that one unique solution to the problem can be guaranteed. This is closely linked to the choice of the discretization of the Lagrange Multipliers fields, which is in general a difficult problem. Nevertheless, the method provided highly satisfactory results for the proposed choice of the Lagrange Multipliers fields discretization.

The augmented Lagrange Multipliers method inherits the properties from both the Lagrange Multipliers and the Penalty method. Therefore, the method results also in a saddle point problem for which the penalty parameters can be used to adjust the accuracy of the results without the necessity of being too large.

Extension of the given DDMs into a parameter-free variational formulation is by this study highly motivated, so that the parameters which are driving the coupled system are kept to minimum and thus minimizing the negative consequences of a possible improper choice.

Acknowledgements The support of the Deutsche Forschungsgemeinschaft (DFG) with grant numbers BL 306/26-1 and BL 306/23-2 is gratefully acknowledged.

References

1. Y. Basar, W. Krätzig, *Mechanik der Flächentragwerke* (Vieweg, Braunschweig, 1985)
2. T. Hughes, J. Cottrell, Y. Bazilevs, Isogeometric analysis: CAD, finite elements, NURBS, exact geometry, and mesh refinement. Comput. Methods Appl. Mech. Eng. **194**, 4135–4195 (2005)
3. K.-J. Bathe, *Finite Element Procedures* (Prentice Hall, Upper Saddle River, 2006)
4. O. Zienkiewicz, R. Taylor, J. Zhu, *The Finite Element Method, Its Basis and Fundamentals*, 6th edn. (Elsevier, Oxford, 2005)
5. L. Piegl, W. Tiller, *The NURBS Book*, 2nd edn. (Springer, New York, 1997)
6. R. Schmidt, J. Kiendl, K.-U. Bletzinger, R. Wüchner, Realization of an integrated structural design process: analysis-suitable geometric modelling and isogeometric analysis. Comput. Vis. Sci. **13**, 315–330 (2010)
7. J. Kiendl, K.-U. Bletzinger, J. Linhard, R. Wüchner, Isogeometric shell analysis with Kirchhoff-Love elements. Comput. Methods Appl. Mech. Eng. **198**, 3902–3914 (2009)
8. R. Schmidt, R. Wüchner, K.-U. Bletzinger, Isogeometric analysis of trimmed NURBS geometries. Comput. Methods Appl. Mech. Eng. **241–244**, 93–111 (2012)
9. M. Dörfel, B. Jüttler, B. Simeon, Adaptive isogeometric analysis based on t-splines. Comput. Methods Appl. Mech. Eng. **199**, 264–275 (2010)
10. T.-K. Uhm, Y. Kim, S.-K. Youn, A locally refinable t-spline finite element method for CAD/CAE integration. Struct. Eng. Mech. **80**, 225–245 (2008)
11. M. Breitenberger, A. Apostolatos, B. Philipp, R. Wüchner, K.-U. Bletzinger, Analysis in computer aided design: nonlinear isogeometric B-Rep analysis of shell structures. Comput. Methods Appl. Mech. Eng. **284**, 401–457 (2015)
12. I. Babuška, The finite element method with penalty. Math. Comput. **27**, 221–228 (1973)
13. I. Babuška, M. Zlámal, Nonconforming elements in the finite element method with penalty. SIAM J. Numer. Anal. **10**(5), 863–875 (1973)
14. J.D. Sanders, J.E. Dolbow, T.A. Laursen, On methods for stabilizing constraints over enriched interfaces in elasticity. Int. J. Numer. Methods Eng. **78**, 1009–1036 (2009)

15. A. Apostolatos, R. Schmidt, R. Wüchner, K.-U. Bletzinger, A Nitsche-type formulation and comparison of the most common domain decomposition methods in isogeometric analysis. Int. J. Numer. Methods Eng. **97**(7), 473–504 (2014)
16. D.N. Arnold, An interior penalty finite element method with discontinuous elements. SIAM J. Numer. Anal. **19**, 742–760 (1982)
17. M.F. Wheeler, An elliptic collocation-finite element method with interior penalties. SIAM J. Numer. Anal. **15**, 152–161 (1978)
18. Y. Guo, M. Ruess, Nitsche's method for a coupling of isogeometric thin shells and blended shell structures. Comput. Methods Appl. Mech. Eng. **284**, 881–905 (2015)
19. J. Kiendl, Y. Bazilevs, M.-C. Hsu, R. Wüchner, K.-U. Bletzinger, The bending strip method for isogeometric analysis of Kirchhoff–Love shell structures comprised of multiple patches. Comput. Methods Appl. Mech. Eng. **199**, 2403–2416 (2010)
20. C. Bernardi, Y. Maday, A. Patera, A new nonconforming approach to domain decomposition: the mortar element method, Technical report, Université Pierre at Marie Curie, Paris (1990)
21. E. Brivadis, A. Buffa, B. Wohlmuth, L. Wunderlich, Isogeometric mortar methods. Comput. Methods Appl. Mech. Eng. **284**, 292–319 (2015)
22. F. Brezzi, K.-J. Bathe, A discourse on the stability conditions for mixed finite element formulations. Comput. Methods Appl. Mech. Eng. **82**, 27–57 (1990)
23. A. Toselli, O. Widlund, *Domain Decomposition Methods-Algorithms and Theory* (Springer, Berlin/Heidelberg, 2005)
24. B.O. Neil, *Elementary Differential Geometry* (Academic, San Diego, 1997)
25. P. Kaufmann, S. Martin, M. Botsch, M. Gross, Implementation of discontinuous Galerkin Kirchhoff-Love shells, Technical report, ETH Zurich, Department of Computer Science (2009)
26. L. Noels, R. Radovitzky, A new discontinuous Galerkin method for Kirchhoff-Love shells. Comput. Methods Appl. Mech. Eng. **197**, 2901–2929 (2008)
27. J. Cottrell, T. Hughes, Y. Bazilevs, *Isogeometric Analysis: Toward Integration of CAD and FEA* (Wiley, Chichester, 2009)
28. K. Sze, X. Liu, S. Lo, Popular benchmark problems for geometric nonlinear analysis of shells. Finite Elem. Anal. Des. **40**, 1551–1569 (2004)
29. S. Timoshenko, S. Woinowsky-Krieger, K. (Firm), *Theory of Plates and Shells*, 2nd edn. (McGraw-Hill, New York, 1959), international edition 1959 – t.p. verso
30. A.C. Scordelis, K.S. Lo, Computer analysis of cylindrical shells. J. Am. Concr. Inst. **61**, 539–561 (1969)

A Reduced Integration for Reissner-Mindlin Non-linear Shell Analysis Using T-Splines

C. Adam, S. Bouabdallah, M. Zarroug, and H. Maitournam

Abstract We propose a reduced shell element for Reissner-Mindlin geometric non-linear analysis within the context of T-spline analysis. The shell formulation is based on the displacements and a first order kinematic in the thickness is used for the transverse shear strains. A total Lagrangian formulation is considered for the finite transformations. The update of the incremental rotations is made using the quaternion algebra. The standard two-dimensional reduced quadrature rules for structured B-spline and NURBS basis functions are extended to the more flexible T-meshes. The non-uniform Gauss-Legendre and patchwise reduced integrations for quadratic shape functions are both presented and compared to the standard full Gauss-Legendre scheme. The performance of the element is assessed with linear and geometric non-linear two-dimensional problems in structural analysis. The effects of mesh distortion and local refinement, using both full and reduced numerical quadratures, are evaluated.

1 Introduction

Isogeometric analysis (IGA) [1] is a recent concept introduced by T.J.R Hughes in 2004 that aims to bridge engineering design to analysis. B-spline and NURBS [2] shape functions, initially used as standard tools in industrial computer aided design (CAD) software, also provide promising properties for analysis. The geometric description in CAD systems needs to be translated in a representation suitable for

C. Adam (✉) • H. Maitournam
Laboratoire de Mécanique des Solides (CNRS UMR 7649), École Polytechnique, Route de Saclay, 91128 Palaiseau, France
e-mail: adam@lms.polytechnique.fr

S. Bouabdallah
Département Mécanique Numérique et Modélisation, École Supérieure d'Ingénieurs Léonard de Vinci, 8 avenue Léonard de Vinci, 92400 Courbevoie, France

M. Zarroug
Direction Scientifique et Technologies Futures, PSA Peugeot Citroën, Route de Gisy, 78140 Vélizy-Villacoublay, France

B. Jüttler, B. Simeon (eds.), *Isogeometric Analysis and Applications 2014*,
Lecture Notes in Computational Science and Engineering 107,
DOI 10.1007/978-3-319-23315-4_5

finite element analysis (FEA) or isogeometric analysis. This process is far from trivial when dealing with complex designs in industrial applications and represents a large amount of the overall process time.

Using different geometries results in several difficulties since both representations need to be updated in the same time, efficient shape optimizations of the CAD geometry require a tight integration between the solver and the optimizer with an automatic CAD geometry to mesh conversion and vice versa. These tasks are complex since meshes are disconnected from CAD geometries. Ideally, having a unique analysis suitable geometric model, which can be used directly from design to simulation, is essential to automatize the overall process.

Engineering designs present some drawbacks for analysis since there are less restrictions when manipulating the CAD geometry than when performing a numerical simulation. Some properties such NURBS trimming or patch merging represent bottlenecks to achieve an automatic process which would join both domains. The structured B-spline and NURBS shape functions are not enough flexible to represent general industrial designs in an analysis suitable fashion.

The notion of T-spline [3] extends the concepts of B-splines and NURBS. They benefit from the flexibility of the point-based splines [3] and the Cartesian structure of B-splines. Many complex geometries can be represented with a single T-spline. As hierarchical B-splines [4], they provide local refinement which is essential from an analysis point of view.

As a generalisation of B-splines, T-splines suffer from numerical locking pathologies and computational efficiency using a full Gauss-Legendre quadrature. Efficient quadrature rule is an essential ingredient to obtain efficient numerical simulations, regardless of the type of basis functions chosen. Full Gauss-Legendre quadrature is optimal for C^0 discretization methods such as standard Lagrange elements. However, it has been shown in [5] that optimal quadrature rules for spline-based analysis are patchwise rules with fewer quadrature points and different locations and weights.

Reduced integration is a controversial subject in computational mechanics which has significantly improved the efficiency of numerical simulations reducing the number of quadrature points and the locking phenomena occurring in thin structures modelled by thick shell elements. Some work on this topic has been initiated in e.g. [6–9] within the context of B-spline and NURBS analysis.

In this paper, we extend the reduced quadrature schemes presented in [10–13] to the more general T-spline analysis for two-dimensional problems. We limit our work to quadratic shape functions. The strengths of T-splines, reduced integration and geometric non-linear Reissner-Mindlin shell formulation are combined to obtain efficient isogeometric elements needed for complex and time consuming industrial simulations.

The paper is outlined as follows. In Sect. 2 we give a brief overview of the T-spline technology introducing the point-based splines and the concept of local knot vectors and refinement. In Sect. 3 we propose a Reissner-Mindlin shell element for geometric non-linear analysis of structures focusing on the update of the normals with arbitrary large rotations. The extension of non-uniform Gauss-Legendre and patchwise reduced quadrature rules is presented in Sect. 4. Both integrations are performed in Bézier elements. The performance of the resulting reduced shell elements is assessed in Sect. 5 using standard linear and non-linear benchmark problems. We draw conclusions in Sect. 6.

2 T-Splines

T-splines generalize the concepts of B-splines and NURBS. Before describing T-splines, we introduce the notion of function subdivision which is a process that enlightens the concept of T-splines.

2.1 Blending Function Subdivision

Let us consider two quadratic B-spline basis $\mathbf{N}$ and $\tilde{\mathbf{N}}$ generated by the two open and global knot vectors $\Xi = \{0,0,0,1,2,3,4,4,4\}$ and $\tilde{\Xi} = \{0,0,0,1,1.5,2,3,4,4,4\}$. We have the relation $\Xi \subset \tilde{\Xi}$ in the sense that the space of all curves that can be built from Ξ is included in the space of all curves that can be built from $\tilde{\Xi}$. In other words, each basis function provided by the knot vector Ξ can be expressed as a linear combination of the basis functions given by $\tilde{\Xi}$ [3]. These relations can be written in the following compact form

$$\mathbf{N} = \mathbf{P}\tilde{\mathbf{N}}, \tag{1}$$

where $\mathbf{P}$ is the operator of knot insertion. In order to describe the same curve $\mathcal{C}(\xi) = \mathbf{N}^{\mathrm{T}}\mathbf{B} = \tilde{\mathbf{N}}^{\mathrm{T}}\tilde{\mathbf{B}}$, a similar relation is obtained for both sets of control points $\mathbf{B}$ and $\tilde{\mathbf{B}}$

$$\tilde{\mathbf{B}} = \mathbf{P}^{\mathrm{T}}\mathbf{B}. \tag{2}$$

Figure 1 represents the two quadratic B-spline basis. On the left, blue curves are provided by the coarse knot vector Ξ while, on the right, red curves represent the new functions obtained from the refinement of the coarse shape functions. The three polynomials represented by dotted blue lines can be expressed as a linear combination of the four new functions denoted by red lines. Further details on the algorithm of the blending function refinement for cubic polynomials are presented in [3].

This example illustrates the notion of function subdivision and local refinement. In fact, knot insertion with B-splines or NURBS is a global process with multi-dimensional problems since it modifies the global knot vectors which generate the basis functions. However, instead of considering refinement of global knot vectors, we can consider the refinement of the basis functions, namely the refinement of the local knot vectors. The Cox-de-Boor formula [2] exhibits a local process to construct

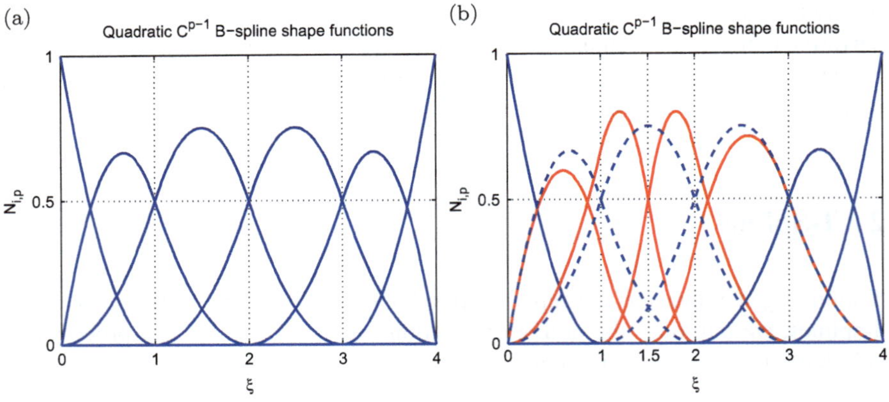

Fig. 1 Univariate B-spline basis functions of order two with both open knot vectors Ξ and $\tilde{\Xi}$. (**a**) Coarse knot vector $\Xi = \{0, 0, 0, 1, 2, 3, 4, 4, 4\}$. (**b**) Refined knot vector $\tilde{\Xi} = \{0, 0, 0, 1, 1.5, 2, 3, 4, 4, 4\}$

Table 1 Relations between the two basis $\mathbf{N}$ and $\tilde{\mathbf{N}}$

Coarse basis	Local knot Ξ_i^{loc}	Refined basis	Local knot $\tilde{\Xi}_i^{loc}$	Linear combination
N_1	[0 0 0 1]	$\tilde{N}_1$	[0 0 0 1]	$N_1 = \tilde{N}_1$
N_2	[0 0 1 2]	$\tilde{N}_2$	[0 0 1 1.5]	$N_2 = \tilde{N}_2 + \frac{1}{4}\tilde{N}_3$
N_3	[0 1 2 3]	$\tilde{N}_3$	[0 1 1.5 2]	$N_3 = \frac{3}{4}\tilde{N}_3 + \frac{3}{4}\tilde{N}_4$
N_4	[1 2 3 4]	$\tilde{N}_4$	[1 1.5 2 3]	$N_4 = \frac{1}{4}\tilde{N}_4 + \tilde{N}_5$
N_5	[2 3 4 4]	$\tilde{N}_5$	[1.5 2 3 4]	$N_5 = \tilde{N}_6$
N_6	[3 4 4 4]	$\tilde{N}_6$	[2 3 4 4]	$N_6 = \tilde{N}_7$
		$\tilde{N}_7$	[3 4 4 4]	

the spline polynomials of order p from the piecewise constant functions defined by the knot vectors. In that sense, to each basis function $N_{i,p}$ can be associated a local knot vector Ξ_i^{loc} of size $p + 2$, necessary and sufficient to construct the spline function.

Table 1 describes the local knot vectors Ξ_i^{loc} and $\tilde{\Xi}_i^{loc}$ as well as the existing relations between the two basis $\mathbf{N}$ and $\tilde{\mathbf{N}}$. The knot insertion $\xi = 1.5$ only impacts $p + 2 = 4$ basis functions.

Finally, knot insertion as subdivision function is a more general concept than the standard refinement for B-splines and NURBS. This notion is more general and exhibits interesting properties such as the local refinement which is a major drawback of B-splines and NURBS. T-splines are constructed from local knot vectors associated with the subdivision function process. Therefore, they permit meshes with the so called T-junctions and they contain the special cases of B-splines and NURBS. Another strength of T-splines is their capability to achieve a high continuity across patch boundaries without introducing new knots which propagate in each direction when merging two patches.

2.2 B-Splines, PB-Splines and T-Splines

The concept of point-based splines [3] has been introduced to benefit from the local construction and refinement of the splines. There is no notion of global knot vector, but only local and independent knot vectors. In that sense, the associated polynomials are called blending functions rather than basis functions since the properties of linear independence and partition of unity are no longer guaranteed. PB-splines are even more flexible than T-splines and have some nice properties for modelling geometries but the lack of structure in the knot vectors is a major deficiency which prevents them from being used in analysis.

T-splines represent a good compromise between the Cartesian structure of B-splines and the flexible PB-splines. They benefit from the strengths of both types of functions. Initially they were defined for bi-cubic surfaces but more recently, they were extended to three-dimensional problems and arbitrary degree p [14].

In this paper, we do not give an in-depth description of the T-spline theory which is very rich and complex. The interested reader can refer to e.g. [15] to have a more detailed study of this theory. We restrict our work to bi-quadratic basis functions which give the possibility of constructing a large part of complex geometries.

2.3 The T-Mesh

Figure 2 represents, on the left, a T-mesh constructed from two NURBS-patches denoted by blue and red lines respectively. It also includes a local refinement strategy represented by a green line. Several T-junctions are thus obtained. On the

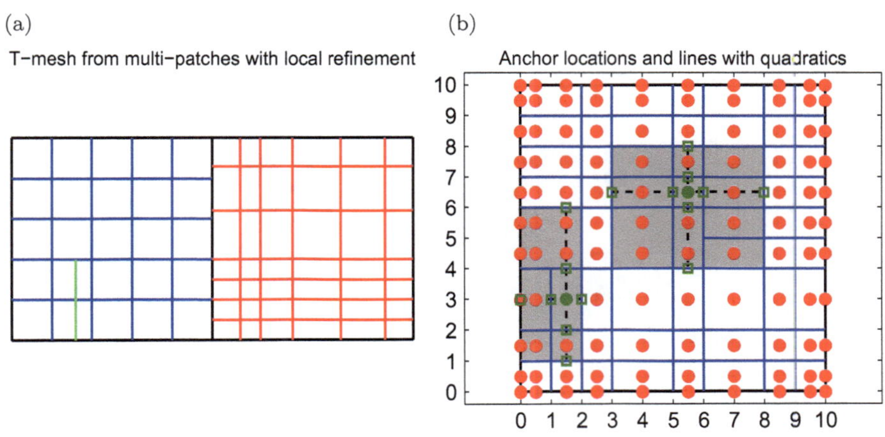

Fig. 2 T-mesh obtained from multi-patches merging (*blue* and *red lines*) with local refinement (*green line*). T-mesh with anchor locations (*red* and *green points*), local knot vectors (*green squares*) and supports of the blending functions (*grey areas*). (**a**) T-mesh with local refinement. (**b**) T-mesh with anchor locations and lines

right, Fig. 2 illustrates the construction of local knot vectors in each parametric direction. Let consider two bi-quadratic functions N_1 and N_2. They are represented in the T-mesh by two anchors (green dots) and their respective support is defined by the two darkened areas. Each anchor (red dots) is associated with one polynomial function and is located at the maximum of the blending function. They are equivalent to the Greville abscissae for B-splines and NURBS meshes. The construction of the anchor locations and lines, namely the local knot vectors defining the supports of the polynomials, is not detailed here but the interested reader can refer to [14]. The corresponding knots vectors are

$$\Xi_1^{\text{loc}} = [0\ 0\ 1\ 2] \text{ and } \mathcal{H}_1^{\text{loc}} = [1\ 2\ 4\ 6], \tag{3}$$

$$\Xi_2^{\text{loc}} = [3\ 5\ 6\ 8] \text{ and } \mathcal{H}_2^{\text{loc}} = [4\ 6\ 7\ 8]. \tag{4}$$

The T-mesh contains all the essential information for defining the local knot vectors and thus the shape functions. Moreover, the structure of the T-mesh defines the properties of linear independence and partition of unity. A T-mesh that guarantees these required properties for the analysis is called an analysis suitable T-mesh. Analysis suitable T-splines are a subset of general T-splines which satisfy several rules described in [16]. We introduce T-junction extensions which correspond to closed line segments that extend from the T-junctions in the opposite direction and cross $\lfloor (p+1)/2 \rfloor$ perpendicular edges or vertices. One may note that T-junction extensions are lines of reduced continuity. In practical terms, a T-mesh is well-characterized and suitable for analysis if no vertical T-junction extension intersects a horizontal T-junction extension. Both T-meshes represented in Fig. 2 are thus suitable for analysis.

The structure of an analysis suitable T-mesh is also flexible enough to describe complex geometries with locally different continuities and holes. Figure 3

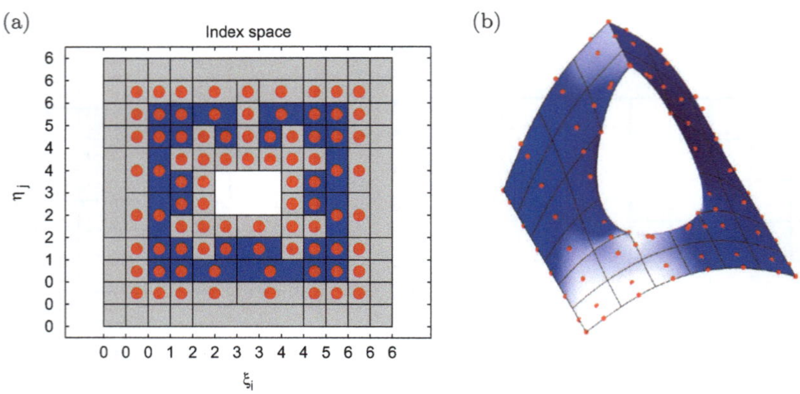

Fig. 3 T-mesh and associated geometry with a hole and local C^0 continuity. *Grey* elements have zero measure while *blue* elements have a positive measure. (**a**) Index space. (**b**) Physical space

represents, on the right, a complex bi-quadratic and three-dimensional surface containing a hole and a C^0 sharp edge which do not propagate within the domain. On the left, the T-mesh corresponding to the geometry is given in the index space so that multiple knots are visible. Grey elements have zero measure, there are thus non-physical, while blue elements have a positive measure. This T-mesh is suitable for analysis.

3 Large Deformations in Thick Shells

This section aims to develop the shell model presented in [10] for large deformations and moderate strains in Reissner-Mindlin shells. Both thick and thin shell structures are generally subjected to arbitrary large deformations and rotations but strains are moderate due to their particular shape ratio between the lengths of the in-plane directions and the thickness.

3.1 Green-Lagrange Strain Tensor

When dealing with moderate strains, the Green-Lagrange strain measure $\mathbf{E}$ [17] is a good candidate to generalize the standard infinitesimal strain tensor $\boldsymbol{\epsilon}$ used for small transformations. The Green-Lagrange tensor is well adapted to the Total Lagrangian Formulation (TLF), in which the discrete equations are formulated with respect to the reference configuration.

To compute $\mathbf{E}$, the current configuration $\mathbf{x}$ is expressed in the initial configuration $\mathbf{X}$ as follows

$$\mathbf{x} = \mathbf{X} + \mathbf{u}. \tag{5}$$

The Green-Lagrange strain measure is defined by the relation

$$\mathrm{d}\mathbf{x}^2 - \mathrm{d}\mathbf{X}^2 = 2\mathrm{d}\mathbf{X}^{\mathrm{T}}[\mathbf{E}]\mathrm{d}\mathbf{X}, \tag{6}$$

with

$$\mathrm{d}\mathbf{x} = \frac{\partial \mathbf{x}}{\partial \mathbf{X}}\mathrm{d}\mathbf{X} = \frac{\partial(\mathbf{X}+\mathbf{u})}{\partial \mathbf{X}}\mathrm{d}\mathbf{X} = \mathbf{F}\mathrm{d}\mathbf{X}, \tag{7}$$

where the deformation gradient tensor $\mathbf{F}$ is related to both reference and current configurations as follows

$$\mathbf{F} = \mathbf{I_3} + \mathbf{D} \text{ and } \mathbf{D} = \frac{\partial \mathbf{u}}{\partial \mathbf{X}} = \mathbf{u}_{,\mathbf{X}}. \tag{8}$$

The Green-Lagrange tensor is thus

$$[\mathbf{E}] = \frac{1}{2}(\mathbf{F}^{\mathrm{T}}\mathbf{F} - \mathbf{I}_3) = \frac{1}{2}(\mathbf{D} + \mathbf{D}^{\mathrm{T}} + \mathbf{D}^{\mathrm{T}}\mathbf{D}). \tag{9}$$

Different measures exist when dealing with finite deformations. The Green-Lagrange strain tensor is a special case of more general strain measures, the Seth-Hill family of strain measures:

$$[\mathbf{E}_m] = \frac{1}{2m}(\mathbf{U}^{\mathbf{2m}} - \mathbf{I}_3) = \frac{1}{2m}((\mathbf{F}^{\mathrm{T}}\mathbf{F})^m - \mathbf{I}_3). \tag{10}$$

For several values of m we obtain the standard Green-Lagrange measure ($m = 1$), the Biot tensor ($m = 1/2$), the Euler-Almansi strain tensor ($m = -1$) or the logarithmic strains ($m = 0$). In the case of small strains, all the measures are similar. The logarithmic strain is a standard measure when dealing with arbitrary finite deformations. However, in structural mechanics, we recall that the displacements and rotations can be large but the strains generally remains moderate so that the Green-Lagrange measure is sufficient. The interested reader can find more details on the strain measures and the different formulations in [17].

3.2 Constitutive Law

We consider an elastic behaviour along with a plane stress hypothesis. The constitutive law relates the second Piola-Kirchhoff stresses $\mathbf{S}$ to the local Green-Lagrange strains $\mathbf{E}$ so that, in Voigt notation, we have

$$\mathbf{S} = \mathbf{C}\mathbf{E}, \tag{11}$$

where $\mathbf{C}$ is the standard constitutive matrix of the isotropic material for elastic shells under plane stress conditions in the case of Reissner-Mindlin's theory. In terms of iterative variation we thus obtain

$$\Delta\mathbf{S} = \mathbf{C}\Delta\mathbf{E}. \tag{12}$$

3.3 Virtual Work

The virtual work of internal forces can be written on the reference geometry V^0 as follows

$$\delta W^{\mathrm{int}} = \int_{V_0} \delta\mathbf{E}^{\mathrm{T}}\mathbf{S}\,\mathrm{d}V_0. \tag{13}$$

The iterative variation of the virtual work is

$$\Delta(\delta W^{\text{int}}) = \int_{V_0} \delta \mathbf{E}^{\mathrm{T}} \Delta \mathbf{S} + \Delta(\boldsymbol{\delta} \mathbf{E})^{\mathrm{T}} \mathbf{S} \, \mathrm{d}V_0. \tag{14}$$

The strain-displacement matrix can be rewritten in the following condensed form

$$\mathbf{E} = \left(\mathbf{H} + \frac{1}{2} \mathbf{A} \left(\mathbf{u}_{,\mathbf{X}} \right) \right) \mathbf{G} \mathbf{U} \tag{15}$$

where **U** is the vector containing the approximate displacements and rotations. The expression of the operators **H**, **A** and **G** is detailed in [17].

Noting that $\delta \mathbf{A} (\mathbf{u}_{,\mathbf{X}}) \mathbf{G} \mathbf{U} = \mathbf{A} (\mathbf{u}_{,\mathbf{X}}) \mathbf{G} \boldsymbol{\delta} \mathbf{U}$, we obtain the following virtual variation for the Green-Lagrange tensor

$$\delta \mathbf{E} = (\mathbf{H} + \mathbf{A} (\mathbf{u}_{,\mathbf{X}})) \mathbf{G} \boldsymbol{\delta} \mathbf{U} = \mathbf{B}_{\text{nl}}(\mathbf{u}) \delta \mathbf{U}. \tag{16}$$

Finally, the virtual iterative variation of **E** is

$$\Delta(\delta \mathbf{E}) = \mathbf{A} (\Delta \mathbf{u}_{,\mathbf{X}}) \mathbf{G} \boldsymbol{\delta} \mathbf{U} + (\mathbf{H} + \mathbf{A} (\mathbf{u}_{,\mathbf{X}})) \mathbf{G} \Delta(\delta \mathbf{U}). \tag{17}$$

The first term $\Delta \mathbf{B}_{\text{nl}}(\mathbf{u}) \delta \mathbf{U}$ is standard in solid element formulations while the second term only appears when large rotations are considered.

3.4 Tangent Stiffness Matrix

The tangent stiffness matrix $\mathbf{K}_t$ is defined using the virtual iterative variation of the work of internal forces such that

$$\Delta(\delta W^{\text{int}}) = \delta \mathbf{U}^{\mathrm{T}} \mathbf{K}_t \Delta \mathbf{U} \tag{18}$$

In this section, we only consider the first part of the virtual iterative variation of **E**, neglecting the effects of large rotations.

Using Eqs. (16) and (18) and noting that $\mathbf{S}^{\mathrm{T}} \Delta \mathbf{B}_{\mathbf{nl}}(\mathbf{u}) = \mathbf{S}^{\mathrm{T}} \mathbf{A} (\Delta \mathbf{u}_{,\mathbf{X}}) \mathbf{G} = \Delta \mathbf{U}^{\mathrm{T}} \mathbf{G}^{\mathrm{T}} \hat{\mathbf{S}} \mathbf{G}$ we obtain the following stiffness matrix

$$\mathbf{K}_t = \mathbf{K}_{t1} + \mathbf{K}_{tff} = \int_{V_0} \mathbf{B}_{\mathbf{nl}}(\mathbf{u})^{\mathrm{T}} \mathbf{C}_t(\mathbf{S}) \mathbf{B}_{\mathbf{nl}}(\mathbf{u}) \, \mathrm{d}V_0 + \int_{V_0} \mathbf{G}^{\mathrm{T}} \hat{\mathbf{S}} \mathbf{G} \, \mathrm{d}V_0 \tag{19}$$

where $\hat{\mathbf{S}}$ is defined by

$$\hat{\mathbf{S}} = \begin{bmatrix} [\mathbf{S}] & \mathbf{0} & \mathbf{0} \\ \mathbf{0} & [\mathbf{S}] & \mathbf{0} \\ \mathbf{0} & \mathbf{0} & [\mathbf{S}] \end{bmatrix} \tag{20}$$

The first term $\mathbf{K}_{\mathbf{t1}}$ corresponds to the "material" part of the tangent stiffness matrix while the second term $\mathbf{K}_{\mathbf{t\sigma}}$ corresponds to the "initial stress" part.

3.5 *Finite Rotations*

Large displacements and rotations are common in structural analysis. In this section, we extend the displacement-based formulation given in [11] to arbitrary large rotations.

3.5.1 Displacement-Based Formulation

We briefly review the shell formulation given in small transformations. The interpolated displacements and the exact geometry of the shell are described by the same shape functions, that is

$$\mathbf{u}(\boldsymbol{\xi}) = \sum_{A=1}^{nm} R_A \left(\mathbf{U}_A + \frac{h}{2}\zeta(\mathbf{n}_A - \mathbf{n}_A^0) \right) = \sum_{A=1}^{nm} R_A \left(\mathbf{U}_A + \frac{h}{2}\zeta\boldsymbol{\theta}_A \times \mathbf{n}_A^0 \right), \qquad (21)$$

where $\boldsymbol{\theta}_A$ is the rotation vector at the control point $\mathbf{X}_A$. It can be highlighted that these displacements are only valid for small rotations if we use a TLF. In this particular case, the simple relation

$$\mathbf{n}_A - \mathbf{n}_A^0 = (\mathbf{R}(\boldsymbol{\theta}_A) - \mathbf{I}_3)\mathbf{n}_A^0 \simeq \boldsymbol{\theta}_A \times \mathbf{n}_A^0 \qquad (22)$$

remains true. However, in the general theory of finite transformations, we use the following relation

$$\mathbf{u}(\boldsymbol{\xi}) = \sum_{A=1}^{nm} R_A \left(\mathbf{U}_A + \frac{h}{2}\zeta(\mathbf{R}(\boldsymbol{\theta}_A) - \mathbf{I}_3)\mathbf{n}_A^0 \right). \qquad (23)$$

Due to the Reissner-Mindlin's hypothesis, the current director vector $\mathbf{n}_A$ is not normal to the mid-surface of the shell and thus need to be correctly rotated to keep accurate results.

3.5.2 Parametrization of the Rotations

Finite rotations can be described by an orthogonal operator $\mathbf{R}$ or using a pseudo-vector of rotation. Different parametrizations are possible but we focus on the axis-angle representation $(\mathbf{e}, \phi)$ where $\mathbf{e}$ is a unit vector.

The orthogonal operator **R** has the three following eigenvalues $\{1, \exp(i\phi), \exp(-i\phi)\}$. The tensor can be represented using the compact form

$$\mathbf{R}(\mathbf{e}, \phi) = \cos(\phi)\mathbf{I} + (1 - \cos(\phi))\mathbf{e} \otimes \mathbf{e} + \sin(\phi)[\mathbf{e}\times], \tag{24}$$

with $||\mathbf{e}|| = 1$ and where the skew-symmetric tensor $[\mathbf{e}\times]$ is

$$\forall \mathbf{b} \in \mathbb{R}^3 : [\mathbf{e}\times]\mathbf{b} = \mathbf{e} \times \mathbf{b}. \tag{25}$$

Noting that, in the Euclidean space, $\mathbf{e} \otimes \mathbf{e} - [\mathbf{e}\times][\mathbf{e}\times] = \mathbf{I}$, we obtain

$$\mathbf{R}(\mathbf{e}, \phi) = \mathbf{I} + (1 - \cos(\phi))[\mathbf{e}\times][\mathbf{e}\times] + \sin(\phi)[\mathbf{e}\times]. \tag{26}$$

The pseudo-vector of rotation $\boldsymbol{\phi}$, of norm ϕ and direction **e**, is introduced using the relation

$$\boldsymbol{\phi} = \phi\mathbf{e} = ||\boldsymbol{\phi}||\mathbf{e}. \tag{27}$$

In the Euclidean space, we have $\boldsymbol{\phi} = (\phi_1, \phi_2, \phi_3)^{\mathrm{T}}$ with $\phi = \sqrt{\phi_1^2 + \phi_2^2 + \phi_3^2}$. The pseudo-vector of rotation is not a vector since it does not satisfy the property of additivity. In fact, finite rotations of the three-dimensional space are not additive in general. The rotation operator can be rewritten as follows

$$\mathbf{R} = \cos(\phi)\mathbf{I} + \frac{1 - \cos(\phi)}{\phi^2}\boldsymbol{\phi} \otimes \boldsymbol{\phi} + \frac{\sin(\phi)}{\phi}[\boldsymbol{\phi}\times], \tag{28}$$

or recalling that, in the Euclidean space, $\boldsymbol{\phi} \otimes \boldsymbol{\phi} - [\boldsymbol{\phi}\times][\boldsymbol{\phi}\times] = \phi^2\mathbf{I}$, we obtain the standard Rodrigues' rotation formula

$$\mathbf{R} = \mathbf{I} + \frac{1 - \cos(\phi)}{\phi^2}[\boldsymbol{\phi}\times][\boldsymbol{\phi}\times] + \frac{\sin(\phi)}{\phi}[\boldsymbol{\phi}\times], \tag{29}$$

which has no singularity in $[0, 2\pi[$.

Using the Taylor series expansions of the functions sine and cosine, the Rodrigues' formula provides a simple algorithm to compute the exponential map

$$\mathbf{R} = \exp([\boldsymbol{\phi}\times]) = \sum_{k=0}^{\infty} \frac{[\boldsymbol{\phi}\times]^k}{k!} \tag{30}$$

so that the variation of the rotation operator is $\delta\mathbf{R} = [\delta\boldsymbol{\phi}\times]\mathbf{R}$. Moreover, the trigonometric relations

$$1 - \cos(\phi) = 2\sin^2\left(\frac{\phi}{2}\right) \text{ and } \sin(\phi) = 2\sin\left(\frac{\phi}{2}\right)\cos\left(\frac{\phi}{2}\right), \tag{31}$$

allows us to define new dependent parameters such that

$$r_0 = \cos\left(\frac{\phi}{2}\right) \text{ and } \mathbf{r} = \sin\left(\frac{\phi}{2}\right)\frac{\boldsymbol{\phi}}{||\boldsymbol{\phi}||}, \tag{32}$$

with the following relation $r_0^2 + \mathbf{r}\cdot\mathbf{r} = 1$.

The Euler parameters, or the unit quaternion $r = (r_0, \mathbf{r})$, give a simple representation of the orthogonal operator $\mathbf{R}$ as follows

$$\mathbf{R} = \exp([\boldsymbol{\phi}\times]) = (2r_0^2 - 1)\mathbf{I} + 2r_0[\mathbf{r}\times] + 2\mathbf{r}\otimes\mathbf{r}. \tag{33}$$

3.5.3 Variation of the Displacements

We recall that for all points Q in the shell we have

$$\mathbf{u}_Q = \mathbf{u}_P + z(\mathbf{R} - \mathbf{I}_3)\mathbf{n}^0 \tag{34}$$

where P is obtained by an orthogonal projection of Q onto the mid-surface of the shell and $\mathbf{n}^0$ is the initial normal. Noting that $\delta\mathbf{n}^0 = 0$, the virtual displacement $\delta\mathbf{u}_Q$ is expressed as

$$\delta\mathbf{u}_Q = \delta\mathbf{u}_P + z\delta\mathbf{R}\mathbf{n}^0 = \delta\mathbf{u}_P + z[\delta\boldsymbol{\phi}\times]\mathbf{R}\mathbf{n}^0 = \delta\mathbf{u}_P + z\delta\boldsymbol{\phi}\times\mathbf{n}. \tag{35}$$

A similar expression is obtained for the iterative displacement $\Delta\mathbf{u}_Q$, knowing that $\Delta\mathbf{n}^0 = 0$

$$\Delta\mathbf{u}_Q = \Delta\mathbf{u}_P + z\Delta\boldsymbol{\phi}\times\mathbf{n}. \tag{36}$$

Moreover, since we have $\Delta(\delta\mathbf{u}_P) = 0$, the variation for the iterative displacement $\Delta(\delta\mathbf{u}_Q)$ is

$$\Delta(\delta\mathbf{u}_Q) = z\delta\boldsymbol{\phi}\times(\Delta\boldsymbol{\phi}\times\mathbf{n}). \tag{37}$$

This term is not standard and does not exist in the classical theory of continuum shell elements. It implies a new contribution in the stiffness matrix which is non-symmetric. Further details about this additional term can be found in [17].

3.5.4 Update of the Rotations and Normals

The quaternion algebra gives simple formulas to update the pseudo-vector of rotation $\boldsymbol{\Phi}$ and the current normal to the shell $\mathbf{n}$.

Given the two rotations $\mathbf{R}_1$ and $\mathbf{R}_2$, respectively represented by the quaternions r^1 and r^2, the resulting composition of the two rotations, namely $\mathbf{R} = \mathbf{R}_2\mathbf{R}_1$, is parametrized by the quaternion r as follows

$$r = r^1 \circ r^2 = (r_0^1 r_0^2 - \mathbf{r}^1 \cdot \mathbf{r}^2, r_0^1 \mathbf{r}^2 + r_0^2 \mathbf{r}^1 + \mathbf{r}^1 \times \mathbf{r}^2), \tag{38}$$

where $r^1 \circ r^2$ denotes the product of the two quaternions r^1 and r^2. One may note that due to the presence of the tensor product $\mathbf{r}^1 \times \mathbf{r}^2$, the composition of two rotations in a three-dimensional space is not, in general, commutative.

Suppose it is known, at the increment i, the total rotation $\mathbf{R}_{i-1}$, that is its representation by the quaternion r^{i-1}, the field $(\mathbf{u}_{i-1}, \boldsymbol{\phi}_{i-1})$, its iterative variation $(\Delta\mathbf{u}_i, \Delta\boldsymbol{\phi}_i)$ and the normal $\mathbf{n}^{i-1}$. The displacement $\mathbf{u}$ is straightforward updated by the additive formula

$$\mathbf{u}_i = \mathbf{u}_{i-1} + \Delta\mathbf{u}_i. \tag{39}$$

However, a similar update can not be applied to the rotation $\boldsymbol{\phi}$ since the property of additivity is only valid for small rotations.

To update the rotation $\boldsymbol{\phi}$ using the computed iterative value $\Delta\boldsymbol{\phi}_i$, we first compute the quaternion Δr^i which parametrizes the iterative rotation

$$\Delta r^i = (\Delta r_0^i, \Delta \mathbf{r}^i) = \left(\cos\left(\frac{||\Delta\boldsymbol{\phi}_i||}{2}\right), \sin\left(\frac{||\Delta\boldsymbol{\phi}_i||}{2}\right)\frac{\Delta\boldsymbol{\phi}_i}{||\Delta\boldsymbol{\phi}_i||}\right). \tag{40}$$

The rotation field is thus updated using the product $r^i = \Delta r^i \circ r^{i-1}$

$$r^i = (r_0^i, \mathbf{r}^i) = (\Delta r_0^i r_0^{i-1} + \Delta\mathbf{r}^i \cdot \mathbf{r}^{i-1}, \Delta r_0^i \mathbf{r}^{i-1} + r_0^{i-1}\Delta\mathbf{r}^i + \Delta\mathbf{r}^i \times \mathbf{r}^{i-1}). \tag{41}$$

Recalling that $r = (\cos(\frac{\phi}{2}), \sin(\frac{\phi}{2})\mathbf{e})$, the pseudo-vector of rotation $\boldsymbol{\phi}$ can be defined in $[0, 2\pi[$ using the relations

$$\phi^i = 2\tan^{-1}\left(\frac{||\mathbf{r}^i||}{r_0^i}\right) \text{ and } \boldsymbol{\phi}^i = \phi^i \frac{||\mathbf{r}^i||}{\mathbf{r}^i}. \tag{42}$$

The director vector $\mathbf{n}^i = \exp([\Delta\boldsymbol{\phi}^i \times])\mathbf{n}^{i-1}$ is easily updated as follows

$$\mathbf{n}^i = (2(\Delta r_0^i)^2 - 1)\mathbf{n}^{i-1} + 2\Delta r_0^i \Delta\mathbf{r}^i \times \mathbf{n}^{i-1} + 2(\Delta\mathbf{r}^i \cdot \mathbf{n}^{i-1})\Delta\mathbf{r}^i. \tag{43}$$

4 Reduced Integration

The reduced quadrature rules for NURBS-based analysis of structural mechanics given in [10, 11] are extended, in this section, to analysis suitable T-meshes.

We recall that due to the high order of the isogeometric method, a full Gaussian integration with $\lceil (p+1)/2 \rceil$ quadrature points in each direction is computationally expensive. Moreover, due to the high continuity of the spline functions, the full integration provides to many quadrature points in the element [5]. The reduced integration presents the double advantage to significantly reduce the computational cost of the method and to alleviate the numerical locking occurring when modelling thin structures [11].

4.1 Bézier Elements and Full Gauss Quadrature

As for B-splines and NURBS, the continuity of T-splines in the physical space is determined by the local knot vectors, defining the blending functions in the parameter space. For polynomials of order p, a multiplicity m of a knot value in the local knot vector implies a regularity C^{p-m} of the blending functions across this knot value. Unlike Cartesian structure of B-splines and NURBS, in which the lines of reduced continuity are global and unambiguous, the lines of reduced continuity in a T-mesh are local to the basis functions and do not generally propagate throughout the domain. Therefore, T-splines may have locally several degrees of continuity within the domain as previously illustrated in Fig. 3.

We have introduced the concept of T-junction extension which, in practical, is a simple and efficient tool to verify whether the T-mesh is suitable for analysis or not. We recall that these T-junction extensions are lines of reduced continuity. Therefore, an extended T-mesh can be constructed adding all continuity reduction lines to the initial T-mesh. The union off all edges and T-junction extensions represents the lines across which the continuity of the blending functions is less than C^{∞}. The extended T-mesh splits the parameter and physical spaces into several regions which define the Bézier elements. This notion of element is essential to perform a correct numerical quadrature.

The numerical integration is thus performed in the regions delimited by the edges of the extended T-mesh which correspond to the Bézier elements. Figure 4 illustrates

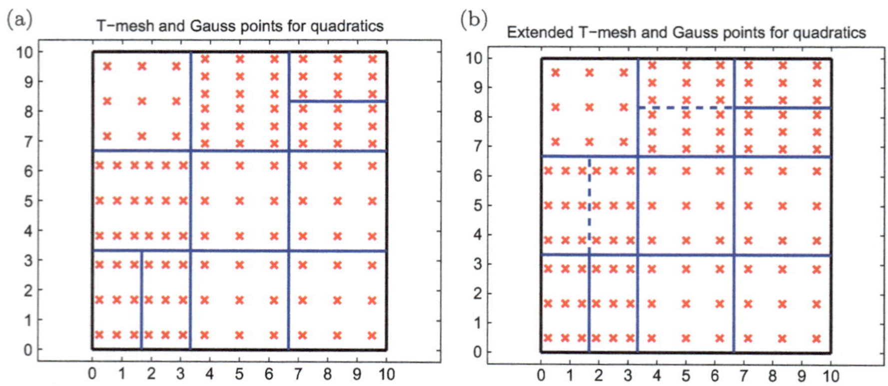

Fig. 4 T-mesh (**a**) and Bézier elements (**b**) with full Gauss quadrature for bi-quadratic blending functions

the full Gauss quadrature for analysis suitable T-splines. The numerical quadrature is performed in each Bézier elements, represented on the right, which are defined adding the lines or reduced continuity (dotted blue lines) to the edges of the T-mesh (solid blue lines).

4.2 Reduced Gauss Quadrature

The non-uniform reduced Gauss quadrature presented in [10] in the context of B-spline meshes, is extended in this work to unstructured analysis suitable T-meshes. The reduced number of quadrature points used in B-spline/NURBS analysis has shown to alleviate numerical locking as well as to improve the computational efficiency, especially as the degree of continuity is high. As with a full Gauss quadrature, the reduced rule is applied in each Bézier element in which the continuity of the blending function is C^{∞}. In practical terms, the local continuity of each Bézier element edge is given by the local knot vectors of the blending functions that are positive on this element.

Figure 5 represents the same analysis suitable T-mesh as previously and the non-uniform reduced Gauss quadrature rule which is constructed locally, namely in a standard element-by-element fashion. Using the concepts introduced in [12] for quadratic polynomials, the target space is $\mathcal{S}^{2p}_{q-1} = \mathcal{S}^4_0$, achieved with a full Gauss or patchwise quadrature, and the approximation space corresponding to this given scheme is $\mathcal{S}^r_s = \mathcal{S}^1_{-1}$.

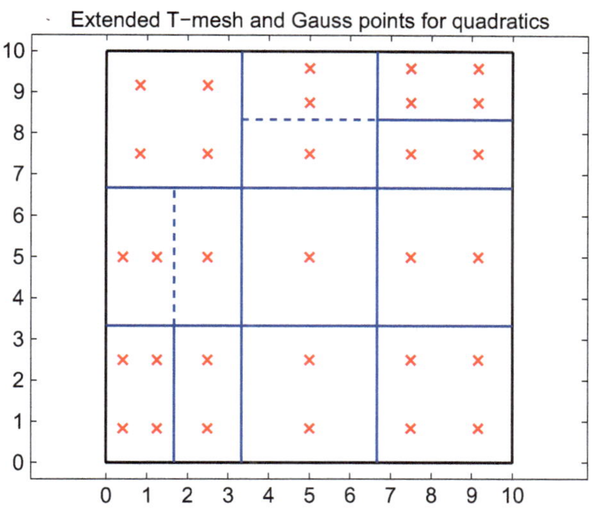

Fig. 5 Extended T-mesh and reduced Gauss quadrature rule

4.3 Reduced Patchwise Quadrature

The reduced patchwise rules given in [12] are applied in the context of T-spline analysis. We recall that, in this work, it is shown that the reduced Gauss rule is not optimal in term of accuracy since it exactly integrates the approximation space $\mathcal{S}^1_{-1}$ which is not included in the target space $\mathcal{S}^4_0$. For C^1 quadratic shape functions, the over-integrated approximation space $\mathcal{S}^2_0$ appears to be optimal when performing a reduced integration in the sense that it significantly reduces the numerical locking and computational effort without introducing Hourglass modes. Moreover, it is the smallest space that guarantees these properties for C^1 quadratics.

The reduced patchwise rules are constructed by tensor products of several one-dimensional rules. We look through all the Bézier elements of the extended T-mesh in both parametric directions since the patchwise rules are not constructed locally, that is from the continuity given by the local knot vectors associated with the positive functions in the element. The complexity of the T-mesh, namely the number of holes or T-junctions, the different degrees of continuity, defines the number of one-dimensional rules to compute in order to construct the two-dimensional quadrature scheme. However, each construction of a one-dimensional rule is independent, that is it can be straightforward performed in parallel. A patchwise one-dimensional rule is defined in a given parametric direction and between two local knots of C^{-1} continuity, namely between two borders of the domain.

Figure 6 represents the two one-dimensional rules that have to be constructed in order to compute the two-dimensional scheme. Again, we use the same T-mesh as with the other presented integrations. We observe that the same number of quadrature points is used as with the non-uniform reduced Gauss quadrature. However, the locations of these points are not standard since the integration rule is not defined locally, it depends on the element in the patch.

Due to the flexible structure of the T-splines, in particular the possibility to perform a local refinement, blending functions of generally distinct supports lie in

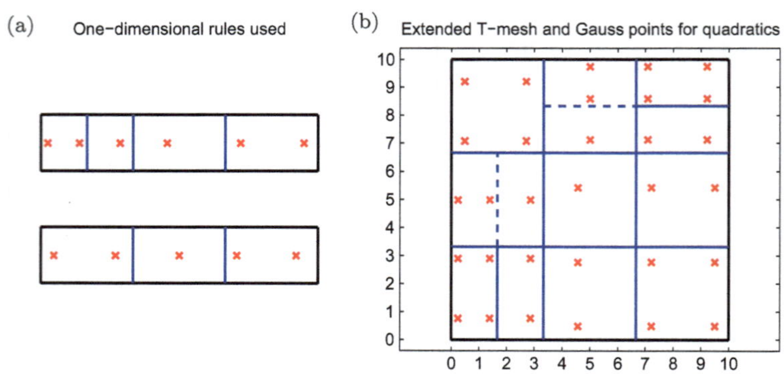

Fig. 6 Extended T-mesh and reduced patchwise quadrature rule. (**a**) One-dimensional patchwise reduced rules used. (**b**) Reduced patchwise quadrature scheme

the element. Therefore, the patchwise rule can not, in principle, exactly integrate all functions within the element. However, we have seen that the refined blending functions can be expressed as a linear combination of the coarse functions when performing a blending function subdivision. Considering the Bézier elements for the construction of the quadrature rules guarantees that all the blending functions in the element, even with a coarse support, are exactly integrated. We note that some shape functions are thus over-integrated near a T-junction, which may imply a slight numerical locking since there are locally more quadrature points than control points. This observation can be made for both quadrature rules presented in this work. For instance, the one-dimensional rule with six quadrature points, presented in Fig. 6, also exactly integrates the blending functions which are exactly integrated by the other one-dimensional rule with only five quadrature points.

5 Numerical Results

In this section, we perform some classical benchmarks for structural analysis. We present numerical solutions for linear elastic shells and geometric non-linear structures. All the calculations are done for the thick shell formulation presented in the previous section. A standard Galerkin formulation is employed. Full Gauss, non-uniform reduced Gauss and reduced patchwise quadratures are used for the numerical integration and compared in several examples. The first example is a bending plate which exhibits strong numerical locking under mesh distortion with full Gauss-Legendre quadrature. The two following examples come from the so-called shell obstacle course: the pinched cylinder and the pinched hemisphere. The relevance of these problems to the assessment of shell analysis have been discussed extensively in the literature [18]. The pinched hemisphere is also considered within the context of geometric non-linear analysis to assess the performance of the shell formulation and the numerical quadrature with large transformations. A set of popular benchmark problems for geometric non-linear analysis of shells is given in [19].

5.1 Bending Plate

The first example, represented in Fig. 7, is a simply supported thin square plate, of length L, with a vertical concentrated loading applied at its center. The case of a uniform pressure loading is not presented in this paper but similar results are obtained. Due to the symmetry of the problem, only a quarter of the plate (blue area in Fig. 7) is simulated. The geometric property $L/h = 10^3$ results in a thin shell structure which is expected to suffer from locking pathologies, in particular transverse shear locking. In this example, no T-junction is considered. We first assess

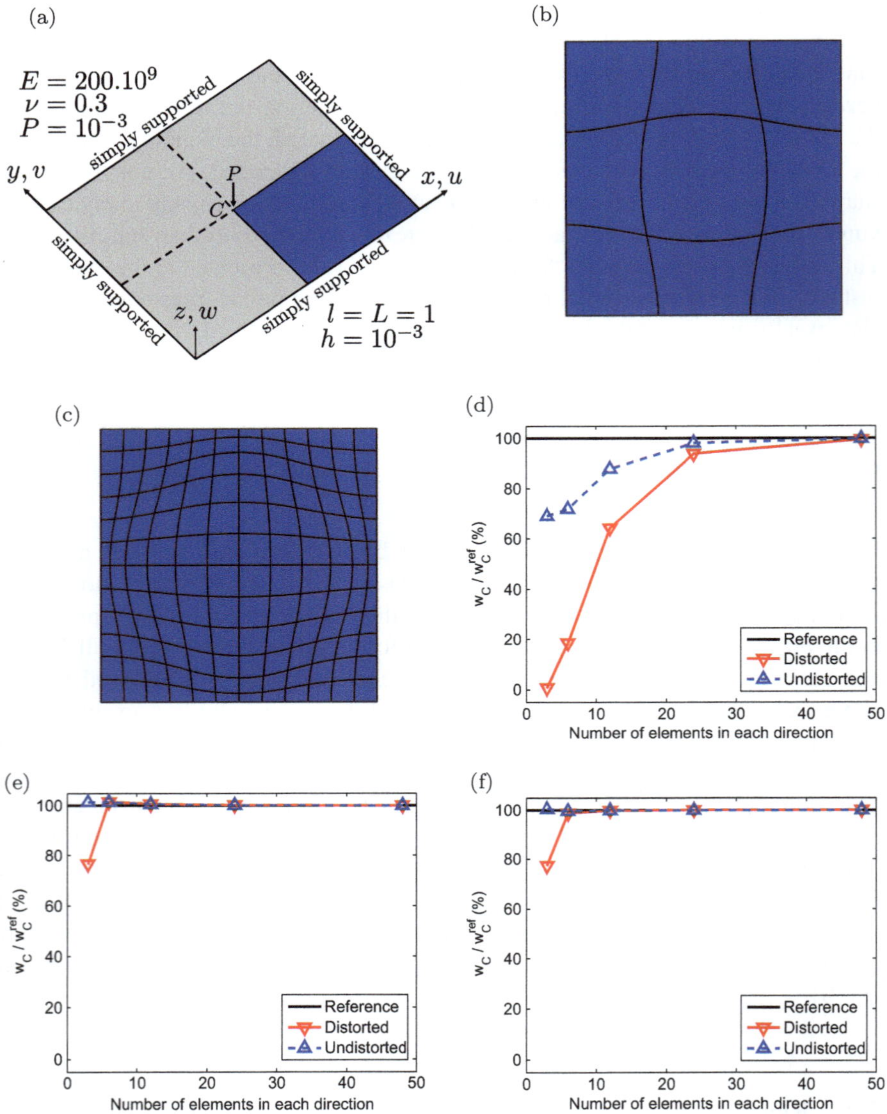

Fig. 7 Central deflection of the bending plate for different numerical quadratures. (**a**) Problem description and geometric data. (**b**) Distorted mesh 3×3. (**c**) Distorted mesh 12×12. (**d**) Full Gauss-Legendre quadrature. (**e**) Reduced Gauss-Legendre quadrature. (**f**) Reduced patchwise quadrature

the performance of both presented reduced rules with respect to mesh quality and numerical locking.

For each integration, namely full, reduced Gauss-Legendre and reduced patchwise, two meshes are compared: a regular mesh and a distorted mesh. The numerical results are compared with the reference central deflection $w_C^{\text{ref}} = 6.3341 \cdot 10^{-6}\,\text{m}$ [11]. An initial mesh distortion is applied on the coarsest mesh with three elements in each direction. A strategy of k-refinement is then applied to generate the refined meshes without modifying the initial mesh distortion.

Figure 7 illustrates that full quadrature is sensitive to mesh distortion in the sense that it strengthens the locking pathologies and decreases the accuracy of the numerical solution. However, in this example, both reduced quadratures are not sensitive to numerical locking neither to mesh distortion. The accuracy of the solution is significantly improved with respect to the full quadrature. Moreover, we have the following relation for the ratio of the number of quadrature points

$$\lim_{n_{\text{elem}} \to +\infty} n_{\text{GP}}^{\text{full}} / n_{\text{GP}}^{\text{reduced}} = 4, \tag{44}$$

which means that the computational effort is also significantly reduced.

5.2 *Pinched Cylinder*

The second example is the so called pinched cylinder subjected to equal and opposite radial forces at the top and bottom mid-spans. The two ends are supported by rigid diaphragms (Fig. 8). The geometry of the cylindrical shell exhibits the shape ratio as in the previous problem. This is a severe test case evaluate the capacity of the shell element to describe the membrane and bending deformations, especially near the application points of the concentrated forces. This problem is characterized by a bending dominant behaviour and results have revealed a tendency towards membrane and shear locking. A severe membrane locking is expected as bending is inextensible. Again, due to the symmetry of the problem, only a quarter of the shell (blue area in Fig. 8a) is modelled. To evaluate the convergence, the vertical displacement w_C^h at the loading points is compared with the reference solution $w_C^{\text{ref}} = 1.8248 \cdot 10^{-7}\,\text{m}$ [11].

Several NURBS meshes are generated using the standard k-refinement procedure. The T-meshes are represented in Fig. 8. We assess the performance of both full and reduced Gauss-Legendre quadratures for NURBS and T-spline meshes. We observe that combining the reduced integration with the locally refined T-meshes significantly improves the accuracy of the numerical solution using less degrees of freedom and integration points. It thus appears as a promising technique to perform complex simulations, involving localized deformations, with a reasonable computational effort.

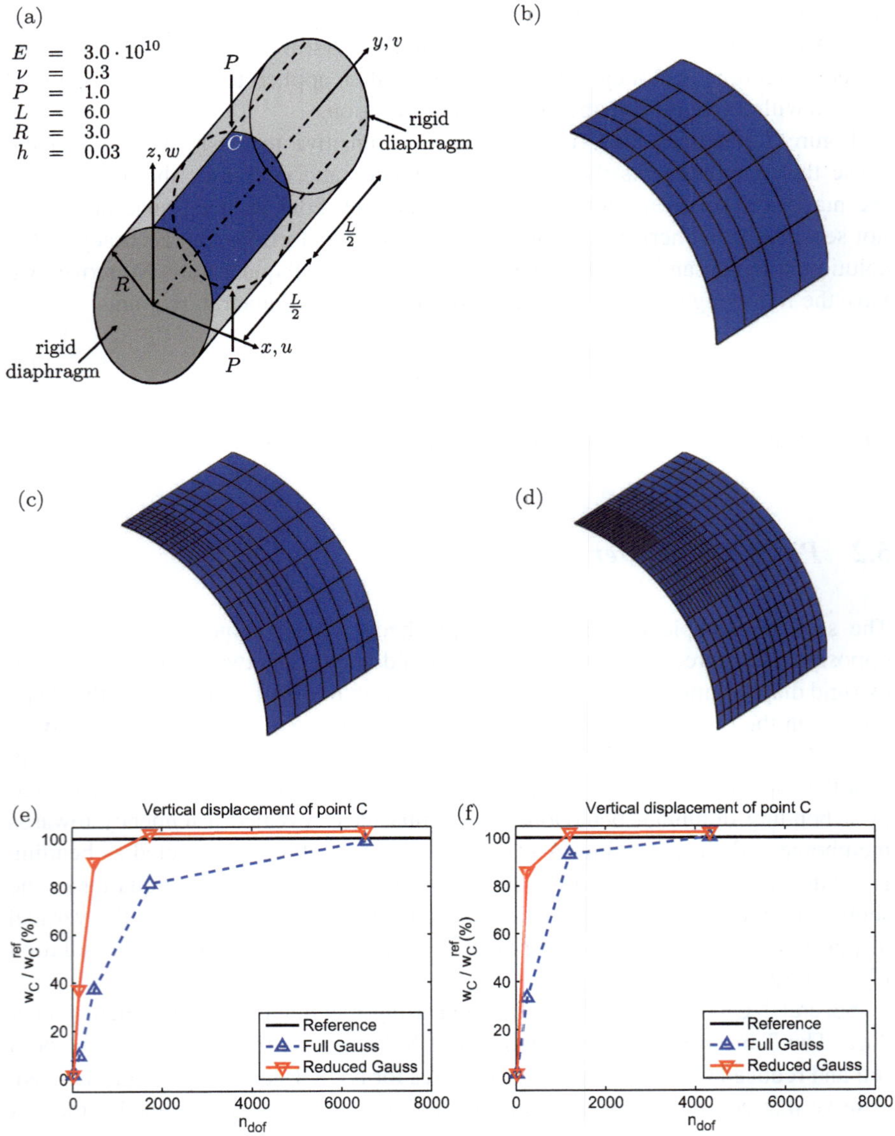

Fig. 8 Vertical displacement at the center of the pinched cylinder for different numerical quadratures and types of meshes. (**a**) Problem description and geometric data. (**b**) $n_{\mathrm{dof}} = 258$. (**c**) $n_{\mathrm{dof}} = 1206$. (**d**) $n_{\mathrm{dof}} = 4326$. (**e**) NURBS meshes. (**f**) T-spline meshes

5.3 Pinched Hemisphere

The last example considered in this work is the pinched hemisphere, with an 18° hole at its pole, which comes from a set of popular benchmark problems for geometric non-linear analysis of shells [19]. The pinched hemispherical shell, represented in Fig. 9, is subjected to equal and opposite concentrated forces P_{max} applied at the four cardinal points of its equator. The two ends are stress free. The rigid body motion is eliminated fixing the vertical displacement at an application point of the forces. The slenderness ratio is $R/h = 250$. This doubly curved shell problem is characterized by inextensible bending modes and large rigid body rotations. Membrane energy is almost zero. This test ensures that the element does not suffer from membrane and shear locking. Due to symmetry, only one quadrant is meshed (blue area in Fig. 9). To assess the accuracy of the shell formulation, the horizontal displacement u_A^h at the loading points is compared with the reference solution using ABAQUS S4R, a four-node reduced shell element, given in [19].

In the non-linear solution procedure, the full Newton-Raphson method is used in a TLF. The convergence criterion employed is the 0.1 % force tolerance. The maximum load P_{max} is subdivided into several load increments which are not necessarily uniform. At the end of each load increment, a converged intermediate solution is obtained. This procedure reduces the degree of non-linearity from an intermediate solution state to another and enhances the chance of obtaining the desired solution corresponding to the maximum load. If the increments are correctly chosen or computed, each step converges in a few iterations.

A standard NURBS mesh with twelve C^1 quadratic elements in each direction is used. Figure 9 illustrates that, as with linear analysis, the reduced Gauss scheme provide a better accuracy than the full Gauss-Legendre quadrature within the context of non-linear analysis. The shell formulation with finite transformations also exhibits a good accuracy. The resulting reduced isogeometric shell element shows

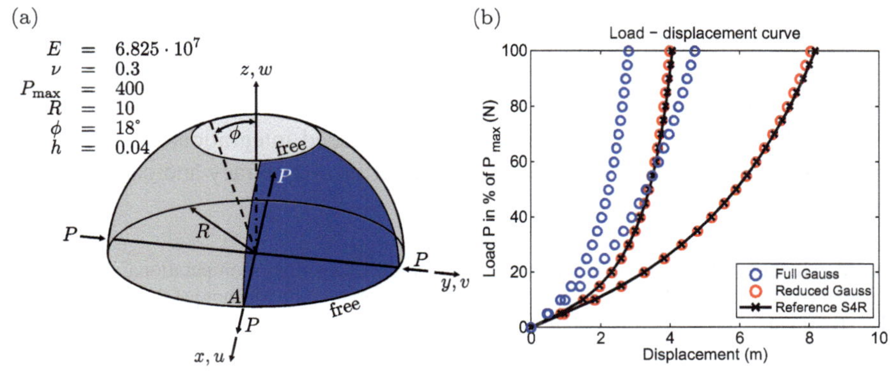

Fig. 9 Horizontal displacement at the center of the pinched cylinder for different numerical quadratures and types of meshes. (**a**) Problem description and geometric data. (**b**) Numerical solution for a 12 × 12 mesh

satisfying performance. If a local refinement, using T-junctions, is added, accuracy and computational efficiency are even more improved, as shown in the previous examples.

6 Conclusions

The concept of T-splines extends the NURBS technology and permits local refinement, holes, which is a solution to the trimmed surface problem, merging of patches without introducing new knots which propagate. These properties are essential in a design context and T-splines represent a promising integrated approach to engineering design and analysis. This technology is attractive for both design and analysis and has recently become available in industrial design software.

In this paper, we explore various integration schemes and adapt the B-spline/NURBS tensor-product rules to T-splines. The uni-dimensional reduced quadrature rules for quadratic shape functions, with non-uniform Gauss-Legendre or patchwise integration, have been extended to bi-dimensional shell problems within the context of unstructured T-splines. A geometric non-linear Reissner-Mindlin shell formulation, based on the displacements, has been proposed. The update of the finite rotations in a TLF is handled using the quaternion algebra.

The non-uniform reduced Gauss-Legendre integration can be straightforward extended to more complex T-meshes. The local continuity of each Bézier element edge, required to construct the reduced element rule, is given by the local knot vectors of the blending functions that are positive on this element. However, we look through all the Bézier elements of the extended T-mesh in both parametric directions to construct the one-dimensional patchwise rules essential for the final two-dimensional tensor-product rule. The latter require a higher computational effort than Gauss rules, e.g. one calculation during the pre-processing phase, but they also offer a better accuracy.

The performance of the resulting reduced shell element is assessed on several linear and geometric non-linear problems. We have observed that, C^1 quadratic blending functions are sensitive to mesh distortion and numerical locking when performing a full integration with a thin structure. They both decrease the accuracy of the numerical solution. However, the reduced integrations alleviate these pathologies and exhibit good performance in terms of accuracy and computational efficiency.

Acknowledgements This work was supported by the "OpenLab Computational Mechanics" created collectively by PSA Peugeot Citroën and the Laboratoire de Mécanique des Solides, École Polytechnique. This support is gratefully acknowledged.

References

1. T.J.R Hughes, J.A. Cottrell, Y. Bazilevs, Isogeometric analysis: CAD, finite elements, NURBS, exact geometry and mesh refinement. Comput. Methods Appl. Mech. Eng. **194**, 4135–4195 (2005)
2. L. Piegl, W. Tiller, *The NURBS Book*, 2nd edn. (Springer, New York, 1997)
3. T.W. Sederberg, J. Zheng, A. Bakenov, A. Nasri, T-splines and T-NURCCSs. ACM Trans. Graph. **22**, 477–484 (2003)
4. A.-V. Vuong, C. Giannelli, B. Jüttler, B. Simeon, A hierarchical approach to adaptive local refinement in isogeometric analysis. Comput. Methods Appl. Mech. Eng. **200**, 3554–3567 (2011)
5. T.J.R. Hughes, A. Reali, G. Sangalli, Efficient quadrature for NURBS-based isogeometric analysis. Comput. Methods Appl. Mech. Eng. **199**, 301–313 (2010)
6. R. Echter, M. Bischoff, Numerical efficiency, locking and unlocking of NURBS finite elements. Comput. Methods Appl. Mech. Eng. **199**, 374–382 (2010)
7. R. Echter, B. Oesterle, M. Bischoff, A hierarchic family of isogeometric shell finite elements. Comput. Methods Appl. Mech. Eng. **254**, 170–180 (2013)
8. D. Schillinger, S.J. Hossain, T.J.R. Hughes, Reduced Bézier element quadrature rules For quadratic and cubic splines in isogeometric analysis. Comput. Methods Appl. Mech. Eng. **277**, 1–45 (2014)
9. J.C. Simo, D. Fox, M.S. Rifai, On the stress resultant geometrically exact shell model. Part III: computational aspects of the nonlinear theory. Comput. Meth. Appl. Mech. Eng. **79**, 21–70 (1990)
10. C. Adam, S. Bouabdallah, M. Zarroug, H. Maitournam, Improved numerical integration for locking treatment in isogeometric structural elements, Part I: beams. Comput. Methods Appl. Mech. Eng. **278**, 1–28 (2014)
11. C. Adam, S. Bouabdallah, M. Zarroug, H. Maitournam, Improved numerical integration for locking treatment in isogeometric structural elements, Part II: plates and shells. Comput. Methods Appl. Mech. Eng. **284**, 106–137 (2015)
12. C. Adam, T.J.R. Hughes, S. Bouabdallah, M. Zarroug, H. Maitournam, Selective and reduced numerical integration for NURBS-based isogeometric analysis. Comput. Methods Appl. Mech. Eng. **284**, 732–761 (2015)
13. C. Adam, A contribution to static and dynamic isogeometric analysis of structures. Ph.D. thesis, École Polytechnique (2015)
14. Y. Bazilevs, V.M. Calo, J.A. Cottrell, J.A. Evans, T.J.R. Hughes, S. Lipton, M.A. Scott, T.W. Sederberg, Isogeometric analysis using T-splines. Comput. Methods Appl. Mech. Eng. **199**, 229–263 (2010)
15. T.W. Sederberg, D.L. Cardon, G.T. Finnigan, N.S. North, J. Zheng, T. Lyche, T-spline simplification and local refinement. ACM Trans. Graph. **23**, 276–283 (2004)
16. M.A. Scott, X. Li, T.W. Sederberg, T.J.R. Hughes, Local refinement of analysis-suitable T-splines. Comput. Methods Appl. Mech. Eng. **213**, 206–222 (2012)
17. M.A. Crisfield, *Non-linear Finite Element Analysis of Solids and Structures* (Wiley, West Sussex, 1996)
18. R. MacNeal, R. Harder, A proposed standard set of problems to test finite element accuracy. Finite Elem. Anal. Des. **1**, 3–20 (1985)
19. K.Y. Sze, X.H. Liu, S.H. Lo, Popular benchmark problems for geometric nonlinear analysis of shells. Finite Elem. Anal. Des. **40**, 1551–1569 (2004)

Multiresolution Shape Optimisation with Subdivision Surfaces

Fehmi Cirak and Kosala Bandara

Abstract We review our recent work on multiresolution shape optimisation and present its application to elastic solids, electrostatic field equations and thin-shells. In the spirit of isogeometric analysis the geometry of the domain is described with subdivision surfaces and different resolutions of the same surface are used for optimisation and analysis. The analysis is performed using a sufficiently fine control mesh with a fixed resolution. During shape optimisation the geometry is updated starting with the coarsest control mesh and then moving on to increasingly finer control meshes. The transfer of data between the geometry and analysis representations is accomplished with subdivision refinement and coarsening operators. Moreover, we discretise elastic solids with the immersed finite element method, electrostatic field equations with the boundary element method and thin-shells with the subdivision finite element technique. In all three discretisation techniques there is no need to generate and maintain an analysis-suitable volume discretisation.

1 Introduction

As widely discussed in isogeometric analysis literature, the geometry representations used in today's computer aided design (CAD) and computational analysis software are inherently incompatible [19]. This is particularly limiting in shape optimisation in which a geometry model is iteratively updated based on the results of a computational analysis [6, 8, 17]. It is tedious and often impossible to repeatedly map the results from an analysis mesh back to the CAD model. To remedy this, the shape optimisation of shells and solids by directly optimising the CAD geometry model using isogeometric analysis has been recently explored [12, 16, 21, 37].

In the present paper we review the multiresolution shape optimisation technique and present its application to linear elastic solids, electrostatic field equations and thin-shell structures [2–4]. We discretise elastostatic boundary value problems using immersed, or embedded, finite elements, see, e.g., [28, 29, 31, 32], which have clear

F. Cirak (✉) • K. Bandara
Department of Engineering, University of Cambridge, Cambridge, CB2 1PZ, UK
e-mail: fc286@cam.ac.uk; kkmb2@cam.ac.uk

B. Jüttler, B. Simeon (eds.), *Isogeometric Analysis and Applications 2014*,
Lecture Notes in Computational Science and Engineering 107,
DOI 10.1007/978-3-319-23315-4_6

advantages when applied to structural shape optimisation [1, 18]. In order to cope with unbounded domains we discretise electrostatic problems with the boundary element method. Crucially, in immersed finite elements and boundary elements the geometry of the domain boundary can be updated without needing to generate or maintain domain meshes. Shell structures are modelled as a surface and are discretised with subdivision finite elements, which use the subdivision surfaces as basis functions [11, 12].

The domain boundaries and the shell mid-surface are represented with subdivision surfaces. Although historically subdivision and related techniques have originated in computer graphics, they recently became available in several CAD software packages, including Autodesk Fusion 360, PTC Creo and CATIA. As will be demonstrated in this paper, subdivision surfaces provide an elegant isogeometric analysis-suitable, bidirectional mapping between the geometry and analysis models. In subdivision a geometry is described using a control mesh and a limiting process of repeated refinement [27, 40]. The refinement rules are usually adapted from knot refinement rules for splines [9, 14, 22]. We consider the Catmull-Clark [9] scheme based on quadrilateral meshes and the Loop [25] scheme based on triangular meshes. The Catmull-Clark scheme is the generalisation of cubic tensor-product b-splines to unstructured meshes and the Loop scheme is the generalisation of quartic box-splines. Both subdivision schemes lead to smooth surfaces even in case of unstructured meshes with extraordinary vertices. The hierarchy of control meshes underlying subdivision surfaces lends itself naturally to multiresolution decomposition of geometries [26, 41]. The size of the geometric region influenced by each vertex depends on the resolution of the control mesh, editing coarser levels leads to large-scale changes while editing finer levels lead to small-scale changes.

The introduced multiresolution optimisation approach relies on subdivision curves/surfaces for the description of boundaries. The multiresolution paradigm allows us to describe the same geometry with control meshes of different resolution for analysis and optimisation purposes. For finite element and boundary element analysis a relatively fine control mesh is used in order to minimise the discretisation errors. In contrast, the degrees of freedom in optimisation (i.e., design variables) are chosen as the vertex coordinates of a coarser control mesh. Perhaps counterintuitively, the use of the same fine control mesh for optimisation and analysis leads to suboptimal optimisation results with possible non-physical geometry oscillations. This behaviour is also known from earlier work on structural optimisation in which the finite element nodes were used as optimisation design variables [8, 17]. This paper will demonstrate that the best optimisation results are achieved when starting with a coarse control mesh increasingly finer control meshes are optimised. During the optimisation iterations the refinement level of the control mesh is increased each time a minimum is reached.

2 Multiresolution Subdivision Surfaces

2.1 Subdivision Refinement of Spline Curves

The refinability property of cubic b-splines can be utilised to derive a corresponding subdivision scheme. To illustrate this, we consider the coarse knot sequence $\xi_i^0 = 0, 1, 2, 3, \ldots$ and the fine knot sequence $\xi_i^1 = 0, 0.5, 1, 1.5, 2, 2.5, 3, \ldots$. We denote the b-splines on the coarse knot sequence with $B_i^0(\xi)$ and the ones on the fine knot sequence with $B_i^1(\xi)$, see Fig. 1. According to the b-spline refinability equation, see, e.g., [7, 40], it is possible to represent the coarse b-splines (in the interior) as a linear combination of the fine b-splines

$$B_i^0(\xi) = \sum_{j=0}^{4} S_{ij} B_{2i+j-2}^1(\xi) \quad \text{with} \quad S_{ij} = \frac{1}{8}\binom{4}{i}, \tag{1}$$

where S_{ij} is the subdivision matrix with its entries given in terms of the usual binomial coefficients. In order to keep the derivations simple we exclude from our discussion the b-splines close to the boundaries. For subsequent derivations (1) is best expressed in matrix notation

$$\boldsymbol{B}^0 = \boldsymbol{S}\boldsymbol{B}^1 . \tag{2}$$

Next, we consider a spline curve defined in terms of the coarse b-splines and the corresponding control vertices, i.e.,

$$\boldsymbol{x}_h(\xi) = \boldsymbol{B}^0 \cdot \boldsymbol{x}^0 \tag{3}$$

with matrix $\boldsymbol{x}^0$ containing the coordinates of control vertices. The number of columns of $\boldsymbol{x}^0$ is equal to space dimension and the number of rows is equal to the number of all control vertices.

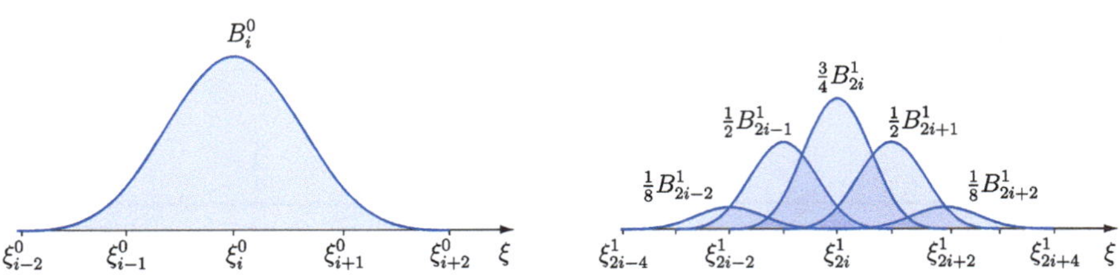

Fig. 1 Refinement relation for cubic b-splines

Introducing the refinement relation (2) into (3) the spline curve can be expressed with

$$x_h(\xi) = \left(SB^1\right) \cdot x^0 = B^1 \cdot \left(S^{\mathsf{T}} x^0\right) . \tag{4}$$

This implies that the control vertex coordinates on the finer level can be computed with the subdivision relation

$$x^1 = S^{\mathsf{T}} x^0 . \tag{5}$$

In subdivision schemes the described refinement approach is applied recursively, i.e.,

$$x^{\ell+1} = S^{\mathsf{T}} x^{\ell} , \tag{6}$$

where $x^{\ell+1}$ and x^{ℓ} are two matrices containing the coordinates of all the vertices at levels $\ell + 1$ and ℓ with $\ell \geq 0$. Although the dimensions of S increase with ℓ we denote all subdivision matrices with S since each row has the same non-zero components only shifted relative to adjacent rows [3]. The successive refinement of a given control polygon using (6) is illustrated in Fig. 2.

For computer implementation and generalisation to bivariate splines, it is instructive to think that the refinement of a control polygon according to (6) consists of a refinement and an averaging step. In the refinement step each segment of the polygon is subdivided into two segments, see Fig. 3a. Subsequently, the vertex

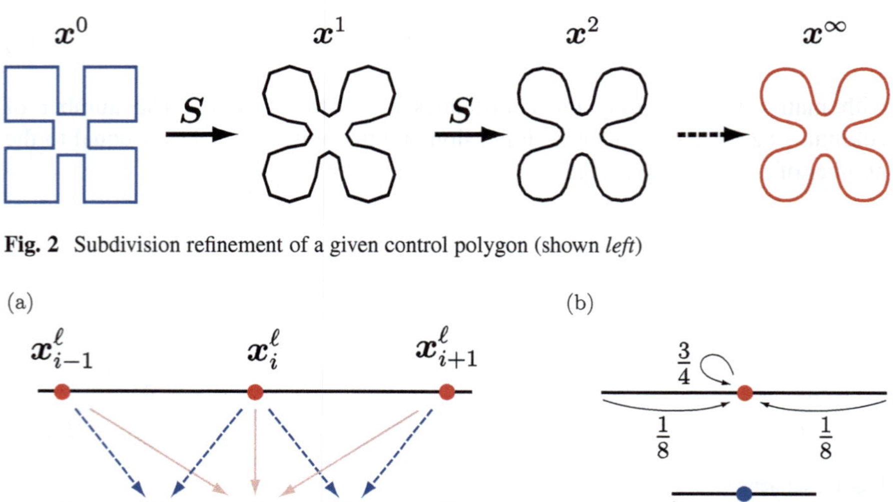

Fig. 2 Subdivision refinement of a given control polygon (shown *left*)

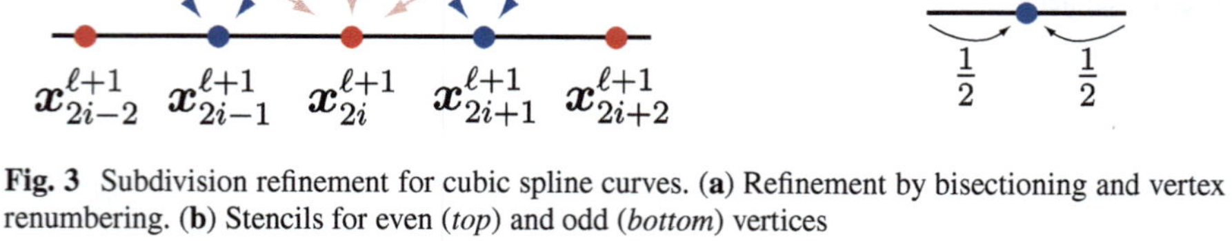

Fig. 3 Subdivision refinement for cubic spline curves. (**a**) Refinement by bisectioning and vertex renumbering. (**b**) Stencils for even (*top*) and odd (*bottom*) vertices

coordinates of the refined polygon are determined by averaging the coarse vertex coordinates with the two stencils shown in Fig. 3b. The even vertex stencil applies to vertices that are already present in the coarse polygon and the odd vertex stencil applies to vertices that are only present in the refined polygon. The naming odd and even is motivated by the consecutive numbering of vertices where newly inserted vertices receive odd numbers. According to Fig. 3b, for a given coarse polygon of level ℓ with vertex coordinates $\boldsymbol{x}_i^\ell$ a refined polygon of level $\ell + 1$ with vertex coordinates $\boldsymbol{x}_i^{\ell+1}$ is computed with

$$\boldsymbol{x}_{2i}^{\ell+1} = \frac{1}{8}\boldsymbol{x}_{i-1}^{\ell} + \frac{3}{4}\boldsymbol{x}_{i}^{\ell} + \frac{1}{8}\boldsymbol{x}_{i+1}^{\ell}\,, \qquad \boldsymbol{x}_{2i+1}^{\ell+1} = \frac{1}{2}\boldsymbol{x}_{i}^{\ell} + \frac{1}{2}\boldsymbol{x}_{i+1}^{\ell}\,. \tag{7}$$

2.2 *Subdivision Refinement of Bivariate Surfaces*

Similar to tensor-product b-splines the tensor products of one-dimensional subdivision stencils yield the corresponding subdivision stencils for surfaces. The tensor product construction works only for quadrilateral meshes and in the refinement step each quadrilateral is subdivided into four quadrilaterals, see Fig. 4a. It is evident that the tensor-product stencils only apply to meshes in which each vertex within the domain is connected to four faces. The number of faces connected to a vertex is referred to as the valence of that vertex. The domain vertices with a valence other than four are known as extraordinary vertices or star-vertices.

For subdivision surfaces using quadrilateral meshes, the original stencils proposed by Catmull and Clark [9], shown in Fig. 5, are used. Note, the vertex stencil in Fig. 5c depends on the valence v of the vertex and reduces to a standard tensor product stencil for structured meshes with valence $v = 4$. There is mathematical theory which shows that the resulting surface is C^2 continuous almost everywhere except at the extraordinary vertices where it is only C^1 continuous [27]. In addition, there are also extended subdivision stencils for vertices on edges, creases and corners, see, e.g., [5, 10, 39]. In this context, a crease is a line on the surface across which the surface is only C^0 continuous. As an illustrative example, Fig. 6

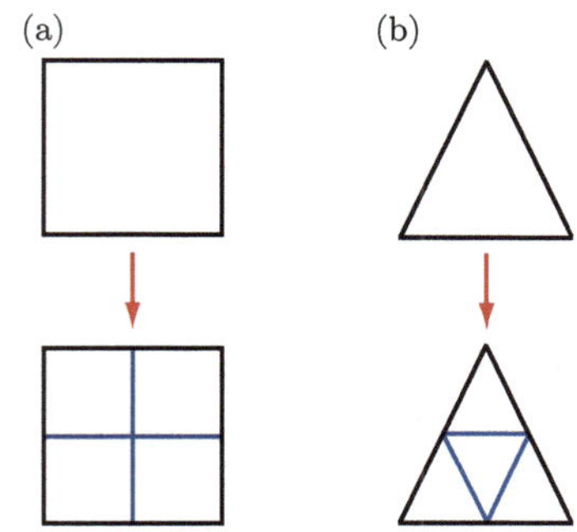

Fig. 4 Refinement by quadrisectioning. (**a**) Quadrilaterals. (**b**) Triangles

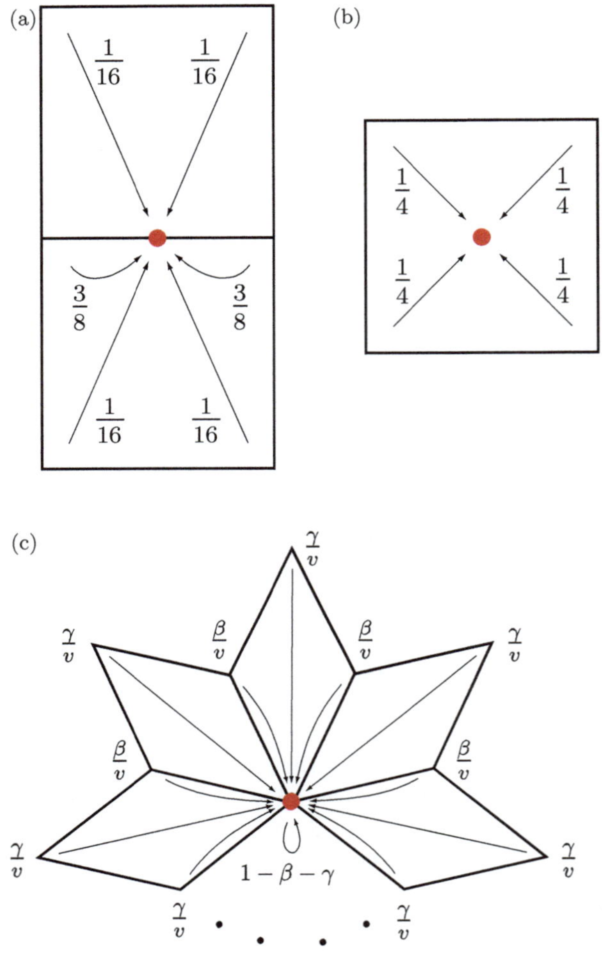

Fig. 5 Subdivision stencils for the Catmull-Clark scheme. Each of the stencils are used for computing the coordinates of vertices of the type indicated by the *red dot*. The weights for the vertex stencil are $\beta = \frac{3}{2v}$ and $\gamma = \frac{1}{4v}$, where v is the valence of the vertex. (**a**) Edge stencil. (**b**) Face stencil. (**c**) Vertex stencil

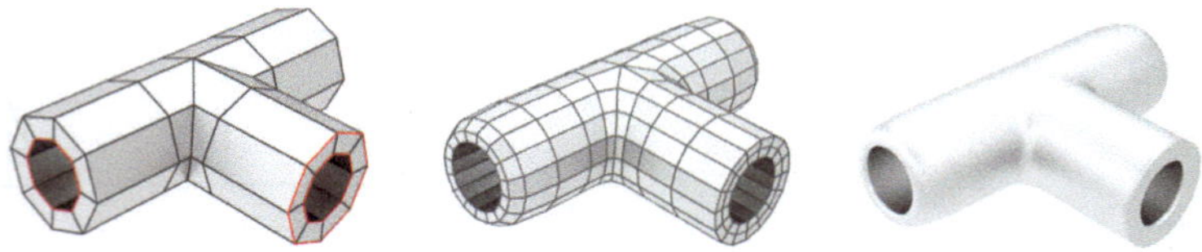

Fig. 6 Refinement of a given quadrilateral control mesh (shown *left*) with Catmull-Clark subdivision. In the vicinity of the edges in *red* modified subdivision stencils are applied in order preserve the sharp edge. The once subdivided mesh is shown in the *middle* and the limit surface on the *right*

shows the subdivision refinement of a control mesh for a T-junction geometry with extraordinary vertices and prescribed crease edges.

In case of triangular control meshes the subdivision scheme introduced by Loop [25] can be used. On three-directional triangular meshes (with each vertex having valence six) the Loop scheme yields quartic box-splines. In this context a vertex is regular when it is inside the domain and incident to six edges, or is on the boundary of the domain and incident to four edges. In the refinement step of the Loop scheme, each triangle of the control mesh is subdivided into four triangles by introducing new vertices at the edge midpoints, as shown in Fig. 4b. Subsequently, the vertex coordinates of the refined mesh are computed with the subdivision stencils given in Fig. 7 [38]. Figure 8 shows a mechanical connector geometry containing extraordinary vertices and sharp features described with the extended subdivision surfaces.

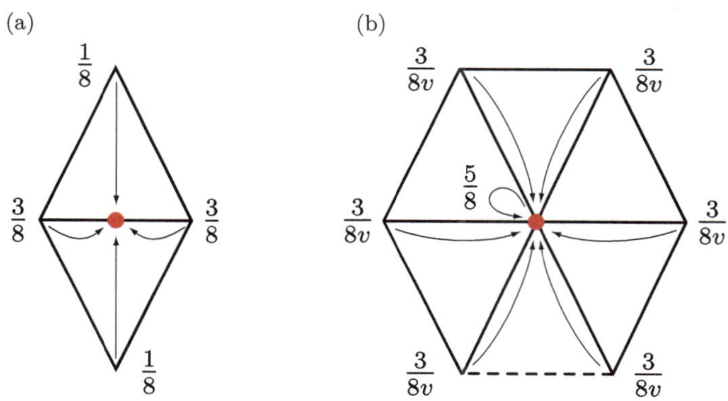

Fig. 7 Subdivision stencils for the Loop scheme. Each of the stencils are used for computing the coordinates of vertices of the type indicated by the *red dot*. (**a**) Edge stencil. (**b**) Face stencil

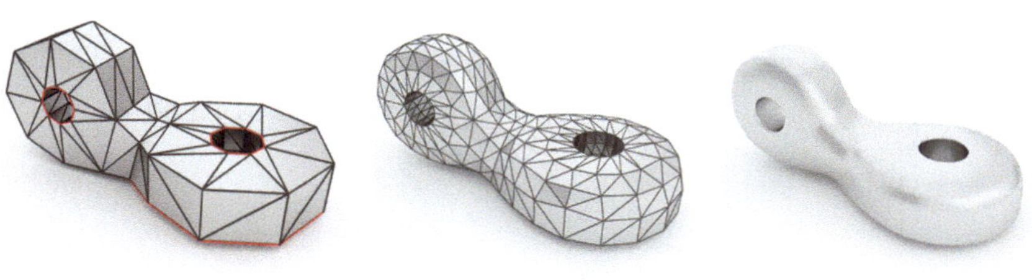

Fig. 8 Subdivision refinement of a given triangular control mesh (shown *left*) with Loop subdivision. In the vicinity of the edges in *red* modified subdivision stencils are applied in order preserve the sharp edge. The once subdivided mesh is shown in the *middle* and the limit surface on the *right*

2.3 Multiresolution Editing

Subdivision surfaces represent a limit surface with a nested hierarchy of control meshes of increasing resolution. As known in computer graphics, this property lends itself to efficient multiresolution editing of surfaces [26, 41]. The basic idea in multiresolution editing is to modify the coarse mesh vertex coordinates to apply large-scale changes (to the limit surface) and to modify the fine mesh vertex coordinates to add localised changes. By way of example, this is illustrated in Fig. 9 for the connector geometry previously introduced in Fig. 8. First the control mesh vertex coordinates $\boldsymbol{x}^0$ at level $\ell = 0$ are modified with $\boldsymbol{x}^0 + \boldsymbol{d}^0$, where $\boldsymbol{d}^0$ can be thought of as a user given perturbation vector. In the considered example, $\boldsymbol{d}^0$ applies displacements only to the vertices placed on one of the hole edges. Subsequent computation of the limit surface (by repeated subdivision) leads to a geometry with rather large scale changes. Alternatively, the edge of the hole can be displaced on level $\ell = 1$, i.e., $\boldsymbol{x}^1 + \boldsymbol{d}^1 = \boldsymbol{S}\boldsymbol{x}^0 + \boldsymbol{d}^1$. This results in a more localised change on the limit surface. It can be shown that the area of influence for each vertex extends over two rings of adjacent triangles.

The multiresolution editing algorithms available in computer graphics allow us to simultaneously edit coarse and fine resolutions, see, e.g., [3, 41]. This is achieved by a wavelet-like decomposition of the geometry into a low resolution part and a collection of wavelet coefficients expressing perturbations from the low resolution part [26, 41]. To compute such a decomposition it is necessary to define, in addition to the subdivision refinement, a coarsening operation

$$\boldsymbol{x}^{\ell} = \boldsymbol{R}\boldsymbol{x}^{\ell+1} . \tag{8}$$

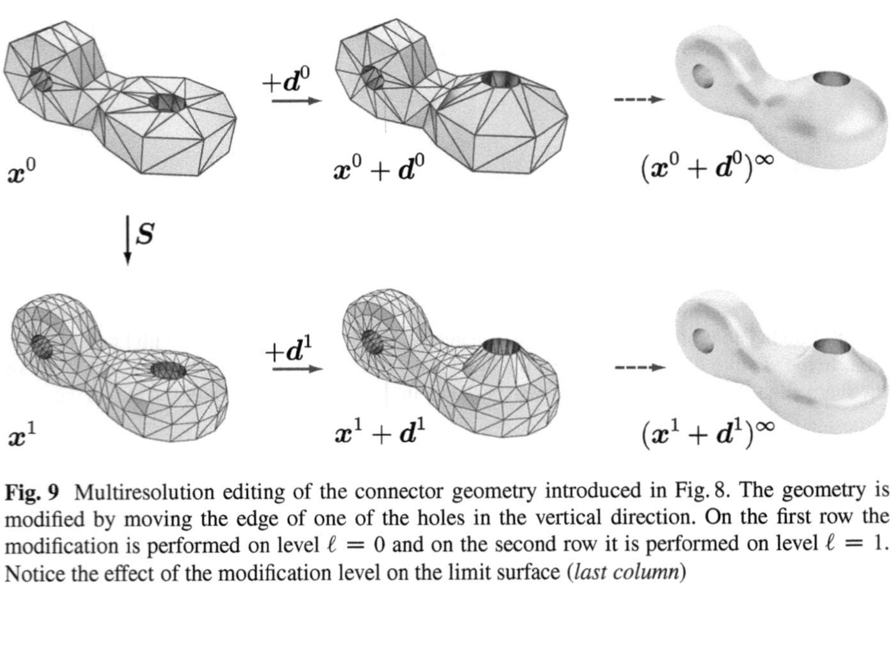

Fig. 9 Multiresolution editing of the connector geometry introduced in Fig. 8. The geometry is modified by moving the edge of one of the holes in the vertical direction. On the first row the modification is performed on level $\ell = 0$ and on the second row it is performed on level $\ell = 1$. Notice the effect of the modification level on the limit surface (*last column*)

The coarsening matrix $\boldsymbol{R}$ enables the computation of the coarse control mesh $\boldsymbol{x}^\ell$ corresponding to a given edited fine control mesh $\boldsymbol{x}^{\ell+1}$. In contrast, recall that the subdivision matrix $\boldsymbol{S}$ enables the computation of a refined mesh from a given coarse mesh, cf. (6). Different choices for the matrix $\boldsymbol{R}$ are possible. For instance, it can be determined with discrete least squares fitting

$$\boldsymbol{x}^\ell = \underset{\boldsymbol{y}^\ell}{\operatorname{argmin}} \|\boldsymbol{x}^{\ell+1} - \boldsymbol{S}\boldsymbol{y}^\ell\|^2 , \tag{9}$$

which leads to

$$\boldsymbol{S}^\mathsf{T}\boldsymbol{S}\boldsymbol{x}^\ell = \boldsymbol{S}^\mathsf{T}\boldsymbol{x}^{\ell+1} . \tag{10}$$

By comparison with (8) we observe that the coarsening matrix has to be $\boldsymbol{R} = (\boldsymbol{S}^\mathsf{T}\boldsymbol{S})^{-1}\boldsymbol{S}^\mathsf{T}$. For more details we refer to [3, 4].

Instead of using least squares fitting the coarsening matrix $\boldsymbol{R}$ can also be defined based on quasi-interpolation [23] or smoothing [41]. On the other hand coarsening by simply subsampling of the fine control mesh usually leads to artefacts in form of oscillations in the coarse control mesh. The proposed least squares fit approach is not very common in computer graphics because of the need for interactivity and fast processing times. Although the least squares matrix in (10) is sparse its solution cannot be found at interactive rates.

Subdivision surfaces do not by themselves provide the possibility of simultaneously editing coarse and fine control meshes. For instance, after a fine control mesh is edited it is not possible to further edit a coarser level in order to apply larger scale changes to the geometry. Simultaneous editing of different levels can be achieved with a wavelet-like multiresolution decomposition of the control meshes, as discussed in [3, 41].

3 Governing Equations of Shape Optimisation

In this section, we revisit the governing equations for shape optimisation of linear elastic solids, electrostatic field equations and thin-shells. The considered cost functions are the structural compliance for elastic solids and thin-shells and the normal flux over the domain boundaries for electrostatic problems. For computing the derivatives of the cost function with respect to the domain boundary perturbations we consider both the continuous and the discrete adjoint formulations. Specifically, for elastic and electrostatic field equations we use the continuous formulation and for thin-shells we use the discrete formulation. In the continuous formulation the differentiation is first performed analytically and subsequently the resulting boundary value problems are discretised with a method of choice. In contrast, in the discrete formulation first the governing equations are discretised and subsequently the resulting algebraic equations differentiated.

3.1 Continuous Shape Sensitivity Formulation

In continuous shape sensitivity analysis it is necessary to differentiate functionals that are defined on varying domains. To this end, borrowing terminology from continuum mechanics, we consider a reference configuration with domain Ω and boundary Γ and a perturbed configuration with domain Ω_t and boundary Γ_t, see Fig. 10. The corresponding deformation $\boldsymbol{\varphi}_t(\boldsymbol{x})$ which maps material points $\boldsymbol{x} \in \Omega$ onto $\boldsymbol{x}_t \in \Omega_t$ is assumed to be of the form

$$\boldsymbol{x}_t = \boldsymbol{\varphi}_t(\boldsymbol{x}) = \boldsymbol{x} + t\boldsymbol{V}\,, \tag{11}$$

where $\boldsymbol{V}$ is a prescribed constant vector field and t is a scalar parameter. It is convenient to interpret $\boldsymbol{V}$ as a velocity field and t as the time parameter.

In the sequel the derivatives of domain and boundary integrals with respect to boundary perturbations are needed. First, we consider the generic domain integral

$$I_1(\Omega_t) = \int_{\Omega_t} \psi(\boldsymbol{x}_t)\, \mathrm{d}\,\Omega_t \tag{12}$$

with a scalar integrand $\psi(\boldsymbol{x}_t)$. At the reference configuration $\boldsymbol{x}$, the derivative of the domain integral in the direction of $\boldsymbol{V}$ is defined with

$$\frac{\partial I_1}{\partial \Omega}\boldsymbol{V} = \frac{d}{dt} I_1(\boldsymbol{x} + t\boldsymbol{V})\Big|_{t=0} = \frac{d}{dt}\int_{\Omega_t} \psi(\boldsymbol{x}_t)\, \mathrm{d}\,\Omega_t\Big|_{t=0}. \tag{13}$$

By applying standard techniques from continuum mechanics and shape calculus, see [3, 13, 33], we obtain

$$\frac{\partial I_1}{\partial \Omega}\boldsymbol{V} = \int_{\Omega} (\nabla\psi(\boldsymbol{x}) \cdot \boldsymbol{V} + \psi(\boldsymbol{x})\nabla \cdot \boldsymbol{V})\, \mathrm{d}\,\Omega = \int_{\Gamma} \psi(\boldsymbol{x})(\boldsymbol{V}\cdot\boldsymbol{n})\, \mathrm{d}\,\Gamma\,, \tag{14}$$

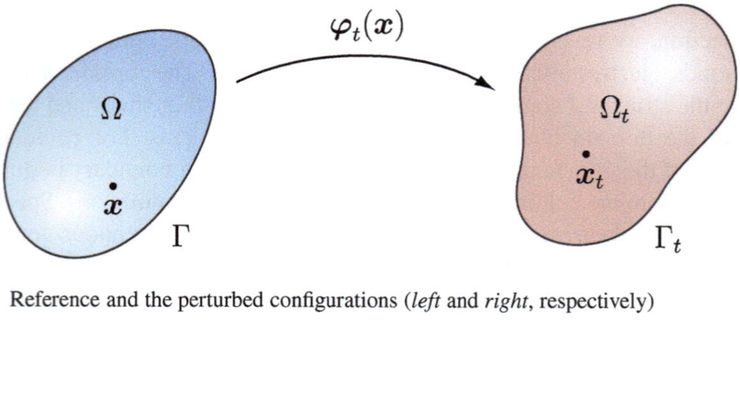

Fig. 10 Reference and the perturbed configurations (*left* and *right*, respectively)

where $\boldsymbol{n}$ is the unit normal to the boundary. As to be expected, this integral is zero when the perturbation direction $\boldsymbol{V}$ is orthogonal to the boundary normal $\boldsymbol{n}$. Perturbations tangential to the boundary do not lead to a change in I_1.

Next, we consider the generic boundary integral

$$I_2(\Gamma_t) = \int_{\Gamma_t} \psi(\boldsymbol{x}_t) \, \mathrm{d}\Gamma_t \tag{15}$$

with a scalar integrand $\psi(\boldsymbol{x}_t)$. The differentiation of this integral at the reference configuration $\boldsymbol{x}$ in the direction of $\boldsymbol{V}$ is more involved and can be found in standard texts on shape calculus, see [13, 33],

$$\frac{\partial I_2}{\partial \Gamma}\boldsymbol{V} = \int_{\Gamma} (\nabla\psi(\boldsymbol{x}) \cdot \boldsymbol{n} + H(\boldsymbol{x})\psi(\boldsymbol{x}))\,(\boldsymbol{V} \cdot \boldsymbol{n}) \, \mathrm{d}\Gamma \,, \tag{16}$$

where $H(\boldsymbol{x})$ is the mean curvature on Γ.

3.1.1 Compliance Optimisation in Elasticity

As the first specific optimisation problem we consider linear elasticity. The equilibrium equation for a solid body with the domain Ω is given by

$$\nabla \cdot \boldsymbol{\sigma}(\boldsymbol{u}) + \boldsymbol{f} = \boldsymbol{0} \qquad \text{in } \Omega \,, \tag{17a}$$

$$\boldsymbol{u} = \boldsymbol{0} \qquad \text{on } \Gamma_D \,, \tag{17b}$$

$$\boldsymbol{\sigma}(\boldsymbol{u})\boldsymbol{n} = \bar{\boldsymbol{t}} \qquad \text{on } \Gamma_N \,, \tag{17c}$$

where $\boldsymbol{\sigma}$ is the stress tensor, $\boldsymbol{u}$ is the displacement vector, $\boldsymbol{f}$ is the external load vector and $\bar{\boldsymbol{t}}$ is the prescribed traction on the Neumann boundary Γ_N with the outward normal $\boldsymbol{n}$. On Γ_D, for simplicity, only homogenous Dirichlet boundary conditions are assumed.

We assume a homogenous linear elastic material model

$$\boldsymbol{\sigma}(\boldsymbol{u}) = \boldsymbol{C} : \boldsymbol{\epsilon}(\boldsymbol{u}) \tag{18}$$

with the linear elastic strain tensor

$$\boldsymbol{\epsilon}(\boldsymbol{u}) = \frac{1}{2}(\nabla \boldsymbol{u} + \nabla^{\mathsf{T}} \boldsymbol{u}) \,. \tag{19}$$

The aim of shape optimisation is to find a shape Ω that minimises a cost functional $J(\Omega, \boldsymbol{u})$ such that

$$\min_{\Omega} J(\Omega, \boldsymbol{u}) \tag{20}$$

where the displacement vector $\boldsymbol{u}$ has to satisfy (17). In practice there are usually additional constraints, e.g., pertaining to the volume of the domain Ω, which are neglected here for brevity. In case of structural compliance, the cost functional $J(\Omega, \boldsymbol{u})$ reads

$$J(\Omega, \boldsymbol{u}) = \int_{\Omega} \boldsymbol{\sigma}(\boldsymbol{u}) : \boldsymbol{\epsilon}(\boldsymbol{u}) \, \mathrm{d}\Omega = \int_{\Omega} \boldsymbol{f} \cdot \boldsymbol{u} \, \mathrm{d}\Omega + \int_{\Gamma_N} \bar{\boldsymbol{t}} \cdot \boldsymbol{u} \, \mathrm{d}\Gamma \, . \tag{21}$$

In adjoint shape sensitivity analysis the minimisation problem (20) with the equilibrium equation (17) as a constraint is expressed with the functional

$$\begin{aligned} L(\Omega, \boldsymbol{u}, \boldsymbol{\lambda}) = J(\Omega, \boldsymbol{u}) &+ \int_{\Omega} \nabla \boldsymbol{\lambda} : \boldsymbol{\sigma}(\boldsymbol{u}) \, \mathrm{d}\Omega - \int_{\Omega} \boldsymbol{\lambda} \cdot \boldsymbol{f} \, \mathrm{d}\Omega \\ &- \int_{\Gamma_D} \boldsymbol{u} \cdot (\boldsymbol{C} : \nabla \boldsymbol{\lambda})\boldsymbol{n} + \boldsymbol{\lambda} \cdot \boldsymbol{\sigma}(\boldsymbol{u})\boldsymbol{n} \, \mathrm{d}\Gamma - \int_{\Gamma_N} \boldsymbol{\lambda} \cdot \bar{\boldsymbol{t}} \, \mathrm{d}\Gamma \, , \end{aligned} \tag{22}$$

where $\boldsymbol{\lambda}$ is a vector-valued Lagrange parameter field. The stationarity condition for the functional $L(\Omega, \boldsymbol{u}, \boldsymbol{\lambda})$, i.e.,

$$\delta L(\Omega, \boldsymbol{u}, \boldsymbol{\lambda}) = 0 \tag{23}$$

provides the complete set of equations for shape optimisation. The variation of $L(\Omega, \boldsymbol{u}, \boldsymbol{\lambda})$ with respect to the Lagrange parameter $\boldsymbol{\lambda}$ yields the original boundary value problem (17). The variation of $L(\Omega, \boldsymbol{u}, \boldsymbol{\lambda})$ with respect to the displacement vector $\boldsymbol{u}$ yields the adjoint boundary value problem

$$\nabla \cdot \boldsymbol{\sigma}(\boldsymbol{\lambda}) - \boldsymbol{f} = \boldsymbol{0} \qquad \text{in } \Omega \, , \tag{24a}$$

$$\boldsymbol{\lambda} = \boldsymbol{0} \qquad \text{on } \Gamma_D \, , \tag{24b}$$

$$\boldsymbol{\sigma}(\boldsymbol{\lambda})\boldsymbol{n} = -\bar{\boldsymbol{t}} \qquad \text{on } \Gamma_N \, . \tag{24c}$$

By comparing the adjoint problem with the original boundary value problem (17) we deduce that $\boldsymbol{\lambda} = -\boldsymbol{u}$. This holds only when the structural compliance (21) is the cost functional.

Next, we consider the variation of $L(\Omega, \boldsymbol{u}, \boldsymbol{\lambda})$ with respect to the domain Ω. In deriving it we make use of the domain deformation $\boldsymbol{x}_t = \boldsymbol{\varphi}_t(\boldsymbol{x}) = \boldsymbol{x} + t\boldsymbol{V}$ introduced in (11). In shape optimisation of linear elasticity problems usually only the parts of the Neumann boundary with zero tractions are allowed to move, i.e.,

$$\begin{aligned} \boldsymbol{V} &= \boldsymbol{0} \quad \text{on } \Gamma_D \, , \\ \boldsymbol{V} &= \boldsymbol{0} \quad \text{on } \Gamma_N \text{ with } \boldsymbol{\sigma n} = \bar{\boldsymbol{t}} \, , \\ \boldsymbol{V} &\neq \boldsymbol{0} \quad \text{on } \Gamma_N \text{ with } \boldsymbol{\sigma n} = \boldsymbol{0} \, . \end{aligned} \tag{25}$$

The variation of the Lagrangian (22) with respect to the domain perturbations yields the shape gradient. For structural compliance (21) as the cost functional (i.e., $\boldsymbol{\lambda} = -\boldsymbol{u}$) and with the result (14) at hand we obtain for the shape gradient

$$\frac{\partial L}{\partial \Omega}\boldsymbol{V} = \int_{\Gamma_N} (2\boldsymbol{u}\cdot\boldsymbol{f} - \nabla\boldsymbol{u} : \boldsymbol{\sigma}(\boldsymbol{u}))\,(\boldsymbol{V}\cdot\boldsymbol{n})\,\mathrm{d}\Gamma = \int_{\Gamma_N} g(\boldsymbol{u})(\boldsymbol{V}\cdot\boldsymbol{n})\,\mathrm{d}\Gamma\,, \tag{26}$$

where $g(\boldsymbol{u})$ is the shape kernel function. It is worth emphasising that without restricting $\boldsymbol{V}$ as stated in (25) the shape derivate would contain several more terms.

In iterative shape optimisation the shape kernel function $g(\boldsymbol{u})$ is used as gradient information. In order to achieve a decrease in the cost functional the boundary perturbation is chosen proportional to

$$\boldsymbol{V} = -g(\boldsymbol{u})\boldsymbol{n} \tag{27}$$

such that

$$\frac{\partial L}{\partial \Omega}\boldsymbol{V} = -\int_{\Gamma_N} g(\boldsymbol{u})^2\,\mathrm{d}\Gamma\,. \tag{28}$$

During the shape optimisation the boundary value problem (17) has to be repeatedly solved on a continuously changing domain. In a conventional finite element setting this would require frequent mesh smoothing or updating. Therefore, immersed, or embedded, finite element approaches that do not require remeshing have clear advantages in shape optimisation. In the present work, we use an immersed finite element technique previously developed in the context of incompressible fluid-structure interaction [28–30]. For more details see [3].

3.1.2 Boundary Flux Optimisation in Electrostatics

The motivation for our second optimisation problem comes from the design of high-voltage electrical devices. Avoidance of electrical breakdown is one of the key design considerations for such devices. In a first approximation, electrical breakdown can be avoided by reducing the electrical field strength on critical components. The aim of shape optimisation is to modify the geometry of components so that the maximum electric field strength on surfaces, i.e., the normal flux, is reduced.

The electrostatic field equation in absence of space charges is given by a Dirichlet boundary value problem for the Laplace equation

$$\begin{aligned} -\Delta u &= 0 \quad \text{in } \Omega\,, \\ u &= 0 \quad \text{on } \Gamma_{D_0}\,, \\ u &= \overline{u} \quad \text{on } \Gamma_{D_f}\,, \end{aligned} \tag{29}$$

where u is the scalar electric potential or voltage, Ω is a multiply connected domain with a Dirichlet boundary consisting of a (to be optimised) free part Γ_{D_f} and a fixed part Γ_{D_0}. It is assumed that the potentials on Γ_{D_f} and Γ_{D_0} are constant. The geometry of the free Dirichlet boundary Γ_{D_f} with the prescribed potential $\overline{u}$ is to be determined with shape optimisation. Notice that the electric potential u is denoted non-bold because it is a scalar field quantity.

In order to pointwise minimise the maximum of the normal flux on the free part of the boundary Γ_{D_f} we chose to minimise the cost functional

$$J(\Omega, u) = \frac{1}{2} \int_{\Gamma_{D_f}} \left(\frac{\partial u}{\partial \boldsymbol{n}} - Q \right)^2 \mathrm{d}\Gamma , \tag{30}$$

where $Q \geq 0$ is a user prescribed constant expected value.

The adjoint sensitivity analysis, as discussed in Sect. 3.1.1 for elasticity, is used for obtaining the adjoint boundary value problem and the shape gradient. The adjoint boundary value problem corresponding to the cost functional (30) reads

$$\begin{aligned} -\Delta \lambda &= 0 && \text{in } \Omega, \\ \lambda &= 0 && \text{on } \Gamma_{D_0}, \\ \lambda &= \frac{\partial u}{\partial \boldsymbol{n}} - Q && \text{on } \Gamma_{D_f} . \end{aligned} \tag{31}$$

Importantly, the adjoint solution λ is different from the solution u of the original problem so that in applications both problems (29) and (31) have to be solved.

Moreover, the shape gradient corresponding to the cost functional (30) reads

$$\frac{\partial L}{\partial \Omega} \boldsymbol{V} = \int_{\Gamma_{D_f}} \left(-\frac{\partial \lambda}{\partial \boldsymbol{n}} \frac{\partial u}{\partial \boldsymbol{n}} - \frac{H}{2} \right) (\boldsymbol{V} \cdot \boldsymbol{n}) \, \mathrm{d}\Gamma = \int_{\Gamma_{D_f}} g(\lambda, u)(\boldsymbol{V} \cdot \boldsymbol{n}) \, \mathrm{d}\Gamma , \tag{32}$$

where the shape kernel function $g(\lambda, u)$ now depends on the electric potential u and adjoint solution λ. In deriving the shape gradient we made use of (16) and H is the mean curvature of the boundary.

As discussed for elasticity, during the iterative shape optimisation the shape kernel function $g(\boldsymbol{u})$ is used as gradient information. In order to achieve a decrease in the cost functional the boundary perturbation is chosen proportional to $\boldsymbol{V} = -g(\lambda, \boldsymbol{u})\boldsymbol{n}$.

For solving the original and the adjoint boundary value problems (29) and (31), respectively, we use the boundary element method, see e.g. [36], which does not require a domain discretisation and is ideal for solving problems with unbounded domains that occur in electrostatic field analysis. For more details see [3].

3.2 Discrete Shape Sensitivity Formulation

3.2.1 Compliance Optimisation of Thin-Shells

We use the Kirchhoff-Love energy functional for modelling the mechanical response of thin shells. Since the related equations are classically formulated in curvilinear coordinates, it is usually easier to compute the shape sensitivities using a discrete approach. To this end, we first discretise the governing equations with finite elements and subsequently differentiate the cost function and the discrete equilibrium equations with respect to control vertex coordinates.

The Kirchhoff-Love energy functional depends on mid-surface displacements and their first and second derivatives. Therefore, the basis functions have to be smooth and their second order derivatives square-integrable. As originally introduced by Cirak et al. [10, 11] Kirchhoff-Love shells can be elegantly discretised with subdivision basis functions. In the resulting discrete equilibrium equations the control vertex displacements are the only degrees of freedom.

To begin with, we consider a shell with the undeformed mid-surface Ω, the position vector $\boldsymbol{x}(\xi_1, \xi_2)$ and the displacement vector $\boldsymbol{u}(\xi_1, \xi_2)$. It is assumed that the mid-surface is parameterised in terms of the curvilinear coordinates (ξ_1, ξ_2). The two corresponding (covariant) surface basis vectors follow from

$$\boldsymbol{a}_\alpha = \boldsymbol{x}_{,\alpha} \,. \tag{33}$$

From here onwards Greek indices take the values 1 and 2 and a comma denotes partial differentiation with respect to curvilinear coordinates.

According to [11] the linearised membrane and bending strains $\alpha_{\alpha\beta}$ and $\beta_{\alpha\beta}$, respectively, are given by

$$\alpha_{\alpha\beta} = \frac{1}{2}(\boldsymbol{a}_\alpha \cdot \boldsymbol{u}_{,\beta} + \boldsymbol{u}_{,\alpha} \cdot \boldsymbol{a}_\beta) \,, \tag{34}$$

$$\begin{aligned} \beta_{\alpha\beta} = &-\boldsymbol{u}_{,\alpha\beta} \cdot \boldsymbol{a}_3 + \frac{1}{\sqrt{a}} \left[\boldsymbol{u}_{,1} \cdot (\boldsymbol{a}_{\alpha,\beta} \times \boldsymbol{a}_2) + \boldsymbol{u}_{,2} \cdot (\boldsymbol{a}_1 \times \boldsymbol{a}_{\alpha,\beta})\right] \\ &+ \frac{\boldsymbol{a}_3 \cdot \boldsymbol{a}_{\alpha,\beta}}{\sqrt{a}} \left[\boldsymbol{u}_{,1} \cdot (\boldsymbol{a}_2 \times \boldsymbol{a}_3) + \boldsymbol{u}_{,2} \cdot (\boldsymbol{a}_3 \times \boldsymbol{a}_1)\right] . \end{aligned} \tag{35}$$

With the above strain expressions at hand, the potential energy of a linear-elastic Kirchhoff-Love shell reads

$$\Pi(\boldsymbol{u}) = \int_\Omega \left(W^m(\alpha_{\alpha\beta}) + W^b(\beta_{\alpha\beta})\right) \mu \,\mathrm{d}\Omega + \Pi^{ext}(\boldsymbol{u}) \,, \tag{36}$$

where Ω is the shell mid-surface, W^m is the membrane energy density, W^b is the bending energy density, μ is the jacobian associated with the integration over the thickness and $\Pi^{ext}(\boldsymbol{u})$ is the potential of the external forces. For a linear elastic material the two energy densities are defined with

$$W^m(\alpha_{\alpha\beta}) = \frac{1}{2}\frac{Et}{1-\nu^2}H^{\alpha\beta\gamma\delta}\alpha_{\alpha\beta}\alpha_{\gamma\delta}\,, \tag{37a}$$

$$W^b(\beta_{\alpha\beta}) = \frac{1}{2}\frac{Et^3}{12(1-\nu^2)}H^{\alpha\beta\gamma\delta}\beta_{\alpha\beta}\beta_{\gamma\delta}\,, \tag{37b}$$

where t is the shell thickness, E is the Young's modulus, ν is the Poisson's ratio and $H^{\alpha\beta\gamma\delta}$ is an auxiliary fourth order tensor

$$H^{\alpha\beta\gamma\delta} = \nu\, a^{\alpha\beta}a^{\gamma\delta} + \frac{1}{2}(1-\nu)\,(a^{\alpha\gamma}a^{\beta\delta} + a^{\alpha\delta}a^{\beta\gamma})\,. \tag{38}$$

The contravariant metric tensor components $a^{\alpha\beta} = \boldsymbol{a}^{\alpha}\cdot\boldsymbol{a}^{\beta}$ are computed from the contravariant basis vectors defined according to $\boldsymbol{a}^{\alpha}\cdot\boldsymbol{a}_{\beta} = \delta^{\alpha}_{\beta}$, where δ^{α}_{β} is the Kronecker delta.

Next, the mid-surface position and displacement vector are discretised with subdivision basis functions

$$\boldsymbol{x}_h(\xi_1,\xi_2) = \sum_i N_i(\xi_1,\xi_2)\boldsymbol{x}_i\,, \qquad \boldsymbol{u}_h(\xi_1,\xi_2) = \sum_i N_i(\xi_1,\xi_2)\boldsymbol{u}_i\,, \tag{39}$$

where $N_i(\xi_1,\xi_2)$ is the basis function, $\boldsymbol{x}_i$ is the position vector and $\boldsymbol{u}_i$ is the displacement of a control vertex with the index i. Recall that on structured meshes the subdivision basis functions are equivalent to quartic box-splines in case of the Loop scheme and equivalent to tensor-product cubic b-splines in case of the Catmull-Clark scheme. As shown by Stam [34, 35] in the vicinity of extraordinary vertices the subdivision surfaces can be evaluated in the same way as regular patches after only few steps of refinement. For implementation details see [10, 11].

In order to obtain the discrete equilibrium equations, first the interpolation (39) is introduced into the energy functional (36) and subsequently the integrals are evaluated with Gauss quadrature. Computing the stationary points of the discretised energy functional yields the discrete equilibrium equations

$$\boldsymbol{K}\boldsymbol{u} = \boldsymbol{f}, \tag{40}$$

where $\boldsymbol{K}$ is the stiffness matrix, $\boldsymbol{f}$ is the array of vertex forces and, with a slight abuse of notation, $\boldsymbol{u}$ is the array of vertex displacements.

In discrete shape sensitivity formulation we aim to minimise a discretised cost function

$$\min_{\boldsymbol{x}} J(\boldsymbol{x},\boldsymbol{u}) \tag{41}$$

which depends on the array of vertex coordinates $\boldsymbol{x}$ and displacements $\boldsymbol{u}$. It is clear that the displacements have to satisfy (40). The compliance cost function corresponding to the discrete equilibrium equation reads

$$J(\boldsymbol{x},\boldsymbol{u}) = \boldsymbol{f}^{\mathsf{T}}\boldsymbol{u} = \boldsymbol{u}^{\mathsf{T}}\boldsymbol{K}\boldsymbol{u} \,. \tag{42}$$

In order to compute the derivatives of the cost function with respect to the vertex coordinates we consider the adjoint formulation

$$L(\boldsymbol{x},\boldsymbol{u},\boldsymbol{\lambda}) = J(\boldsymbol{x},\boldsymbol{u}) + \boldsymbol{\lambda}^{\mathsf{T}}[\boldsymbol{f} - \boldsymbol{K}\boldsymbol{u}], \tag{43}$$

where $\boldsymbol{\lambda}$ is a Lagrange parameter vector. The stationarity condition for $L(\boldsymbol{x},\boldsymbol{u},\boldsymbol{\lambda})$ with respect to the vertex displacements leads to the adjoint problem

$$\boldsymbol{K}\boldsymbol{\lambda} = \frac{\partial J(\boldsymbol{x},\boldsymbol{u})}{\partial \boldsymbol{u}} \,. \tag{44}$$

Here, we made use of the symmetry of the stiffness matrix $\boldsymbol{K}$. The stationarity condition for $L(\boldsymbol{x},\boldsymbol{u},\boldsymbol{\lambda})$ with respect to the vertex coordinates leads to the discrete shape gradients

$$\frac{\partial L(\boldsymbol{x},\boldsymbol{u},\boldsymbol{\lambda})}{\partial \boldsymbol{x}} = \frac{\partial J(\boldsymbol{x},\boldsymbol{u})}{\partial \boldsymbol{x}} + \boldsymbol{\lambda}^{\mathsf{T}}\left[\frac{\partial \boldsymbol{f}}{\partial \boldsymbol{x}} - \frac{\partial \boldsymbol{K}}{\partial \boldsymbol{x}}\boldsymbol{u}\right] . \tag{45}$$

By introducing the cost function (42) and the adjoint solution with $\boldsymbol{\lambda} = \boldsymbol{u}$ we obtain

$$\frac{\partial L(\boldsymbol{x},\boldsymbol{u},\boldsymbol{\lambda})}{\partial \boldsymbol{x}} = 2\boldsymbol{u}^{\mathsf{T}}\frac{\partial \boldsymbol{f}}{\partial \boldsymbol{x}} - \boldsymbol{u}^{\mathsf{T}}\frac{\partial \boldsymbol{K}}{\partial \boldsymbol{x}}\boldsymbol{u} \,. \tag{46}$$

The derivatives of the stiffness matrix with respect to the vertex coordinates can be determined by consecutively differentiating the discretised Kirchhoff-Love energy functional (36) first with respect to vertex displacements and then with respect to the vertex positions. In gradient-based optimisation, in order achieve a decrease in the cost function the perturbation of the control vertex positions is chosen proportional to

$$\boldsymbol{V} = -2\boldsymbol{u}^{\mathsf{T}}\frac{\partial \boldsymbol{f}}{\partial \boldsymbol{x}} + \boldsymbol{u}^{\mathsf{T}}\frac{\partial \boldsymbol{K}}{\partial \boldsymbol{x}}\boldsymbol{u} \,. \tag{47}$$

Note that, with a slight abuse of notation, the perturbation vectors for the continuous and discrete shape sensitivity formulations are denoted with the same symbol $\boldsymbol{V}$. For more details we refer to [2].

4 Multiresolution Shape Optimisation

The introduced subdivision multiresolution editing technique enables to use two different resolutions of the same geometry for optimisation and analysis. A simplified two-level version of the proposed optimisation algorithm is shown Fig. 11. The optimisation and analysis meshes correspond to different refinement levels in a multiresolution hierarchy. The optimisation level is $\ell_o = 0$ and the analysis level is $\ell_c = n$. In our actual implementation the optimisation level is successively increased until $\ell_o = \ell_c$ and the analysis level ℓ_c is fixed. Crucially, in the spirit of isogeometric analysis the control meshes for analysis and optimisation represent the same geometry.

In shape optimisation it is usually necessary to use a coarse control mesh for geometry updating and a relatively fine control mesh for analysis. As is known, unwanted geometry oscillations may appear when the analysis and geometry representations have similar resolutions [8]. These geometry oscillations are usually a numerical artefact or a result of the ill-posedness of the considered optimisation problem. Moreover, in practical applications it might be desirable to optimise only a very coarse representation out of aesthetic or manufacturability reasons.

A more detailed description of the employed multiresolution technique is given in Algorithm 1. In the presented computations, different from Algorithm 1 we use instead of the steepest descent update on line 9 the method of moving asymptotes (MMA) as implemented in the NLopt library [20]. In addition, in practical computations there are usually constraints, such as bounds on vertex positions or area/volume constraints, which are not mentioned in Algorithm 1. For more details we refer to [3, 4].

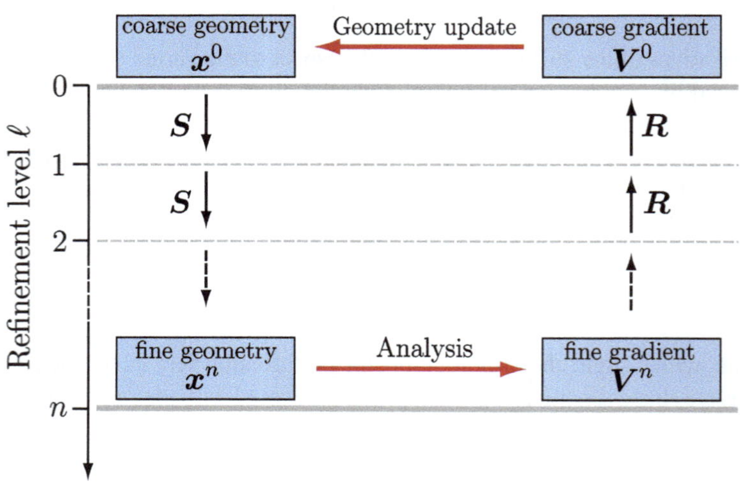

Fig. 11 Flowchart of the simplified multiresolution optimisation algorithm

Algorithm 1 Multiresolution shape optimisation

// choose computational level ℓ_c
// read coarse input control mesh $\boldsymbol{x}^0$
// Initialise optimisation level
$\ell_o = 0$
// Initialise cost functional
$J = \infty$
// iterate over optimisation levels
while $\ell_o \leq \ell_c$ **do**
  // update vertex coordinates $\boldsymbol{x}^{\ell_o}$ *while the cost functional decreases*
  repeat
    // subdivide optimisation level ℓ_o *up to analysis level* ℓ_c
    for $\ell \leftarrow \ell_o$ to ℓ_c **do**
      $\boldsymbol{x}^\ell \leftarrow \boldsymbol{S}\boldsymbol{x}^\ell$
    // compute cost functional $J = J(\boldsymbol{x}^{\ell_c}, \boldsymbol{u}(\boldsymbol{x}^{\ell_c}))$ *and the descent direction* $\boldsymbol{V}^{\ell_c}$
    // project the descent direction $\boldsymbol{V}^{\ell_c}$ *to the optimisation level*
    for $\ell \leftarrow \ell_c$ to ℓ_o **do**
      $\boldsymbol{V}^\ell \leftarrow \boldsymbol{R}\boldsymbol{V}^\ell$
    // update vertex coordinates of the optimisation level
    $\boldsymbol{x}^{\ell_o} \leftarrow (\boldsymbol{x}^{\ell_o} + \alpha \boldsymbol{V}^{\ell_o})$ with $\alpha > 0$
  until $J < J_{\text{previous}}$
  // increment optimisation level
  $\ell_o \leftarrow (\ell_o + 1)$

5 Examples

5.1 *Immersed Finite Element Discretised Elasticity Problems*

5.1.1 Simply Supported Plate with a Hole

This example highlights the advantages of multiresolution optimisation over classical approaches that use only one or two representation levels. The problem consists of a square plate with an edge length $L = 2$ and a circular hole with diameter $D = 1$, see Fig. 12. The plate is loaded with a line load of length 1. The Young's modulus and Poisson's ratio of the plate are $E = 100$ and $\nu = 0.4$, respectively. During optimisation the shape of the hole is to be modified so that the structural compliance of the plate is minimised.

Initially, at level $\ell = 0$ the hole is represented with a cubic spline with 8 control points. The immersed finite element grid has 100×100 cells of uniform size. Three cases referred to as C1, C2 and C3 with different geometry and analysis resolutions are studied:

- In C1 only one level with $\ell_o = \ell_c = 0$ is used for analysis and optimisation.
- In C2 a four times subdivided control mesh at refinement level $\ell_o = \ell_c = 4$ is used for analysis and optimisation.

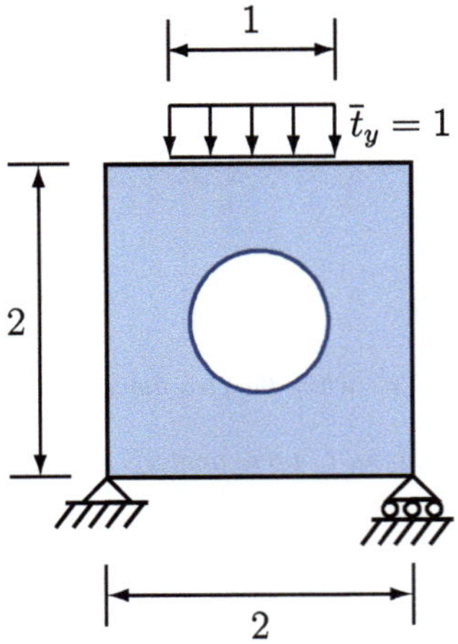

Fig. 12 Simply supported plate with a hole. Problem description

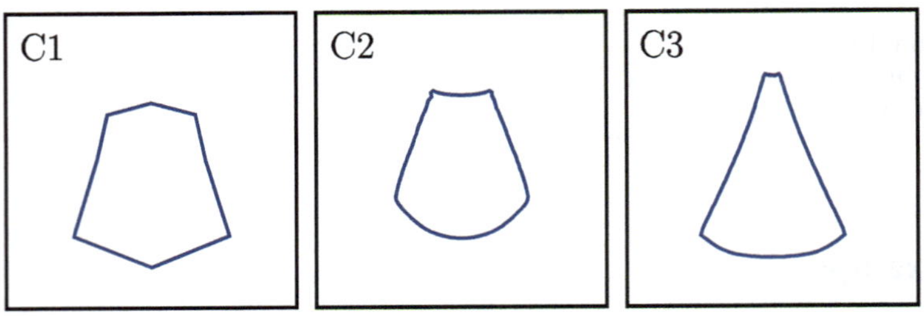

Fig. 13 Simply supported plate with a hole. Optimised hole shapes for cases C1, C2 and C3

- In C3 the optimisation level starts with $\ell_o = 0$ and increases until $\ell_o = \ell_c = 4$ is reached. Throughout the computations the analysis level is fixed to $\ell_c = 4$.

In case C1 the control mesh that is visible by the immersed finite element grid contains 8 elements and in cases C2 and C3 it contains 128 elements. It is clear that in case C1 the hole geometry is poorly resolved on the immersed finite element grid.

In Fig. 13 the optimised final hole shapes for the three cases are shown. In particular, the difference in optimal shapes for cases C2 and C3 which use the same analysis level $\ell_c = 4$ is striking. The case C1 is different from the other two cases because of the mentioned inadequately coarse analysis control mesh with $\ell_c = 0$. During optimisation the optimisation level ℓ_o is successively increased only for case C3. The optimisation level is always incremented when a minimum is reached, cf. Algorithm 1. For the three cases the reduction of the relative cost function over the number of iterations is shown in Fig. 14. The case C2 with fixed fine resolution achieves the smallest cost reduction while the case C3 with multiresolution achieves the largest cost reduction. The strong dependence of the optimisation results on geometry parameterisation is well known in structural optimisation and is often

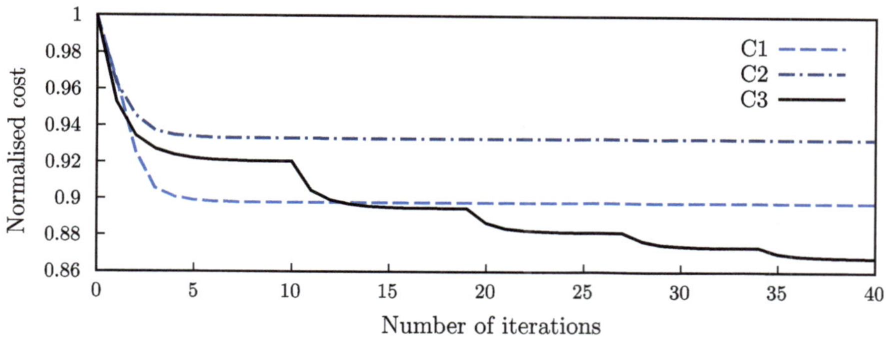

Fig. 14 Simply supported plate with a hole. Reduction of the normalised cost over the number of optimisation iterations. The initial cost for case C1 is 0.073 and for cases C2 and C3 is 0.065

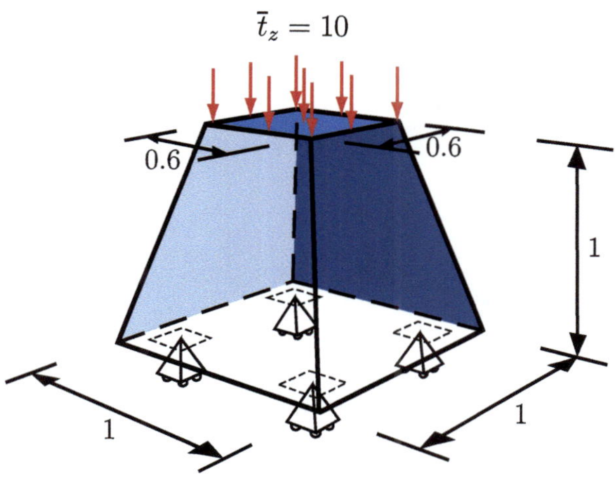

Fig. 15 Three-dimensional stool. Problem description. Roller supports are applied to all finite element nodes inside the regions of size 0.2 × 0.2 marked by *dashed squares*

associated with the non-convexity of the considered optimisation problem. We conjecture that by initially using a coarse control mesh for optimisation the possible number of local minima is significantly reduced which reduces the possibility of landing in a non-optimal local minimum. It appears that in case C2 the optimisation problem is caught in a local minimum which is significantly higher than the global minimum.

5.1.2 Three-Dimensional Stool

In this example we present the combined topology and shape optimisation of a three-dimensional solid, see Fig. 15. Our algorithm for combined topology and shape

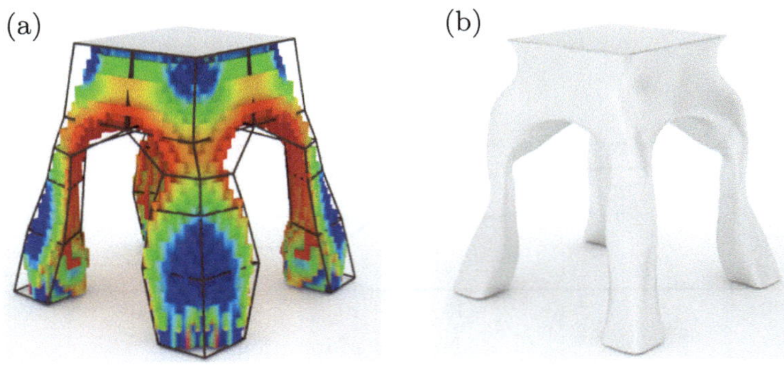

Fig. 16 Three-dimensional stool. First topology and subsequent shape optimisation step. (**a**) Topology optimised. (**b**) Shape optimised

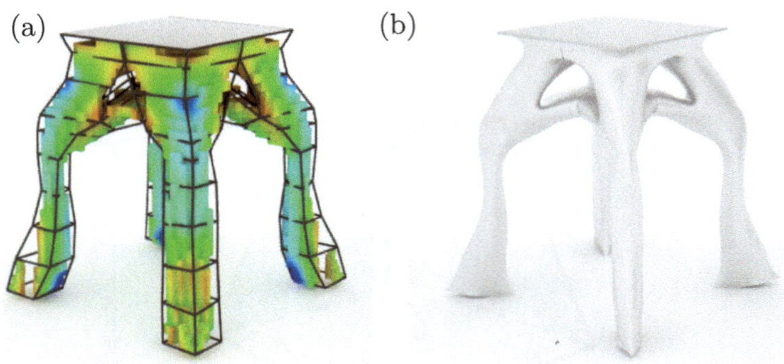

Fig. 17 Three-dimensional stool. Second topology and subsequent shape optimisation step. (**a**) Topology optimised. (**b**) Shape optimised

optimisation is described in [3]. The initial domain is a truncated pyramid and is at its top loaded with a uniform distributed load $\bar{t}_z = 10$. At its bottom it is supported by four distributed roller supports each of size 0.2×0.2. The Young's modulus and Poisson's ratio are chosen with $E = 100$ and $\nu = 0.4$, respectively. In the optimisation study only one quarter of the domain is considered and appropriate bounds and geometry tags are applied at the two planes of symmetry. The corresponding immersed finite element grid is of size $0.7 \times 0.7 \times 1$ and consists of $30 \times 30 \times 30$ cells.

The sequence of the performed topology and shape optimisation steps are shown in Figs. 16 and 17. In total two topology and two shape optimisation steps are performed. During each topology optimisation step we remove in one go a relatively large amount of material by deleting computational cells with the topology derivative below a given threshold. Figure 16a shows the result of the first topology optimisation step and the semi-manually generated coarse control mesh for representing the new topology. In the following shape optimisation step, see Fig. 16b, the generated control mesh serves as the optimisation level $\ell_o = 0$ and the computation level is chosen with $\ell_c = 2$. During the shape optimisation the volume of the domain is constraint to remain constant. The result of the second topology optimisation step and the generated control mesh are shown in Fig. 17a. This is followed by the final shape optimisation step shown Fig. 17b.

5.2 Boundary Element Discretised Electrostatic Problems

5.2.1 Box in a Sphere

We optimise the shape of a box placed inside a sphere, see Fig. 18, with the expected normal flux density Q in (30) set to 20. It can be shown that the optimal shape for the inner box is a sphere with half the diameter of the outer sphere [15]. The box, representing the to be optimised boundary Γ_{D_f}, is of size $0.16 \times 0.2 \times 0.24$ and the outer sphere, representing the fixed boundary Γ_{D_0}, has radius 0.2. The coarse mesh for the box contains 48 elements which increases to 768 elements in the twice subdivided fine mesh at level $\ell_c = 2$. During the subdivision refinement, the creases in the coarse mesh are maintained as creases using the extended subdivision stencils mentioned in Sect. 2.2, see also [5, 10]. With the extended subdivision stencils the limit surface corresponding to the coarse box mesh is a box of the same geometry. Note that on the limit surface the creases are only C^0-continuous and not at least C^1-continuous. The resolution of the outer sphere remains fixed with 3840 elements. Hence, the meshes for the boundary element analysis of the cube and sphere consist of 768 and 3840 elements, respectively.

Figure 18a shows the initial coarse geometry yielding a cost functional value of $J(\Omega^{\ell_c}, u^{\ell_c}) = 15.38$. First we select this coarse geometry as optimisation level, i.e., $\ell_o = 0$. After consecutively selecting $\ell_o = 1$ and $\ell_o = 2$ and optimising we obtain the final optimised geometry shown in Fig. 18b. The final shape of the initial box is nearly a sphere of diameter 0.215 and the cost functional value is $J(\Omega^{\ell_c}, u^{\ell_c}) = 8.46 \cdot 10^{-3}$, which represents a reduction of 99.95 %. As to be

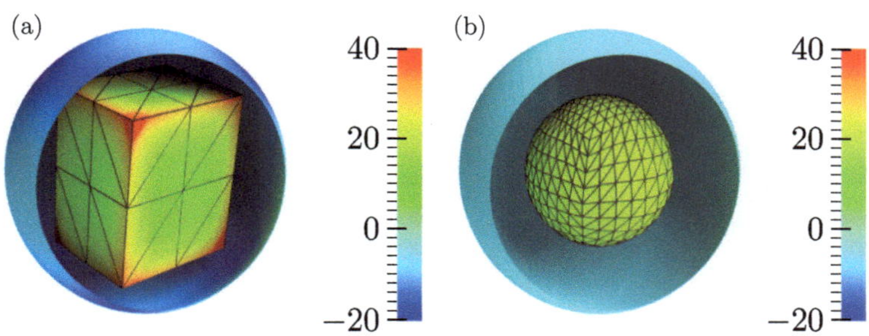

Fig. 18 Box in a sphere. Initial and optimised geometries with isocontours of the normal flux. The meshes indicate the optimisation level ℓ_o. The isocontours belong to the fine computational mesh at level $\ell_c = 2$. **(a)** Initial geometry at $\ell_o = 0$. **(b)** Final optimised geometry at $\ell_o = 2$

expected, the optimisation leads to a geometry with nearly uniform distribution of normal flux as seen in Fig. 18b. Since the extended subdivision scheme was used in this case, the marked creases were not smoothed out by the subdivision itself, but rather by the shape optimisation procedure.

5.2.2 Gas Insulated Switchgear

In this example we apply the proposed shape optimisation approach to the design of an electrode in a gas insulated switchgear, see Fig. 19a. Such devices are widely used as circuit-breakers in high-voltage power transmission. The objective of shape optimisation is to reduce the propensity for electrical breakdown with the ultimate aim of enabling more compact device geometries. This can be achieved by modifying the electrode geometries such that the cost functional $J(\Omega^{\ell_c}, u^{\ell_c})$ in (30) is minimised.

In Fig. 19a the gas insulated switchgear is shown with the electrode in the form of a primitive cylinder. The cylinder represents the electrode geometry to be optimised. The initial coarse mesh of the cylinder contains 264 elements. The creases on the cylinder are not tagged. Therefore, the geometry becomes smoother while it is refined by subdivision, see Fig. 19b. As a design constraint, the inner surface of the cylinder is required to have a constant radius for a bolt passing through it. Geometric constraints on the positions of vertices lying on the inner surface are applied to prevent any movement that would violate this design requirement.

In this example, a once subdivided electrode mesh with 1056 elements is chosen as the computational level ($\ell_c = 1$) and the geometry at level $\ell_o = 0$ is used for optimisation. In the initial design, Fig. 19b, the maximum normal flux is 81.63 before optimisation and reduces to 66.99 in the optimised shape shown in Fig. 20, corresponding to a reduction of 17.94 %. However, the reduction in the cost function $J(\Omega^{\ell_c}, u^{\ell_c})$ is much higher with 38.24 %. In Fig. 20b two electrode

(a) (b)

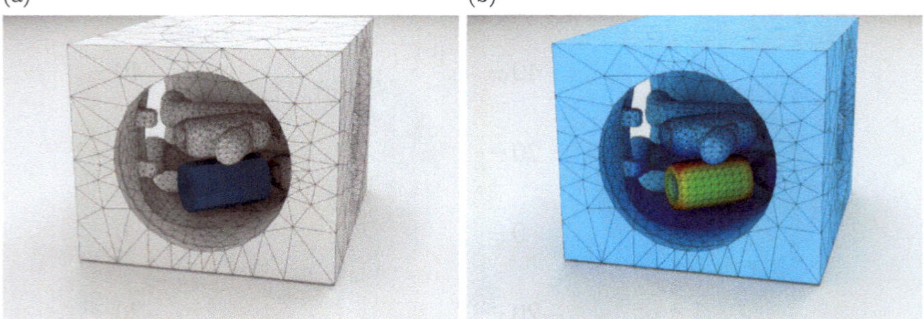

Device with coarse electrode mesh. Device with refined electrode mesh.

Fig. 19 Gas insulated switchgear with coarse and refined electrode meshes. The considered electrode is shown in *dark blue* in (**a**). The isocontours in (**b**) represent the normal flux

(a) (b)

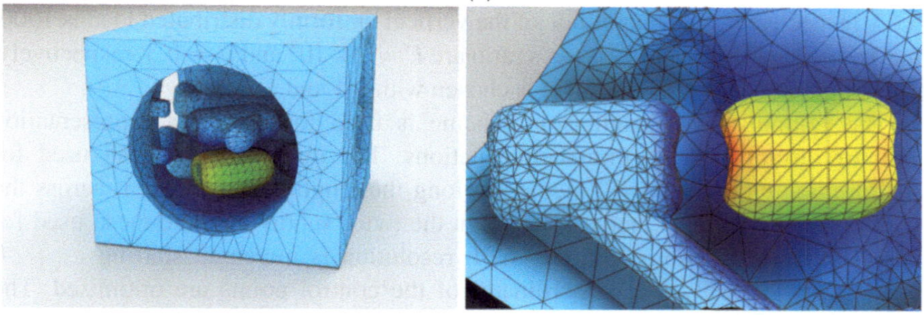

Fig. 20 Gas insulated switchgear with the optimised electrode geometry. The isocontours represent the normal flux. (**a**) Isometric view. (**b**) Top view

geometries are visible. The one in blue is an electrode geometry that has been obtained over the years by combining engineering intuition with simple calculations and testing. The similarities between the systematically shape optimised and the electrode geometry in production are striking. Notice in particular the saddle shapes at the two ends of the both electrodes, which help to reduce the large normal fluxes that are present at the sharp edges of the inner hole.

5.3 Subdivision Finite Element Discretised Thin-Shell Problems

5.3.1 Inverted Catenary Arch

In this verification example we optimise a thin strip pinned at both ends and subjected to a vertical distributed load. The optimal shape for such a strip is (in the limit of zero thickness) a catenary because for shells it is energetically more favourable to carry loads through membrane action instead of bending action. The catenary is the shape assumed by a loose string hung freely from two fixed points [24], its equation reads

$$y = \alpha \cosh\left(\frac{x+\beta}{\alpha}\right) + \gamma\,, \tag{48}$$

where the y-axis is parallel to the applied load vector and the three constants α, β and γ are determined based on the location of the supports and the string length l. Two cases are studied, one where the supports are at equal height and the other where one support is higher than the other. In both cases the horizontal distance between the supports is 1. The vertical offset between the supports in the second case is 0.2. Initially, the strip is a narrow flat plate connecting the supports with length equal to the distance between the supports. Width and thickness are constant with 0.05 and 0.02, respectively. The magnitude of the vertical uniformly distributed load is 1000, the Young's modulus and Poisson's ratio are $E = 2 \times 10^8$ and $\nu = 0.3$, respectively. The length l of the optimised strip is chosen with 1.1 and 1.3.

The Catmull-Clark subdivision scheme is used for geometry representation and for discretising the thin-shell equations. The initial coarse mesh used for optimisation contains only 3 elements along the length and 1 element across the width of the strip. This increases to 48 in the twice subdivided fine mesh used for analysis. During optimisation the mesh resolution is increased starting $\ell_o = 0$ up to $\ell_o = 2$. Only the y coordinates of the control points are optimised. The comparison of the optimisation results with the corresponding catenary curve for different curve lengths and support positions is shown in Fig. 21. The reduction of the objective function is more than 99.9 % for all cases and the results show good visual agreement with catenary curves. The slight deviations from the catenary are possibly due to the finite width of the strip, which leads during optimisation to some curvature generation across the width of the shell (not visible in Fig. 21).

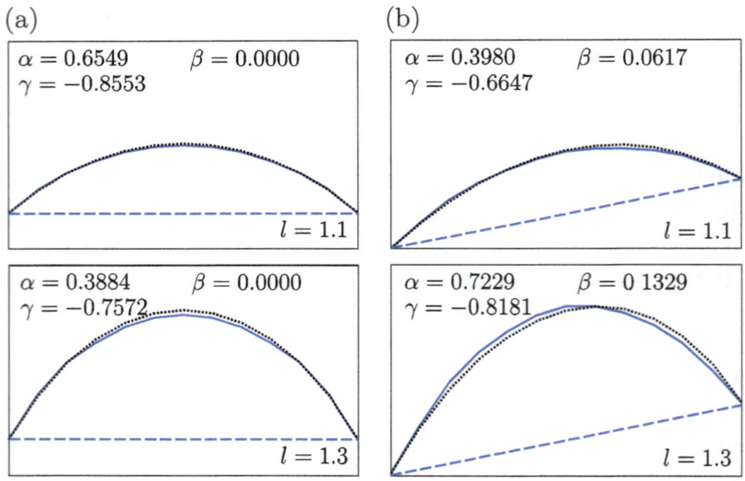

Fig. 21 Inverted catenary arch. The *dashed* and *solid blue lines* show the centre lines of the strip before and after optimisation, respectively. The *dotted black line* is the catenary curve corresponding to length l. The α, β and γ values are the parameters in (48). In the chosen coordinate system the left support has the coordinates $(-0.5, 0.0)$. (**a**) Supports at same height. (**b**) Supports at different heights

5.3.2 Architectural Roof

We consider the optimisation of an initial roof design shown in Fig. 22 with approximate dimensions of 2.31 × 6.27 × 0.75. This design was created by an architect using the Autodesk Maya software. A vertical uniformly distributed load of 1000 is chosen as the design load and the shell thickness is $t = 0.02$. The Young's modulus and the Poisson's ratio are $E = 1 \times 10^{10}$ and $\nu = 0.2$, respectively.

During the shape optimisation only the vertical coordinates of control points are chosen as design variables. Moreover, an area constraint is applied in order to restrict the size of the optimised shape. An important architectural feature of the roof is the ridge profile, which is preserved by applying special extended subdivision stencils at the corner vertices and crease edges, see Figs. 22 and 23.

The coarse mesh at optimisation level $\ell_o = 0$ contains 26 nodes which is twice subdivided to obtain the computation mesh at level $\ell_c = 2$ with 272 nodes. The second optimisation stage is done at level $\ell_o = 1$ but no optimisation is done at level $\ell_o = 2$ as this results in oscillatory surface features. The optimised roof design is shown in Fig. 24. The initial value of the compliance cost function is 31.36 and is reduced by 79.13 % through optimisation.

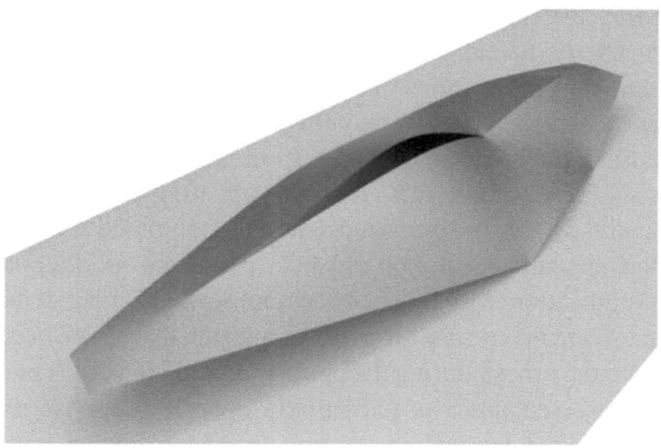

Fig. 22 Limit surface of the initial architectural roof design. The model has three supports, a central opening and a roof ridge. The corresponding coarse control mesh and selected vertex and edge tags are shown in Fig. 23

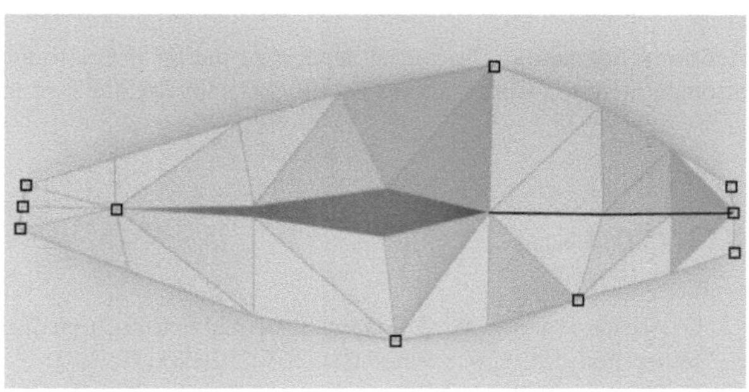

Fig. 23 Top view of the initial control mesh of the architectural roof. The nodes indicated by *black squares* are tagged as corner and the edges highlighted by the *black solid line* as crease

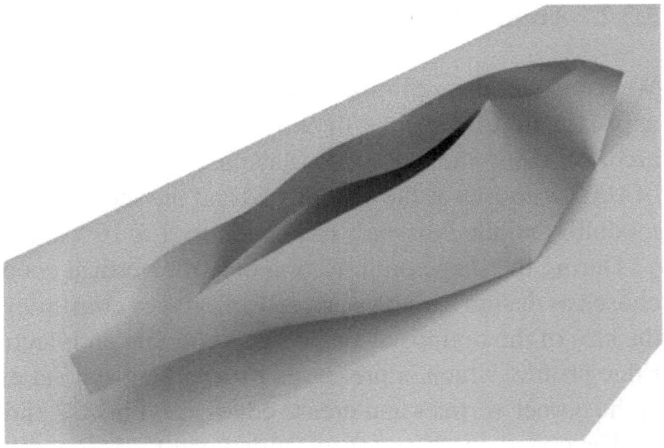

Fig. 24 Limit surface of the optimised architectural roof design

6 Summary and Conclusions

We reviewed the multiresolution optimisation technique based on subdivision surfaces and presented its application to elasticity, electrostatics and thin-shell problems. For discretising the governing partial differential equations we employed three different discretisation techniques, namely immersed finite elements for elasticity, boundary elements for electrostatics and subdivision finite elements for thin-shells. The inherent hierarchy of the subdivision surfaces allows us to consider the same surface at different resolutions and to take advantage of multiresolution editing techniques. Starting from the coarsest control mesh increasingly finer meshes are used for geometry updating and always a fine mesh is used for analysis. As demonstrated with the computed examples, this effectively inhibits the appearance of non-physical geometry oscillations that may occur in shape optimisation. Moreover, any pathological element distortions on the computational mesh are practically avoided. As a result, there is no need to regenerate or smooth the boundary mesh during the optimisation. Finally, it is evident that the multiresolution editing semantics is particularly appealing for isogeometric analysis because it enables the decoupling of geometry and analysis representations.

Acknowledgements The partial support of the EPSRC through grant # EP/G008531/1 and EC through the Marie Curie Actions (IAPP) program CASOPT project are gratefully acknowledged.

References

1. G. Allaire, F. Jouve, A.M. Toader, Structural optimization using sensitivity analysis and a level-set method. J. Comput. Phys. **194**, 363–393 (2004)
2. K. Bandara, F. Cirak, Multiresolution shape optimisation of thin-shell structures with subdivision finite elements. Technical report, University of Cambridge (2015)

3. K. Bandara, T. Rüberg, F. Cirak, Shape optimisation with multiresolution subdivision surfaces and immersed finite elements. arXiv:1510.02719 (2015)
4. K. Bandara, F. Cirak, G. Of, O. Steinbach, J. Zapletal, Boundary element based multiresolution shape optimisation in electrostatics. J. Comput. Phys. **297**, 584–598 (2015)
5. H. Biermann, A. Levin, D. Zorin, Piecewise smooth subdivision surfaces with normal control, in *SIGGRAPH 2000 Conference Proceedings*, New Orleans, 2000, pp. 113–120
6. K.U. Bletzinger, S. Kimmich, E. Ramm, Efficient modeling in shape optimal design. Comput. Syst. Eng. **2**, 483–495 (1991)
7. P.B. Bornemann, F. Cirak, A subdivision-based implementation of the hierarchical b-spline finite element method. Comput. Methods Appl. Mech. Eng. **253**, 584–598 (2013)
8. V. Braibant, C. Fleury, Shape optimal design using b-splines. Comput. Methods Appl. Mech. Eng. **44**, 247–267 (1984)
9. E. Catmull, J. Clark, Recursively generated b-spline surfaces on arbitrary topological meshes. Comput.-Aided Des. **10**, 350–355 (1978)
10. F. Cirak, Q. Long, Subdivision shells with exact boundary control and non-manifold geometry. Int. J. Numer. Methods Eng. **88**, 897–923 (2011)
11. F. Cirak, M. Ortiz, P. Schröder, Subdivision surfaces: a new paradigm for thin-shell finite-element analysis. Int. J. Numer. Methods Eng. **47**, 2039–2072 (2000)
12. F. Cirak, M. Scott, E. Antonsson, M. Ortiz, P. Schröder, Integrated modeling, finite-element analysis, and engineering design for thin-shell structures using subdivision. Comput.-Aided Des. **34**, 137–148 (2002)
13. M. Delfour, J.P. Zolésio, *Shapes and Geometries: Analysis, Differential Calculus, and Optimization*. Advances in Design and Control (SIAM, Philadelphia, 2001)
14. D. Doo, M. Sabin, Behavior of recursive division surfaces near extraordinary points. Comput.-Aided Des. **10**, 356–360 (1978)
15. M. Flucher, M. Rumpf, Bernoulli's free-boundary problem, qualitative theory and numerical approximation. Journal für die reine und angewandte Mathematik **486**, 165–204 (1997)
16. D. Fußeder, B. Simeon, A.V. Vuong, Fundamental aspects of shape optimization in the context of isogeometric analysis. Comput. Methods Appl. Mech. Eng. **286**, 313–331 (2015)
17. R.T. Haftka, R.V. Grandhi, Structural shape optimization – a survey. Comput. Methods Appl. Mech. Eng. **57**, 91–106 (1986)
18. J. Haslinger, R.A.E. Mäkinen, *Introduction to Shape Optimization: Theory, Approximation, and Computation* (SIAM, Philadelphia, 2003)
19. T. Hughes, J. Cottrell, Y. Bazilevs, Isogeometric analysis: CAD, finite elements, NURBS, exact geometry and mesh refinement. Comput. Methods Appl. Mech. Eng. **194**, 4135–4195 (2005)
20. S.G. Johnson, The NLopt nonlinear-optimization package. http://ab-initio.mit.edu/nlopt
21. J. Kiendl, R. Schmidt, R. Wüchner, K.U. Bletzinger, Isogeometric shape optimization of shells using semi-analytical sensitivity analysis and sensitivity weighting. Comput. Methods Appl. Mech. Eng. **274**, 148–167 (2014)
22. J.M. Lane, R.F. Riesenfeld, A theoretical development for the computer generation and display of piecewise polynomial surfaces. IEEE Trans. Pattern Anal. Mach. Intell. **PAMI-2**, 35–46 (1980)
23. N. Litke, A. Levin, P. Schröder, Fitting subdivision surfaces, in *IEEE Visualization 2001*, San Diego, 2001, pp. 319–324
24. E.H. Lockwood, *A Book of Curves* (Cambridge University Press, Cambridge, 1961)
25. C. Loop, Smooth subdivision surfaces based on triangles. Master's thesis, Department of Mathematics, University of Utah (1987)
26. M. Lounsbery, T.D. DeRose, J. Warren, Multiresolution analysis for surfaces of arbitrary topological type. ACM Trans. Graph. **16**, 34–73 (1997)
27. J. Peters, U. Reif, *Subdivision Surfaces*. Springer Series in Geometry and Computing (Springer, Berlin/Heidelberg, 2008)
28. T. Rüberg, F. Cirak, An immersed finite element method with integral equation correction. Int. J. Numer. Methods Eng. **86**, 93–114 (2011)

29. T. Rüberg, F. Cirak, Subdivision-stabilised immersed b-spline finite elements for moving boundary flows. Comput. Methods Appl. Mech. Eng. **209–212**, 266–283 (2012)
30. T. Rüberg, F. Cirak, A fixed-grid b-spline finite element technique for fluid–structure interaction. Int. J. Numer. Methods Fluids **74**(9), 623–660 (2014)
31. R. Sanches, P. Bornemann, F. Cirak, Immersed b-spline (i-spline) finite element method for geometrically complex domains. Comput. Methods Appl. Mech. Eng. **200**, 1432–1445 (2011)
32. D. Schillinger, L. Dede, M.A. Scott, J.A. Evans, M.J. Borden, E. Rank, T.J.R. Hughes, An isogeometric design-through-analysis methodology based on adaptive hierarchical refinement of NURBS, immersed boundary methods, and T-spline CAD surfaces. Comput. Methods Appl. Mech. Eng. **249–253**, 116–150 (2012)
33. J. Sokolowski, J.P. Zolesio, *Introduction to Shape Optimization: Shape Sensitivity Analysis* (Springer, Berlin/Heidelberg, 1992)
34. J. Stam, Exact evaluation of Catmull-Clark subdivision surfaces at arbitrary parameter values, in *SIGGRAPH 1998 Conference Proceedings*, Orlando, 1998, pp. 395–404
35. J. Stam, Evaluation of Loop subdivision surfaces, in *SIGGRAPH 1999 Course Notes*, Los Angeles (1999)
36. O. Steinbach, *Numerical Approximation Methods for Elliptic Boundary Value Problems. Finite and Boundary Elements* (Springer, New York, 2008)
37. W. Wall, M. Frenzel, C. Cyron, Isogeometric structural shape optimizisation. Comput. Methods Appl. Mech. Eng. **197**, 2976–2988 (2008)
38. J. Warren, H. Weimer, *Subdivision Methods for Geometric Design: A Constructive Approach* (Morgan Kaufmann, San Francisco, 2001)
39. L. Ying, D. Zorin, Nonmanifold subdivision, in *IEEE Visualization 2001*, San Diego, 2001, pp. 325–332
40. D. Zorin, P. Schröder, Subdivision for modeling and animation, in *SIGGRAPH 2000 Course Notes*, New Orleans (2000)
41. D. Zorin, P. Schröder, W. Sweldens, Interactive multiresolution mesh editing, in *Proceedings of the 24th annual conference on computer graphics and interactive techniques (SIGGRAPH)*, Los Angeles, (ACM Press/Addison-Wesley, New York, 1996), pp. 259–268

Atom Simplification and Quality T-mesh Generation for Multi-resolution Biomolecular Surfaces

Tao Liao, Guoliang Xu, and Yongjie Jessica Zhang

Abstract In this paper, we present an algorithm to simplify low-contributing atoms and generate quality T-meshes for multi-resolution biomolecular surfaces. The structure of biomolecules is first simplified using an error-bounded atom elimination method. An extended cross field-based parameterization method is then developed to adapt the parametric line spacings to different surface resolutions. Moreover, an anisotropy defined from an input scalar field can also be achieved. From the parameterization results, we extract adaptive and anisotropic T-meshes for the further T-spline surface construction. Finally, a gradient flow-based method is developed to improve the T-mesh quality, with the anisotropy preserved in the quadrilateral elements. The effectiveness of the presented algorithm has been verified using several large biomolecular complexes.

1 Introduction

Geometric modeling of biomolecules plays an important role in various applications [10, 22, 36, 37, 45]. New challenges arise in simulations involving extremely large biomolecular complexes. The computational cost for biomolecular surface construction increases as the atom number becomes progressively larger. Generally, biomolecular complexes are formed by a binding of several biomolecules, and a specific biomolecule can be contained in different complexes. For example in human cardiac muscle, there are many different types of filament lattice complexes, and they all contain the same biomolecule called filament protein. Simplifying the

T. Liao • G. Xu • Y.J. Zhang (✉)
Department of Mechanical Engineering, Carnegie Mellon University, 5000 Forbes Ave, Pittsburgh, PA 15213, USA

Institute of Computational Mathematics, Academy of Mathematics and Systems Science, Chinese Academy of Sciences, Beijing 100190, China
e-mail: taoliao@andrew.cmu.edu; xuguo@lsec.cc.ac.cn; jessicaz@andrew.cmu.edu

B. Jüttler, B. Simeon (eds.), *Isogeometric Analysis and Applications 2014*, Lecture Notes in Computational Science and Engineering 107, DOI 10.1007/978-3-319-23315-4_7

structure of filament protein can significantly improve the modeling efficiency in building different types of filament lattice complexes.

Various methods have been developed for biomolecular surface generation [11, 12]. Some of them construct biomolecular surface in an explicit way, such as alpha-shapes [1, 16], MSMS [35], advancing front and generalized Delaunay approaches [24], NURBS approximation [3], and PDE-based methods [19, 42, 49]. Other methods implicitly represent biomolecular geometry using a Gaussian density map, and then extract surface or volumetric meshes from it [44, 48]. Techniques were developed to improve the computational efficiency, such as PDE transform based on Fast Fourier Transform (FFT) [19, 49], Non-uniform Fast Fourier Transform (NFFT) [4], coarse-grained modeling technique [2], as well as CPU and GPU parallel computation based on neighboring search, KD-tree and Bounding Volume Hierarchy (BVH) [26].

In the context of isogeometric analysis [20, 38, 39], the T-spline surface provides a powerful basis for computation in different applications, which can also greatly benefit the analysis of biomolecules. Due to the high efficiency of the T-spline control mesh, namely T-mesh, the multi-resolution feature on the biomolecular surface can be represented efficiently. Various methods have been developed for T-mesh generation [28, 40, 41]. In recent years, the cross field-based global parameterization methods were introduced in surface quadrangulation [6], which capture surface features based on the principal curvature directions [7, 21, 23] or eigenfunctions of the Laplace-Beltrami operator [25]. Various methods were introduced to reduce distortion in the parameterization. For example, the distortion in the cross field was minimized in [30, 31], the mapping of the input triangle meshes was optimized in [27], and the local bijectivity was achieved in [29]. More contributions have been made in different aspects such as controlling the scale of captured features [15] and achieving the anisotropy in parametric lines, which has been fulfilled based on the metric tensor [23], adapted cross field [25] or surface deformation [32]. The integer-grid map [5] was introduced to achieve quadrangulation in the surface parameterization, and a robust method was developed to extract quadrilateral meshes [14]. Other than the cross field, a novel parameterization method was introduced based on the N-poly vector field [13], which can avoid the mixed-integer problem. These techniques provide a nice basis for us to build T-meshes for biomolecules with multi-resolution features preserved efficiently.

During Gaussian density map construction, faraway atoms have little contribution to the bimolecular surface due to the decay of Gaussian kernel function. In this paper, an atom simplification method is developed to eliminate these atoms from the biomolecular structure. Based on the simplified structure, a multi-resolution biomolecular surface can be built for quality T-mesh generation.The main contributions in this paper include:

1. An error-bounded atom simplification algorithm is introduced for biomolecular complexes to improve the computational efficiency of density map construction;
2. An extended cross field-based method is developed for adaptive and anisotropic parameterization, which adapts the parametric line spacings to multi-resolution surface features; and
3. A new gradient flow-based method is introduced for T-mesh quality improvement, preserving the anisotropy in quadrilateral elements robustly.

The remainder of this paper is organized as follows. Section 2 describes our atom simplification method for biomolecular complexes. Section 3 explains the extended parameterization algorithm together with the new quality improvement method. Section 4 shows the results. Finally, Sect. 5 draws conclusions and points out the future work.

2 Algorithm Overview and Atom Simplification

Figure 1 shows the main steps of our algorithm. Biomolecular structures from the Protein Data Bank (PDB) are simplified to improve the modeling efficiency, which are used for multi-resolution triangular surface generation. Then T-meshes are generated based on surface parameterization. The two main components of our work (atom simplification and T-mesh construction) are marked in orange and green, respectively.

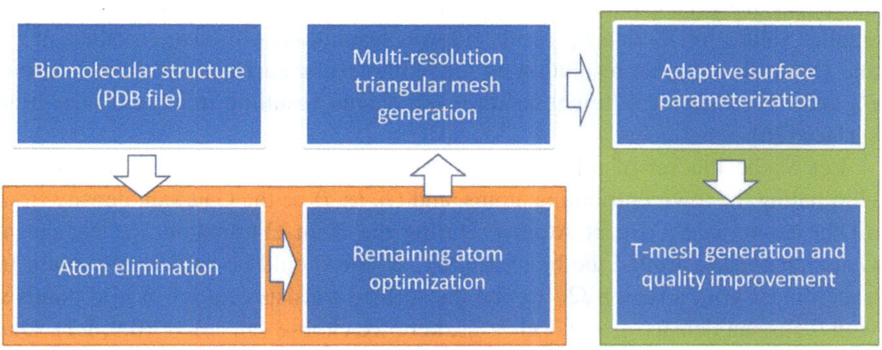

Fig. 1 The main steps of T-mesh generation for multi-resolution biomolecular surfaces. Note that the simplification process in the *orange* area is not required if the simplified structure is already prepared

In this paper, we simplify the structure of biomolecular complexes by eliminating low-contributing atoms. The efficiency of density map construction can be improved since the atom number is reduced. There are two main steps in our atom simplification: (1) eliminating the low-contributing atoms; and (2) optimizing the center location, the radius and the decay rate of the remaining atoms to minimize the error in the density map around the biomolecular surface.

2.1 Atom Elimination

Biomolecules, especially proteins, have hierarchical structures. They consist of several chains, each chain is formed by a set of residues, and each residue contains a group of atoms. The Gaussian density can be computed using a multi-level summation of Gaussian kernel functions [26, 48],

$$G(\mathbf{x}) = \sum_{k \in U_C} \left(\sum_{j \in U_R^k} \left(\sum_{i \in U_A^j} e^{-\kappa_i \left(\|\mathbf{x}-\mathbf{x}_i\|^2 - r_i^2 \right)} \right)^{P_R^{k,j}} \right)^{P_C^k}, \tag{1}$$

where U_C is the set of chains in the biomoleucle, U_R^k is the set of residues in Chain k, U_A^j is the set of atoms in Residue j, $\left(P_R^{k,j}, P_C^k\right)$ are coefficients controlling the surface resolution, and $(\mathbf{x}_i, r_i, \kappa_i)$ are the atom center, the radius and the decay rate of Atom i. The biomolecular surface is represented by a Gaussian density map implicitly, which can be extracted as an isosurface $\{M : G(\mathbf{x}) = 1\}$ in a rectilinear grid. In this paper, the grid spacings in x, y and z directions are set to be 0.3Å. Generally, smaller values for P_R and P_C result in lower resolution, while larger values of P_R and P_C result in higher resolution on the biomolecular surface. We choose various P_R and P_C values for different residues and chains, resulting in a multi-resolution surface [26].

Let S_A be the set of all the atoms, S_R be the set of remaining atoms after eliminating the low-contributing atoms, and $G_R^-(\mathbf{x}, i)$ be the Gaussian density from all the atoms in S_R except Atom i. During the atom elimination, the change in Gaussian density around the biomolecular surface should be sufficiently small. In a rectilinear grid, a region Ω_S around the surface consists of a set of grid points $\mathbf{x}$ with Gaussian density $g_l < G(\mathbf{x}) < g_u$, here we choose $(g_l, g_u) = (0.9, 1.1)$. Due to the decay of Gaussian kernel functions, eliminating Atom i only influences the

Gaussian density map in its neighboring region Ω_i, where all its grid points $\mathbf{x}$ satisfy

$$\|\mathbf{x} - \mathbf{x}_i\|^2 \leq -\frac{\ln \epsilon_r}{\kappa_i} + r_i^2, \tag{2}$$

and ϵ_r is a pre-defined threshold (e.g. $\epsilon_r = 10^{-6}$). The Gaussian density error e_G^i of ignoring Atom i can be defined as

$$e_G^i = \max_{\mathbf{x}\in\Omega_i\cap\Omega_S} \left| \frac{G(\mathbf{x}) - G_R^-(\mathbf{x}, i)}{G(\mathbf{x})} \right|. \tag{3}$$

In a rectilinear grid, the overall contribution C_i of Atom i to the surface is defined as

$$C_i = \left(\sum_{\mathbf{x}\in\Omega_i\cap\Omega_S} \left(\frac{G^-(\mathbf{x}, i)}{G(\mathbf{x})} \right)^2 \right)^{-1/2}, \tag{4}$$

where $G(\mathbf{x})$ and $G^-(\mathbf{x}, i)$ are the Gaussian density with and without considering Atom i, respectively. We sort all the atoms based on their overall contributions, and keep eliminating the one with the lowest contribution until we find an atom with the Gaussian density error defined in Eq. (3) greater than ϵ_G, where ϵ_G is an input threshold. After atom elimination, we build the Gaussian density map with the remaining atoms and extract a triangle mesh to represent the simplified surface. Let M be the triangle mesh. The surface error in M can be measured as

$$e_G^M = \max_{\mathbf{x}\in M} \left| \frac{G(\mathbf{x}) - 1}{G(\mathbf{x})} \right|, \tag{5}$$

where $\mathbf{x}$ is a vertex in the mesh. See Algorithm 1 for the details.

During Gaussian density map construction, a neighboring search and BVH-based GPU parallel method is employed for efficient computation [26]. In this paper biomolecules are represented by their PDB IDs. Figure 2 shows the simplification results of 2O53 with different ϵ_G values and coefficients (P_R, P_C). As ϵ_G increases, the Gaussian density error on the surface e_G^M tends to increase faster in concave and valley areas than the rest of the surface. As shown in Table 1, as the surface resolution is improved with larger P_R, less atoms can be eliminated. This is because more atoms are exposed on the high resolution surface than the low resolution one.

Algorithm 1 Atom elimination

$S_R := S_A$;
for each grid point **x** **do**
  Compute Gaussian density value $G(\mathbf{x})$;
end for
Identify the region Ω_S around the biomolecular surface, where $g_l < G(\mathbf{x}) < g_u$;
for each atom $i \in S_A$ **do**
  Identify its neighboring region Ω_i using Eq. (2);
  Compute the overall contribution C_i to the surface in $\Omega_i \cap \Omega_S$;
end for
Sort all the atoms in S_A based on C_i;
for the atom with the lowest contribution in S_R **do**
  Set this atom to be Atom i;
  Compute the Gaussian density error e_G^i using Eq. (3);
  if $e_G^i > \epsilon_G$ **then**
    STOP;
  else
    $S_R := S_R \setminus \{i\}$;
    Continue;
  end if
end for
Output the surface error e_G^M according to Eq. (5).

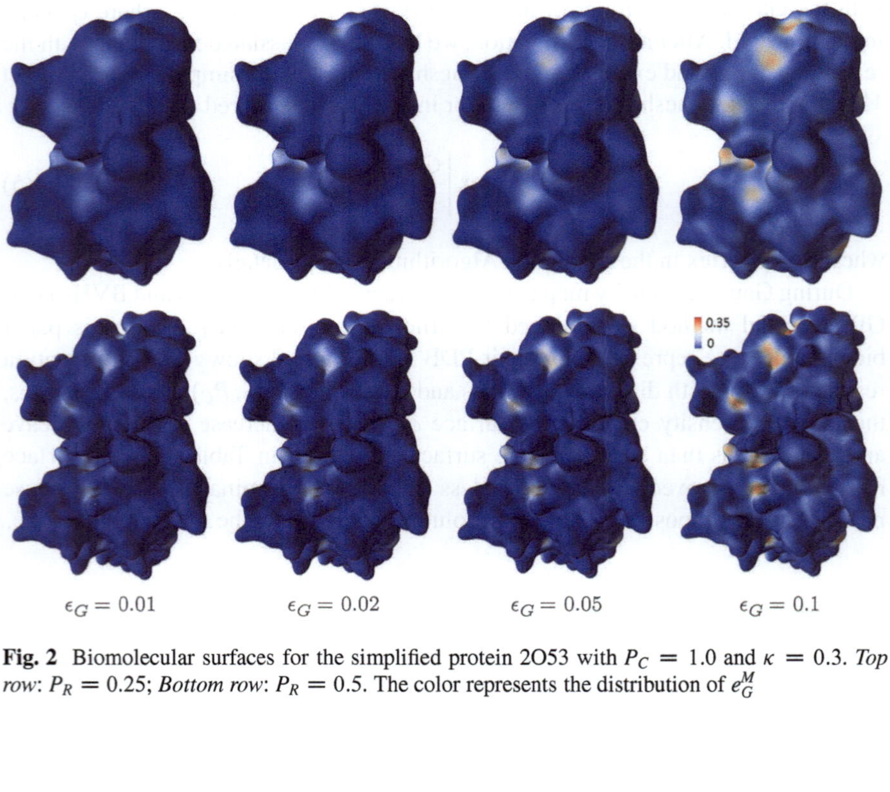

$\epsilon_G = 0.01$ $\epsilon_G = 0.02$ $\epsilon_G = 0.05$ $\epsilon_G = 0.1$

Fig. 2 Biomolecular surfaces for the simplified protein 2O53 with $P_C = 1.0$ and $\kappa = 0.3$. *Top row*: $P_R = 0.25$; *Bottom row*: $P_R = 0.5$. The color represents the distribution of e_G^M

Table 1 Atom Elimination Results for 2O53

		$\epsilon_G = 0.0$	$\epsilon_G = 0.01$	$\epsilon_G = 0.02$	$\epsilon_G = 0.05$	$\epsilon_G = 0.1$
Case 1	N_R	4,912(100 %)	3,316(67.5 %)	3,011(61.3 %)	2,810(57.2 %)	2,557(52.1 %)
	e_G^M	0.0	0.021	0.032	0.057	0.221
	Optimized e_G^M	–	0.019	0.030	0.053	0.215
	T_G	0.92	0.55	0.52	0.47	0.42
Case 2	N_R	4,912(100 %)	3,662(74.6 %)	3,513(71.5 %)	3,362(68.4 %)	3,102(63.2 %)
	e_G^M	0.0	0.023	0.028	0.060	0.210
	Optimized e_G^M	–	0.021	0.024	0.057	0.208
	T_G	0.98	0.81	0.79	0.76	0.72

Note: Case 1 – $P_R = 0.25$; Case 2 – $P_R = 0.5$; N_R – the number of remaining atoms; atoms; e_G^M – the surface error defined in Eq. (5); and T_G – time for Gaussian density map computation (unit: second). The percentage of remaining atoms is shown in parentheses

2.2 Optimization of Remaining Atoms

To reduce the surface error of the simplified biomolecule, the parameters of the remaining atoms can be optimized. Let N_R be the number of remaining atoms. The atoms in S_R are sorted, and their parameters are optimized using a gradient flow method. The objective function is defined as

$$\mathbf{F}(\mathbf{f}) = \int_{\mathbb{R}^3} (G(\mathbf{x}) - G_R(\mathbf{f}, \mathbf{x}))^2 \mathrm{d}\mathbf{x}, \tag{6}$$

and in a rectilinear grid we have

$$\mathbf{F}(\mathbf{f}) = \sum_{l=1}^{N_G} (G(\mathbf{x}_l) - G_R(\mathbf{f}, \mathbf{x}_l))^2, \tag{7}$$

where N_G is the number of grid points, and $\mathbf{f} = (x_1, y_1, z_1, r_1, \kappa_1, \ldots, x_{N_R}, y_{N_R}, z_{N_R}, r_{N_R}, \kappa_{N_R})^T$. The initial value $\mathbf{f}^0$ is given by the original centers, radii and decay rates of the remaining atoms. For the kth iteration of the gradient flow-based optimization, we have

$$\mathbf{f}^{k+1} = \mathbf{f}^k - \tau \nabla \mathbf{F}, \tag{8}$$

where τ is the step length, and

$$\nabla \mathbf{F} = \left(\frac{\partial \mathbf{F}}{\partial x_1}, \frac{\partial \mathbf{F}}{\partial y_1}, \frac{\partial \mathbf{F}}{\partial z_1}, \frac{\partial \mathbf{F}}{\partial r_1}, \frac{\partial \mathbf{F}}{\partial \kappa_1}, \ldots, \frac{\partial \mathbf{F}}{\partial x_{N_R}}, \frac{\partial \mathbf{F}}{\partial y_{N_R}}, \frac{\partial \mathbf{F}}{\partial z_{N_R}}, \frac{\partial \mathbf{F}}{\partial r_{N_R}}, \frac{\partial \mathbf{F}}{\partial \kappa_{N_R}} \right)^T.$$

As $\mathbf{F}$ is minimized step by step, the error in the Gaussian density map is minimized. In each step, the parameters of the remaining atoms are updated until the change in the parameters is less than a threshold ϵ_f,

$$\left\| \mathbf{f}^{k+1} - \mathbf{f}^k \right\|_\infty < \epsilon_f. \tag{9}$$

During the iteration in Eq. (8), the variation in $\mathbf{f}$ should be bounded. Let $\mathbf{x}_i^0$, r_i^0 and κ_i^0 be the original location, radius and decay rate of Atom i. If $\left\| \mathbf{x}_i^k - \mathbf{x}_i^0 \right\| > \rho_x$ (e.g., $\rho = 0.5$), then

$$\mathbf{x}_i^k = \mathbf{x}_i^0 + \rho_x \frac{\mathbf{x}_i^k - \mathbf{x}_i^0}{\left\| \mathbf{x}_i^k - \mathbf{x}_i^0 \right\|},$$

where ρ_x defines a sphere around $\mathbf{x}_i^0$ and the updated atom center should be inside the sphere. Similarly, the modifications in the radius $|r_i^k - r_i^0|$ and the decay rate $|\kappa_i^k - \kappa_i^0|$ are bounded to ρ_r and ρ_κ (e.g., $\rho_r = 0.5$ and $\rho_\kappa = 0.2$), respectively. As shown in Table 1, the surface error e_G^M can be improved after the optimization of remaining atoms.

3 Multi-resolution T-mesh Construction

For the analysis of large biomolecular complexes, usually only a specific component is essential for accuracy. A multi-resolution surface can be used to represent the biomolecular surface, maintaining high resolution details on the surface for the emphasized components while providing low resolution for the rest. In this paper, the emphasized components are defined as specific chains in the biomolecular complexes based on their biological functions. For example in Fig. 3a, the chains belonging to actin (blue and orange) in the human Thin Filament protein should be emphasized because they are receptors for some inhibitors. For the emphasized chains, we keep all the atoms and also use larger coefficients (P_R, P_C) in Eq. (1) to obtain higher resolution; while for the rest of the biomolecule, we simplify atoms and choose smaller (P_R, P_C). Sometimes, sharp noises may happen in the Gaussian density map around the connection region of the lower and higher resolution surface, which can be removed by applying a low-pass filter based on the fast Fourier transform [17]. Then, the multi-resolution surface is extracted from the constructed Gaussian density map using the dual contouring method [47, 48], and adaptive triangular meshes are obtained. In the following, we will talk about how to construct surface parameterization and quality T-meshes using these triangular meshes.

3.1 Surface Parameterization and T-mesh Construction

In this paper, we extend the cross field-based parameterization method [7, 21, 23, 25, 33, 34] to T-mesh generation for biomolecular surfaces. Firstly, a cross field is built on the triangle mesh guided by the principal curvature directions. Then an adaptive parameterization is computed based on the cross field. Anisotropy can also be achieved during surface parameterization. Finally, T-meshes are constructed by connecting the nodes with integer parametric coordinates.

Review of Cross Field-based Parameterization Here we briefly review the cross field generation and parametric coordinate computation, more details can be found in [7]. A cross field is defined in each triangle T_i with four perpendicular vectors, which can be represented as an angle θ_i referring to an edge $\mathbf{e}_i$ of the triangle, namely the reference edge. These vectors can be initialized using different inputs, such as the principal curvature directions. Then the cross field is smoothed by minimizing the smoothness energy [7]

$$\Gamma^S = \sum_{e_{ij} \in E} (\theta_i + \varphi_{ij} + \frac{\pi}{2} p_{ij} - \theta_j)^2, \tag{10}$$

where e_{ij} is the edge shared by triangles T_i and T_j, E_T is the set of edges in the mesh, φ_{ij} is the angle between the reference edges of triangle T_i and T_j, and p_{ij} is the integer valued period jump of the cross field across e_{ij}. The principal curvature directions in a set of sparsely distributed triangles are chosen as the guidance for the cross field. By minimizing the smoothness energy, θ_i and p_{ij} are updated in each triangle T_i with the constraints from the guidance, and a smooth cross field is obtained, see Fig. 3b. Singularities of the cross field (red dots) can be located with the optimized θ_i and p_{ij} in each triangle [7]. Then, the surface is cut into a disk-like planar region with all the singularities on its boundary (black lines). The parametric coordinates (u, v) of

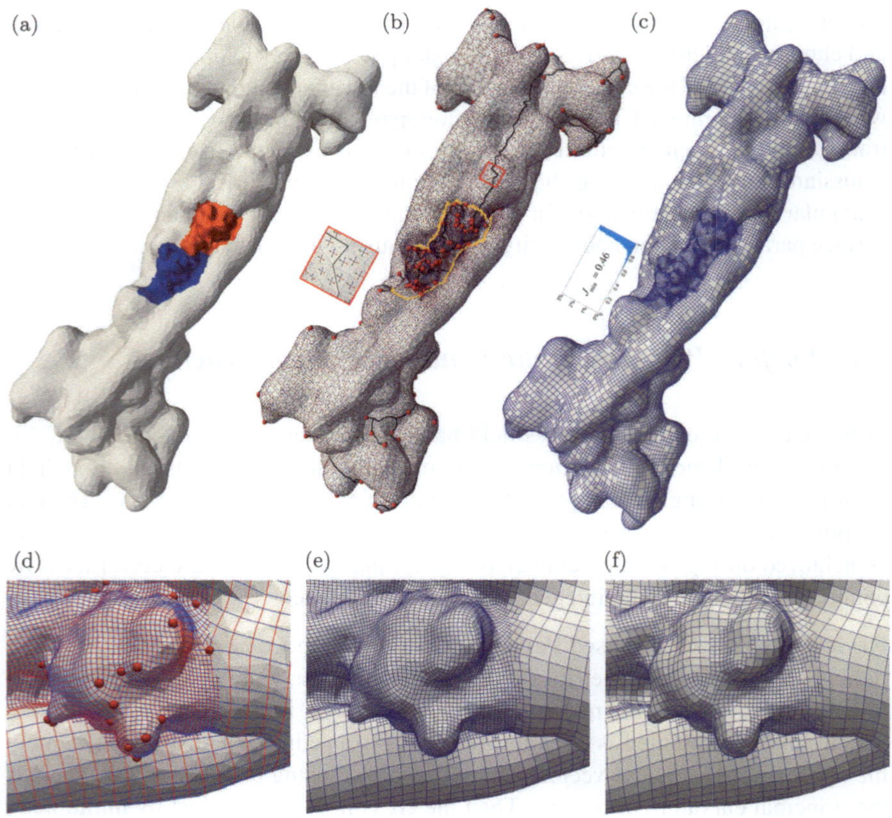

Fig. 3 Adaptive parameterization for the multi-resolution surface of 2W4U. (**a**) The multi-resolution triangle surface; (**b**) the cross field; (**c**) the adaptive T-mesh; (**d**) the parametric lines; and (**e, f**) the T-meshes corresponding to (**d**) before and after the removal of redundant vertices

each vertex behave as two piece-wise linear scalar fields, which can be obtained by minimizing an orientation energy [7, 25, 33],

$$\Gamma^O = \sum_{i=1}^{N_T} A_i \left(\left\| h_i^u \nabla_{T_i} u - \mathbf{u}_i \right\|^2 + \left\| h_i^v \nabla_{T_i} v - \mathbf{v}_i \right\|^2 \right), \tag{11}$$

where N_T is the number of triangles, A_i is the area of triangle T_i, $\left(h_i^u, h_i^v\right)$ are parameters controlling the spacings of the parametric lines, $(\nabla_{T_i} u, \nabla_{T_i} v)$ are gradients of (u, v) in T_i, and $(\mathbf{u}_i, \mathbf{v}_i)$ are two perpendicular vectors chosen from the cross field (red and blue arrows in Fig. 3b). Details about the cutting process and the choice of $(\mathbf{u}_i, \mathbf{v}_i)$ directions can be found in [7]. The parametric line spacings in u and v directions equal to $1/h_i^u$ and $1/h_i^v$, respectively. Integer constraints are set on the planar region boundary to ensure consistent parametric lines, enabling a valid quadrangulation of the surface. Minimizing the smoothness energy in Eq. (10) and the orientation energy in Eq. (11) are two mixed-integer problems, which can be solved using the greedy mixed-integer optimizer introduced in [8].

Adaptive T-mesh Generation For the multi-resolution surface in Fig. 3a, we need to generate denser elements in the high resolution patches (blue and orange) to capture the detailed features. This can be achieved by adapting the line spacings in the orientation energy in Eq. (11): smaller (h_i^u, h_i^v) values are set for the higher resolution regions, while larger values are set for others. Different from [7], the multi-resolution surface is first cut along the patch boundaries. Note that different patches are not completely independent, and each patch is connected with one of its neighbors on the boundary. Similar to [7], each patch is cut into a disk-like region based on a Dijkstra tree, and all the singularities are on the boundary of the disk-like region. Then the $(\mathbf{u}_i, \mathbf{v}_i)$ directions are rotated to keep consistent inside each patch. As shown in Fig. 4 when a higher resolution patch i (blue) is connected with a lower resolution patch j (yellow), we restrict $h_j = 2^m h_i$, where m is a positive integer. In

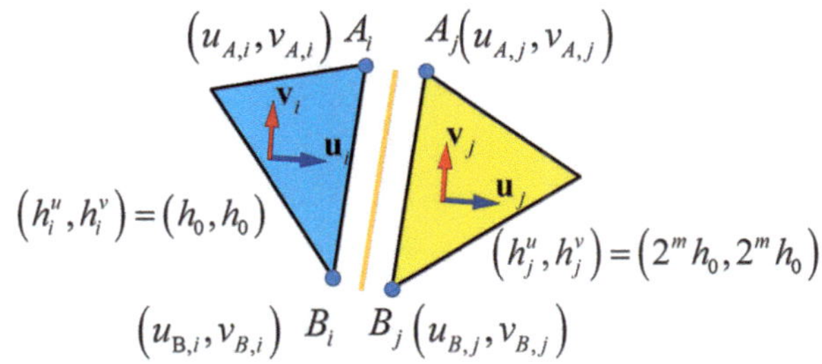

Fig. 4 Two triangles across the patch boundary (*orange*)

addition, the parametric coordinates at the patch boundaries should satisfy

$$\begin{cases} (u_{A,i}, v_{A,i})^T = 2^m \mathbf{R}_{AB}\big(u_{A,j}, v_{A,j}\big)^T + 2^m (I_u, I_v)^T, \\ (u_{B,i}, v_{B,i})^T = 2^m \mathbf{R}_{AB}\big(u_{B,j}, v_{B,j}\big)^T + 2^m (I_u, I_v)^T, \end{cases} \tag{12}$$

where $\mathbf{R}_{AB}$ is the rotation matrix across edge AB, (I_u, I_v) are integer-valued shift in the parametric coordinates, and m controls the difference of the parametric line spacing across the patch boundary. The parametric coordinates are computed by minimizing the orientation energy Γ^O with the constraints in Eq. (12). By connecting vertices with integer-valued parametric coordinates, a quadrilateral mesh can be built with T-junctions on the edges. To ensure the strongly-balanced structure in the T-mesh ($m \leq 1$ for any two neighboring elements), some quadrilateral elements in the transition region need to be subdivided.

Figure 3d, e shows the adaptive parameterization result for the multi-resolution surface (a local region) of Thin filament (2W4U). High resolution is set for two emphasized chains (Chains 18 and 20) with denser elements. To have a smooth transition from a higher resolution to a lower one, some T-mesh elements need to be modified. Figure 5 shows four different connections across the patch boundary (orange curve). Here $m = 2$, so some quadrilateral elements are subdivided to ensure a strongly-balanced structure. The obtained T-meshes may be too fine in some regions, therefore we identify redundant vertices based on their surface error and remove them from the T-meshes. Figure 3e, f shows a comparison between the T-meshes before and after removing redundant vertices.

Anisotropic T-mesh Generation Similar with [25], an anisotropy can be defined from an input scalar field f. Instead of using the principal curvatures, here we choose eigenfunction based cross-field guidance for anisotropic T-mesh generation.

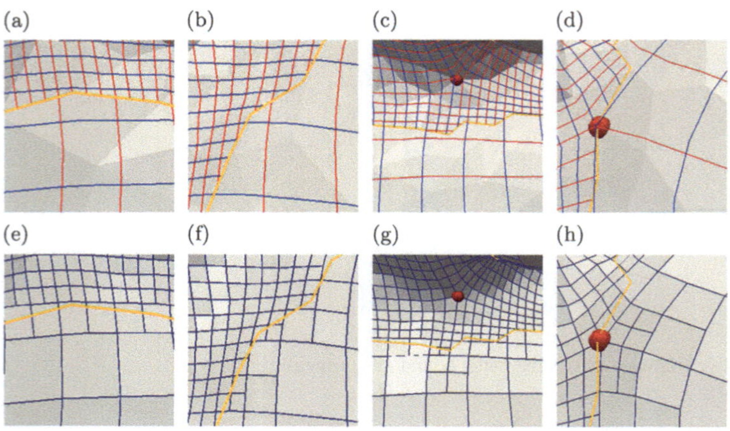

Fig. 5 Various parametric lines across the patch boundary and their resulting T-meshes. (**a–d**) The parametric lines; and (**e–h**) the corresponding T-meshes

In anisotropic T-meshes, quadrilateral elements are stretched in certain directions. The principal curvature directions usually vary intensively on the surface especially for regions with many detailed features, yielding a lot of singularities in the cross field, which makes it difficult to define the directions for stretching elements. On the contrary, low-mode eigenfunctions are very smooth and yield much fewer singularities. Their gradient directions in a set of sparsely distributed triangles are taken as the constraint for minimizing the smoothness energy in Eq. (10). After the cross field is built, singularities can be located using the same method in [7]. For the parametric lines following the gradient direction ∇f, h_i is set to be an uniform value h^0. For the perpendicular direction to ∇f, the line spacing $h_i^{\perp}$ is determined by the gradient magnitude,

$$\frac{1}{h_i^{\perp}} = \frac{\alpha \cdot \max\left(\|\nabla f\|\right)}{h^0 \, \|\nabla f_i\|}, \tag{13}$$

where α controls the minimum line spacing. If $\|\nabla f_i\| < \alpha \cdot \max\left(\|\nabla f\|\right)$, the line spacing is set to be $h_i^{\perp} = h^0$. As such, the quadrilateral elements are stretched along the direction perpendicular to the gradient, and dense elements are generated in the gradient direction. In Fig. 6, Mode 2 eigenfunction of the Laplace-Beltrami operator is used as the input scalar field. Similar with the isotropic parameterization in Fig. 3d, we adapt the parametric line spacings to the surface resolution by quadrupling the line spacings across the patch boundary.

3.2 T-mesh Quality Improvement

As shown in Fig. 7a, for each T-junction we extend it to make the surrounding elements form a local unstructured quadrilateral mesh. For a Vertex $\mathbf{x}_i$ in Element j, the other vertices in the element can be represented as $\mathbf{x}_{i+1}^j$, $\mathbf{x}_{i+2}^j$ and $\mathbf{x}_{i+3}^j$ in the counter-clockwise order, see Fig. 7b. The normal direction at Vertex $\mathbf{x}_i$ is computed as

$$\mathbf{n}_i = \sum_{j \in Q_i} \left(\mathbf{x}_{i+1}^j - \mathbf{x}_i\right) \times \left(\mathbf{x}_{i+3}^j - \mathbf{x}_i\right),$$

where Q_i is the set of neighboring elements of Vertex $\mathbf{x}_i$ and "$\times$" is the cross product of vectors. $\mathbf{n}_i$ is then normalized as $\mathbf{n}_i / \|\mathbf{n}_i\|$. In this way, $\mathbf{n}_i$ is estimated as the weighted average of normal directions in the neighboring elements, and the areas of these elements are taken as the weights. The *scaled Jacobian* [18, 46] at $\mathbf{x}_i$ in Element j equals to $\|\mathbf{v}_1 \times \mathbf{v}_2\| \cdot s$, where

$$\mathbf{v}_1 = \frac{\mathbf{x}_{i+1}^j - \mathbf{x}_i}{\left\|\mathbf{x}_{i+1}^j - \mathbf{x}_i\right\|}, \ \mathbf{v}_2 = \frac{\mathbf{x}_{i+3}^j - \mathbf{x}_i}{\left\|\mathbf{x}_{i+3}^j - \mathbf{x}_i\right\|}, \ \text{and } s = \operatorname{sign}\left((\mathbf{v}_1 \times \mathbf{v}_2) \cdot \mathbf{n}_i\right).$$

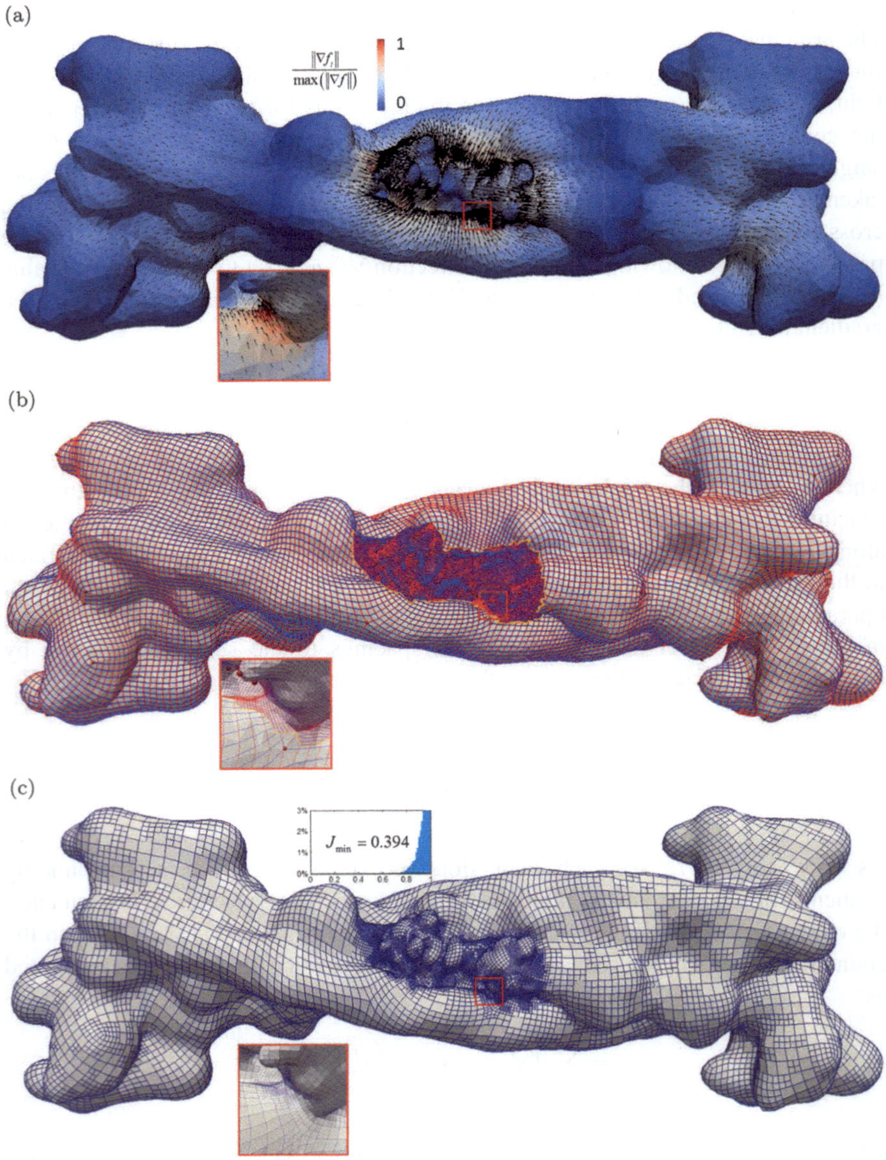

Fig. 6 Anisotropic T-mesh generation from the Mode 2 eigenfunction of LBO ($\alpha = 0.1$). (**a**) The gradient and gradient magnitude of the input field; (**b**) the adaptive and anisotropic parameterization; and (**c**) the corresponding T-mesh

The quality of Element j can be measured by the minimum Jacobian at its four vertices. Generally, the overall quality of the T-mesh from surface parameterizaion is good in term of Jacobian, except a few elements. To improve the Jacobian of

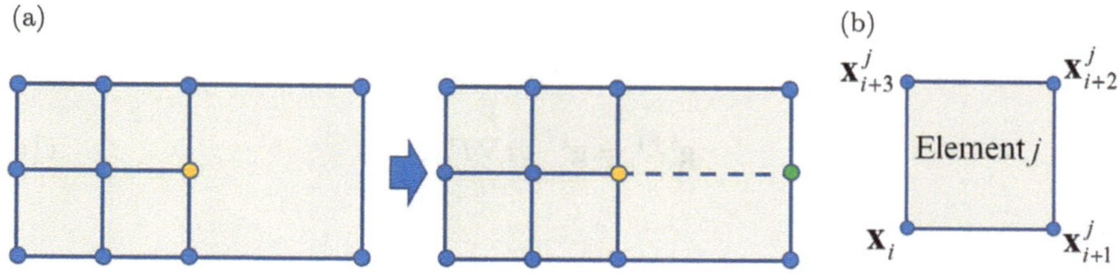

Fig. 7 (**a**) Extending the T-junction (*orange*) to form a local unstructured mesh; and (**b**) A neighboring element j surrounding Vertex $\mathbf{x}_i$

the T-mesh elements, we apply a new gradient flow-based quality improvement method, which considers both the original shape of elements and the orthogonality of edges. Contrast to other existing approaches [9, 43], our method can preserve the anisotropy of the T-mesh.

Suppose Element j in Fig. 7b is a neighboring element surrounding Vertex $\mathbf{x}_i$, the *cross distortion* at $\mathbf{x}_i$ is defined as

$$\Gamma^C(\mathbf{x}_i) = \sum_{j \in Q_i} \frac{\left\| \frac{\mathbf{x}_{i+1}^j - \mathbf{x}_i}{a_j} \right\|^2 + \left\| \frac{\mathbf{x}_{i+3}^j - \mathbf{x}_i}{b_j} \right\|^2}{\sigma_{ij}^2 + \sqrt{\sigma_{ij}^2 + 4\delta^2}}, \tag{14}$$

where $\sigma_{ij} = \mathbf{n}_i \cdot \left(\frac{\mathbf{x}_{i+1}^j - \mathbf{x}_i}{a_j} \times \frac{\mathbf{x}_{i+3}^j - \mathbf{x}_i}{b_j} \right)$, δ is an arbitrary small value (e.g. $\delta = 0.1$), and

$$\left(a_j, b_j\right) = \left(\frac{\left\|\mathbf{x}_{i+1}^j - \mathbf{x}_i\right\| + \left\|\mathbf{x}_{i+2}^j - \mathbf{x}_{i+3}^j\right\|}{2}, \frac{\left\|\mathbf{x}_{i+3}^j - \mathbf{x}_i\right\| + \left\|\mathbf{x}_{i+2}^j - \mathbf{x}_{i+1}^j\right\|}{2} \right)$$

are the *feature lengths* of Element j. The total cross distortion energy of the T-mesh is defined as

$$\Gamma_M^C = \sum_{i=1}^{N} \Gamma^C(\mathbf{x}_i), \tag{15}$$

where N is the number of vertices. Minimizing Γ_M^C not only makes the corners of each quadrilateral element close to 90°, but also makes the edge length adapt to the local surface feature, which improves the element quality and preserves its stretched shape. The minimization can be fulfilled via a gradient flow method.Let

$\mathbf{g} = (x_1, y_1, z_1, \ldots, x_N, y_N, z_N)^T$ be the vector containing the coordinates of all the vertices in the T-mesh.For the kth step of the gradient flow, we have

$$\mathbf{g}^{k+1} = \mathbf{g}^k - \tau \nabla \Gamma_M^C, \tag{16}$$

where

$$\nabla \Gamma_M^C = \left(\frac{\partial \Gamma_M^C}{\partial x_1}, \frac{\partial \Gamma_M^C}{\partial y_1}, \frac{\partial \Gamma_M^C}{\partial z_1}, \ldots, \frac{\partial \Gamma_M^C}{\partial x_N}, \frac{\partial \Gamma_M^C}{\partial y_N}, \frac{\partial \Gamma_M^C}{\partial z_N} \right)^T.$$

Note that in each step, vertices can only move on the tangent plane. Therefore, we update $\mathbf{x}_i^k$ with $\mathbf{x}_i^{k+1} - \left(\mathbf{n}_i \cdot \mathbf{p}_i^k\right) \mathbf{n}_i$, where $\mathbf{n}_i$ is the surface normal and $\mathbf{p}_i^k = \mathbf{x}_i^{k+1} - \mathbf{x}_i^k$. Our quality improvement method can also handle flat and concave elements although so far they were not observed in our T-mesh results.

We also implemented an isotropic T-mesh quality improvement method based on the triangle optimization [9, 43] and compared our method with it. For an isotropic T-mesh, the ideal shapes of the elements are squares or rectangles, as shown in Fig. 8, which can be decomposed into several *ideal triangles*. For a general quadrilateral element, we decompose it into several *real triangles* following the same splitting format. An affine mapping between a real triangle j and its corresponding ideal triangle j_I can be defined as

$$\begin{aligned} \mathbf{f_K} &: j_I \to j, \\ \tilde{\mathbf{x}} &\mapsto \mathbf{x} = \mathbf{K}_j \tilde{\mathbf{x}} + \mathbf{v}, \end{aligned} \tag{17}$$

where $\mathbf{x}$ is a vertex in the real triangle, $\tilde{\mathbf{x}}$ is the corresponding vertex in the ideal triangle, and $\mathbf{v}$ is a constant term. The distortion of the T-mesh is defined as

$$\Gamma_M^T = \sum_{i=1}^{N_Q} \sum_{j \in R_i} \frac{\|K_j\|_F^2}{\det\left(\mathbf{K}_j\right) + \sqrt{\left(\det\left(\mathbf{K}_j\right)\right)^2 + 4\delta^2}}, \tag{18}$$

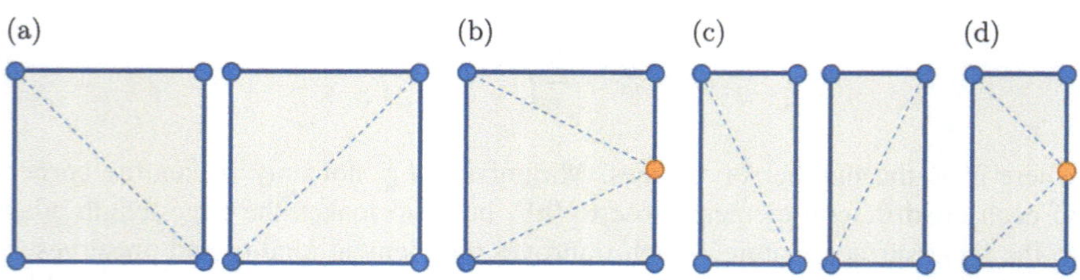

Fig. 8 Ideal triangles from a square (**a–b**) or rectangle (**c–d**). (**a, c**) No T-junction; and (**b, d**) with a T-junction. The *orange dots* are T-junctions

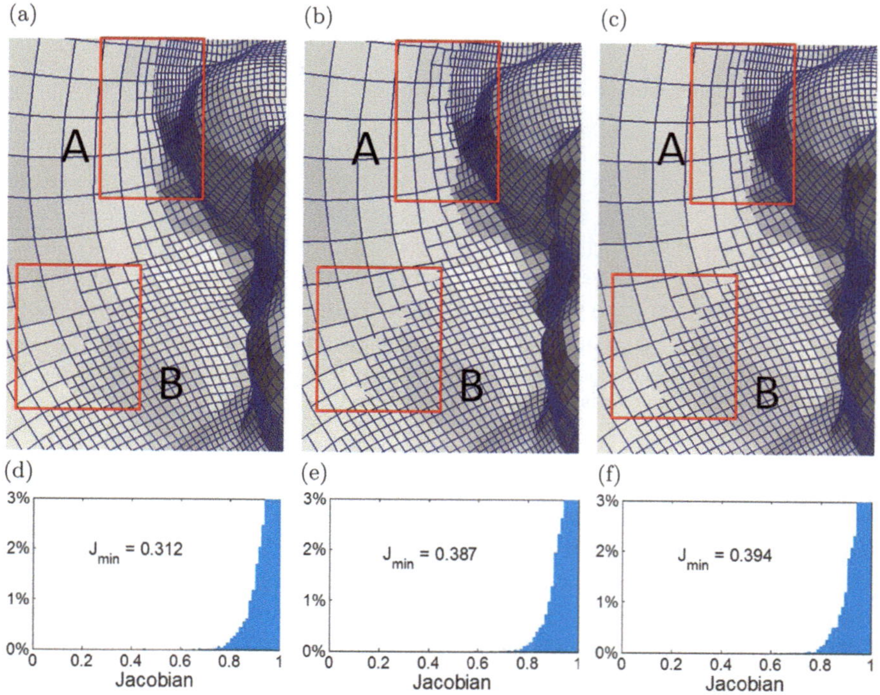

Fig. 9 Quality improvement results for the T-mesh in Fig. 6c using two different methods. (**a**) The original T-mesh; (**b–c**) the improved T-meshes using the triangle optimization method and our method, respectively; and (**d–f**) the Jacobian distribution corresponding to (**a–c**)

where N_Q is the number of quadrilateral elements, R_i is the set of real triangles generated from Element i, and $\|K_j\|_F = \sqrt{\text{tr}\left(\mathbf{K}_j{}^T\mathbf{K}_j\right)}$ is the Frobenius norm of $\mathbf{K}_j$. Similar with Eq. (16), a gradient flow method can be applied to minimize Γ_M^T.

Figure 9 shows a comparison between our method and the triangle optimization method. We can observe that both methods can improve the mesh quality. As shown in (b–c), for the isotropic elements in Region B, both methods yield similar results. But for the anisotropic elements in Region A, only our method preserves the anisotropic rectangle shape of the elements.

4 Results and Discussion

In this section, the biomolecule simplification and multi-resolution T-mesh generation algorithms are applied to various biomolecular complexes. All the results are generated from a computer with an Intel Xeon E5-1620 CPU, a Nvidia GeForce GTX680 graphic card, and 16GB of memory.

Table 2 Simplification results for biomolecules

	No simplification		Atom simplification (no emphasized chains)			Atom simplification (with emphasized chains)		
PDB ID	N_A	T_G	N_R	T_G	e_G^M	N_R	T_G	e_G^M
2O53	4,912	0.92	3,316 (67.5 %)	0.55	0.019	4,114 (83.8 %)	0.87	0.018
4KYT	7,908	1.35	5,377 (68.0 %)	0.97	0.018	5,501 (69.6 %)	0.99	0.018
4N78	22,843	3.26	14,436 (63.2 %)	2.13	0.023	15,107 (66.1 %)	2.33	0.021
4A7F	33,500	3.88	20,134 (60.1 %)	2.42	0.026	20,465 (62.7 %)	2.47	0.026
2W4U	0.14M	16.21	74,962 (54.4 %)	10.13	0.032	78,853 (57.2 %)	10.98	0.031
2KU2	1.23M	33.21	0.51M (39.3 %)	15.81	0.055	0.64M (49.3 %)	17.48	0.051

Note: N_A – the number of all the atoms; N_R – the number of remaining atoms; and T_G – time for Gaussian density map computation (unit: second). $\epsilon_G = 0.01$ for the atom elimination. The percentage of remaining atoms is shown in the parentheses

Table 2 shows the results of atom simplification for proteins with various sizes. Among them, 2O53 is a protein in human brain, and one of its two identical components is emphasized. 4KYT, 4N78, 4A7F and 2W4U are involved in the heart contractile process. Components with important biological functions are chosen to be emphasized. 2KU2 is one of the largest proteins in the PDB with 1.23M atoms. One of its seven symmetric components is chosen to be emphasized. As shown in Table 2, the percentage of remaining atoms varies from 39.3 % to 69.6 %, and tends to decrease as the protein size increases because most atoms are buried inside the biomolecular surface. The number of remaining atoms also depends on the size of emphasized components. For example in 2O53, when the emphasized components are considered, the percentage of remaining atoms increases by 16.3 %. Contrast to the other proteins, the emphasized components in 4A7F are exposed to the surface, and most of their atoms are kept during simplification. Therefore the number of remaining atoms does not change much when considering these emphasized components.

Figures 3, 10 and 11 show adaptive T-mesh generation results for the five proteins in Table 2, which can be used directly as the control mesh to build rational T-spline surfaces [41]. The principal curvatures are used to guide the parametric line directions, therefore the obtained T-meshes follow the local surface features. On the contrary in Fig. 12, eigenfunctions are used to define the input vector field. Therefore, the obtained anisotropic T-mesh follows the gradient direction of those eigenmodes. The variation of different modes is reflected by both the element orientations and the stretched shapes. We can also set various resolutions for different emphasized components. In Fig. 13, the parametric line spacings are $1 : 2 : 4$ for Chain 20 (orange), Chain 18 (blue) and the rest of the surface. Table 3 shows statistics of all the T-mesh generation results. We can observe that the number of T-junctions and the number of singularities vary for different proteins due to their complex surface features. In addition, the generated T-meshes are in good quality with $J_{\min} \geq 0.37$.

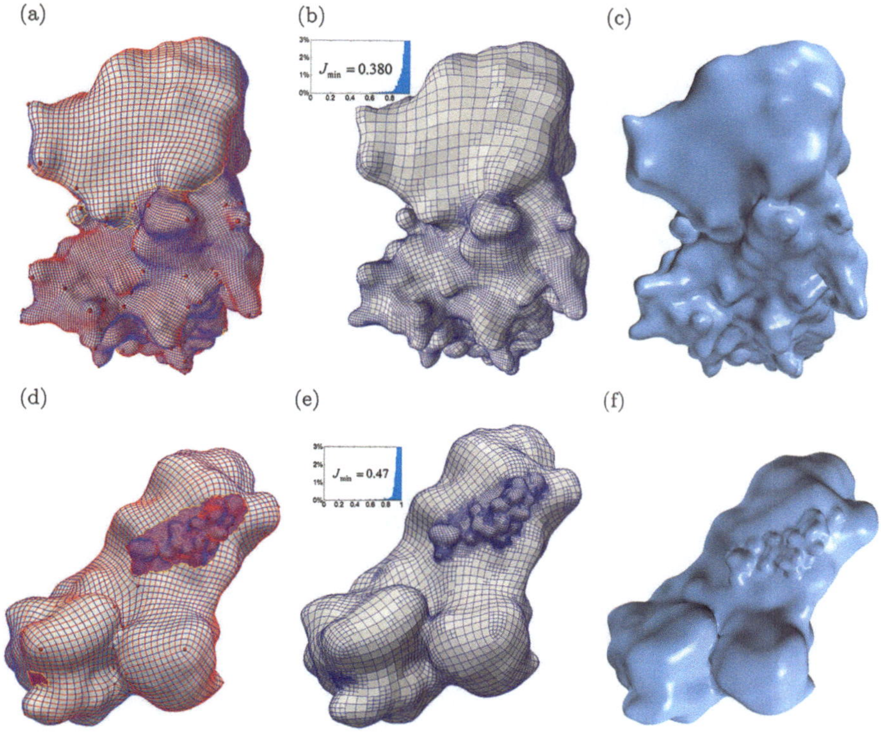

Fig. 10 Multi-resolution surfaces for 2O53 and 4KYT. (**a–c**) 2O53; and (**d–f**) 4KYT. *Left column*: adaptive parameterization; *Middle column*: T-mesh; *Right column*: T-spline surface

5 Conclusion and Future Work

In this paper, we have introduced a new approach to simplify low-contributing atoms and generate quality T-meshes for multi-resolution biomolecular surfaces. An error-bounded atom elimination algorithm is designed to reduce the atom number and preserve multi-resolution surface feature at the same time. An extended cross field-based parameterization is introduced to generate adaptive and anisotropic T-meshes, which can be used further for T-spline surface construction. In addition, a new gradient flow-based method is introduced for T-mesh quality improvement, preserving the anisotropy in the input T-mesh.

Isogeometric analysis has been applied in a lot of engineering fields, it also has a great potential for applications in computational biology to study biomolecular complexes or proteins. In the future we intend to explore along this direction.

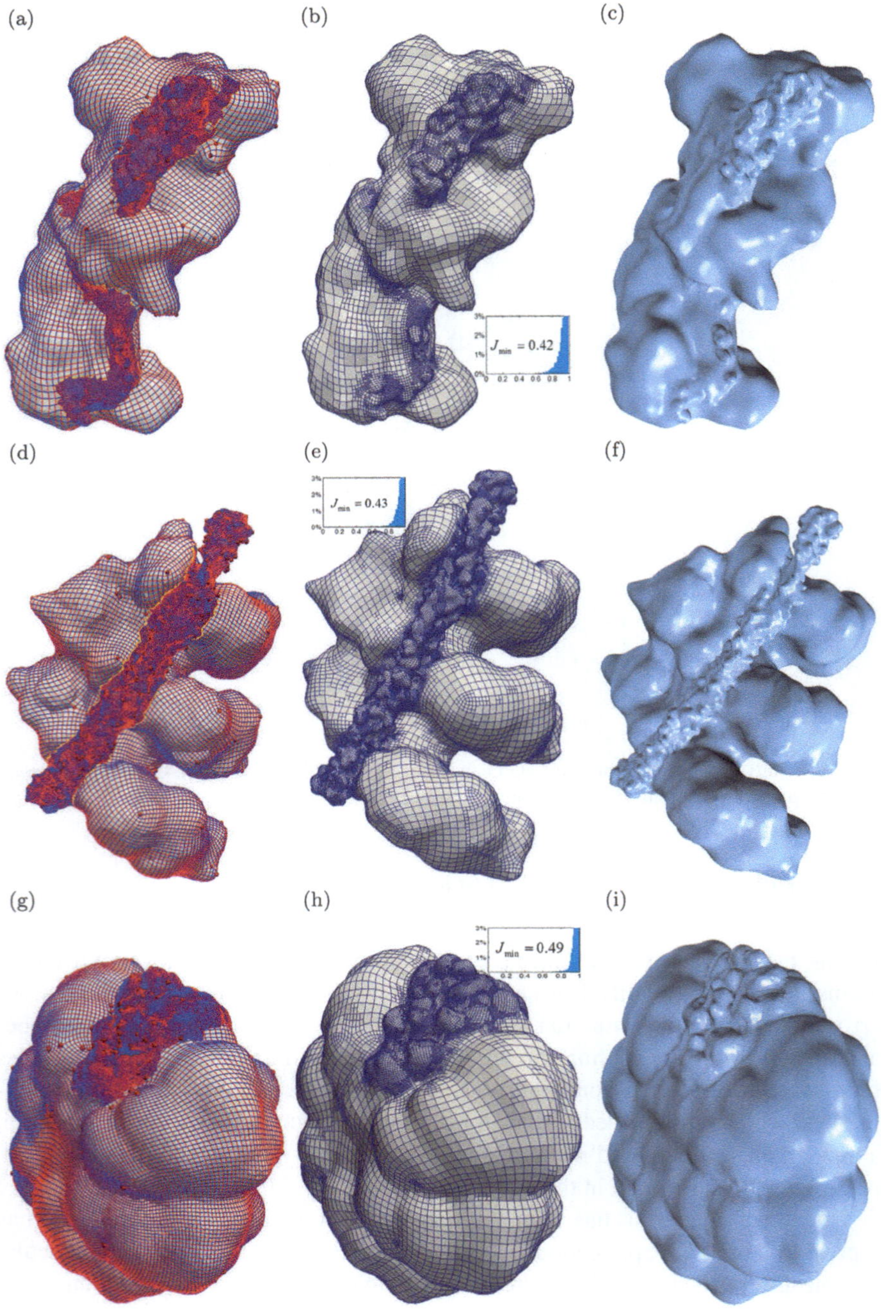

Fig. 11 Multi-resolution surfaces for 4N78, 4A7F and 2KU2. (**a–c**) 4KYT; (**d–f**) 4N78; and (**g–i**) 2KU2. *Left column*: adaptive parameterization; *Middle column*: T-mesh; *Right column*: T-spline surface

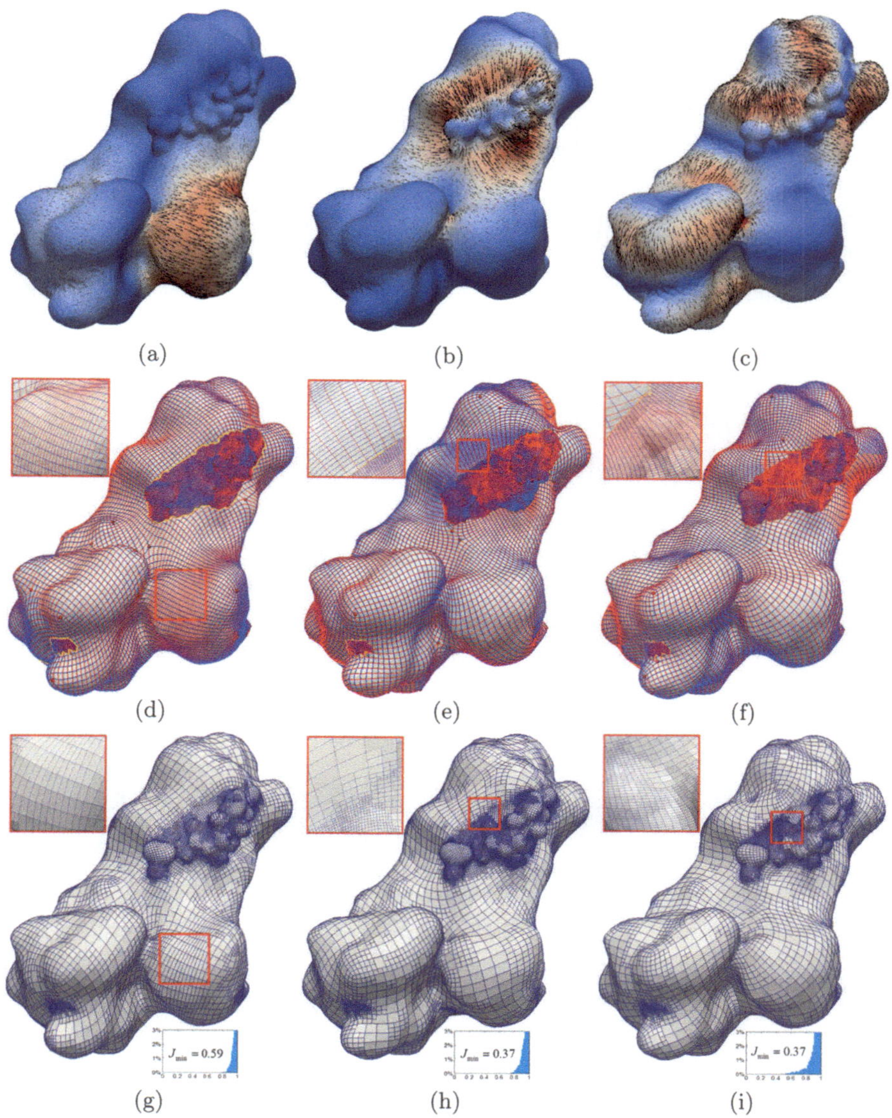

Fig. 12 Adaptive and anisotropic T-mesh construction of 4KYT from three different eigenmodes. (**a–c**) The gradient direction and magnitude of the eigenfunctions; (**d–f**) surface parameterization; and (**g–i**) adaptive and anisotropic T-meshes. *Left column*: results from Mode 2; *Middle column*: results from Mode 3; and *Right column*: results from Mode 6

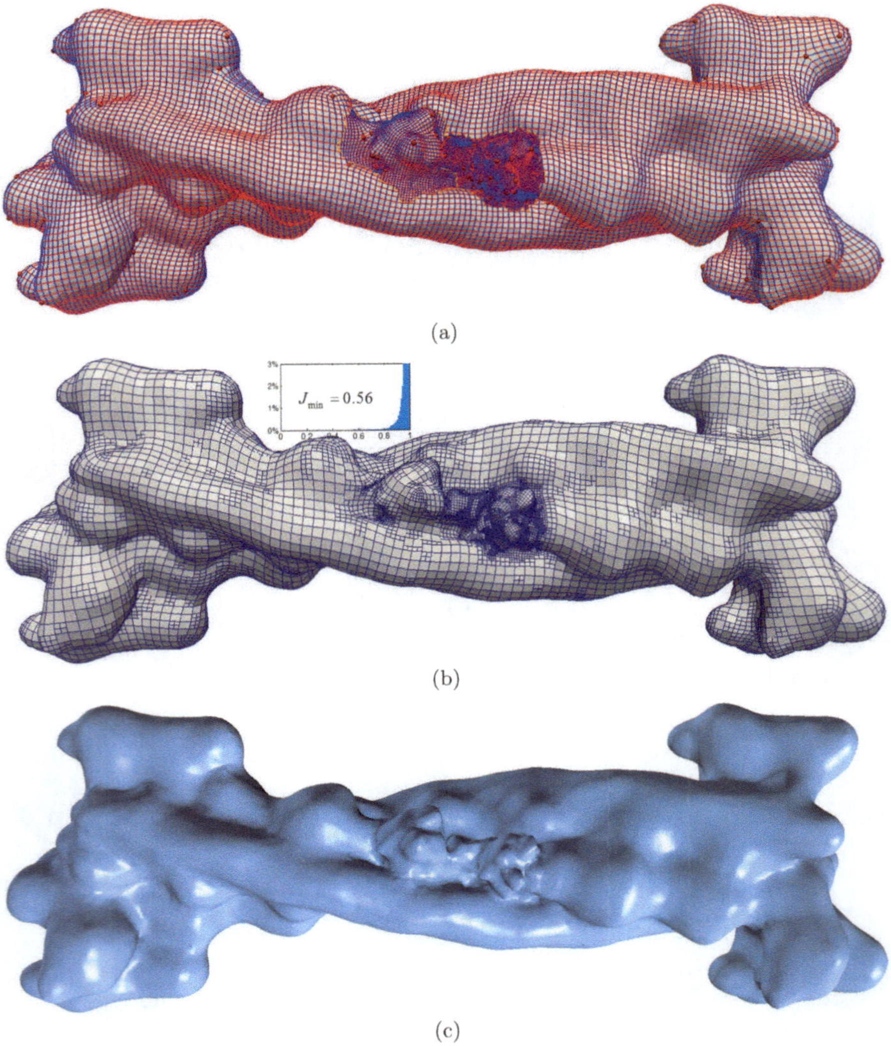

Fig. 13 Biomolecular surfaces for 2W4U emphasizing Chains 18 and 20 with different resolutions. (**a**) Adaptive parameterization; (**b–c**) the corresponding T-mesh and T-spline surface

Table 3 T-mesh generation results for biomolecules

PDB ID	(Vertex No., Element No.)	Number of T-junctions	Number of Singularities	Number of Levels	Number of J_{min}	min_Jacobian (s)	Time
2O53	Fig. 10b	(17,235 15,494)	3,766	156	4	0.47	0.38
4KYT	Fig. 10e	(11,815, 10,576)	3,272	136	4	0.47	13.3
	Fig. 12g	(13,557, 12,251)	2,770	56	4	0.59	13.6
	Fig. 12h	(16,432, 15,045)	2,270	62	4	0.37	13.9
	Fig. 12i	(13,212, 11,942)	2,582	56	4	0.37	13.5
4N78	Fig. 11b	(14,484, 12,843)	4,782	122	4	0.42	14.2
4A7F	Fig. 11e	(42,560 38,423)	3,272	172	4	0.43	15.6
2W4U	Fig. 3c	(30,142, 26,782)	5,214	240	4	0.46	16.3
	Fig. 6c	(28,993, 26,227)	2,956	226	4	0.39	16.6
	Fig. 13b	(21,908 19,298)	5,210	168	4	0.56	15.2
2KU2	Fig. 11h	(25,759, 23,774)	5,433	120	4	0.49	18.3

Acknowledgements T. Liao and Y. Zhang were supported in part by PECASE Award N00014-14-1-0234 and NSF CAREER Award OCI-1149591. G. Xu was supported in part by NSFC under the grants (11101401, 81173663), and NSFC Fund for Creative Research Groups of China (grant No. 11321061).

References

1. L. Albou, B. Schwarz, O. Poch, J. Wurtz, D. Moras, Defining and characterizing protein surface using alpha shapes. Proteins **76**(1), 1–12 (2009)
2. C. Bajaj, S. Chen, G. Xu, Q. Zhang, W. Zhao, Hierarchical molecular interfaces and solvation electrostatics, in *SIAM/ACM Joint Conference on Geometric and Physical Modeling*, San Francisco, 2009, pp. 283–288
3. C.L. Bajaj, V. Pascucci, A. Shamir, R.J. Holt, A.N. Netravali, Dynamic maintenance and visualization of molecular surfaces. Discret. Appl. Math. **127**(1), 23–51 (2003)
4. C.L. Bajaj, V. Siddavanahalli, Fast error-bounded surfaces and derivatives computation for volumetric particle data. Tech. rep., ICES 06-03, 2006
5. D. Bommes, M. Campen, H. Ebke, P. Alliez, L. Kobbelt, Integer-grid maps for reliable quad meshing. ACM Trans. Graphics **32**(4), 98:1–98:12 (2013)
6. D. Bommes, B. Lévy, N. Pietroni, E. Puppo, C. Silva, D. Zorin, Quad-mesh generation and processing: a survey. Comput. Graphics Forum 32(6), 51–76 (2013)
7. D. Bommes, H. Zimmer, L. Kobbelt, Mixed-integer quadrangulation. ACM Trans. Graphics **28**(3), 1–10 (2009)
8. D. Bommes, H. Zimmer, L. Kobbelt, Practical mixed-integer optimization for geometry processing, in *Curves and Surfaces*, ed. by J.-D. Boissonnat, P. Chenin, A. Cohen, C. Gout, T. Lyche, M.-L. Mazure, L. Schumaker (Springer, Berlin/Heidelberg, 2012), pp. 193–206
9. M. Brovka, J.I. López, J.M. Escobar, J.M. Cascón, R. Montenegro, A new method for T-spline parameterization of complex 2D geometries. Eng. Comput. **30**(4), 457–473 (2014)
10. Y. Cheng, C.A. Chang, Z. Yu, Y. Zhang, M. Sun, T.S. Leyh, M.J. Holst, J.A. Mccammon, Diffusional channeling in the sulfate activating complex: combined continuum modeling and coarse-grained Brownian dynamics studies. Biophys. J. **95**(10), 4659–4667 (2008)
11. M.L. Connolly, Analytical molecular surface calculation. J. Appl. Crystallogr. **16**(5), 548–558 (1983)
12. M.L. Connolly, *Molecular Surface: A Review* (Network Science, 1996). www.netsci.org
13. O. Diamanti, A. Vaxman, D. Panozzo, O. Sorkine-Hornung, Designing N-polyvector fields with complex polynomials. Comput. Graphics Forum **33**(5), 1–11 (2014)
14. H. Ebke, D. Bommes, M. Campen, L. Kobbelt, Qex: robust quad mesh extraction. ACM Trans. Graphics **32**(6), 168–177 (2013)
15. H. Ebke, M. Campen, D. Bommes, L. Kobbelt, Level-of-detail quad meshing. ACM Trans. Graphics **33**(6), 184:1–184:11 (2014)
16. Edelsbrunner, E.P. H., Mücke, Three-dimensional alpha shapes. ACM Trans. Graphics **13**(1), 43–72 (1994)
17. M. Frigo, S. Johnson, The design and implementation of FFTW3. Proc. IEEE **93**(2), 216–231 (2005)
18. R. Garimella, M. Shashkov, P. Knupp, Triangular and quadrilateral surface mesh quality optimization using local parametrization. Comput. Methods Appl. Mech. Eng. **193**(9), 913–928 (2004)
19. L. Hu, D. Chen, G. Wei, High-order fractional partial differential equation transform for molecular surface construction. Mol. Based Math. Biol. **1**, 1–25 (2013)
20. B. Jüttler, M. Kapl, D. Nguyen, Q. Pan, M. Pauley, Isogeometric segmentation: the case of contractible solids without non-convex edges. Comput.-Aided Des. **57**, 74–90 (2014)

21. F. Kälberer, M. Nieser, K. Polthier, Quadcover-surface parameterization using branched coverings. Comput. Graphics Forum **26**, 375–384 (2007)
22. P. Kekenes-Huskey, T. Liao, A. Gillette, J. Hake, Y. Zhang, A. Michailova, A. McCulloch, J. McCammon, Molecular and subcellular-scale modeling of nucleotide diffusion in the cardiac myofilament lattice. Biophys. J. **105**(9), 2130–2140 (2013)
23. D. Kovacs, A. Myles, D. Zorin, Anisotropic quadrangulation. Comput. Aided Geom. Des. **28**(8), 449–462 (2011)
24. P. Laug, H. Borouchaki, Molecular surface modeling and meshing. Eng. Comput. **18**, 199–210 (2002)
25. T. Liao, G. Xu, Y. Zhang, Structure-aligned guidance estimation in surface parameterization using eigenfunction-based cross field. Graphical Models **76**, 691–705 (2014)
26. T. Liao, Y. Zhang, P. Kekenes-Huskey, Y. Cheng, A. Michailova, A. McCulloch, M. Holst, J. McCammon, Multi-core CPU or GPU-accelerated multiscale modeling for biomolecular complexes. Mol. Based Math. Biol. **1**, 164–179 (2013)
27. Y. Lipman, Bounded distortion mapping spaces for triangular meshes. ACM Trans. Graphics **31**(4), 108:1–108:13 (2012)
28. A. Myles, N. Pietroni, D. Kovacs, D. Zorin, Feature-aligned T-meshes. ACM Trans. Graphics **29**(4), 98:1–98:12 (2010)
29. A. Myles, N. Pietroni, D. Zorin, Robust field-aligned global parametrization. ACM Trans. Graphics **33**(4), 135 (2014)
30. A. Myles, D. Zorin, Global parametrization by incremental flattening. ACM Trans. Graphics **31**(4), 109:1–109:11 (2012)
31. A. Myles, D. Zorin, Controlled-distortion constrained global parametrization. ACM Trans. Graphics **32**(4), 105:1–105:13 (2013)
32. D. Panozzo, E. Puppo, M. Tarini, O. Sorkine-Hornung, Frame fields: anisotropic and non-orthogonal cross fields. ACM Trans. Graphics **33**(4), 134–144 (2014)
33. N. Pietroni, M. Tarini, O. Sorkine, D. Zorin, Global parametrization of range image sets. ACM Trans. Graphics **30**(6), 149:1–149:10 (2011)
34. N. Ray, W. Li, B. Lévy, A. Sheffer, P. Alliez, Periodic global parameterization. ACM Trans. Graphics **25**(4), 1460–1485 (2006)
35. M.F. Sanner, A.J. Olson, J.C. Spehner, Reduced surface: an efficient way to compute molecular surfaces. Biopolymers **38**(3), 305–320 (1996)
36. Y. Song, Y. Zhang, C.L. Bajaj, N.A. Baker, Continuum diffusion reaction rate calculations of wild-type and mutant mouse acetylcholinesterase: adaptive finite element analysis. Biophys. J. **87**(3), 1558–1566 (2004)
37. Y. Song, Y. Zhang, T. Shen, C.L. Bajaj, J.A. McCammon, N.A. Baker, Finite element solution of the steady-state Smoluchowksi equation for rate constant calculations. Biophys. J. **86**(4), 2017–2029 (2004)
38. D. Thomas, M. Scott, J. Evans, K. Tew, E. Evans, Bézier projection: a unified approach for local projection and quadrature-free refinement and coarsening of NURBS and T-splines with particular application to isogeometric design and analysis. Comput. Methods Appl. Mech. Eng. **284**, 55–105 (2015)
39. A. Vuong, C. Giannelli, B. Jüttler, B. Simeon, A hierarchical approach to adaptive local refinement in isogeometric analysis. Comput. Methods Appl. Mech. Eng. **200**(49), 3554–3567 (2011)
40. W. Wang, Y. Zhang, M. Scott, T. Hughes, Converting an unstructured quadrilateral mesh to a standard T-spline surface. Comput. Mech. **48**(4), 477–498 (2011)
41. W. Wang, Y. Zhang, G. Xu, T. Hughes, Converting an unstructured quadrilateral/hexahedral mesh to a rational T-spline. Comput. Mech. **50**(1), 65–84 (2012)
42. G. Wei, Y. Sun, Y. Zhou, M. Feig, Molecular multiresolution surfaces. arXiv math-ph/0511001 (2005)
43. T. Wilson, J. Sarrate, X. Roca, R. Montenegro, J. Escobar, Untangling and smoothing of quadrilateral and hexahedral meshes, in *Proceedings of the 8th International Conference on Engineering Computational Technology*, Stirlingshire, Paper 36 (2012)

44. Z. Yu, M.J. Holst, Y. Cheng, J.A. McCammon, Feature-preserving adaptive mesh generation for molecular shape modeling and simulation. J. Mol. Graphics Model. **26**(8), 1370–1380 (2008)
45. D. Zhang, J. Suen, Y. Zhang, Y. Song, Z. Radic, P. Taylor, M.J. Holst, C.L. Bajaj, N.A. Baker, J.A. McCammon, Tetrameric mouse acetylcholinesterase: continuum diffusion rate calculations by solving the steady-state Smoluchowski equation using finite element methods. Biophys. J. **88**(3), 1659–1665 (2005)
46. Y. Zhang, C. Bajaj, Adaptive and quality quadrilateral/hexahedral meshing from volumetric data. Comput. Methods Appl. Mech. Eng. **195**(9), 942–960 (2006)
47. Y. Zhang, C.L. Bajaj, B. Sohn, 3D finite element meshing from imaging data. Comput. Methods Appl. Mech. Eng. **194**(48–49), 5083–5106 (2005)
48. Y. Zhang, G. Xu, C.L. Bajaj, Quality meshing of implicit solvation models of biomolecular structures. Comput. Aided Geom. Des. **23**(6), 510–530 (2006)
49. Q. Zheng, S. Yang, G. Wei, Biomolecular surface construction by PDE transform. Int. J. Numer. Methods Biomed. Eng. **28**(3), 291–316 (2012)

Algorithmic Aspects of Isogeometric Shape Optimization

Daniela Fußeder and Bernd Simeon

Abstract Shape optimization is concerned about finding optimal designs under the aspect of some cost criteria often involving the solution of a partial differential equation (PDE) over the afore said unknown shape.

In general, industrial cases involve a geometric model from Computer Aided Design (CAD). However, solving PDEs requires an analysis suitable working model, typically a Finite Element (FEM) triangulation. Hence, some of the geometric properties known from the CAD model may be lost during this format change. Therefore, we employ isogeometric analysis (IGA) instead, which has a tighter connection between geometry, simulation and shape optimization.

In this paper, we present a self-contained treatment of gradient based shape optimization method with isogeometric analysis, focusing on algorithmic and practical aspects like computation of shape gradients in an IGA formulation and updating B-spline and NURBS geometries.

1 Introduction

Isogeometric analysis was invoked in the seminal paper [8] to serve as a bridge between Computer Aided Design (CAD) and the Finite Element Method (FEM). With IGA, so the hope, a seamless work flow from designing CAD models to simulation is possible, saving man power from (manually) adapting FEM meshes to CAD models. Moreover, in IGA there is no information loss as is for such mesh adaptations of classical FEM. Shape optimization was targeted as an application which might benefit from IGA already in that first introduction to IGA. The reasons

D. Fußeder (✉) • B. Simeon
Technische Universität Kaiserslautern, Paul-Ehrlich-Str. Geb. 31, 67633 Kaiserslautern, Germany
e-mail: fusseder@mathematik.uni-kl.de; simeon@mathematik.uni-kl.de

B. Jüttler, B. Simeon (eds.), *Isogeometric Analysis and Applications 2014*,
Lecture Notes in Computational Science and Engineering 107,
DOI 10.1007/978-3-319-23315-4_8

are the same as above, namely that

- there is no loss of information between optimization and analysis model, and
- B-splines and NURBS practically are the toolbox of designing because of their nice approximation properties, so why not use them also for finding optimal designs numerically.

A general formulation of shape optimization problems, see for instance [22], reads

$$\min J(u, \Omega) \quad \text{s.t. } E(u, \Omega) = 0 \text{ and } g(\Omega) \leq 0, \quad h(\Omega) = 0 \tag{1}$$

where J is a real valued objective or cost function depending on a domain Ω and the solution u of a PDE, which is given by the term $E(u, \Omega) = 0$. In g and h we collect additional inequality and equality constraints on the shape, but, in this work, not on the state u.

A gradient-based optimization method is employed to solve the discretized version of the minimization problem (1). Typical gradient-based optimization, as in [16], are for instance a sequential quadratic programming method (SQP), method of moving asymptotes (MMA), or an interior point method. Using such a black box gradient-based optimization routine, e.g. in the form of MATLAB solver `fmincon`, we illustrate the steps towards an optimal shape in Algorithm 1.

Algorithm 1 Basic Black Box Shape Optimization Algorithm

Require: Initial geometry Ω
Require: PDE(Ω) ▷ PDE solver on Ω which yields solution u_h
Require: OBJECTIVE(Ω, u) ▷ evaluate objective function $J(\Omega, u)$ for domains Ω
Require: CONSTRAINTS(Ω) ▷ evaluate constraint functions $g(\Omega)$, $h(\Omega)$ for domains Ω
Require: SHAPE GRAD(Ω, u) ▷ shape gradient computation ∇J, $\nabla C = (\nabla g, \nabla h)^T$
Require: UPDATE(Ω, s) ▷ geometry update routine Ω_{new} = "$\Omega + s$" with a descent direction s

Black Box Optimization, for instance with `fmincon`

1: **repeat**
2: $u_h \leftarrow$PDE(Ω)
3: $J \leftarrow$ OBJECTIVE(Ω, u_h)
4: $C \leftarrow$CONSTRAINTS(Ω)
5: $(\nabla J, \nabla C) \leftarrow$SHAPE GRAD($\Omega, u_h$)
6: **if** Ω is not optimal, i.e. does not satisfy a stopping criteria **then**
7: compute a descent direction s involving ∇J, ∇g, ∇h
8: $\Omega \leftarrow$ UPDATE(Ω, s)
9: **end if**
10: **until** Ω is optimal

Of course, in this work we use a B-spline or NURBS representation of $\Omega \subset \mathbb{R}^2$ and solve the state equation PDE(Ω) in Line 2 with IGA. Here, $E(u, \Omega) = 0$ describes a linear elasticity or Poisson equation. One focus will be on the practical computation of shape gradients ∇J, ∇g and ∇h in SHAPE GRAD(Ω, u) in Line 5 which are formulated in an isogeometric way using classical methods from [13, 22]. For that, we take an *optimize first–then discretized view* as opposed to *discretize first–optimize then*, see [7] for the first and [22] for the second approach. However, we delegate finding a descent direction in Line 7 to a black box nonlinear optimization routine, MATLAB `fmincon`, [10] or [26], which vitally depends on the quality of said shape gradients. Although checking the stopping criteria in Line 6 is part of the black-box optimizer we need to address it also for possible pitfalls. Last but not least updating the geometry in UPDATE(Ω) Line 8 falls again to our lot and consequently needs our attention to avoid infeasible representations.

A basic reference for IGA is [8]; B-splines and NURBS are treated in [17]. Isogeometric analysis has been applied to several fields of shape optimization problems already, for instance [14, 15] with application to electromagnetism and [2, 3, 20, 27] with application to solid mechanics and also shells [11]. As one of the basic references in the classical shape optimization field, we refer to [18], and a rigorous mathematical treatment is provided by [22]. Especially the definition of shape gradients is mainly attributed to [13, 18, 22]. Among many others, the introduction to optimization methods [16] treats SQP and interior-point methods. A popular optimizer in structural shape optimization is MMA by [23]. For our model problems, we use the packages [10] which includes for instance SQP and MMA, and [26], an interior-point solver. Also, SQP and interior-point method both are options for `fmincon` in [12].

Algorithm 1 provides the structure for this work: we introduce the core of isogeometric shape optimization – NURBS and B-splines – in Sect. 2 and apply it subsequently to obtain representations of domains in Sect. 2.2, find solutions of PDEs with IGA in Sect. 3, and formulate derivatives w.r.t. geometries in Sect. 4. For that, the space of admissible domains is provided with a vector space structure which allows us to define "summation of domains", i.e. updating shapes in Sect. 5. Finally in Sect. 6, we apply the theory from the previous sections to obtain the required procedures in Algorithm 1 for the abstract shape optimization problem, and realize them for some model examples. We conclude this work with the numerical solutions of these model problems in Sect. 7, a short summary, and an outlook.

2 Introduction to B-Splines, NURBS and Geometries in IGA

Isogeometric analysis starts and ends with B-splines and NURBS as basis functions for geometric modeling, as basis functions for the Galerkin projection in simulation and as design variables for shape optimization.

Therefore, we present the basic definitions subsequently in the first part, Sect. 2.1, but refer to various sources like [4, 21] for B-splines and for instance to the monograph [17] for a more elaborate treatment of NURBS than we can provide here. Equipped with the notation from this introduction we apply it to geometry representation in IGA in the following in Sect. 2.2.

2.1 Introduction to B-Splines and NURBS

From a given knot vector $\Xi = (\xi_1, \ldots, \xi_{n+p+1})$ we form n B-spline basis functions of degree p recursively

$$\text{for } p = 0: \quad N_{i,0}(\hat{x}) = \begin{cases} 1, & \text{if } \xi_i \leq \hat{x} < \xi_{i+1} \\ 0, & \text{otherwise} \end{cases} \tag{2}$$

$$\text{for } p > 0: \quad N_{i,p}(\hat{x}) = \frac{\hat{x} - \xi_i}{\xi_{i+p} - \xi_i} N_{i,p-1}(\hat{x}) + \frac{\xi_{i+p+1} - \hat{x}}{\xi_{i+p+1} - \xi_{i+1}} N_{i,p-1}(\hat{x}) \; . \tag{3}$$

Here, we always assume an open knot vector where the first and last knots are repeated $p+1$ times, i.e. $\xi_i = a$ for $i = 1, \ldots, p+1$ and $\xi_i = b$ for $i = n, \ldots, p+1$. This way, $N_{1,p}$ and $N_{n,p}$ are interpolative. We obtain non-uniform rational B-spline basis functions, in short NURBS, by weighting the B-spline basis

$$R_{i,p,W} = \frac{\omega_i N_{i,p}(\hat{x})}{\sum_{j=1}^{n} \omega_j N_{j,p}(\hat{x})} \tag{4}$$

with positive weights $\omega_i > 0$ collected in a weight vector $W = (\omega_1, \ldots, \omega_n)$.

A set of B-spline basis functions spans a spline space $\mathcal{S}(\Xi, p)$ of degree p which is determined uniquely by the underlying knot vector,

$$\mathcal{S}(\Xi, p) = \text{span}\{N_i(\hat{x}) := N_{i,p}(\hat{x}),\; i = 1, \ldots, n,\; \hat{x} \in [0, 1]\}. \tag{5}$$

We reach higher space dimensions by forming tensor products: given two knot vectors Ξ_1, Ξ_2 and corresponding sets of basis functions $N_{i_1,p_1}(\hat{x}_1)$ and $N_{i_2,p_2}(\hat{x}_2)$ a tensor product space

$$\mathcal{S}(\Xi_1, p_1) \otimes \mathcal{S}(\Xi_2, p_2) = \text{span}\{N_i(\hat{x})\colon i = 1, \ldots, n_1 n_2,\; \hat{x} \in [0, 1]^2\} \tag{6}$$

is obtained with $n = n_1 n_2$ basis functions $N_i = N_{i_1,p_1}(\hat{x}_1) N_{i_2,p_2}(\hat{x}_2)$ for $i_1 = 1, \ldots, n_1$, $i_2 = 1, \ldots, n_2$.

Taking a NURBS basis by exchanging N_{i_1,p_1} with R_{i_1,p_1,W_1} and N_{i_2,p_2} with R_{i_2,p_2,W_2} leads to a NURBS space

$$\mathcal{S}(\Xi, p, W) = \text{span}\{N_i(\hat{x}) := R_{i,p,W}(\hat{x}),\; i = 1, \ldots, n,\; \hat{x} \in [0, 1]\} \tag{7}$$

and analogously to (6) we have $\mathcal{S}(\Xi_1, p_1, W_1) \otimes \mathcal{S}(\Xi_2, p_2, W_2)$ with basis functions $N_i := R_{i_1,p_1,W_1} R_{i_2,p_2,W_2}$.

Of course, the tensor product can be extended to arbitrary space dimensions, however, for our examples two dimensions are sufficient. Since B-splines are a special case of NURBS with all weights equal to one, we from now on use the same symbols N_i for uni- and bivariate NURBS and B-splines, also dropping the product notation and the indices for fixed degrees p and weight vectors W. When we speak of NURBS, we mean both NURBS and B-splines, except where we distinguish them explicitly.

With such B-splines and NURBS we are equipped for designing geometric forms, performing simulation and shape optimization in the following sections.

2.2 *Geometries in Isogeometric Analysis*

One key ingredient of isogeometric analysis is the representation G of geometries by linear combinations of NURBS with fixed weights. Moreover, in isogeometric shape optimization we search for designs also in such NURBS spaces, i.e. the linear coefficients of G are the optimization variables. In addition, one could also look for shapes where the weights are variable as well. However, for the definition of shape gradients we need a linear space which we can provide for NURBS with variable weights in homogeneous coordinates only.

In the following, we first introduce geometries from the easier case of B-spline and NURBS spaces with fixed weights, and then geometries built from NURBS with variable weights.

2.2.1 B-Splines and NURBS with Fixed Weights

A NURBS geometry in $\mathbb{R}^d$ is a linear combination of NURBS basis functions

$$G(\hat{x}) = \sum_{i=1}^{n} X_i N_i(\hat{x}) \in \mathcal{S}^d \text{ with control points } X_i \in \mathbb{R}^d \,, \tag{8}$$

typically $d = 1, 2, 3$. In our examples, G is defined over a parameter space $\hat{\Omega} = [0, 1]^d$.

For NURBS *curves*, $d = 1$, we have univariate NURBS $N_i \in \mathcal{S} = \mathcal{S}(\Xi, p, W)$ and $\hat{x} \in [0, 1]$. For *surfaces*, $d = 2$ and the parameter domain is $[0, 1]^2$; we use bivariate NURBS $N_i \in \mathcal{S} = \mathcal{S}(\Xi_1, p_1, W_1) \otimes \mathcal{S}(\Xi_2, p_2, W_2)$.

In Algorithm 1 we assume that the geometries Ω are surfaces given by a NURBS representation as in (8), $\Omega = G(\hat{\Omega})$.

2.2.2 NURBS Space with Variable Weights

For the shape gradient derivation in the variational setting we have to restore linearity of the NURBS space with variable weights. For that, we resort to homogeneous coordinates and a perspective map, see [17]. A homogeneous coordinate vector $\tilde{X}$ in

$\mathbb{R}^{d+1}$ with $\tilde{X} := (X^w, X_{d+1})$ and $X^w \in \mathbb{R}^d$ is projected to $\mathbb{R}^d$ by

$$H(\tilde{X}) = \begin{cases} X^w/X_{d+1}, & X_{d+1} \neq 0 \\ X^w/|X^w|, & \text{else.} \end{cases}$$

A $d + 1$ dimensional B-spline geometry $(G^w, w) \in \mathcal{S}^{d+1}$ with $G^w \in \mathcal{S}^d$ and $0 < w \in \mathcal{S}$ is mapped to a rational representation in $\mathbb{R}^d$ by

$$H: (G^w, w) \mapsto G := G^w/w\,. \tag{9}$$

Finally, $\mathcal{S}^d \oplus \mathcal{S}$ has a basis $\{N_i e_k : N_i \in \mathcal{S},\ i = 1, \ldots, n,\ k = 1, \ldots, d + 1\}$ and is a linear space which is isomorphic to $\mathbb{R}^{(n+1)d}$.

The global geometry representations with fixed weights are a key ingredient of isogeometric analysis in Sect. 3. The case of variable weights is only important for shape optimization in Sect. 4.

3 Simulation in Isogeometric Analysis

Solving PDEs with isogeometric analysis works with the variational form and Galerkin projection the same way as in FEM. However, the PDE is transformed with a geometry function G such that it is posed over the parameter domain $\hat{\Omega} = [0, 1]^d$.

Given a PDE in its strong form

$$Lu = f \tag{10}$$

with a linear elliptic partial differential operator L, its relation to the weak form $a(u, v) = l(v)$ is given by the dual pairing $a(u, v) = \langle Lu, v\rangle$ and by $l(v) = (f, v)$. As in FEM, we use the variational form of (10) to find its numerical solution. In general, we say that u solves the state equation $E(u, \Omega) = 0$ if it solves the variational form of (10) over the domain $\Omega \subset \mathbb{R}^d$

$$u \in H_0^1(\Omega)^m: \quad a(u, v) = l(v) \quad \forall v \in H_0^1(\Omega)^m\,. \tag{11}$$

We now first introduce the Galerkin projection according to the isogeometric paradigm and then state the PDE solver for Algorithm 1.

3.1 Galerkin Projection in IGA

The approximation space for the state equations in the form of (11) is in isogeometric analysis

$$\mathcal{V}_h := \{v_h \in H_0^1(\Omega)^m \colon v_h \circ G \in \mathcal{M}^m\} \tag{12}$$

where m is the dimension of $u \colon \Omega \to \mathbb{R}^m$ and $\mathcal{M}$ is a NURBS space of space dimension d. Typically in IGA, it is the same NURBS space as for the geometric representation in (8), i.e. $\mathcal{M} = \mathcal{S}$, but that is not mandatory.

The important point also for shape optimization later is, that we move the dependency on the domain from the Sobolev space to the bilinear and linear form:

$$a_G(\hat{v}, \hat{w}) := \int_{\hat{\Omega}} L_G \hat{v}\hat{w} |\det J_G| \, d\hat{\Omega} \,, \quad L_G := L \circ G \tag{13}$$

$$l_G(\hat{v}) := \int_{\hat{\Omega}} f \circ G \cdot \hat{v} |\det J_G| \, d\hat{\Omega} \,. \tag{14}$$

The variational form which we solve in IGA then is

$$\hat{u}_h \in \mathcal{M}_0^m \colon a_G(\hat{u}_h, \hat{v}_h) = l_G(\hat{v}_h) \quad \forall \hat{v}_h \in \mathcal{M}_0^m \,, \tag{15}$$

which is equivalent to (11) for $\hat{u}_h = u_h \circ G$. The NURBS space $\mathcal{M}_0^m := \mathcal{M}^m \cap H_0^1(\hat{\Omega})^m$ contains only those NURBS $\hat{v}_h \in \mathcal{M}^m$ which respect (possible) Dirichlet boundary conditions. Note, that through the transformed differential operator L_G we also introduce J_G^{-T} into the forms a_G and l_G besides $\det J_G$.

All in all, the solution of (15) has the form

$$\hat{u}_h = \sum_{i=1}^{n} Q_i M_i(\hat{x}) \text{ with solution coefficients } Q_i \in \mathbb{R}^m \text{ and } M_i \in \mathcal{M} \,. \tag{16}$$

The Poisson equation is an example for such a transformed notation. In that case, the dimension of the solution is $m = 1$ and

$$a_G(\hat{v}, \hat{w}) = \int_{\hat{\Omega}} \nabla \hat{v}^T J_G^{-1} J_G^{-T} \nabla \hat{w} |\det J_G| \mathrm{d}\hat{x} \,.$$

A vector field example with $m = 2$ is linear elasticity for which a weak IGA form is found in (52) in Sect. 6 about model problems.

3.2 PDE Solver

Going back to the Basic Shape Optimization Algorithm 1, line 2 corresponds to assembling and solving equation (15). This leads to the linear system

$$KQ = F, \quad K_{iljk} := a_G(M_i e_l, M_j e_k), \quad F_{il} := l_G(M_i e_l) \tag{17}$$

for the model problems with stiffness matrix K, right hand side F and coefficients of the solution $\hat{u} = \sum_{i=1}^{n} \sum_{l=1}^{m} Q_{il} M_i e_l(\hat{x})$. We silently assume that the coefficients $Q_{il} \in \mathbb{R}$ can be ordered lexicographical in a solution vector Q, for instance like

$$Q = \left(Q_{11}\ Q_{12}\ \dots\ Q_{n1}\ \dots\ Q_{nm}\right)^T$$

and in the same breath we write

$$Q_i = \left(Q_{i1}\ \dots\ Q_{im}\right)^T.$$

Hence, for any admissible domain Ω given by a NURBS parameterization $G = \sum_{i=1}^{n_N} X_i N_i \in \mathcal{S}^d$ and a test function space of NURBS $\mathcal{M}^m$ the procedure PDE(Ω) is given by Algorithm 2.

Algorithm 2 PDE(Ω)

for all basis functions $M_i e_l$ **do**
    for all basis functions $M_j e_k$ **do**
        $K_{iljk} \leftarrow a_G(M_i e_l, M_j e_k)$
    end for
    $F_{il} \leftarrow l_G(M_i e_l)$
end for
Apply boundary conditions of state to K and F
Solve $KQ = F$
$u_h \leftarrow \sum_{i=1}^{m} Q_i M_i$.

4 Shape Gradients in Isogeometric Analysis

Gradient based optimization routines use gradients of the objective function and of the constraints to find a descent direction towards a (local) minimum and also as first order optimality condition. In case of shape optimization, the objective and constraints depend on a domain from the space of admissible domains and mostly also on a state u defined on that domain. Hence, a gradient would be w.r.t. a shape.

In classical shape optimization such gradients are obtained by defining perturbations of identity

$$F_t(x) = id(x) + th(x) \tag{18}$$

where a perturbed domain results from $\Omega_t := \{x_t = F_t(x) : x \in \Omega\}$. The vector fields h must be chosen in a way that respects certain regularity conditions, see for instance [22]. Then, for a function $\phi(\Omega)$ the derivative in direction of h is understood as

$$\phi'(\Omega; h) := \lim_{t \to 0+} \frac{1}{t} \big(\phi(\Omega_t) - \phi(\Omega)\big) = \mathrm{d}_t \phi(\Omega_t)|_{t=0} \,. \tag{19}$$

Next, we derive the corresponding formulas in IGA for NURBS with fixed weights in Sect. 4.1 and then extend them to variable weights in 4.2.

4.1 Isogeometric Shape Sensitivities

In isogeometric analysis we have for $x \in \Omega$ that $x = G(\hat{x})$ for some $\hat{x} \in \hat{\Omega}$. Applying this transformation to the perturbation of identity (18) yields

$$G_t(\hat{x}) := F_t \circ G(\hat{x}) = G(\hat{x}) + th \circ G(\hat{x}) \,. \tag{20}$$

Choosing h true to the isogeometric paradigm $\theta := h \circ G \in \mathcal{S}^d$, as shown in Fig. 1, this results in the isogeometric perturbation of identity

$$G_t(\hat{x}) = G(\hat{x}) + t\theta(\hat{x}) \text{ and } \Omega_t = G_t(\hat{\Omega}) \,. \tag{21}$$

Therefore it makes sense to notate $\phi(G)$ and the derivative in direction of θ as

$$\phi'(G; \theta) = \lim_{t \to 0+} \frac{1}{t} \big(\phi(G_t) - \phi(G)\big) = \mathrm{d}_t \phi(G_t)|_{t=0} \,. \tag{22}$$

The perturbations in isogeometric analysis fulfill the regularity conditions such that the directional derivatives constitute the shape gradient. As we have shown in [6]

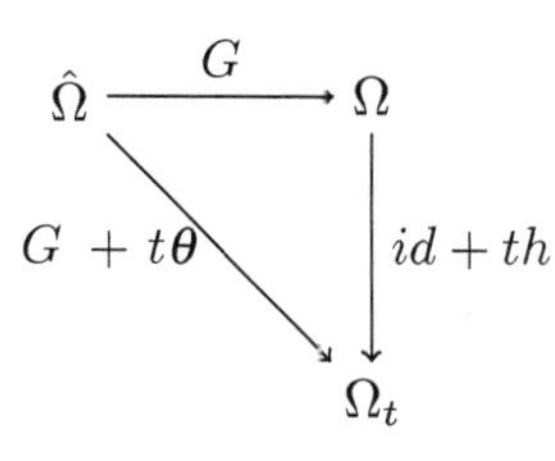

Fig. 1 Perturbation of identity with IGA

the directional derivatives of integrals in isogeometric analysis of the form

$$\phi(G) := \int_{\Omega} j_1(x)\mathrm{d}x + \int_{\partial\Omega} j_2(s)\mathrm{d}s \tag{23}$$

have the expression

$$\phi'(G;\theta) = \int_{\hat{\Omega}} \big(\nabla j_1 \circ G \cdot \theta + j_1 \circ G\,\mathrm{trace}(J_G^{-1} D\theta)\big)|\det J_G|\mathrm{d}\hat{x} \tag{24}$$

$$+ \int_{\partial\hat{\Omega}} \Big(\nabla j_2 \circ G \cdot \theta |J_G^{-T}\hat{n}| + j_2 \circ G\,\mathrm{trace}(J_G^{-1} D\theta) \tag{25}$$

$$- j_2 \circ G \frac{\hat{n}^T J_G^{-1} J_G^{-T} D\theta^T J_G^{-T}\hat{n}}{|J_G^{-T}\hat{n}|}\Big)|\det J_G|\mathrm{d}\hat{s}\ . \tag{26}$$

The isogeometric shape sensitivities in Eqs. (24), (25), and (26) can be obtained by using existing classical results and the transformation rule, or can be developed by solving at $t = 0$

$$d_t\phi(G_t) = d_t \int_{\hat{\Omega}} j_1 \circ G_t |\det J_{G_t}|\mathrm{d}\hat{x} + d_t \int_{\partial\hat{\Omega}} j_2 \circ G_t \|d_s G_t\|_2 \mathrm{d}\hat{s}\ . \tag{27}$$

Also in [6], we proved for linear elliptic PDEs of second order that the directional derivative from solving (24), (25), and (26) for $\theta = N_i e_k$ for a B-spline $N_i e_k \in \mathcal{S}^d$ is equal to the standard derivative $d_{X_{ik}}\phi$, i.e. taking the derivative of ϕ w.r.t. a control point component k of X_i

$$d_{X_{ik}}\phi(G) = d_{X_{ik}} \int_{\hat{\Omega}} j_1 \circ G|\det J_G|\mathrm{d}\hat{x} + d_{X_{ik}} \int_{\partial[0,1]^2} j_2 \circ G|\det J_G|\mathrm{d}\hat{s}\ . \tag{28}$$

Reversing the order of differentiation and integration shows that in (28) we, too, need $\mathrm{d}_{X_{ik}}G$ and $\mathrm{d}_{X_{ik}}|\det J_G|$. An explicit, efficient term has been derived in [20] and [27].

The two notions of derivatives stem from two ways of approaching optimal control problems: in the first case, the optimality conditions are derived for the continuous problem in an infinite dimensional function space setting, then all occurring function spaces and operators are approximated by finite dimensional ones, leading to the directional derivatives in (23). In the second case, the continuous problem is discretized by substituting all infinite dimensional function spaces with finite approximations, and then the optimality criteria are derived. It so happens, that the spline spaces are isomorphic to real vector spaces and therefore the directional derivatives are the standard derivatives in isogeometric shape optimization.

The above derivative (24) is a domain representation. In literature, one can also find a boundary representation by using the so called Hadamard structure theorem

[5, Theorem 3.6]. Basically, it says that a shape derivative in direction h of the form

$$\int_{\Omega} \operatorname{div}(j_1(x)h(x))\mathrm{d}x = \int_{\partial\Omega} j_1(x)h(x)\mathrm{d}s \tag{29}$$

with Stokes' divergence theorem. In this work, we only consider domain representations, but all steps can be applied to Hadamard sensitivities as well. In [2], such Hadamard representations have been used together with IGA in linear elasticity compliance minimization. In both cases, the directions θ of the derivatives are chosen such that θ is a basis function living on the *moving* boundary. This is, of course, natural for a boundary representation, but it also reduces the degrees of freedom for the domain representation significantly. By doing so, we speak of such basis functions $\theta = N_i e_k \neq 0$ on the moving boundary as well as their corresponding control points X_{ik} as *design variables* and denote their number by n_{des}.

4.2 Weight Optimization

For the following argument, we explicitly denote the NURBS representation of G with NURBS from Eq. (7) as

$$G(\hat{x}) = \sum_{i=1}^{n} X_i \frac{\omega_i N_i(\hat{x})}{\sum_{j=1}^{n} \omega_j N_j(\hat{x})} \,. \tag{30}$$

In the light of taking the derivative w.r.t. a control point in Eq. (28), it is not far fetched to get the derivative w.r.t. a weight ω_i for rational geometries G as in (30), which implies the need for $\mathrm{d}_{\omega_i} G$ and $\mathrm{d}_{\omega_i} |\det J_G|$, an explicit derivation can again be found in [20]. In [6] the authors have shown how to embed weight optimization in the function space setting using homogeneous coordinates. We emphasize here, that in the case of variable weights, the formulas for the two approaches of directional derivatives are not the same. One could say, that in the function space setting a term is missing. This is compensated by a different update rule, though. See also Sect. 5 for that.

We use this in the following sensitivity formula. For a shape functional ϕ like (23), we obtain the isogeometric shape sensitivities in direction $\tilde{\theta} \in \mathcal{S}^{d+1}$ for $G = H \circ \tilde{G}$ and $\theta = H \circ \tilde{\theta}$ as

$$\begin{aligned} &\phi'(G;\theta) \\ &= \int_{\hat{\Omega}} \left(\nabla j_1 \circ G \cdot DH \circ \tilde{G}\tilde{\theta} + j_1 \circ G \operatorname{trace}(J_G^{-1}\dot{D})\right) |\det J_G| \mathrm{d}\hat{x} \end{aligned} \tag{31}$$

$$+\int_{\hat{\Gamma}} \left(\nabla j_2 \circ G \cdot DH \circ \tilde{G}\tilde{\theta} |J_G^{-T}\hat{n}| + j_2 \circ G |J_G^{-T}\hat{n}| \operatorname{trace}(J_G^{-1}\dot{D}) + \right. \tag{32}$$

$$\left. -j_2 \circ G \frac{\hat{n}^T J_G^{-1} J_G^{-T} \dot{D}^T J_G^{-T} \hat{n}}{|J_G^{-T}\hat{n}|} \right) |\det J_G| \mathrm{d}\hat{x} \tag{33}$$

with $\dot{D} := \mathrm{d}_t D(H \circ \tilde{G}_t)|_{t=0} = D^2 H \circ \tilde{G} \odot \tilde{\theta} D\tilde{G} + DH \circ \tilde{G} D\tilde{\theta}$, where

$$D^2 H \circ \tilde{G} \odot \tilde{\theta} = \frac{1}{w^2} \left((0_d, -e_1)\tilde{\theta}, \ldots, (0_d, -e_d)\tilde{\theta}, (-I_d, 2G)\tilde{\theta} \right) \tag{34}$$

$$= \begin{cases} \frac{1}{w^2} \left(0_d, -\theta^w \right) & , \text{if } \tilde{\theta}_{d+1} = 0 \\ \frac{-\tilde{\theta}_{d+1}}{w^2} \left(I_d, \frac{1}{\tilde{\theta}_{d+1}} \theta^w - 2G \right) & , \text{if } \tilde{\theta}_{d+1} \neq 0 \,. \end{cases} \tag{35}$$

$$DH \circ \tilde{G} = \frac{1}{w} \left(I_d, -G \right). \tag{36}$$

In this formulation $\tilde{\theta} = N_i e_k$ for e_k a unit vector of $\mathbb{R}^{d+1}$ and $k = 1, \ldots, d$ corresponds to varying control point components X_{ik} and $k = d+1$ varies weights ω_i. However note, that $\phi'(G;\theta) \neq \mathrm{d}_X \phi(G)$ anymore, where $X = X_{ik}$ or ω_i.

4.3 Regularization

Though both interpretations of directional derivatives, (27) and (28), lead to the same system for the *directional derivative*, in the functional space setting the gradient is a member of a Hilbert space H and thus depending on the underlying scalar product $(\cdot,\cdot)$. For a Fréchet-differentiable real valued function ϕ, the gradient denoted by $\nabla\phi \in H$ is defined as the solution to

$$(\nabla\phi, \theta) = \phi'(G;\theta) \quad \forall \theta \in H \,. \tag{37}$$

This leaves some freedom in choosing the scalar product $(\cdot,\cdot)$ and hence the regularity of $\nabla\phi$. According to [19] such a regularization has the effect of preconditioning the optimization process.

As yet we have not used regularization in any of the model problems, however, in [2] an example can be found that combines regularization with extending the gradient to the inside, or what we call mesh movement and treat later in Sect. 5.1.

4.4 With State Equation

Typically, the cost functional J depends on the domain Ω but also on the solution u of a PDE over the domain, $u = u(\Omega)$. So formally, a shape derivative of J employs the chain rule and yields

$$\mathrm{d}_\Omega J = \partial_\Omega J + \partial_u J \mathrm{d}_\Omega u \,. \tag{38}$$

This sloppy notation can be justified by both approaches, *discretize first–optimize then* and vice versa, and realized with the adjoint state. All in all, we have the following necessary optimality system for an optimal shape $\Omega = G(\hat{\Omega})$ in IGA.

1. State equation (PDE): $u \in \mathcal{V}_h$: $u = \sum Q_i M_i$:

$$a_G(u, v) = l_G(v) \quad \forall v \in \mathcal{V}_h \qquad \Leftrightarrow \qquad KQ = F \tag{39}$$

2. Adjoint state (chain rule): $z \in \mathcal{V}_h$: $z = \sum P_i M_i$:

$$a_G(v, z) = \mathrm{d}_u J(\Omega, v) \quad \forall v \in \mathcal{V}_h \qquad \Leftrightarrow \qquad K^T P = \big(\mathrm{d}_u J(\Omega, M_k)\big)_k \tag{40}$$

3. Stationary point (shape gradient): For all design variables $N_k e_l$ and $X_{k,l}$ respectively

$$\begin{aligned} J'(\Omega, \hat{u}; N_k e_l) + l'_G(\hat{z}; N_k e_l) - a'_G(\hat{u}, \hat{z}; N_k e_l) = 0 \qquad \Leftrightarrow \\ \partial_{X_{k,l}} J + p^T(\partial_{X_{k,l}} F - \partial_{X_{k,l}} KQ) = 0 \end{aligned} \tag{41}$$

with $l'_G(\hat{v}; \theta) = \mathrm{d}_t l_{G+t\theta}(\hat{v})|_{t=0}$ and $a'_G(\hat{y}, \hat{v}; \theta) = \mathrm{d}_t a_{G+t\theta}(\hat{y}, \hat{v})|_{t=0}$.

An example of such shape derivatives in IGA notation can be found in Sect. 6.

4.5 Shape Gradient Computation

In this section we summarize in Algorithm 3 the previous points in one procedure for Algorithm 1. Moreover, we make a practical consideration on the relation between the (possible) two discretizations, simulation and optimization mesh. When using two meshes, we have shape sensitivities like $a'_G(u, z; N_k e_l)$ which involve integrals over basis functions $N_k e_l$ from geometry and $M_i e_j$ from simulation. Therefore, the domain of integration will be the mutual support $\mathrm{supp}(N_k) \cap \mathrm{supp}(M_j)$. For an easy implementation, it is favorable when the meshes match, i.e. if one mesh is the h-refinement of the other. We then can assemble over the finer mesh and the mutual support is found straight forward.

Algorithm 3 SHAPE GRAD(Ω, u)

Shape gradient of objective function ▷ **compute terms in** (41) **for objective function**

for all basis functions $M_i e_l$ **do** ▷ assemble right hand side of adjoint equation
  $B_{i,l} \leftarrow \mathrm{d}_u J(\Omega, M_i e_l)$
end for
Solve $K^T P = B$: adjoint state $z = \sum_{i=1}^{m} P_i M_i$ ▷ solve adjoint
for all design variables $N_k e_l$ **do** ▷ directional derivatives of objective function
  $\nabla J_{k,l} \leftarrow J'(\Omega, u; N_k e_l) + l'_G(z; N_k e_l) - a'_G(u, z; N_k e_l)$
end for

Shape gradient of constraint functions

for all design variables $N_k e_l$ **do** ▷ directional derivatives of constraints
  $\nabla C_{k,l} \leftarrow (g'(G; N_k e_l), h'(G; N_k e_l))$ derived from Eqs. (24), (25), and (26)
end for

Regularization ▷ **if needed**

for all design variables $N_k e_l$ **do** ▷ assemble system matrix
  for all design variables $N_r e_s$ **do**
    $R_{klrs} \leftarrow (N_k e_l, N_r e_s)$ with scalar product $(\cdot, \cdot)$ of Hilbert space H of wanted regularity
  end for
end for
Solve $Ry = \nabla J$, $\nabla J \leftarrow y$
Solve $Ry = \nabla C$, $\nabla C \leftarrow y$

5 Geometry Update

In Algorithm 1 we need a means how to update shapes. This is also given by the perturbation of identity (18) $\Omega_t = F_t(\Omega) = (id + th)(\Omega)$ or in IGA by (21)

$$\Omega_t = G_t(\hat{\Omega}) = G(\hat{\Omega}) + t\theta(\hat{\Omega}) \ . \tag{42}$$

Since in IGA θ is from the same NURBS space as the geometry function G, the update rule in IGA consists of adding control points.

A special case is updating geometries from NURBS spaces with variable weights. There we have

$$\Omega_t = H(\tilde{G} + t\tilde{\theta})(\hat{\Omega}) \tag{43}$$

where $\tilde{G}$ and $\tilde{\theta}$ are NURBS geometries in homogeneous coordinates.

Before we cast the update rule into a procedure, we have to consider another issue. Often, large steps result in infeasible parameterizations. However, it can partly be avoided with the right mesh movement algorithm.

5.1 Mesh Movement

In case of large deformation at the boundary of a geometry due to updates by shape gradients it frequently leads to infeasible parameterizations, for instance, when the control polygon overlaps with inner control lines, as in Fig. 2. Such tangled meshes are no phenomenon of IGA nor of shape optimization alone. Several techniques have been successfully applied in IGA:

- minimizing the Winslow functional in electromagnetic shape optimization, which aims for a positive determinant of the Jacobian of G [15],
- solving a linear elasticity problem with the displacement of control points on the boundary as Neumann boundary condition [2],
- Analysis-aware IGA meshes [28], which also optimize the magnitude of the determinant of the Jacobian of G.

We write here very generally that a mesh movement function $\Phi\colon \mathbb{R}^{n_{des}} \to \mathbb{R}^{nd}$ propagates a descent direction s to all control points, where s results from the sensitivity analysis and has the size n_{des} corresponding to the number ob design variables as in Sect. 4.1 The focus is here on an algorithmic framework for shape optimization, therefore we assume that Φ is given by a mesh movement method, for instance from above or the one we used in [6]. In the end, the update in IGA is carried out by changing the control points, with Φ respecting fixed boundaries and the movement on the optimization boundary, i.e.

$$\begin{cases} y_\ell = s_\ell, & \text{if } N_{i_\ell} e_{k_\ell} \text{ is design variable,} \\ y_\ell = 0, & \text{if } N_{i_\ell} e_{k_\ell} \text{ is on fixed boundary .} \end{cases} \tag{44}$$

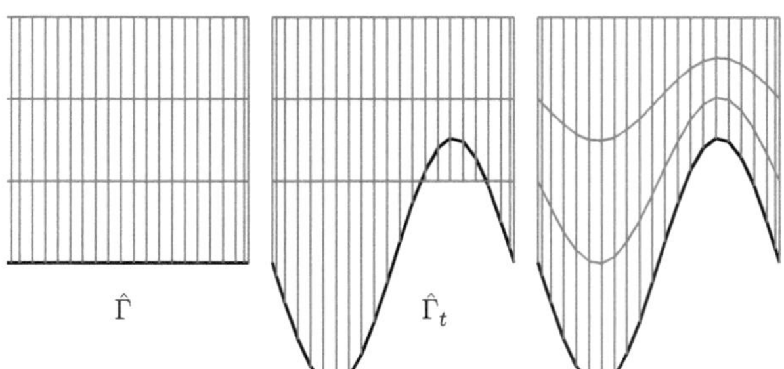

Fig. 2 Moving a boundary which results in non-convex domains can cause irregular parameterizations by overlapping control polygons, hence a suitable method to also move inner control points must be applied

5.2 Computation of New Geometry

Here, we give the procedure for updating a geometry in Algorithm 1. The descent direction s depends on what the optimization routine returns on the basis of ∇J and ∇C but it is up to the user how to update the geometry with it and possibly to move the mesh. We summarize the whole procedure in Algorithm 4. As mentioned above, the update rules differ for NURBS with variable weights in the discretize first and optimize first ansatz. For a comparison with line 6, we state the update rule of discretize first in the following code snippet

$X_i \leftarrow X_i + (s_{i,1}, \ldots, s_{i,d})^T$
$\omega_i \leftarrow \omega_i + s_{i,d+1}$.

Hence, for $S_i := (s_{i,1}, \ldots, s_{i,d})^T$ and $W_i := s_{i,d+1}$ an update rule with homogeneous coordinates results in a geometry

$$G_{new} = \sum_{i=1}^{n} (X_i \omega_i + S_i) \frac{N_i}{\sum_{j=1}^{n} (\omega_j + W_j) N_j} \tag{45}$$

as opposed to the new geometry from the discretize first approach

$$G_{new} = \sum_{i=1}^{n} (X_i + S_i)(\omega_i + W_i) \frac{N_i}{\sum_{j=1}^{n} (\omega_j + W_j) N_j} . \tag{46}$$

Algorithm 4 UPDATE(Ω, s)

Require: stiffness matrix A of mesh moving problem
Require: geometry function G of initial domain Ω_0
Require: descent direction s from black-box gradient descent optimizer
1: apply mesh movement function $y = \Phi(s)$ to propagate the change also to the inside control points, $s \leftarrow y$
2: **if** G is a B-spline geometry or a NURBS geometry with fixed weights **then**
3:     **for all** geometry basis functions N_i **do**
4:         $X_i \leftarrow X_i + (s_{i,1}, \ldots, s_{i,d})^T$
5:     **end for**
6: **else** G is a NURBS geometry with variable weights
7:     **for all** geometry basis functions N_i **do**
8:         $X_i \leftarrow X_i + (s_{i,1}, \ldots, s_{i,d})^T / s_{i,d+1}$
9:         $\omega_i \leftarrow \omega_i + s_{i,d+1}$
10:    **end for**
11: **end if**

6 Model Problems

In this section we apply the complete Algorithm 1 with the procedures 2–8 to two model problems. These examples, though simple, illustrate the isogeometric shape optimization process and moreover, expose possible weaknesses, which have to be considered. Both examples were also treated by the authors in [6].

We treat two optimization problems: maximizing an area and increasing stiffness. The first one involves no PDE, but demonstrates the influence of quadrature errors in optimizing weights and control points simultaneously. We review it here under this aspect which sheds new light on using weights as optimization variables.

The second problem is well known in shape optimization and has been treated already with isogeometric analysis, also by others. We include it here, for two reasons. Firstly, it is posed in the discretize first setting in IGA literature, so here we view it also in the optimize first picture. Secondly, in this example we show practically the use of two meshes.

6.1 Area Maximization

Since NURBS are able to represent conics exactly, we treat first a simple example without PDE to show the use and also pitfalls of isogeometric shape optimization with variable weights.

Example 1 (Area maximization)

$$\begin{aligned} &\min J(\Omega) = -\int_{\Omega} 1 \mathrm{d}x \quad \text{subject to} \\ &h(\Omega) = \int_{\partial\Omega} 1 \mathrm{d}s - P_0 = 0 \text{ with a constant } P_0 \,. \end{aligned} \tag{47}$$

Shape gradients for J and h without variable weights are given by Eqs. (24), (25), and (26), where $j_1 = 1$ covers the volume and $j_2 = 1$ the perimeter functionals.

$$J'(\Omega;\theta) = -\int_{\hat{\Omega}} \operatorname{trace}(J_G^{-1} D\theta) |\det J_G| \mathrm{d}\hat{x} \tag{48}$$

$$h'(\Omega;\theta) = \int_{\partial\hat{\Omega}} \Big(\operatorname{trace}(J_G^{-1} D\theta) - \frac{\hat{n}^T J_G^{-1} J_G^{-T} D\theta^T J_G^{-T} \hat{n}}{|J_G^{-T}\hat{n}|}\Big) |\det J_G| \mathrm{d}\hat{s} \,. \tag{49}$$

Considering also variable weights, the shape derivative equations (31), (32), and (33) yield

$$J'(\Omega;\theta) = -\int_{\hat{\Omega}} \left(\operatorname{trace}(J_G^{-1}\dot{D})\right) |\det J_G| \mathrm{d}\hat{x} \tag{50}$$

$$h'(\Omega;\theta) = \int_{\hat{\Gamma}} \Bigg(|J_G^{-T}\hat{n}| \operatorname{trace}(J_G^{-1}\dot{D}) + \\ - \frac{(J_G^{-T}\hat{n})^T J_G^{-T}\dot{D}^T J_G^{-T}\hat{n}}{|J_G^{-T}\hat{n}|} \Bigg) |\det J_G| \mathrm{d}\hat{x}\,. \tag{51}$$

6.1.1 Results

We start the optimization process with a square $[-1, 1]^2$ given by B-splines with degree $p = 1$ to 4 over knot vectors $(0, \dots, 0, 1, \dots, 1)$ in both space dimensions. The first and last knot are repeated $p + 1$ times and $V_0 = 4$ and $P_0 = 8$. A convergence study of Fig. 3 reveals, two things. First, we see that non-rational B-splines behave as one expects. Namely, since higher degree B-splines can approximate conics better, they should perform better, i.e. giving smaller errors than lower degree approximations. Second, one anticipates h-convergence for a fixed degree on the same grounds. Both is indicated in the plot. For $p = 4$ the flat behaviour after two refinements is due to reaching machine precision. A first

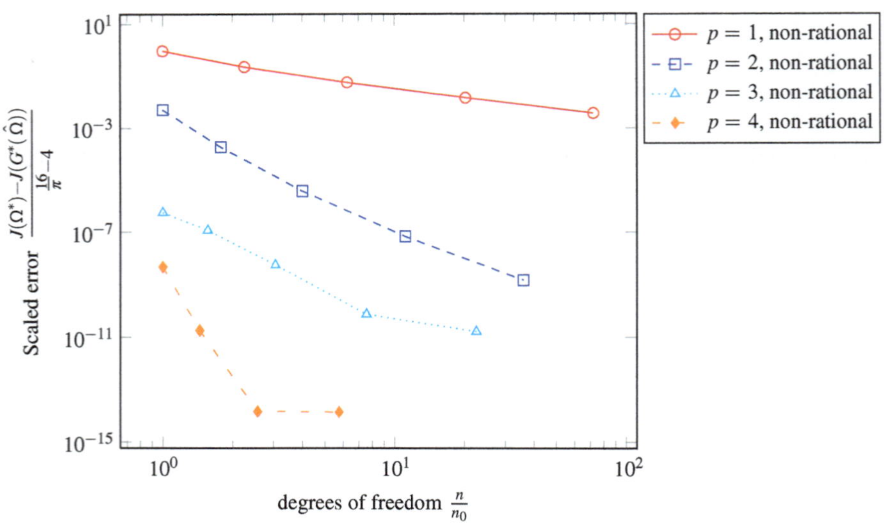

Fig. 3 Convergence study of Example 1 with non-rational B-splines of different degrees p

guess would assign the same behaviour to rational B-splines, even expect a much lower error for $p = 2$ already, because the disk has a representation in this space. We use again the knot vectors $(0, 0, 0, 1, 1, 1)$ in both space direction with 3×3 rational B-spline functions and 9 weights, all equal to 1; control points are such that we have the same initial square as above. Since rational B-splines are rational polynomials we have to take quadrature errors into account. We pursue this in the following before starting optimization with NURBS.

6.1.2 Influence of Quadrature Errors

When picking efficient quadrature rules for the assembly of a_G and l_G in IGA, transformation terms $\det J_G$ and J^{-T} are neglected in this search by [9] on the basis that these terms change slowly in comparison to other factors. This is in general valid since the geometry is fixed at a coarse level with degree p_0 and mesh size parameter h_0; for analysis a (several times) p- and h-refined version is used. Hence, G is a polynomial with degree $p_0 \leq p$ and almost constant on elements of sizes $h \ll h_0$. This carries over to the transformation terms. However, this assumption does not hold anymore when the geometry G has the same refinement level or worse, for cost functionals or for constraints on Ω without state, as for instance a volume constraint in the form of $h(\Omega) = \int_\Omega 1 \mathrm{d}x - const.$ In IGA, this yields $h(G) = \int_{\hat{\Omega}} |\det J_G| \mathrm{d}\hat{x}$.

Using rational B-splines, this simple constraint already exhibits a need for higher order quadrature rules. That is, if we enforce the volume constraint by a tolerance smaller than 10^{-6} in the optimization routine, Table 1 shows that we need at least more than 5 quadrature points in each direction to get rid of integration errors.

6.1.3 Optimization with NURBS

For Example 1 with NURBS of degree $p = 2$ and variable weights the exact disk Ω^* is in this NURBS space, so we expect an error $J(\Omega^*) - J(G^*(\hat{\Omega}))$ dominated by quadrature errors and optimization parameters. In Fig. 4 a visually good result is obtained for the following setting. A Gauß quadrature with 10 quadrature points in each space direction is used. The perimeter constraint is enforce within a tolerance of 10^{-8} and as stopping criteria for the SQP optimizer in the package NLOPT [10] the same tolerance is chosen. In Fig. 4 no difference between exact and numerical

Table 1 Quadrature errors in area and perimeter of a circle with radius 1 given by exact representation with NURBS of degree $p = 2$

#quad. points	5	9	15
Area	$\approx 8 \cdot 10^{-06}$	$\approx 3 \cdot 10^{-11}$	$\approx 1 \cdot 10^{-15}$
Perimeter	$\approx 9 \cdot 10^{-07}$	$\approx 2 \cdot 10^{-12}$	$\approx 9 \cdot 10^{-16}$

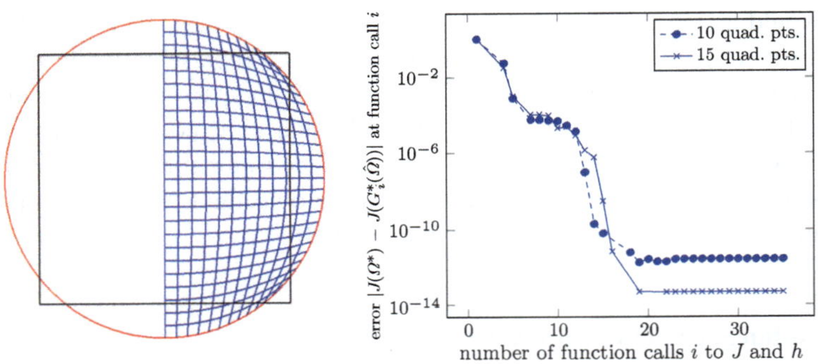

Fig. 4 Result for Example 1 with initial domain (*square*): The optimal numerical shape (plotted as IGA mesh on half disk) is a NURBS approximation of a disk (exact disk plotted as boundary *circle*). On the right, the history of (feasible) SQP optimization steps is shown

optimum can be detected. In fact, the true error in the cost functional is $|J(\Omega^*) - J(G^*(\hat{\Omega}))| \approx 2.1 \cdot 10^{-10}$ after 34 function calls, i.e. iterations. For an even smaller error we increase the quadrature rule to 15 points in each direction and require that the perimeter may not deviate more than 10^{-12} from $P_0 = 8$ and as stopping tolerance we give 10^{-9}. Then, the error $|J(\Omega^*) - J(G^*(\hat{\Omega}))|$ drops to $\approx 8.9 \cdot 10^{-15}$ after 34 function calls. It is unreasonable to expect anything less since from Table 1 we learn that the error in the objective is then dominated by quadrature errors.

Comparison with Non-rational B-Splines

We compare the results from rational and non-rational B-spline optimization and how many design variables we need in both cases for an error in the cost functional of order 10^{-15}. In case of NURBS with weight optimization we have 9 control points with 8 of them corresponding to basis functions with support on the boundary. Since they have two components, we end up with 16 optimization variables plus the 8 weights for each basis, amounting to 24 design variables in total. In case of B-splines without weights Fig. 3 shows that for the desired error we need B-splines of degree $p = 4$ which have been h-refined twice. This leaves us with 64 degrees of freedom of which 32 are nonzero at the boundary. Although we have less design variables using NURBS we pay for it by using higher quadrature rules. Together with a PDE constraint it might therefore be computationally more efficient to use B-splines of higher degrees if the optimal shape is expected to be conic.

6.2 Compliance Minimization

This prominent example has been treated already in [2, 6, 20, 27] with IGA. Our linear elasticity problem for plane stress has dimension $m = 2$ and the strong form

$$\begin{cases} -\operatorname{div}\sigma(u) &= 0 \text{ in } \Omega, \\ u &= 0 \text{ on } \Gamma_D, \\ \sigma(u)\cdot n &= g \text{ on } \Gamma_N, \\ \sigma(u)\cdot n &= 0 \text{ on } \Gamma, \end{cases} \tag{52}$$

with strain $\varepsilon(u) = \frac{1}{2}(\nabla u + \nabla u^T)$ and stress $\sigma(u) = 2\mu\varepsilon(u) + \lambda(\nabla\cdot u)I_2$. We use a constant vector field g in the Neumann boundary conditions on Γ_N, the domain is fixed on the boundary part Γ_D and allowed to move on Γ, where $\partial\Omega$ is the union of all closed boundaries $\bar{\Gamma}_D \cup \bar{\Gamma}_N \cup \bar{\Gamma}$. For this problem $E(u,\Omega) = 0$ is equivalent to

$$u \in H_0^1(\Omega)^2\colon \quad \int_\Omega \underline{\varepsilon}(u)^T C \underline{\varepsilon}(v) \mathrm{d}\Omega = \int_{\Gamma_N} g\cdot v \mathrm{d}\Gamma_N \quad \forall v \in H_0^1(\Omega)^2, \tag{53}$$

where $\underline{\varepsilon}(u) = \left(\varepsilon_{11}, \varepsilon_{22}, \frac{1}{2}(\varepsilon_{12}+\varepsilon_{21})\right)^T$ and elasticity tensor C characterized by the linearized stress-strain relation. Isogeometric Galerkin projection of the plane stress problem in Eq. (52) yields

$$a_G(M_i e_l, M_j e_k) = \int_{\hat{\Omega}} \underline{\varepsilon}_G(M_i e_l)^T C \underline{\varepsilon}_G(M_j e_k) |\det J_G| \mathrm{d}\hat{x}\,. \tag{54}$$

with $\underline{\varepsilon}_G$ obtained from the modified gradient $\nabla u \circ G = J_G^{-T}\nabla\hat{u}$. The linear form is transformed to $l_G(M_i e_l) = \int_{\hat{\Gamma}_N} g \circ G \cdot M_i e_l \|\mathrm{d}_{\hat{s}} G\| \mathrm{d}\hat{s}$ where $\hat{\Gamma}_N$ is the preimage of Neumann boundary Γ_N under G.

Example 2 (Compliance minimization) The shape optimization problem takes the form

$$\min J(\Omega, u) = \int_{\Gamma_N} g\cdot u \mathrm{d}x \tag{55}$$

$$\text{s.t. } h(\Omega) = \int_\Omega 1 \mathrm{d}x - V_0 = 0\,. \tag{56}$$

In this example we have a volume constraint on the domain with V_0 a constant, such as initial volume. Moreover, our domain of interest is a square plate with a hole, where the form of the hole is to be optimized. In [24] it is argued that a circular hole has maximal stiffness, i.e. minimal compliance. Since the problem is symmetric we consider only a quarter of the plate which is depicted in Fig. 5 together with its boundary conditions.

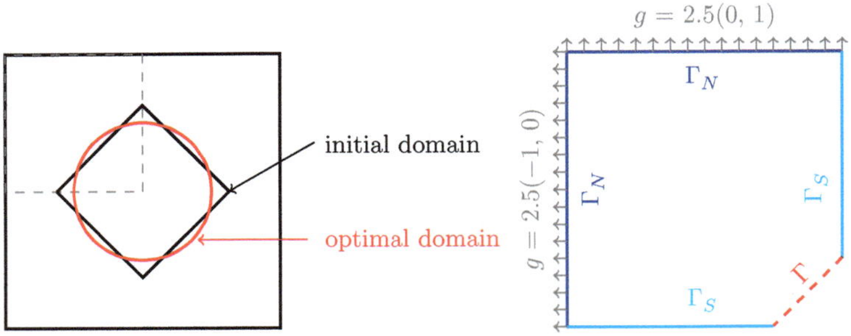

Fig. 5 Configuration of Example 2 where the quarter plate Ω is connected at boundary Γ_S to congruent plates symmetrically to give a closed hole, the moving boundary is given by Γ

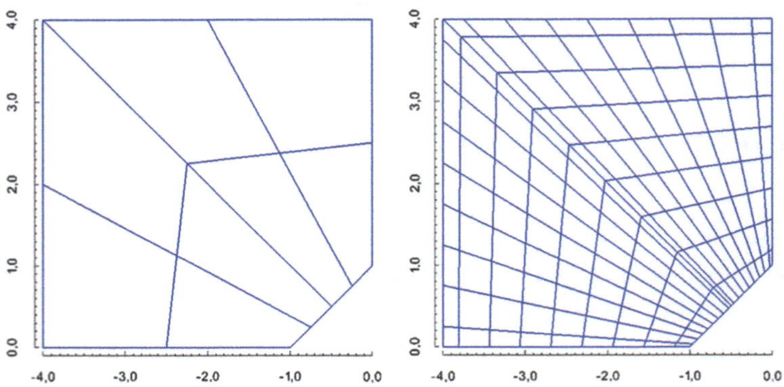

Fig. 6 *Left*: optimization mesh, *right*: simulation mesh, both degree 2, simulation mesh is 3 times h-refined, both depicted as control mesh

From Eq. (41) we deduce the directional shape derivative for the objective function in case of fixed weights using either a domain $J(\Omega, u) = a(u, u)$ or boundary representation $J(\Omega, u) = l(u)$ of the cost functional. This problem is self-adjoint, hence we have for the domain representation of the shape derivative of J

$$J'(\Omega, u; \theta) = \mathrm{d}_t \int_{\hat{\Omega}} \underline{\varepsilon}_{G+t\theta}(M_i e_l)^T C \underline{\varepsilon}_{G+t\theta}(M_j e_k) |\det J_{G+t\theta}| \mathrm{d}\hat{x} \bigg|_{t=0} . \tag{57}$$

The shape derivative of the volume constraint h is given by Eq. (48). For this example, we assume two separate levels of h-refinement in the simulation and optimization mesh, the parameterizations are shown in Fig. 6. As predicted, the optimization process leads to a circular hole in Fig. 7 and stress peaks are much alleviated.

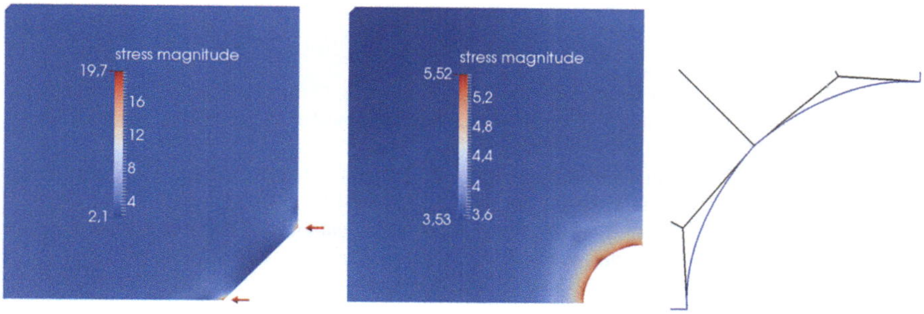

Fig. 7 *Left*: stress on initial physical domain with stress peaks at arrow tips, *middle*: stress on numerical optimal physical domain, *right*: control polygon and physical boundary of optimization mesh after end of optimization

7 Conclusion

In this work, we have treated shape optimization in isogeometric analysis in an algorithmic way. In particular, we formulated its pieces on a discretized level. Model examples verified the implementation. We relied on third party libraries for optimization like [10]. Of course, stopping tolerances of such optimization routines only make sense if we can provide reasonably good approximations of all occurring terms. Usually, quadrature errors go without saying in this discussion as they are easily diminished. Instead of neglecting them here, in the first problem we show that optimization with NURBS requires more carefully chosen quadrature rules than with B-splines. However, in future we mostly are interested in the less obvious influence of interpolation errors for state and controls, which we want to address separately. For that, the second example showcased the use of two different meshes originated from one initial parameterization. From a priori error estimates, also valid for IGA [1], interpolation power of state ansatz spaces are known. With such a detached simulation grid, the error in the state equation can be controlled by finer grids and, together with a posteriori error estimates, locally adaptive refinement [25] for simulation is possible. Lacking at the moment is a means to determine the refinement level of the optimization grid and the influence of its approximation error in the cost term. Our next steps involve first hierarchical B-splines for locally refined state solutions on a static optimization mesh. Next, a successive refinement at the boundary of interest of the optimization mesh is intended.

Acknowledgements We thank Utz Wever from Siemens AG, Corporate Technology, for many helpful discussions on the subject.

This work was supported by the European Union within the Project 284981 “TERRIFIC” (7th Framework Program).

References

1. Y. Bazilevs, L. Beirão da Veiga, J. Cottrell, T.J.R. Hughes, G. Sangalli, Isogeometric analysis: approximation, stability and error estimates for h-refined meshes. Math. Models Methods Appl. Sci. **16**(7), 1031–1090 (2006)
2. L. Blanchard, R. Duvigneau, A.V. Vuong, B. Simeon, Shape gradient for isogeometric structural design. J. Optim. Theory Appl. **161**(2), 1–7 (2013)
3. P.B. Bornemann, F. Cirak, A subdivision-based implementation of the hierarchical B-spline finite element method. Comput. Methods Appl. Mech. Eng. **253**, 584–598 (2013)
4. C. De Boor, *A Practical Guide to Splines*, vol. 27 (Springer, New York, 2001)
5. M.C. Delfour, J.P. Zolésio, *Shapes and Geometries: Metrics, Analysis, Differential Calculus, and Optimization*. Advances in Design and Control, 2nd edn. (SIAM, Philadelphia, 2011)
6. D. Fußeder, A.V. Vuong, B. Simeon, Fundamental aspects of shape optimization in the context of isogeometric analysis. Comput. Methods Appl. Mech. Eng. **286**, 313–331 (2015)
7. J. Haslinger, P. Neittaanmäki, *Finite Element Approximation for Optimal Shape Design: Theory and Applications* (Wiley, Chichester/New York, 1988)
8. T.J.R. Hughes, J.A. Cottrell, Y. Bazilevs, Isogeometric analysis: CAD, finite elements, NURBS, exact geometry and mesh refinement. Comput. Methods Appl. Mech. Eng. **194**(39), 4135–4195 (2005)
9. T.J.R. Hughes, A. Reali, G. Sangalli, Efficient quadrature for NURBS-based isogeometric analysis. Comput. Methods Appl. Mech. Eng. **199**, 301–313 (2010)
10. S.G. Johnson, The NLOPT nonlinear-optimization package (2014), http://ab-initio.mit.edu/nlopt. [Online; Accessed 10 Oct 2014]
11. J. Kiendl, R. Schmidt, R. Wüchner, K.U. Bletzinger, Isogeometric shape optimization of shells using semi-analytical sensitivity analysis and sensitivity weighting. Comput. Methods Appl. Mech. Eng. **274**, 148–167 (2014)
12. MATLAB, Release 2012a (2012), "The MathWorks Inc."
13. F. Murat, J. Simon, Etude de problèmes d'optimal design, in *Optimization Techniques Modeling and Optimization in the Service of Man Part 2* (Springer, Berlin/Heidelberg, 1976), pp. 54–62
14. D.M. Nguyen, Isogeometric analysis and shape optimization in electromagnetism. Ph.D. thesis, Technical University of Denmark, Feb 2012
15. D.M. Nguyen, A. Evgrafov, J. Gravesen, Isogeometric shape optimization for electromagnetic scattering problems. Prog. Electromagn. Res. B **45**, 117–146 (2012)
16. J. Nocedal, S.J. Write, *Numerical Optimization*. Springer Series in Operations Research and Financial Engineering, 2nd edn. (Springer, New York, 2006)
17. L. Piegl, W. Tiller, *The NURBS Book* (Springer, New York, 1995)
18. O. Pironneau, *Optimal Shape Design for Elliptic Systems* (Springer, New York, 1983)
19. B. Protas, T.R. Bewley, G. Hagen, A computational framework for the regularization of adjoint analysis in multiscale PDE systems. J. Comput. Phys. **195**, 49–89 (2004)
20. X. Qian, Full analytical sensitivities in NURBS based isogeometric shape optimization. Comput. Methods Appl. Mech. Eng. **199**, 2059–2071 (2010)
21. L.L. Schumaker, *Spline Functions: Basic Theory* (Wiley, New York, 1981)
22. J. Sokolowski, J.P. Zolésio, *Introduction to Shape Optimization: Shape Sensitivity Analysis*. Springer Series in Computational Mathematics, vol. 16 (Springer, Berlin/New York, 1992)
23. K. Svanberg, A class of globally convergent optimization methods based on conservative convex separable approximations. SIAM J. Optim. **12**(2), 555–573 (2002)
24. S. Timoshenko, J.N. Goodier, *Theory of Elasticity* (McGraw-Hill, New York, 1951)
25. A.V. Vuong, *Adaptive Hierarchical Isogeometric Finite Element Methods* (Springer, Wiesbaden, 2012)

26. A. Wächter, L.T. Biegler, On the implementation of a primal-dual Interior point filter line search algorithm for large-scale nonlinear programming. Math. Program. **106**(1), 25–57 (2006)
27. W.A. Wall, M.A. Frenzel, C. Cyron, Isogeometric structural shape optimization. Comput. Methods Appl. Mech. Eng. **197**, 2976–2988 (2008)
28. G. Xu, B. Mourrain, R. Duvigneau, A. Galligo, Optimal analysis-aware parameterization of computational domain in 3D isogeometric analysis. Comput. Aided Des. **45**(4), 812–821 (2013)

Effective Integration of Sophisticated Operators in Isogeometric Analysis with `igatools`

Nicola Cavallini, Oliver Weeger, M. Sebastian Pauletti, Massimiliano Martinelli, and Pablo Antolín

Abstract `igatools` is a newly released library for operators assembly in isogeometric analysis. The library, which is object oriented designed and written in `C++11`, is general purpose, therefore it is not devoted to any specific application. In this paper we show that such a design makes `igatools` an effective tool in assembling isogeometric discretizations of sophisticated differential operators. This effectiveness will be demonstrated showing code snippets relating one-to-one with the operators written on paper. To embrace a wide audience, applications from nonlinear incompressible solid and fluid mechanics will be addressed. In both cases we are going to deal with mixed isogeometric formulations. The applicative nature of this paper will be stressed solving industrially relevant tests cases.

1 Introduction

Scientific computing is an area where expertise from several backgrounds such as computer science, engineering, mathematics, and physics come together. Contributions to this scientific field range from most theoretical to the most applied ones.

N. Cavallini (✉)
Scuola Internazionale Superiore di Studi Avanzati, Via Bonomea 265, Trieste, Italy
e-mail: nicola.cavalli@sissa.it

O. Weeger
Faculty of Mathematics, TU Kaiserslautern, P.O. Box 3049, 67653 Kaiserslautern, Germany
e-mail: weeger@rhrk.uni-kl.de

M.S. Pauletti
Instituto de Matemática Aplicada del Litoral (IMAL), Consejo Nacional de Investigaciones científicas y técnicas (CONICET), Santa Fe, Argentinia

M. Martinelli
Istituto di Matematica Applicata e Tecnologie Informatiche (IMATI), Consiglio Nazionale delle Ricerche (CNR), Pavia, Italy

P. Antolín
Dipartimento di Ingegneria Civile ed Architettura, Università degli Studi di Pavia, Via Ferrata 3, 27100 Pavia, Italy

B. Jüttler, B. Simeon (eds.), *Isogeometric Analysis and Applications 2014*,
Lecture Notes in Computational Science and Engineering 107,
DOI 10.1007/978-3-319-23315-4_9

With this work we address the numerical solution of nonlinear partial differential equations arising from industrial applications in solid and fluid mechanics. The aim of the paper is to demonstrate that the newly released software library `igatools` allows an effective implementation of isogeometric finite element discretizations of rather complex differential operators, here in particular mixed formulations of nonlinear incompressible elasticity and Navier-Stokes equations, defined on complex domains.

As the software prefix `iga` suggests, we are dealing with isogeometric type of spaces. Isogeometric analysis has been introduced in [24] with the aim of bridging scientific computing and computational geometry. The basic idea here, is to use the shape functions that describe the geometry, as basis functions for a Galerkin method. Several aspects in computational modeling benefit from this linking. In particular, an appealing aspect of this approach is the higher inter element continuity that characterizes the shape functions describing the geometry and numerical solution. Among other contributions we would mention [13] for structural vibration, [27] for shell structures and [35] in non-linear vibration analysis. In fluid dynamics we would like to mention [6] for a variational multi scale modeling approach to turbulent flows and [19] for divergence conforming B-splines.

`igatools` is an object oriented library for discretization of partial differential equations, using isogeometric type of spaces. It has been first presented in [29] and it is available under GPL conditions at www.igatools.org. The idea that underlays the software design is encapsulating mathematical concepts used in isogeometric method into objects, namely classes. Mapping the interaction between classes we get methods used in the actual integration. The software is implemented in `C++11` [7, 26] and it makes an extensive use of generic programming, templates in `C++`. In this context, generic programming is particularly useful to obtain dimension independent code, a very interesting feature of the software.

This paper is devoted to the community of scientists curious about numerical solution implementation of partial differential equations. In this respect we thought at two model readers. Those who might be skeptic about object oriented programming in scientific computing, and to those who already have an acquaintance with this programming style. We aim at involving the first class of readers showing that a carefully designed software can help in implementing non elementary operators. In the second case we would like to capture their attention in a collaborative way. To pursue this twofold objective, we think there is no better way rather than presenting a simple way of implementing sophisticated operators. The implementation ease will be demonstrated providing code snippets, that prove a one-to-one correspondence between the way we write the operators on paper and the way these operators are coded in assembly loops.

Validation of the software will be provided with respect to literature test cases and, industrial applications will be presented to prove feasibility of the software in applicative contexts. In order to embrace the widest possible audience, we present two systems of partial differential equations, one arising from solid mechanics and one from fluid dynamics. In both cases the problems are nonlinear. Though this paper is not focused on numerical treatment of nonlinear systems, we would like to remark that the nonlinearities treatment is a matter of linear algebra strategies, that do not interfere with operators assembly.

One of the goals of this paper is to show that `igatools` can be an effective tool in industrially relevant applications. `igatools` development has been supported by several institutions and projects. In the context of this paper, a special mention goes to the TERRIFIC project, as the presented applications are partial fulfillment of the project itself. TERRIFIC (Towards Enhanced Integration of Design and Production in the Factory of the Future through Isogeometric Technologies) is a project within the seventh framework program of the European Union, that lasted from 1st of September 2011 to the 31st of August 2014. The aim of the project was to inject isogeometric analysis in an industrial context. The geometries presented are courtesy of ALENIA aeronautica, in particular the authors are grateful to L. Morrone and G. Mirra. The other supporting institutions and projects will be listed in the acknowledgments.

In Sect. 2 we continue with a brief summary of `igatools` design. As the applications we are going to deal with require mixed formulations, we recall the isogeometric formulation of Taylor-Hood elements in Sect. 3. In Sect. 4 we detail our computational mechanics applications and in Sect. 5 our fluid dynamics applications are addressed. Finally, in Sect. 6 we conclude the paper.

2 `igatools` Design Description

The software prefix `iga` restricts the discussion to isogeometric methods. Isogeometric methods commonly address Galerkin approximations, characterized by high inter element continuity of basis functions. Originally, in historical sense, basis functions for isogeometric methods are B-Splines and NURBS, the interested reader can refer to [30, 31] for a precise definition, and to [29] for `igatools` related notation. Other than these, new reference spaces have been introduced lately. The reader can refer to [5, 11, 15, 32] for T-splines, to [14] for locally refined splines, and to [20, 22, 34] for hierarchical splines. In this paper, we are going to use B-splines. B-splines of degree d and regularity r ($0 \leq r < d$) will be addressed as Splines S_r^d. In this section we briefly review the objects, or classes, that build the backbone of the software. As extensively explained in [29], these objects are designed to resemble the key mathematical concepts in isogeometric methods. A visual sketch of the major players is outlined in Fig. 1.

2.1 Reference Space (`BSplineSpace` and `NURBSSpace`)

The classes `BSPlineSpace` and `NURBSSpace` represent the shape functions defined on the parametric domain (usually the unit square in two dimensions and the unit cube in three dimensions). These spaces can be non-homogeneous, in the

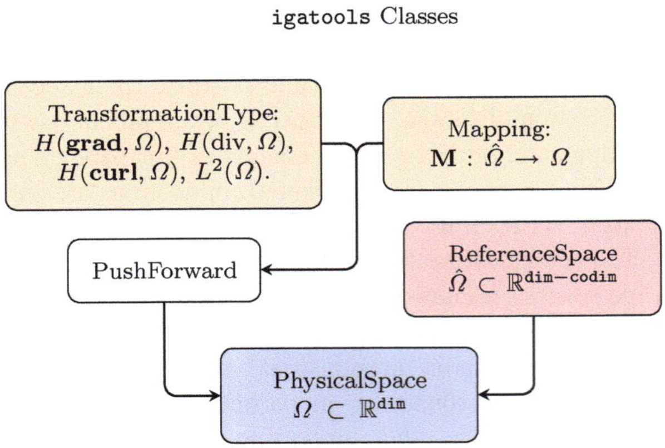

Fig. 1 In this sketch we picture the major object we isolated in isogeometric analysis, together with their mathematical symbols. We outlined that in general the dimension of the reference space can differ from the physical space, this difference is addressed as codimension `codim`

sense that they can be characterized by different degrees and regularities in each dimension.

2.2 *Mapping* (`Mapping`)

The mathematical mapping is the object that maps the parametric domain into the physical one. Notice that, in principle, there is no reason why this mapping should be restricted to isogeometric type of maps. Moreover, in the desired case of an isogeometric type of map, `Mapping` is not supposed to be refined together with the reference space to achieve convergence.

2.3 *Push Forward* (`PushForward`)

The `PushForward` object combines the `Mapping` together with the `TransformationType`. The transformation type defines how to transform functions. Depending on the different transformations, different operators can be preserved throughout the transformation itself. Example transformations are $H(\mathbf{grad}, \Omega)$, $H(\mathrm{div}, \Omega)$, $H(\mathbf{curl}, \Omega)$ and $L^2(\Omega)$.

2.4 *Physical Space* (`PhysicalSpace`)

The combination of the `ReferenceSpace` with the `PushForward` gives the `PhysicalSpace`. This class contains all the information necessary to recover point values of functions and derivatives on the physical domain. This point values are the ones used to assemble the desired operators.

Element Iterator (`ElementIterator`) One of the similarities between standard finite elements and the isogeometric method, is the decomposition of the domain in a collection of elements where a small number of functions have support. The global matrix, is built *iterating* through all the elements in the grid. In object oriented programming the mechanism that points to an element, and modifies itself to point to the next element is called iterator. In [29] we pointed out that we consider the previously mentioned classes as *grid-li ke* containers, this why the element level information is accessed via the `ElementIterator` class.

For an immediate flavor on how a simple $\int \nabla u \nabla v$ operator is assembled, we sketch a sort of "Hello World" program, in Listing 1. In the first four lines we define the presented classes. We start with a `CartesianGrid`, namely the knots without repetitions. This class is used to build the corresponding `BSplineSpace`. The transformation type (`h_grad`, in this case) couples with a `Mapping` to build the `PushForward`. Finally it only remains to instantiate the `PhysicalSpace` as the combination of `BSplineSpace` and `PushForward`. The mechanism that implies the initialization of the cache `elem->init_values()` and its filling is detailed in [29]. In this paragraph we limit ourselves to emphasize that we access

```
    auto grid = CartesianGrid<dim>::create(box, n_knots);
    auto ref_space = BSplineSpace<dim>::create(grid, deg);
    map        = BallMapping<dim>::create(grid);
    space      = PhysicalSpace<BSplineSpace<dim>,
        PushForward<h_grad, dim>>::create(ref_space,
         PushForward<h_grad, dim>::create(map));

    auto elem = space->begin();
    const auto elem_end = space->end();
    ValueFlags fill_flags = ...
    elem->init_values(fill_flags, elem_quad);

    for (; elem != elem_end; ++elem){
        elem->fill_values();
        for (int i = 0; i < n_basis; ++i){
            for (int j = 0; j < n_basis; ++j){
                for (int qp = 0; qp < n_qp; ++qp){
                    loc_mat(i,j) += scalar_product(
                        elem->get_basis_gradient(i,qp),
                        elem->get_basis_gradient(j,qp))*
                        w_meas[qp];}}}
```

Listing 1 A first "Hello World" like code. Intended to assemble a simple $\int \nabla u \nabla v$ operator the shape function values, using `PhysicalSpace` as a grid-like container, to assemble the local matrix.

3 Mixed Methods with Isogeometric Finite Elements

The mathematical description of physical phenomena involves several mechanisms interplaying at the same time. It often happens that different independent variables are necessary to describe a given mechanism. When such independent variables are discretized using different solution spaces we have a mixed method.

A typical case is an equation of motion coupled with an incompressibility constrain. In the applications we are going to address, namely incompressible elasticity (Sect. 4) and incompressible Navier-Stokes (Sect. 5), the independent variables are two. The first one will be referred to as $\mathbf{u}$ and will be used to describe velocities or displacements. The second one is going to be p that represents the Lagrangian multiplier of the incompressibility condition. In both mechanical and fluid dynamical contexts the discretization and linearization of the continuous operators will lead to a block linear system of the form:

$$\begin{pmatrix} A & B^T \\ B & 0 \end{pmatrix} \begin{pmatrix} \mathbf{u} \\ p \end{pmatrix} = \begin{pmatrix} \mathbf{f} \\ 0 \end{pmatrix}.$$

If we consider the two discrete spaces $\mathcal{V}_h$ and $\mathcal{Q}_h$ for $\mathbf{u}$ and p respectively, it is well known that these two spaces have to satisfy the inf-sup condition, see [1, 8, 9]. In standard finite elements, an established choice is the Taylor-Hood element. Here the discrete spaces for $\mathbf{u}$ and p share the same triangulation and the inf-sup condition is fulfilled when the degree of the shape functions for $\mathbf{u}$ is $d+1$, being d the shape functions degree for p.

The isogeometric counterpart of standard Taylor-Hood elements is constructed choosing $\mathcal{V}_h = (S_{d-1}^{d+1})^{\mathtt{dim}}$ and $\mathcal{Q}_h = S_{d-1}^{d}$, being `dim` the dimension. This type of element has been first used in [4] and a stability proof can be found in [10]. For sake of clearness we represent in one dimension the shape functions corresponding to S_1^3/S_1^2 pair in Fig. 2.

As mentioned in Sect. 2, the software design should resemble the mathematical concepts needed to integrate the boundary value problem we are interested in, in a one-to-one correspondence. As different spaces are required to solve for $\mathbf{u}$ and p, the same situation will be reproduced in the code.

Listing 2 reports the code necessary to define $\mathcal{Q}_h$ and $\mathcal{V}_h$. We first focus on the solution space for p, namely `BSplineSpace<dim>`. As previously mentioned, one of the most appealing features of `igatools` is its code dimension independence. In the present case of a scalar space, we only need to define the first template parameter as `dim`, the other template parameters are default defined to give a scalar space. A bit different is the situation for the vector space $\mathcal{V}_h$. In this case mathematics is elegantly resembled by specifying the second template

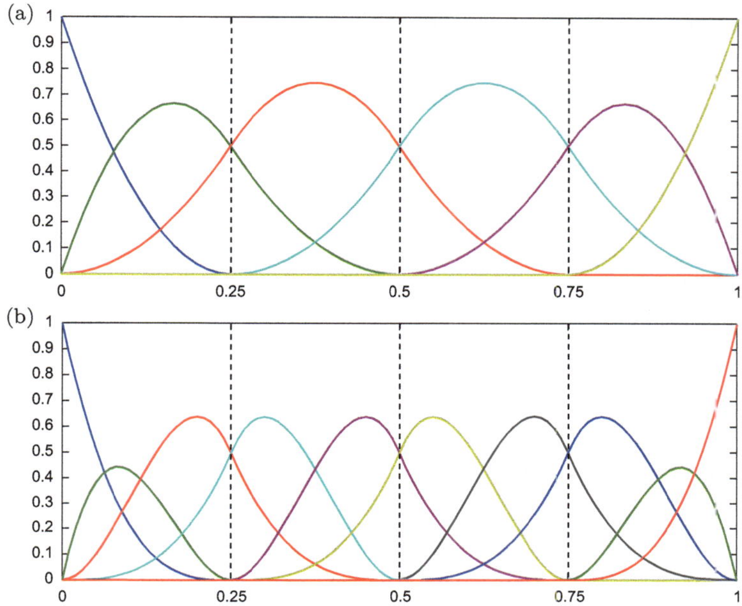

Fig. 2 In this figure we sketch one dimensional shape functions for the Taylor-Hood type of spaces. In particular we represent S_1^3 splines for the velocity, and S_1^2 splines for the pressure. The corresponding knot vectors are: $\Xi_{\mathbf{u}} = \{0, 0, 0, 0, 0.25, 0.25, 0.5, 0.5, 0.75, 0.75, 1, 1, 1, 1\}$ for the velocity and $\Xi_p = \{0, 0, 0, 0.25, 0.5, 0.75, 1, 1, 1\}$ for the pressure. (**a**) Space $\mathcal{Q}_h = \mathcal{S}_1^2$ for the Lagrangian multiplier p. (**b**) Space $\mathcal{V}_h = \mathcal{S}_1^3$ of primal variable $\mathbf{u}$

parameter, the equations range is the same as the problem dimension, this is why we get `BSplineSpace<dim,dim>`.

In code snippet in Listing 2 the reader can also notice how two different vectors, `mult_u` and `mult_p`, are defined to assign the right multiplicity to each knot. Following the Taylor-Hood element definition, the two spaces must have the same regularity and different degree, and we need to set the knots multiplicity to obtain compatible spaces.

4 Computational Mechanics: Nonlinear Incompressible Elasticity

As a first application for the `igatools` implementation of a mixed method, we present the isogeometric discretization of nonlinear elasticity, i.e. large deformation kinematics and a hyperelastic constitutive law, subject to an incompressibility constraint.

```
using PreSpace = BSplineSpace<dim>;
using VelSpcae = BSplineSpace<dim, dim>;

deg_u = deg_p + 1;

shared_ptr<BSplineSpace<dim>> pre_space;
shared_ptr<BSplineSpace<dim, dim>> vel_space;

vector<int> mult_p(n_knots, deg_p - reg);
vector<int> mult_u(n_knots, deg_u - reg);
mult_p[0] = mult_p[n_knots-1] = deg_p + 1;
mult_u[0] = mult_u[n_knots-1] = deg_u + 1;

pre_mult.fill(mult_p);
vel_mult.fill(mult_u);

pre_deg.fill(deg_p);
vel_deg.fill(deg_u);

auto grid = CartesianGrid<dim>::create(n_knots);
pre_space = PreSpace::create(grid, pre_mult, pre_deg);
vel_space = VelSpace::create(grid, vel_mult, vel_deg);
```

Listing 2 Definition of to different spaces for the variables $\mathbf{u}$ and p. In this listing we detail the construction of spaces with different degrees and multiplicites

Incompressible elasticity has been studied already by a few researchers in the context of isogeometric formulations. Elguedj et al. investigated the use of $\bar{B}$- and $\bar{F}$-projection methods for linear and nonlinear incompressible elasticity. Taylor [33] and Mathisen et al. [28] employed a three-field mixed approach for nonlinear incompressible elasticity with independent approximation of $\mathbf{u}$, p and θ. Here we use a classical mixed method with independent variables $\mathbf{u}$ and p, as presented in the previous section, and outline its implementation using `igatools`. The developed solver is validated using a well-known benchmark example and then applied to an industrial test case.

In this paper we employ the Langrangian or material description of the large deformation elasticity problem in $\mathtt{dim} = 2, 3$ dimensions, as for instance introduced in [1, 36]. Thus, a body in its undeformed, Lagrangian configuration is given by its reference or material domain $\Omega \in \mathbb{R}^{\mathtt{dim}}$ and its deformed, current configuration is expressed with the mapping $\hat{\varphi} : \Omega \to \mathbb{R}^{\mathtt{dim}}$:

$$\hat{\varphi}(\mathbf{X}) = \mathbf{X} + \hat{\mathbf{u}}(\mathbf{X}). \tag{1}$$

Following the notation of [1] from the displacement field $\hat{\mathbf{u}}$ we can derive the kinematic quantities such as the deformation gradient

$$\hat{\mathbf{F}} = \mathbf{F}(\hat{\mathbf{u}}) = \mathbf{I} + \nabla\hat{\mathbf{u}}, \tag{2}$$

and the right Cauchy-Green strain tensor

$$\hat{\mathbf{C}} = \mathbf{C}(\hat{\mathbf{u}}) = \hat{\mathbf{F}}^T\hat{\mathbf{F}}. \tag{3}$$

$\hat{J} = \det\hat{\mathbf{F}}$ is the determinant of the deformation gradient and for a purely incompressible material it must hold $\hat{J} = 1$.

Using Lamé parameter μ and a pressure-like variable $\hat{p}$ we can introduce the elastic energy functional for a purely incompressible Neo-Hookean material:

$$\Pi_{elast}(\hat{\mathbf{u}}, \hat{p}) = \int_\Omega \left\{ \frac{1}{2}\mu\left[\mathbf{I} : \hat{\mathbf{C}} - d\right] - \mu \ln \hat{J} + \hat{p}\ln(\hat{J}) \right\} d\Omega - \mathcal{F}(\hat{\mathbf{u}}, \gamma). \tag{4}$$

Here $\mathcal{F}(\hat{\mathbf{u}}, \gamma)$ denotes the external energy functional and has the form

$$\mathcal{F}(\hat{\mathbf{u}}, \gamma) = \gamma \left(\int_\Omega \mathbf{f}\cdot\hat{\mathbf{u}} d\Omega + \int_{\Gamma_N} \mathbf{p}\cdot\hat{\mathbf{u}} d\Omega \right), \tag{5}$$

where $\mathbf{f}$ is a body load, $\mathbf{p}$ a traction force acting on the Neumann boundary of the domain $\Gamma_N \subset \delta\Omega$ and $\gamma \in \mathbb{R}$ a load parameter.

For the minimization of the energy functional (4) it follows from the calculus of variations that the total differential of Π_{elast} must be zero,

$$\mathrm{d}\Pi_{elast}(\hat{\mathbf{u}}, \hat{p})[\mathbf{v}, q] = 0,$$

for all generic virtual displacement fields $\mathbf{v}$ and pressure fields q. Explicitly this reads:

$$\begin{cases} \mu \int_\Omega [\hat{\mathbf{F}} - \hat{\mathbf{F}}^{-T}] : \nabla\mathbf{v} + \int_\Omega \hat{p}\,\hat{\mathbf{F}}^{-T} : \nabla\mathbf{v} - \mathcal{F}(\mathbf{v}; \gamma) = 0 \quad \forall\mathbf{v}, \\ \int_\Omega \ln(\hat{J})q = 0 \quad \forall q. \end{cases} \tag{6}$$

The nonlinear system of equations (6) is then discretized using isogeometric Taylor-Hood elements in order to obtain displacements $\hat{\mathbf{u}} \in \mathcal{V}_h$ and pressures $\hat{p} \in \mathcal{Q}_h$. We also need to evaluate the second derivative of the energy functional (4), in order to setup a Newton's method for our solver:

$$\mathrm{d}^2\Pi_{elast}(\hat{\mathbf{u}}, \hat{p})[(\mathbf{u}, p), (\mathbf{v}, q)] = a_\gamma(\mathbf{u}, \mathbf{v}) + b_\gamma(\mathbf{v}, p) + b_\gamma(\mathbf{u}, q), \tag{7}$$

where

$$\begin{cases} a_\gamma(\mathbf{u},\mathbf{v}) := \mu \int_\Omega \nabla\mathbf{u} : \nabla\mathbf{v} + \int_\Omega [\mu - \hat{p}]\,(\hat{\mathbf{F}}^{-1}\nabla\mathbf{u})^T : \hat{\mathbf{F}}^{-1}\nabla\mathbf{v}, \\ b_\gamma(\mathbf{v},q) := \int_\Omega q\,\hat{\mathbf{F}}^{-T} : \nabla\mathbf{v}. \end{cases} \tag{8}$$

This formulation is a good example of what the authors intend as "sophisticated formulation". If we were to give a precise definition of "sophisticated" in this context, we would define it as an open scale, where grade zero is the classical $\int_\Omega \nabla u \cdot \nabla v$ arising from a Poisson's problem. Of course we do not aim grading every possible operator, but we think that a good measure of complexity is the number of manipulations beyond the classical grad – grad. It is in such cases that `igatools` can help.

In particular we focus on the term

$$\int_\Omega q\,\hat{\mathbf{F}}^{-T} : \nabla\mathbf{v}$$

from (7), which serves as a good example for the difficulties in this formulation. Its implementation is sketched in Listing 3. We first start by noticing that this is a mixed element. As we have two different spaces for $\mathbf{u}$ and p we need to iterate trough both of them and therefor we need two different iterators: the fist one is `defo_elem`, which iterates over the elements of the deformation space, and the second one is `prex_elem`, which iterates over the elements of the pressure like space. The single terms such as $\hat{\mathbf{F}}$, $\nabla\mathbf{v}$, or q are defined as tensors, and we refer to `igatools` documentation of the `Tensor` class for details [25]. Two classical for loops are used to evaluate $\nabla\mathbf{u}$ and p at quadrature points. Then $\hat{\mathbf{F}}^{-T}$ can be easily evaluated combing the inverse and the transpose functions. Once all the variables have been evaluated, they can be combined to obtain the local contribution to the tangent matrix evaluation. The remaining terms of the residual and tangent matrix are assembled in an analogous way. Once all the operators are in place, the setup of a Newton's method is a matter of linear algebra.

For validation of the implementation we chose the well-known Cook's membrane problem as benchmark [12]. This is a well established test in the context of finite element and also isogeometric methods [16, 28]. It is used by a number of researchers to validate solvers and benchmark performance of discretizations. Cook's membrane is a 2D panel, clamped on its side and subject to a shear load on the right side, causing combined bending and shear deformation. Geometry, dimensions and values of parameters can be found in Fig. 3. The quantity of interest is the horizontal deformation of the top right corner of the panel u_y, which is used to study p/k- and h-convergence and locking-free behavior of mixed elements.

In Fig. 4 convergence behavior of u_y with respect to the total number of degrees of freedom is shown. We compare the isogeometric Taylor-Hood elements with pure displacement isogeometric formulations. The Poisson's ratio ν is set to 0.4998

```
for (; defo_elem != defo_end ; ++defo_elem, ++prex_elem)
{
   ...
   for (Index q = 0; q < n_qp; ++q)
   {
     for(Index i = 0; i < defo_loc_ndofs; ++i)
       defo_grad_q += defo_vec(dof) * grad_phi_q[i];

     for(Index i = 0; i < prex_loc_ndofs; ++i)
       prex_q += prex_vec(dof) * prex_phi_q[i];

     defgrad_q = unit_defgrad + defo_grad_q;

     inverse (defgrad_q,defgrad_inv_q);
     defgrad_invT_q = transpose(defgrad_inv_q);

     for (Index i = 0; i < defo_loc_ndofs; ++i)
       defo_loc_res(i) += mat_mu * prex_q[0] *
                   scalar_product(defgrad_invT_q,
                                   grad_phi_q[i]) *w_meas[q];
     }
}
```

Listing 3 Assembly of the mixed term in the second variation of the energy functional

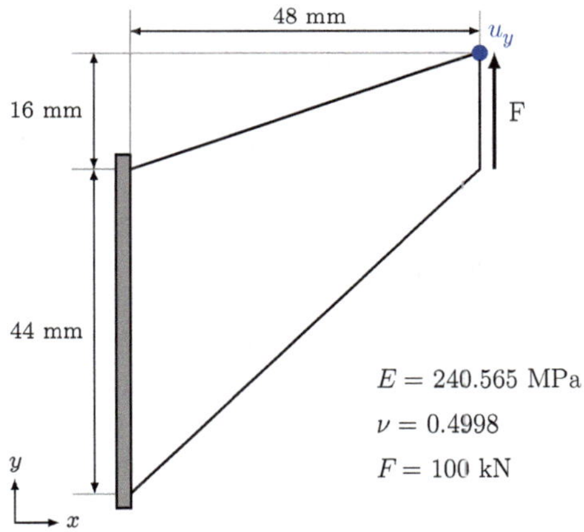

Fig. 3 Cook's membrane problem. A 2D panel clamped on the left side, subject to shear load on right side

that correspond to nearly incompressible material. The displacement-based show a rather slow convergence gradient due to locking. On the other hand, as expected, the Taylor-Hood elements converge quickly for p/k- and h-refinement. These results also match very well with the ones obtained by [16], where an isogeometric $\bar{F}$-method was used instead of a mixed formulation.

Even though the purely displacement formulation also converges towards the correct value of u_y for higher polynomial degree and number of degrees of freedom, the approximation of stresses exhibits spurious oscillations, which are typical for

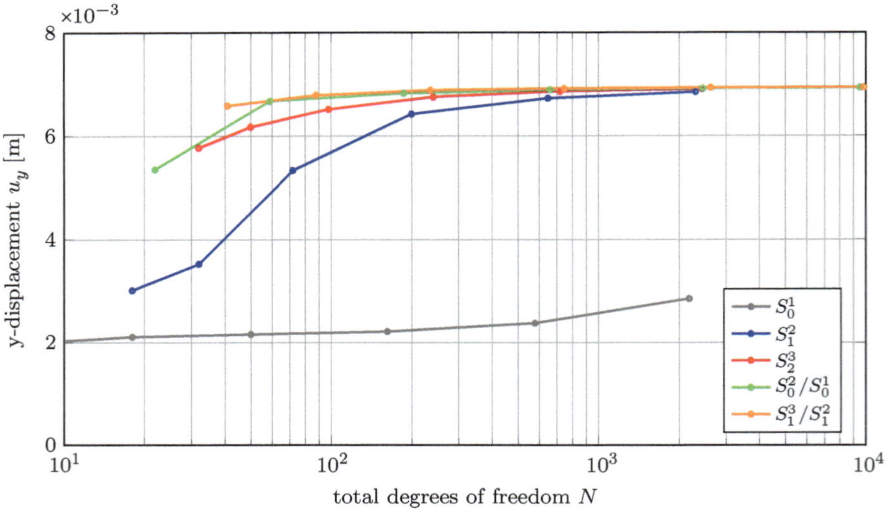

Fig. 4 Convergence test for the Cook's membrane. The reader can notice the fast convergence the Taylor Hood elements and a slower convergence for pure displacements formulation. In the following, the plots of the stresses, will demonstrate that volumetric locking is preventing from a fast convergence

locking phenomena. In Fig. 5 we plot the stress component σ_{xx}. The non mixed elements show a highly oscillatory behavior of volumetric stress components – a typical sign of volumetric locking – while the stress distributions for mixed elements are smooth.

The geometry of a pneumatic pipe, used in aircraft manufacturing, is one of the industrial applications within the TERRIFIC project. Using symmetry boundary conditions we can parametrize only half of the pipe as a B-Spline volume with degrees $d = (1, 2, 2)$ and $n = (2, 34, 3)$ control points, i.e. 204 control points of the volume. The first parameter direction is the thickness of the pipe, the second the circumferential direction and the third its length. For numerical computations we degree-elevate to $d = (2, 2, 2)$ and perform four steps of uniform h-refinement in the length-direction. Thus we have the following Taylor-Hood spaces for pressure and displacements:

$$\begin{aligned} d_p &= (2,2,2), & n_p &= (3,34,18), & N_p &= 1836, \\ d_{\mathbf{u}} &= (3,3,3), & n_{\mathbf{u}} &= (4,66,34), & N_{\mathbf{u}} &= 8976. \end{aligned} \tag{9}$$

here n stands for degrees of freedom per direction, while N indicates the total number of degrees of freedom. The material parameters of the pipe are:

$$E = 240.565\,\text{MPa}, \qquad \nu = 0.5, \qquad \rho = 1.0 \cdot 10^3\,\text{kg/m}^3. \tag{10}$$

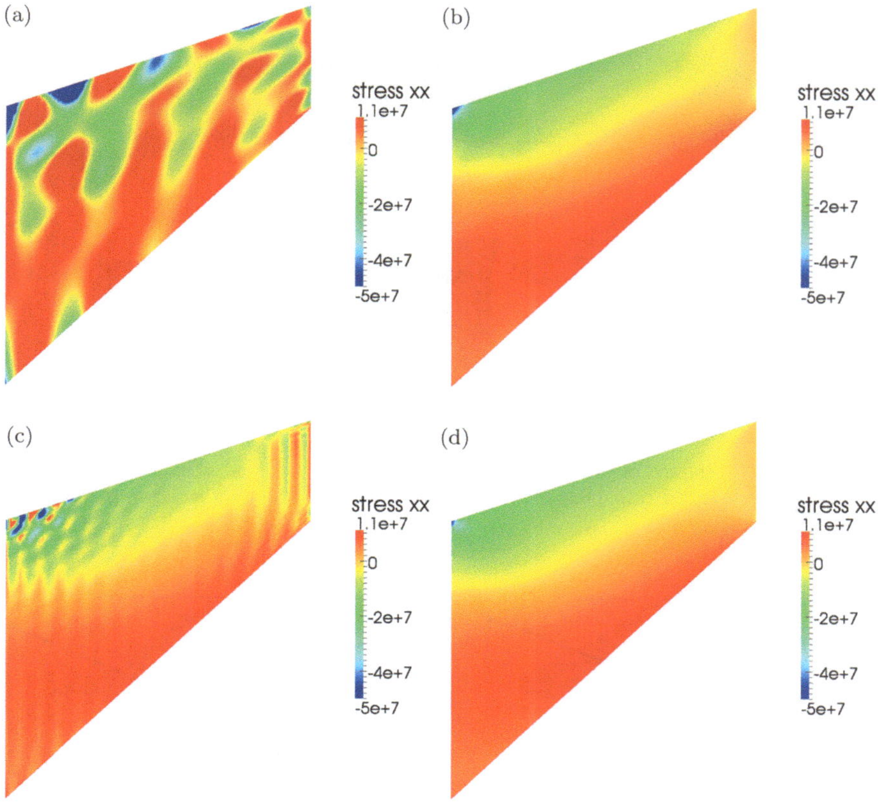

Fig. 5 Stress σ_{xx} for purely displacement-based and mixed formulations. The reader can notice how purely displacement based formulations are affected from volumetric locking, while, as expected, mixed methods are not. (**a**) S_2^3 with 4 elements per side. (**b**) S_1^3/S_1^2 with 4 elements per side. (**c**) S_2^3 with 16 elements per side. (**d**) S_1^3/S_1^2 with 16 elements per side

Boundary conditions are symmetry, zero displacement at long ends and an outwarded-directed Neumann load. We run 10 load steps increasing the magnitude of the surface load from 1.0 to 10.0 kN.

Figure 6 shows the deformed pipe after 10 load steps. Large displacements occur and no volume change is visible in the figure. In Fig. 7 the evolution of displacements u_x, u_y and pressure-like Lagrangian multiplier p, evaluated at the center point where maximum x-displacement and pressure occur, is plotted over the 10 load steps. The nonlinear behaviour can be noticed in the values of the displacements and pressure.

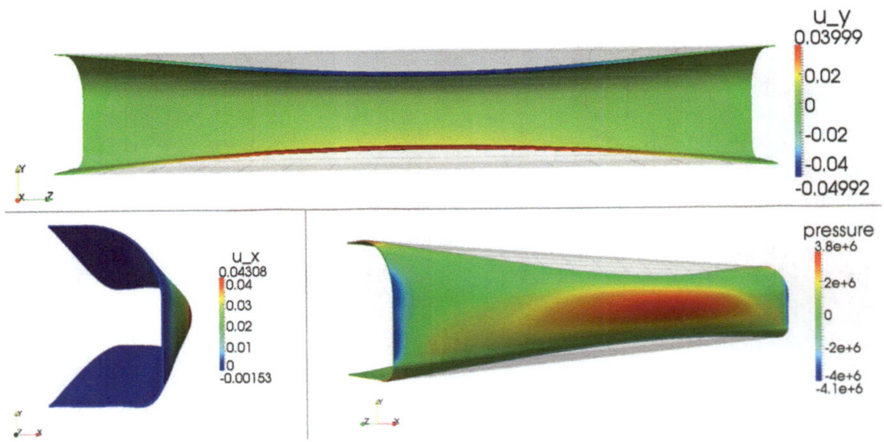

Fig. 6 Displacement and pressure of the pneumatic pipe after 10 load steps. Deformed pipe is colored by displacement resp. pressure, initial un-deformed configuration is shown in translucent grey

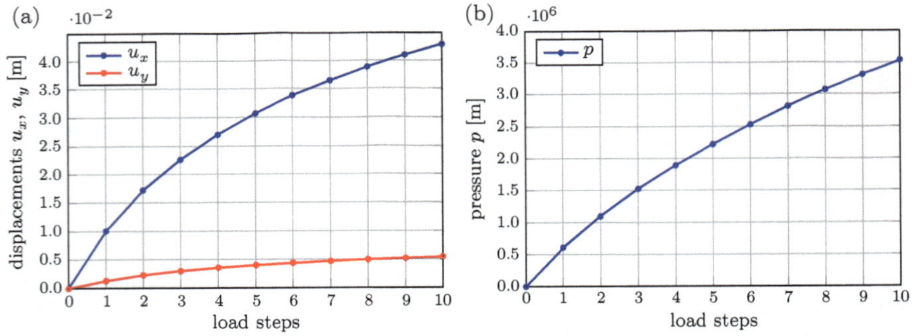

Fig. 7 Evaluation of point values over load steps for pneumatic pipe provided by ALENIA. (**a**) Displacements u_x, u_y. (**b**) Pressure-like Langrangian multiplier p

5 Computational Fluid Dynamics

The laminar motion of an incompressible fluid is modeled by the well known Navier-Stokes equations. We first write the momentum conservation equation:

$$\partial_t \mathbf{u} + (\nabla \mathbf{u})\, \mathbf{u} = -\nabla p + \nu \nabla^2 \mathbf{u} + \mathbf{f},$$

then the mass preservation one:

$$\nabla \cdot \mathbf{u} = 0.$$

Here $\mathbf{u}$ is the fluid velocity, p is the fluid pressure and ν is the kinematic viscosity. The external forces are denoted by $\mathbf{f}$. As a matter of clearness, since it is going to be useful for the method implementation, we explicitly write the component wise formulation of the momentum equation:

$$\partial_t \begin{pmatrix} u_x \\ u_y \end{pmatrix} + \begin{pmatrix} \partial_x u_x & \partial_y u_x \\ \partial_x u_y & \partial_y u_y \end{pmatrix} \begin{pmatrix} u_x \\ u_y \end{pmatrix} = - \begin{pmatrix} \partial_x p \\ \partial_y p \end{pmatrix} + \nu \begin{pmatrix} \partial_{xx} u_x + \partial_{yy} u_x \\ \partial_{xx} u_y + \partial_{yy} u_y \end{pmatrix} + \begin{pmatrix} f_x \\ f_y \end{pmatrix}.$$

In the following, we focus on the stationary version of the problem:

$$(\nabla \mathbf{u})\,\mathbf{u} - \nu \nabla^2 \mathbf{u} + \nabla p = \mathbf{f}, \tag{11}$$

The spatial integration of equation (11) is again performed using the Galerkin approach. Given the computational domain $\Omega \subset \mathbb{R}^{\mathtt{dim}}$, we consider $\mathcal{V}_h \subset (H^1(\Omega))^{\mathtt{dim}}$ the solution space for the velocity, and $\mathcal{Q}_h \subset L^2(\Omega)$ the solution space for the pressure. We consider all the $\mathbf{v}_h \in \mathcal{V}_h$ the test functions for the velocity, then the assembled viscous term $\nabla^2 \mathbf{u}$ correspond to the matrix A such that:

$$A_{ij} = \int_\Omega (\nabla \mathbf{u}_j) : (\nabla \mathbf{v}_i) = \int_\Omega \begin{pmatrix} \partial_x u_x & \partial_y u_x \\ \partial_x u_y & \partial_y u_y \end{pmatrix}_j : \begin{pmatrix} \partial_x v_x & \partial_y v_x \\ \partial_x v_y & \partial_y v_y \end{pmatrix}_i.$$

The convective term is a nonlinear one. For a given value of the velocity field we obtain the matrix N:

$$N_{ij} = \int_\Omega (\nabla \mathbf{u}_j)\,\mathbf{u} \cdot \mathbf{v}_i = \int_\Omega \begin{pmatrix} \partial_x u_x & \partial_y u_x \\ \partial_x u_y & \partial_y u_y \end{pmatrix}_j \begin{pmatrix} u_x \\ u_y \end{pmatrix} \cdot \begin{pmatrix} v_x \\ v_y \end{pmatrix}_i. \tag{12}$$

The operator that couples the momentum equation and the preservation of mass is B^T:

$$B^T_{ij} = \int_\Omega p_j \nabla \cdot \mathbf{v}_i = \int_\Omega p_j (\partial_x v_x + \partial_y v_y)_i.$$

These operators when combined together result in the following block nonlinear system:

$$\begin{pmatrix} A + N(\mathbf{u}) & B^T \\ B & 0 \end{pmatrix} \begin{pmatrix} \mathbf{u} \\ p \end{pmatrix} = \begin{pmatrix} \mathbf{f} \\ 0 \end{pmatrix}.$$

The nonlinearity is treated with a classical, residual-tangent strategy. Given an initial guess for the velocity $\mathbf{u}_0$ we first obtain the tentative solution $\mathbf{x}_k = (\mathbf{u}, p)_k$ by solving:

$$\begin{pmatrix} A + N(\mathbf{u}_0) & B^T \\ B & 0 \end{pmatrix} \begin{pmatrix} \mathbf{u} \\ p \end{pmatrix}_k = \begin{pmatrix} \mathbf{f} \\ 0 \end{pmatrix}.$$

For $k > 0$ we need to form the residual $\mathbf{F}(\mathbf{x}_k)$:

$$\mathbf{F}(\mathbf{x}_k) = \begin{pmatrix} A + N(\mathbf{u}_k) & B^T \\ B & 0 \end{pmatrix} \begin{pmatrix} \mathbf{u} \\ p \end{pmatrix}_k - \begin{pmatrix} \mathbf{f} \\ 0 \end{pmatrix},$$

and correct the tentative solution with an approximation of the tangent. The correction at the k-th step will be denoted as $(\delta\mathbf{u}, \delta p)_k$. The most popular strategies to approximate the tangent are the fixed point and the Newton approximation. In the first case we approximate the tangent with the function itself:

$$\begin{pmatrix} A + N(\mathbf{u}_k) & B^T \\ B & 0 \end{pmatrix} \begin{pmatrix} \delta\mathbf{u} \\ \delta p \end{pmatrix}_k = -\mathbf{F}(\mathbf{x}_k).$$

In the second case we consider a first order expansion of the operator:

$$\begin{pmatrix} A + N(\mathbf{u}_k) + D(\mathbf{u}_k) & B^T \\ B & 0 \end{pmatrix} \begin{pmatrix} \delta\mathbf{u} \\ \delta p \end{pmatrix}_k = -\mathbf{F}(\mathbf{x}_k).$$

In this case we need to assemble the function Jacobian as well:

$$D_{ij} = \int_\Omega (\nabla\mathbf{u})\,\mathbf{u}_j \cdot \mathbf{v}_i = \int_\Omega \begin{pmatrix} \partial_x u_x & \partial_y u_x \\ \partial_x u_y & \partial_y u_y \end{pmatrix} \begin{pmatrix} u_x \\ u_y \end{pmatrix}_j \cdot \begin{pmatrix} v_x \\ v_y \end{pmatrix}_i. \qquad (13)$$

Finally, the solution is updated with its correction:

$$\begin{pmatrix} \mathbf{u} \\ p \end{pmatrix}_{k+1} = \begin{pmatrix} \mathbf{u} \\ p \end{pmatrix}_k + \begin{pmatrix} \delta\mathbf{u} \\ \delta p \end{pmatrix}_k,$$

the procedure stops when the residual satisfies a given tolerance.

Now we intend to focus on the two terms involved in the nonlinearity treatment. Looking at the convective term, we have the velocity gradient applied to the fluid velocity. This mathematical formulation has a pretty simple implementation in `igatools`. In Listing 4 we have the gradient of the trial function that acts, through the function `action`, on a point evaluation of the velocity `vel_q`. The result of this `action` is scalar multiplied with the test functions, by using a `scalar_product` function. In this way we obtain the local contribution to the global matrix. Almost the same procedure applies to the evaluation of the operator (13). In this case the quantity that has to be evaluated is the velocity gradient `grad_vel_q`. We evaluate the `action` of the velocity gradient on the trial functions, and then we perform the `scalar_product` with the test functions to get the local contribution to the global matrix, see Listing 5. Both `vel_q` and `grad_vel_q` are considered as tensors. We refer to the `Tensor`'s class documentation for further details.

```
for (dof_index q = 0; q < quad.get_num_points(); q++)
{
   for (dof_index i = 0; i <local_ndofs; i ++){
      auto phi = element->get_value(i,q);
      vel_q = vel_q+vel[local_dofs[i]]*phi;}

   for (dof_index i = 0; i <local_ndofs; i ++){
      for (dof_index j = 0; j <local_ndofs; j ++){
         adv_ij = scalar_product(action(
                                         element->get_gradient(j,q),
                                         vel_q),
                                         element->get_value(i,q) )*
                                         element->get_w_measures()[q];]}}
}
```

Listing 4 Code snippet for the advection operator assembly. In this case the reader can notice the point evaluation of the velocity, the action of the velocity gradient on the velocity value, and the scalar product with the velocity test functions

```
for (dof_index q = 0; q < quad.get_num_points(); q++)
{

   for (dof_index i = 0; i <local_ndofs; i++){
      auto grad_phi = element->get_gradient(i,q);
      grad_vel_q = grad_vel_q + vel[local_dofs[i]] * grad_phi;}

   for (dof_index i = 0; i <local_ndofs; i++){
      for (dof_index j = 0; j <local_ndofs; j++){
         jac_ij = scalar_product(action(grad_vel_q,
                                         element->get_value(j,q)),
                                         element->get_value(i,q) )*
                                         element->get_w_measures()[q];}}
}
```

Listing 5 Code snippet for the Jacobian assembly. In this case we have the point evaluation of the velocity gradient. Then the velocity gradient acts on the trial functions. The result of this multiplication is scalar multiplied with the test functions, to get the local contribution to the global matrix

The solver is tested using the cavity flow test case. This is a typical test case for stationary Navier-Stokes equations [21, 23]. The horizontal velocity u_x is imposed at the top of a square domain. Its value is:

$$u_x(x) = \begin{cases} \sin(\pi x/2a) & \text{if } 0 \leq x \leq a \\ 1 & \text{if } a \leq x \leq 1-a \\ \sin(\pi(1-x)/2a) & \text{if } 1-a \leq x \leq 1 \end{cases}$$

with $a = 1/32$. No-slip boundary conditions are imposed elsewhere. It is known that isogeometric shape functions are not interpolatory to the degrees of freedom. It is

```
class CavityVel  : public Function
{
public:
    CavityVel  ()  : Function() {}

    void evaluate(const ValueVector & points,
                  ValueVector & values)      const;
}
...
CavityVel cv;
project_boundary_values(cv,velocity_space,face_q,
direchelet_id,dof_values);
apply_boundary_values(dof_values,matrix,rhs,solution)
```

Listing 6 In this code snippet we present how to impose non constant boundary conditions for our problem. The user needs to create a derived class that represents the function to be evaluated. Implementation for the `evaluate` virtual function has to be provided. `igatools` will take care of L^2 projecting the function on the velocity boundary trace space. The resulting values will be applied to the global matrix, right hand side and solution

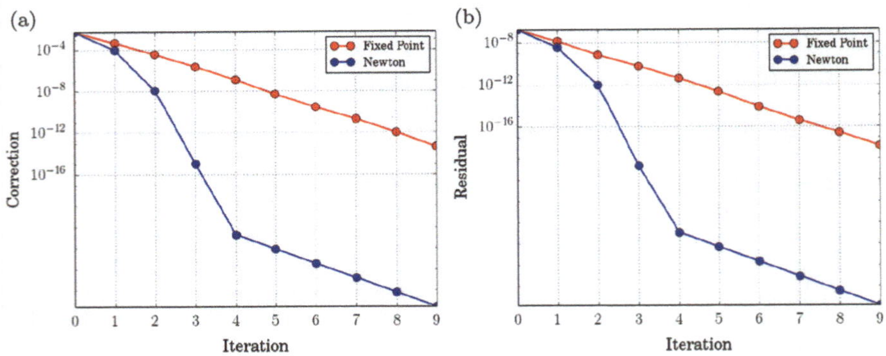

Fig. 8 Correction and residual for the cavity flow test case with Re = 100. (**a**) Solution correction. (**b**) Residual

then necessary to define values for the imposed degrees of freedom. In `igatools`, Listing 6, this is done defining a `CavityVel` class that inherits form the base class `Function` and providing implementation for the virtual function function `evaluate`. At this point the user is let to define an instance of his own class `CavityVel` and `igatools` will take care of evaluating an L^2 projection of the boundary function on the velocity trace space. The resulting `dof_values` will be applied to global matrix, right hand side and solution.

Once the solver is implemented we start with the experiments. In Fig. 8 we represent the residual and the correction for Reynolds's number Re $=$ 100 for each iteration of the nonlinear solver. As expected, the residual and correction drop linearly with the fixed point type of iterations, and quadratically with the Newton's solver.

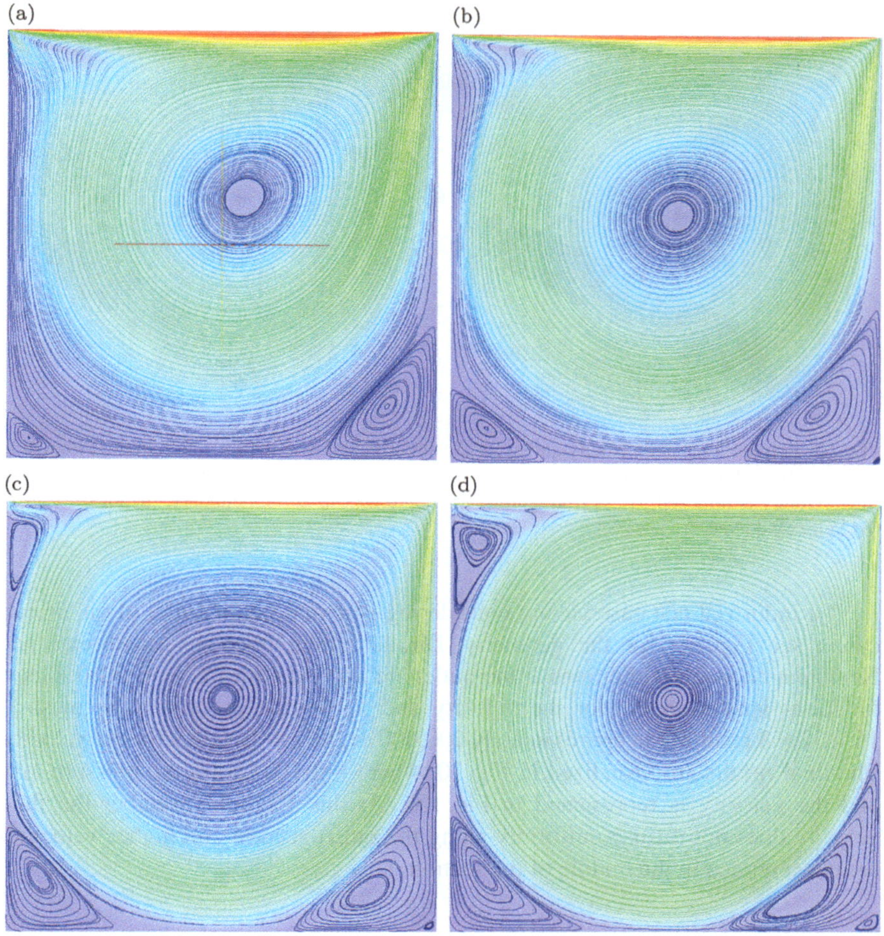

Fig. 9 Cavity flow test case for the Navier-Stokes solver. Colormap represents the velocity magnitude in a scale the ranges from zero to one. (**a**) Re = 400. (**b**) Re = 1000. (**c**) Re = 3200. (**d**) Re = 5000

In the present case of the cavity flow experiment we used S_1^3 splines for the velocity and S_1^2 splines for the pressure with mesh size $h = 1/32$. Efficient linear algebra solvers for this system is still an active research area, the reader can refer to [17, 18] as example references. In this experiment and in the ones that follow case we set for a standard direct solver.

In a second set of experiments we consider increasing Re numbers. The solution space is still S_3^1/S_2^1, but the mesh size is 64×64 elements. In Fig. 9 we present the velocity streamlines. The colormap pictures the velocity magnitude that ranges from zero to one. The agreement with results from [21] is good and the solver can be considered as validated.

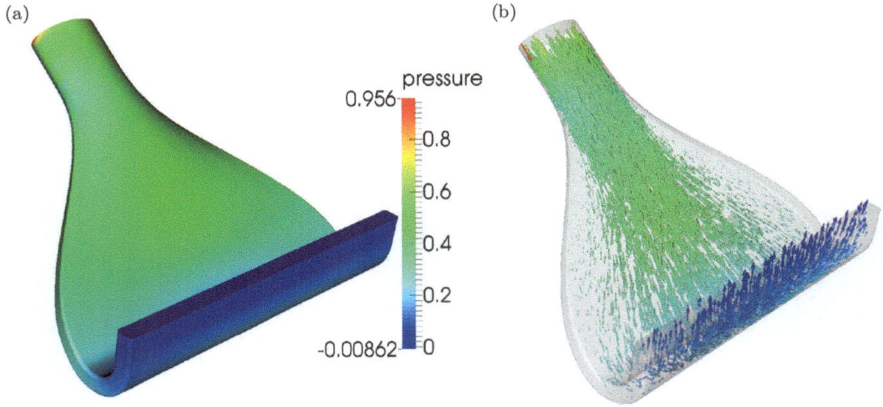

Fig. 10 Pressure map and velocity quiver for the aircraft cabin ventilation outlet. In this simulation Re = 10,000, shape functions are S_1^3/S_1^2, the number of knots, without repetitions, is $17 \times 33 \times 9$. Inlet velocity is imposed, stress free on the outlet, and no slip at the boundary. (**a**) Pressure map. (**b**) Velocity quiver

The industrial test case we are considering is a ventilation outlet of an aircraft cabin, see Fig. 10a for the geometry. Dirichlet inlet boundary conditions and Neumann boundary conditions at the outlet are applied with $\mathrm{Re} = 10,000$. Shape functions spaces are S_1^3 for the velocity and S_1^2 for the pressure. The mesh is $16 \times 32 \times 8$ resulting in approximately 200,000 degrees of freedom for the velocity. The final results are visualized in Fig. 10, where Fig. 10a shows the pressure map and the velocity is represented in Fig. 10b. At this Reynolds number the fluid velocity does not diffuse uniformly along the outlet section and higher velocities are concentrated in the middle of the section.

6 Conclusions

In this paper we showed how `igatools` is an effective tool for assembling isogeometric discretizations of sophisticated operators. `igatools` has been recently released and its design has been presented in [29]. Only time will say if its design will be capable to attract new users and developers. We intended to demonstrate that we are on a good track in this direction. We picked sophisticated operators arising from industrial applications, and showed that a one to one relationship exists between how we write operators on paper, and how we implement them using `igatools`. We believe this attempt has been successful both in computational mechanics and computational fluid dynamics. In order to demonstrate a realistic effectiveness of the software we solved industrially relevant applications.

Acknowledgements The authors are grateful to C. Lovadina for the useful discussion. N. Cavallini and O. Weeger have been supported by the TERRIFIC project, European Community's Seventh Framework Programme, Grant Agreement 284981 Call FP7-2011-NMP-ICT-FoF. In all the experiments in this paper we used linear algebra packages from `deal.II` [2, 3].

References

1. F. Auricchio, L. Beirão da Veiga, C. Lovadina, A. Reali, R.L. Taylor, P. Wriggers, Approximation of incompressible large deformation elastic problems: some unresolved issues. Comput. Mech. **52**(5), 1153–1167 (2013)
2. W. Bangerth, R. Hartmann, G. Kanschat, deal.II – a general purpose object oriented finite element library. ACM Trans. Math. Softw. **33**(4), 24/1–24/27 (2007)
3. W. Bangerth, T. Heister, G. Kanschat, *`deal.II` Differential Equations Analysis Library, Technical Reference*, http://www.dealii.org
4. Y. Bazilevs, L. Beirão da Veiga, J.A. Cottrell, T.J.R. Hughes, G. Sangalli, Isogeometric analysis: approximation, stability and error estimates for h-refined meshes. Math. Models Methods Appl. Sci. **16**(7), 1031–1090 (2006)
5. Y. Bazilevs, V.M. Calo, J.A. Cottrell, J.A. Evans, T.J.R. Hughes, S. Lipton, M.A. Scott, T.W. Sederberg, Isogeometric analysis using T-splines. Comput. Methods Appl. Mech. Eng. **199**(5–8), 229–263 (2010)
6. Y. Bazilevs, C. Michler, V.M. Calo, T.J.R. Hughes, Isogeometric variational multiscale modeling of wall-bounded turbulent flows with weakly enforced boundary conditions on unstretched meshes. Comput. Methods Appl. Mech. Eng. **199**(13–16), 780–790 (2010)
7. P. Becker, Working draft, standard for programming language C++. Technical Report N3242=11-0012, ISO/IEC JTC 1, Information Technology, Subcommittee SC 22, Programming Language C++, Feb 2011
8. D. Boffi, F. Brezzi, M. Fortin, *Mixed Finite Element Methods and Applications*. Springer Series in Computational Mathematics (Springer, London, 2013)
9. D. Boffi, C. Lovadina, Analysis of new augmented lagrangian formulations for mixed finite element schemes. Numerische Mathematik **75**(4), 405–419 (1997)
10. A. Bressan, G. Sangalli, Isogeometric discretizations of the Stokes problem: stability analysis by the macro element technique. IMA J. Numer. Anal. (2012)
11. A. Buffa, D. Cho, G. Sangalli, Linear independence of the T-spline blending functions associated with some particular T-meshes. Comput. Methods Appl. Mech. Eng. **199**(23–24), 1437–1445 (2010)
12. R.D. Cook, Improved two-dimensional finite element. J. Struct. Div. **100**, 1851–1863 (1974)
13. J.A. Cottrell, A. Reali, Y. Bazilevs, T.J.R. Hughes, Isogeometric analysis of structural vibrations. Comput. Methods Appl. Mech. Eng. **195**(41–43), 5257–5296 (2006)
14. T. Dokken, T. Lyche, K.F. Pettersen, Locally refinable splines over box-partitions. Technical report, SINTEF, Feb 2012
15. M. Dörfel, B. Jüttler, B. Simeon, Adaptive isogeometric analysis by local h-refinement with T-splines. Comput. Methods Appl. Mech. Eng. **199**(5–8), 264–275 (2009)
16. T. Elguedj, Y. Bazilevs, V.M. Calo, T.J.R. Hughes, B-bar and f-bar projection methods for nearly incompressible linear and non-linear elasticity and plasticity based on higher-order NURBS elements. Comput. Methods Appl. Mech. Eng. **197**, 2732–2762 (2008)
17. H.C. Elman, D.J. Silvester, A.J. Wathen, Performance and analysis of saddle point preconditioners for the discrete steady-state Navier-Stokes equations. Numer. Math. **90**, 665–688 (2000)
18. H.C. Elman, D.J. Silvester, A.J. Wathen, *Finite Elements and Fast Iterative Solvers: With Applications in Incompressible Fluid Dynamics*. Numerical Mathematics and Scientific Computation (Oxford University Press, Oxford, 2005)

19. J.A. Evans, T.J.R. Hughes, Isogeometric divergence-conforming B-spline for the steady Navier-Stokes equations. Math. Models Methods Appl. Sci. **23**(8), 1421–1478 (2013)
20. D.R. Forsey, R.H. Bartels, Hierarchical B-spline refinement, in *Proceedings of the 15th Annual Conference on Computer Graphics and Interactive Techniques (SIGGRAPH'88)*, Atlanta, 1988, pp. 205–212
21. U. Ghia, K.N. Ghia, C.T. Shin, High-Re solutions for incompressible flow using the Navier-Stokes equations and a multigrid method. J. Comput. Phys. **48**, 387–411 (1982)
22. C. Giannelli, B. Jüttler, H. Speleers, THB–splines: the truncated basis for hierarchical splines. Comput. Aided Geom. D. **29**, 485–498 (2012)
23. R. Glowinski, P.G. Ciarlet, J.L. Lions, *Numerical Methods for Fluids*. Handbook of Numerical Analysis, vol. 3 (Elsevier, Amsterdam, 2002)
24. T.J.R. Hughes, J.A. Cottrell, Y. Bazilevs, Isogeometric analysis: CAD, finite elements, NURBS, exact geometry and mesh refinement. Comput. Methods Appl. Mech. Eng. **194**(39–41), 4135–4195 (2005)
25. igatools 0.3.0, An isogeometric analisys tool library – documentation and manual, Oct 2014
26. ISO/IEC 14882:2011 – Information technology – programming languages – C++, 2011
27. J. Kiendl, Y. Bazilevs, M.-C. Hsu, R. Wüchner, K.-U. Bletzinger, Kirchhoff-Love shell structures comprised of multiple patches. Comput. Methods Appl. Mech. Eng. **199**, 2403–2416 (2010)
28. K.M. Mathisen, K.M. Okstad, T. Kvamsdal, S.B. Raknes, Isogeometric analysis of finite deformation nearly incompressible solids. Rakenteiden Mekaniikka (J. Struct. Mech.) **44**(3), 260–278 (2011)
29. M.S. Pauletti, M. Martinelli, N. Cavallini, P. Antolin, Igatools: an isogeometric analysis library. *I.M.A.T.I.-C.N.R.*, 2014, pp. 1–27
30. L.A. Piegl, W. Tiller, *The NURBs Book*. Monographs in Visual Communication Series (Springer, Berlin/New York, 1997)
31. L.L. Schumaker, *Spline Functions: Basic Theory*. Cambridge Mathematical Library, 3rd edn. (Cambridge University Press, Cambridge, 2007)
32. M.A. Scott, X. Li, T.W. Sederberg, T.J.R. Hughes, Local refinement of analysis-suitable T-splines. Comput. Methods Appl. Mech. Eng. **213–216**, 206–222 (2012)
33. R. Taylor, Isogeometric analysis of nearly incompressible solids. Int. J. Numer. Methods Eng. **87**(1–5), 273–288 (2010)
34. A.-V. Vuong, C. Giannelli, B. Jüttler, B. Simeon, A hierarchical approach to adaptive local refinement in isogeometric analysis. Comput. Methods Appl. Mech. Eng. **200**, 3554–3567 (2011)
35. O. Weeger, U. Wever, B. Simeon, Isogeometric analysis of nonlinear Euler Bernoulli beam vibrations. Nonlinear Dyn. **72**(4), 813–835 (2013)
36. P. Wriggers, *Nonlinear Finite Element Methods* (Springer, Berlin, 2008)

Two-Sided Robust and Sharp a Posteriori Error Estimates in Isogeometric Discretization of Elliptic Problems

Stefan K. Kleiss and Satyendra K. Tomar

Abstract We present two-sided a posteriori error estimates for isogeometric discretization of elliptic problems. These estimates, derived on functional grounds, provide robust, guaranteed and sharp two-sided bounds of the exact error in the energy norm. Moreover, since these estimates do not contain any unknown/generic constants, they are fully computable, and thus provide quantitative information on the error. The numerical realization and the quality of the computed error distribution are addressed. The potential of the proposed estimates are illustrated using several computational examples.

1 Introduction

Interchange of geometry information between the engineering design process, e.g., computer aided design (CAD), and simulation/analysis, e.g., using finite element methods (FEM), is generally not only costly, but also prone to approximation errors, and may also require manual input. *Isogeometric analysis* (IGA), introduced by Hughes et al. [22], see also [13] and the references therein, aims at facilitating the transformation of geometry data between these two rather separately evolved directions. In IGA, instead of transforming the geometry data to a conventional FEM representation, the original geometry description is used directly, and the functions representing the geometry (typically non-uniform rational B-splines, abbreviated hereinafter as NURBS) are also used as basis for the discrete solution. This way, the geometry obtained from CAD is not changed, and the need for data transformation is eliminated. Furthermore, the exact representation from the coarsest mesh is preserved throughout the refinement process. IGA has been thoroughly studied and

S.K. Kleiss (✉)
Johannes Kepler University, Altenbergerstrasse 69, 4040 Linz, Austria
e-mail: stefan.kleiss@jku.at; Stefan_Ken.Kleiss@jku.at

S.K. Tomar
Cardiff School of Engineering, Queen's Buildings, The Parade, Cardiff CF24 3AA, UK
e-mail: tomar.sk@iitkalumni.org

B. Jüttler, B. Simeon (eds.), *Isogeometric Analysis and Applications 2014*,
Lecture Notes in Computational Science and Engineering 107,
DOI 10.1007/978-3-319-23315-4_10

its potential has been shown by successful applications to a wide range of problems, see, e.g., [3–5, 7, 9, 12, 18, 23, 30].

Since naive mesh refinement in a tensor-product setting (which is the hallmark of NURBS meshes) has global effects, the development of local refinement strategies for isogeometric analysis is a subject of current active research. Such local refinement techniques include, e.g., T-splines [6, 29, 36–38], truncated hierarchical B-splines (THB-splines) [20, 21], polynomial splines over hierarchical T-meshes (PHT-splines) [15, 40], and locally-refinable splines (LR-splines) [16, 25]. These approaches are mainly guided by the a-priori knowledge of the solution behavior or geometry and does not take in to account any information from the computed solution. To the best of the authors' knowledge, the issue of a-posteriori error estimation in isogeometric analysis is still in an infancy stage, and the only published results are [14, 17, 24, 27, 39–43].

The first (and often only) practical application of a posteriori error estimation is the issue of adaptive local refinement (see, e.g., [1, 35] for a general overview on error estimators). A good a posteriori error estimation method can accurately identify local area(s) where further refinement is needed due to the local error being significantly larger than in the rest of the domain. Thereby, computing resources can be more efficiently used than uniform global distribution.

A posteriori error estimates based on hierarchical bases, proposed by Bank and Smith [2], have been used in [17, 39]. The reliability and efficiency of this approach is subjected to the saturation assumption on the (enlarged) underlying space and the constants in the strengthened Cauchy inequality. As the authors remarked, the first assumption is critical and its validity depends on the considered example. Moreover, an accurate estimation of constants in the strengthened Cauchy inequality requires the solution of generalized minimum eigenvalue problem. As noted in [24, Page 41], this approach delivers *less than satisfactory* results. Residual-based a posteriori error estimates have been used in [24, 40–42]. This approach requires the computation of constants in Clement-type interpolation operators. Such constants are mesh (element) dependent, often generic/unknown or incomputable for general element shape; and the global constant often over-estimates the local constants, and thus the exact error. This fact has been explicitly stated by the authors in [24, Pages 42–43] and in [40, Remark 1]. Goal-oriented error estimation approach has been studied in [14, 27, 43]. The results presented in these studies show that neither the estimates of this approach are *guaranteed* to be an upper bound, nor the efficiency indices of the estimates are sharp. Moreover, this approach also requires the solution of an adjoint problem, the cost of which can not be entirely neglected. The approach of Zienkiewicz-Zhu type a posteriori error estimates is based on post-processing of approximate solutions, and depends on the superconvergence properties of the underlying basis. The superconvergence properties for B-splines (NURBS) functions are not yet well-known. Summarily, in general situations, the reliability and efficiency of these methods often depend on undetermined constants, which is not suitable for quality assurance purposes.

A posteriori error estimates have another important application, namely the issue of *quality assurance*, i.e., to quantify the error in the computed solution with certain degree of *guarantee*. IGA has shown great potential in many real life engineering applications. Therefore, even if in a niche domain of applications where one can obtain excellent a posteriori error estimates, authors believe that the issue of *quality assurance* of the computed solution deserves more attention than it is given today. In [26], authors addressed the issue of *guaranteed, fully computable and sharp* a posteriori error estimates for the isogeometric discretizations of elliptic problems. The estimates presented in [26] are for the upper bound of the error in the energy norm. In this paper we extend this study to cover the lower bound of the error in the energy norm. These *functional-type* a posteriori error estimates, which were introduced in [32, 33], and have been studied for various fields (see [35] and the references therein), provide guaranteed, sharp and fully computable bounds without any generic undetermined constants. These estimates are derived on purely functional grounds (based on integral identities or functional analysis) and are thus applicable to any conforming approximation in the respective space.

For elliptic problems considered in this paper and in [26], the weak solution of the problem lies in $H_0^1(\Omega)$. For such problems, the proposed upper bound of the error involve computing an auxiliary function $y \in H(\Omega, \mathrm{div})$, and the proposed lower bound of the error involve computing an auxiliary function $w \in H_0^1(\Omega)$. To get sharp two-sided estimates, the functions y and w are computed by solving global problems. As briefly explained above, we reiterate that our interest in a posteriori error estimates is not only for adaptivity, but also on quantifying the error in the computed solution (and thus guaranteeing the quality of the computed solution). To the best of authors' knowledge, there is no other, particularly cheaper, method available which can fulfill these objectives in general situations.

Two aspects motivate the application of functional-type error estimates in IGA. Firstly, unlike the standard Lagrange basis functions, NURBS basis functions of degree p are, in general, globally C^{p-1}-continuous. Hence, NURBS basis functions of degree $p \geq 2$ are, in general, at least C^1-continuous, and therefore, their gradients are automatically in $H(\Omega, \mathrm{div})$. Thereby, a sharp upper error bound can be computed without the need of constructing complicated functions in $H(\Omega, \mathrm{div})$, in particular for higher degrees (see, e.g., [8, 10, 11, 19]). Secondly, since the considered problems are solved in an isogeometric setting, an efficient implementation of NURBS basis functions is readily available, which can be used to construct the above mentioned functions y and w. Hence, applying the technique of functional-type a posteriori error estimation in a setting that relies only on the use of already available NURBS basis functions is greatly appealing.

The remainder of this paper is organized as follows. In Sect. 2, we define the model problem, and recall the definition of B-spline and NURBS basis functions. In Sect. 3, we recall functional-type two-sided a posteriori error estimates. Numerical examples are presented in Sect. 4, and some conclusions are drawn in Sect. 5.

2 Notation and Problem Setting

Let $\Omega \subset \mathbb{R}^2$ be an open, bounded and connected Lipschitz domain with boundary $\partial\Omega$. We shall consider the following model problem:

Find the scalar function $u : \overline{\Omega} \to \mathbb{R}$ such that

$$\begin{aligned} -\mathrm{div}(A\nabla u) &= f && \text{in } \Omega, \\ u &= u_D && \text{on } \Gamma_D = \partial\Omega, \end{aligned} \tag{1}$$

where A, f and u_D are given data. We assume that A is a symmetric positive definite matrix and has a positive inverse A^{-1}, and that there exist constants $c_1, c_2 > 0$ such that

$$c_1|\xi|^2 \le A\xi \cdot \xi \le c_2|\xi|^2, \quad \forall \xi \in \mathbb{R}^2. \tag{2}$$

Then, the norms

$$\|v\|_A^2 = \int_\Omega Av \cdot v \, dx, \quad \|v\|_{\bar{A}}^2 = \int_\Omega A^{-1}v \cdot v \, dx, \tag{3}$$

are equivalent to the L^2-norm $\|v\|^2 = \int_\Omega v \cdot v \, dx$. The weak form of problem (1) can be written as follows:

Find $u \in V_g$, such that

$$a(u, v) = f(v), \quad \forall v \in V_0, \tag{4}$$

where $V_0 \subset H^1(\Omega)$ contains the functions which vanish on Γ_D, and $V_g \subset H^1(\Omega)$ contains the functions satisfying the Dirichlet boundary conditions $u = u_D$ on Γ_D. We assume that the problem data A, f and u_D are given such that the bilinear form $a(\cdot,\cdot)$ is bounded, symmetric and positive definite, and that $f(\cdot)$ is a bounded linear functional. The energy norm of a function v is given by $\|\nabla v\|_A = \sqrt{a(v,v)}$. Note that we have considered the Dirichlet problem only for the sake of simplicity. Functional-type error estimates can be easily generalized to problems with mixed boundary conditions, see, e.g., [28, 35].

The problem (4) is discretized by choosing a finite-dimensional manifold $V_h \subset V_g$ and looking for an *approximate solution* $u_h \in V_h$. This leads to a linear system of equations of the form

$$\underline{K}_h \underline{u}_h = \underline{f}_h, \tag{5}$$

where $\underline{K}_h$ is the stiffness matrix induced by the bilinear form $a(\cdot,\cdot)$, $\underline{f}_h$ is the load vector, and $\underline{u}_h$ is the coefficient vector of the discrete solution u_h.

For brevity reasons, we recall only the IGA related notations which are used in this paper. For detailed discussions of NURBS basis functions, geometry mappings and their properties, we refer to, e.g., [13, 22, 31] and the references therein. Let p be a non-negative *degree* and let $s = (s_1, \ldots, s_m)$ be a *knot vector* with $s_i \leq s_{i+1}$ for all i. We consider only *open knot vectors*, i.e., knot vectors s where the multiplicity of a knot is at most p, except for the first and last knot which have multiplicity $p+1$. For simplicity, we assume that $s_1 = 0$ and $s_m = 1$, which can be easily achieved by a suitable scaling. For $i = 1, \ldots, n$, and $n = m - p - 1$, we denote the univariate *B-spline basis functions* by $B^s_{i,p} : (0, 1) \to \mathbb{R}$. For open knot vectors, the first and the last basis functions are interpolatory at the first and the last knot, respectively. B-spline basis functions of degree p are, in general, globally C^{p-1}-continuous. In the presence of repeated knots, the continuity reduces according to the multiplicity, i.e., if a knot appears k times, the continuity of a B-spline basis function of degree p at that knot is C^{p-k}. Let $\{B^s_{i,p}\}_{i=1}^{n_1}$ and $\{B^t_{j,q}\}_{j=1}^{n_2}$ be two families of B-spline basis functions defined by the degrees p and q, and the open knot vectors

$$s = (s_1, \ldots, s_{n_1+p+1}),\ t = (t_1, \ldots, t_{n_2+q+1}),$$

respectively. We denote the set of all double-indices (i,j) by

$$\mathcal{I}_R = \{(i,j) :\ i \in \{1, \ldots, n_1\}, j \in \{1, \ldots, n_2\}\}.$$

By $R_{(i,j)}(\xi_1, \xi_2)$, $(i,j) \in \mathcal{I}_R$ we denote the *bivariate NURBS basis functions* constructed from the above-mentioned families of B-spline basis functions and positive *weights* $w_{(i,j)}$, $(i,j) \in \mathcal{I}_R$ (see, e.g., [13, 22] for details). The continuity of the B-spline basis functions is inherited by the NURBS basis functions. Note that B-splines can be seen as a special case of NURBS with all weights being equal to one. Hence, we will not distinguish between these two and we will only use the term *NURBS* in the remainder of the paper.

Associated with the *parameter domain* $\hat{\Omega} = (0, 1)^2$, the set of functions

$$\hat{V}_h = \text{span}\{R_{(i,j)},\ (i,j) \in \mathcal{I}_R\},$$

is uniquely determined by the degrees p and q, the knot vectors s and t, and the weights w. Given the set of functions $\hat{V}_h$ and a *control net* of *control points* $P_{(i,j)} \in \mathbb{R}^2$, where $(i,j) \in \mathcal{I}_R$, the two-dimensional *NURBS-surface* $G : \hat{\Omega} \to \Omega$ is defined by

$$G(\xi_1, \xi_2) = \sum_{(i,j) \in \mathcal{I}_R} R_{(i,j)}(\xi_1, \xi_2)\, P_{(i,j)}. \tag{6}$$

We refer to $\Omega = G(\hat{\Omega})$ as the *physical domain*. We assume that the geometry mapping is continuous and bijective (i.e., not self-penetrating), which are natural assumptions for CAD-applications.

In IGA, the isoparametric principle is applied by using the same basis functions for the discrete solution u_h which are used for representing the geometry. For detailed discussion, we refer the reader to, e.g., [13, 22]. The discrete solution u_h on the physical domain Ω is represented as follows:

$$u_h(x) = \sum_{(i,j)\in\mathcal{I}_R} u_{(i,j)} \left(R_{(i,j)} \circ G^{-1}\right)(x), \tag{7}$$

where $u_{(i,j)} \in \mathbb{R}$ are real-valued coefficients which form the coefficient vector $\underline{u}_h$. The discrete function space is thus defined by

$$V_h = \text{span}\{R_{(i,j)} \circ G^{-1},\ (i,j) \in \mathcal{I}_R\}.$$

The initial mesh, and thereby the basis functions on this initial mesh, are assumed to be given via the geometry representation of the computational domain, i.e., the initial discretization is already determined by the problem domain. The exact representation of the geometry on the initial (coarsest) level is preserved in the process of mesh refinement.

As mentioned in the introduction, the straightforward definition of NURBS basis functions leads to a tensor-product structure of the discretization, which is the focus of this paper. Nevertheless, since the presented error estimators do not exploit any information from the underlying discretization method, the beauty of them is that they are also applicable to local refinement techniques (e.g., T-splines, THB-splines, PHT-splines, LR-splines, see Sect. 1) without any additional work.

3 Guaranteed Two-Sided Bounds of the Error

In this section, we recall the well-known theoretical bounds for the error in the energy norm (see, e.g., [32–35]). For a detailed theoretical exposition the reader is referred to [35].

For an upper bound of the error in the energy norm we have the following estimate.

Theorem 1 *Let C_Ω be the constant in the inequality $\|v\| \leq C_\Omega \|\nabla v\|_A$ (Friedrich's type), for all $v \in V_0$, where $\|\cdot\|$ denotes the standard L_2-norm. Let u be the exact solution of the problem* (4)*, and let $u_h \in V_h$ be an approximate solution. Then, the following estimate holds:*

$$\|\nabla u - \nabla u_h\|_A \leq \|A\nabla u_h - y\|_{\bar{A}} + C_\Omega \|\text{div}\, y + f\|, \tag{8}$$

where y is an arbitrary vector-valued function in $H(\Omega, \text{div})$, and the norms are as defined in (3)*.*

The constant C_Ω depends only on the domain Ω and the coefficient matrix A (but not on the underlying mesh), see, e.g., [26, 28, 35]. Note that C_Ω can be computed either numerically or, if one can find a domain $\Omega_\square \supset \Omega$, where $\Omega_\square$ is a square domain with side-length ℓ, then $C_\Omega \leq c_2 \frac{\ell}{\pi\sqrt{d}}$, where d is the dimension and c_2 is the constant in (2). Therefore, the estimate given in Theorem 1 is fully computable upper bound for any conforming approximation $u_h \in V_g$. Note that, if we choose y via the (unknown) exact solution $y = A\nabla u$, both sides of (8) coincide. Hence, the estimate is sharp in the sense that, for any fixed u_h, we can find a function y such that the upper bound is as close to the exact error as desired.

To obtain a sharp estimate (and not just an indicator), one can find a function y which minimizes the right-hand-side of (8). To achieve this, we first rewrite the estimate in the following form

$$\|\nabla u - \nabla u_h\|_A^2 \leq (1+\beta)\|A\nabla u_h - y\|_{\bar{A}}^2 + (1+\tfrac{1}{\beta})C_\Omega^2\|\operatorname{div} y + f\|^2 =: M_\oplus^2(y,\beta), \tag{9}$$

where $\beta > 0$ is a free parameter, see e.g. [26, 28, 35]. Note that the upper bound in (9) holds true for *any* fixed $y \in H(\Omega, \operatorname{div})$ and $\beta > 0$. Hereinafter, for simplicity, we will refer to $M_\oplus^2(y,\beta)$ as the *majorant*. Introducing

$$\begin{aligned} a_1 &= 1 + \beta, & a_2 &= (1 + \tfrac{1}{\beta})C_\Omega^2, \\ B_1 &= \|A\nabla u_h - y\|_{\bar{A}}^2, & B_2 &= \|\operatorname{div} y + f\|^2, \end{aligned} \tag{10}$$

we can briefly write the majorant as

$$M_\oplus^2(y,\beta) = a_1 B_1 + a_2 B_2. \tag{11}$$

Therefore, a *sharp* estimate can be obtained by finding $y \in H(\Omega, \operatorname{div})$ and $\beta > 0$ as solutions to the global minimization problem

$$\min_{y \in H(\Omega,\operatorname{div}),\ \beta > 0} M_\oplus^2(y,\beta). \tag{12}$$

As explained in detail in [26], to find the minimizing parameter y, we set $M_\oplus^2(y)' = 0$, and solve

$$a_1 \int_\Omega A^{-1} y \cdot \tilde{y}\, dx + a_2 \int_\Omega \operatorname{div} y \operatorname{div} \tilde{y}\, dx = a_1 \int_\Omega \nabla u_h \cdot \tilde{y}\, dx - a_2 \int_\Omega f \operatorname{div} \tilde{y}\, dx \tag{13}$$

for a solution $y_h \in Y_h \subset H(\Omega, \operatorname{div})$. Note that the solution of the linear system of equations (13) incurs non-negligible cost as we need to assemble the matrix on the left hand side and the vector on the right hand side, and then solve the linear system. Minimization with respect to β results in $\beta = C_\Omega \sqrt{\frac{B_2}{B_1}}$, which requires only the evaluation of the integrals, and thus involves negligible cost.

For a lower bound of the error in the energy norm we have the following estimate, which can be found in [35].

Theorem 2 *Let u be the exact solution of the problem* (4). *Then, the following estimate holds.*

$$M_{\ominus}^2(w) \leq \|\nabla u - \nabla u_h\|_A^2, \tag{14}$$

where the minorant $M_{\ominus}^2(w)$ *is defined as*

$$M_{\ominus}^2(w) := \int_{\Omega} (2fw - 2A\nabla u_h \cdot \nabla w)dx - \|\nabla w\|_A^2, \tag{15}$$

and where w *is an arbitrary function in* V_0.

Note that, in contrast to the upper bound $M_{\oplus}^2(y, \beta)$, we have only one free parameter in the lower bound, namely the scalar-valued function w. To maximize $M_{\ominus}^2(w)$ with respect to w, we proceed as follows (see, e.g., [28]). Let $M_{\ominus}^2(w)'$ denote the derivative of $M_{\ominus}^2(w)$ with respect to w. Setting $M_{\ominus}^2(w)' = 0$, we obtain

$$\int_{\Omega} A\nabla w \cdot \nabla \tilde{w} dx = \int_{\Omega} (f\tilde{w} - A\nabla u_h \cdot \nabla \tilde{w})dx. \tag{16}$$

To solve (16), we choose a finite-dimensional subspace W_h of V_0 and search for a solution $w_h \in W_h$.

Together with the minorant $M_{\ominus}^2$ and the majorant $M_{\oplus}^2$, we have a two-sided estimate of the true error in the form

$$M_{\ominus}^2 \leq \|\nabla u - \nabla u_h\|_A^2 \leq M_{\oplus}^2. \tag{17}$$

Thereby, it is possible to provide a guaranteed, fully computable, and as sharp as desired interval containing the true error. To measure the efficiency of the computed lower bound and upper bound, the *efficiency indices* are defined by

$$I_{\ominus} = \frac{M_{\ominus}(w)}{\|\nabla u - \nabla u_h\|_A}, \qquad I_{\oplus} = \frac{M_{\oplus}(y, \beta)}{\|\nabla u - \nabla u_h\|_A}. \tag{18}$$

Clearly, the closer $I_{\ominus}$ or $I_{\oplus}$ is to 1, the better the estimate is.

Remark 1 To obtain sharp bounds it is necessary to choose the spaces Y_h and W_h sufficiently rich such that they have better approximation properties than V_h.

4 Numerical Results

In this section, we present numerical results of several examples with varying complexity. These examples include simple Poisson problem with solution as a sinusoidal function on a unit square domain, also with varying continuity of basis functions, a problem with non-identity coefficient matrix where the PDE operator coefficients are of exponential form, a problem with curved boundary domain, and a problem with solution having two sharp peaks.

To reflect the associated polynomial degrees in respective dimensions, we will use the notation $\mathcal{S}_h^{p,q}$ for $\hat{V}_h$, which denotes the NURBS function of degrees p and q in the first and second coordinate, respectively, and where the parameter h is the characteristic cell size (non-vanishing knot-span) of the mesh for $\hat{V}_h$. Unless indicated otherwise, we assume that the NURBS functions in $\mathcal{S}_h^{p,q}$ are of C^{p-1}-continuity in the first coordinate, and C^{q-1}-continuity in the second coordinate. Since for $p \geq 2$, we have $(C^{p-2})^d \subset H(\Omega, \text{div})$, we choose the following choice for $\hat{Y}_h$ (which is the *Case 1* studied in detail in [26])

$$\hat{Y}_h = \mathcal{S}_h^{p+1,p+1} \otimes \mathcal{S}_h^{p+1,p+1}. \tag{19}$$

Thereby, we choose a function space $\hat{Y}_h$ on the parameter domain and, analogously to the relation of $\hat{V}_h$ and V_h, we define the function space Y_h by the push-forward

$$Y_h = \hat{Y}_h \circ G^{-1}.$$

Cost efficient computation of the majorant are discussed in [26]. We shall not discuss a cost-efficient computation of the minorant in detail, and only present first results as a proof-of-concept. We consider a straightforward approach where W_h is obtained from V_0 by applying one step of p-refinement (degree elevation while keeping the continuity at the knots). In this setting, the size of the discretized problem for computing w_h is larger than the original problem for u_h. Since this set-up is costlier, the presented results are not computed on very fine meshes.

Example 1 (Sinus function in a unit square) In this numerical example, the computational domain is the unit square $\Omega = (0, 1)^2$ and $u_h \in \mathcal{S}_h^{2,2}$, i.e., a piecewise quadratic function in both directions. The coefficient matrix is the identity matrix, i.e., $A = I$, and the exact solution is given by

$$u = \sin(6\pi x) \sin(3\pi y).$$

The right-hand-side f and the (homogeneous) boundary conditions u_D are determined by the prescribed exact solution u.

The computed efficiency indices for this example are presented in Table 1. We see that with successively refined meshes the efficiency indices approaches the academic value of 1, which means that the obtained upper and lower bounds are practically same as the true error. Note that the lack of sharpness of the bounds on coarser meshes is due to the oscillations in the solution of the problem (with 3 full wavelengths in the x direction and 1.5 wavelengths in the y direction).

Example 2 (Reduced continuity of basis functions) We consider the same exact solution and the same physical domain as in Example 1. However, we now use

Table 1 Efficiency indices for Example 1, $\hat{V}_h = \mathcal{S}_h^{2,2}$

$1/h$	$I_\ominus$	$I_\oplus$
16×16	0.9918	1.7887
32×32	0.9966	1.3762
64×64	0.9990	1.1883
128×128	0.9997	1.0947
256×256	0.9999	1.0476

Table 2 Efficiency indices for Example 2, $\hat{V}_h = \mathcal{S}_h^{4,4}$, with only C^1-continuity at $x = 0.5$ and $y = 0.5$

$1/h$	$I_\ominus$	$I_\oplus$
16×16	0.9984	2.3040
32×32	0.9974	1.5029
64×64	0.9987	1.2399
128×128	0.9996	1.1188
256×256	0.9999	1.0595

B-splines of degree $p = q = 4$ to represent Ω, and we add a triple knot at the coordinates $x = 0.5$ and $y = 0.5$. The initial knot vectors are thus given by

$$s = t = (0, 0, 0, 0, 0, 0.5, 0.5, 0.5, 1, 1, 1, 1, 1),$$

and the geometry mapping is only C^1-continuous at the coordinate 0.5. This is to mimic a multi-patch situation where the solution has full continuity everywhere except along certain lines with reduced continuity. Note that, in this case, we also reduced the continuity of $\hat{Y}_h$ at $x = 0.5$ and $y = 0.5$ by one order.

The computed efficiency indices for this example are presented in Table 2. We see that the results are qualitatively same as the results of Example 1, and with successively refined meshes the obtained upper and lower bounds are practically same as the true error.

Example 3 (Varying coefficients of PDE operator) Let the matrix A be of the form of

$$\begin{pmatrix} e^{b_{11}x+b_{12}y} & 0 \\ 0 & e^{b_{21}x+b_{22}y} \end{pmatrix},$$

which is positive definite for $b_{ij} \in \mathbb{R}^+$, $i, j = \{1, 2\}$. This will result in the PDE operator to be of the form of

$$e^{b_{11}x+b_{12}y}\frac{\partial^2}{\partial x^2} + e^{b_{21}x+b_{22}y}\frac{\partial^2}{\partial y^2} + b_{11}e^{b_{11}x+b_{12}y}\frac{\partial}{\partial x} + b_{22}e^{b_{21}x+b_{22}y}\frac{\partial}{\partial y}.$$

To have this PDE operator with full generality, we take $b_{11} = 0.1, b_{12} = 0.8, b_{21} = 0.4, b_{22} = 0.7$. With this generality, to have a good comparison of the efficiency indices with the examples considered so far, we again choose the exact solution to

Table 3 Efficiency indices for Example 3, $\hat{V}_h = \mathcal{S}_h^{2,2}$

$1/h$	$I_{\ominus}$	$I_{\oplus}$
16×16	0.9918	5.9954
32×32	0.9966	2.4968
64×64	0.9990	1.7377
128×128	0.9997	1.3699
256×256	0.9999	1.1856

be $u = \sin(6\pi x)\sin(3\pi y)$. The domain and the polynomial space for this example are the same as of Example 1. The right hand side function is accordingly calculated and the solution has homogeneous Dirichlet boundary values.

Note that in this case the constant C_Ω has to be accordingly modified. For the unit square domain, and the matrix A given above, its value is taken as $\frac{c_2}{\pi\sqrt{2}}$, where

$$c_2 = \max\{e^{b_{11}x+b_{12}y}, e^{b_{21}x+b_{22}y}\}.$$

The computed efficiency indices for this example are presented in Table 3. We see that the proposed estimator is robust with respect to the non-trivial PDE coefficient matrix A, and its performance is asymptotically similar to the case with the matrix A being identity. Note that some deviation could be attributed to the fact that we used same number of quadrature points for the evaluation of the matrices in both the cases, which is not sufficient when the PDE coefficients are of exponential form.

Example 4 (Domain with curved boundary) Consider the domain of a quarter annulus. In polar coordinates, Ω is defined by $(r, \phi) \in (1, 2) \times (0, \frac{\pi}{2})$. The circular parts of the domain boundary are represented exactly by the NURBS geometry mapping of degree 2, i.e., we have $p = q = 2$. We set $A = I$, and we prescribe the exact solution

$$u = (r-1)(r-2)\phi(\phi - \tfrac{\pi}{2})e^{-\alpha(r\cos\phi - 1)^2}.$$

We test our method with two values of α, namely,

$$\text{Example 4.a:} \quad \alpha = 20, \qquad \text{Example 4.b:} \quad \alpha = 50.$$

In both examples, this function has zero Dirichlet boundary values and a peak at $x = 1$, the sharpness of which is determined by the value of α. The exact solutions are depicted in Fig. 1.

The computed efficiency indices for this example are presented in Table 4. We again observe a similar pattern as in earlier examples with the unit square domain, with sharper peak demanding more resolution. This confirms the robustness of the two-sided bounds with respect to the boundary shape. From Figs. 2 and 3, we see that the marking based on the majorant (9) shows an excellent agreement with the marking based on the true error.

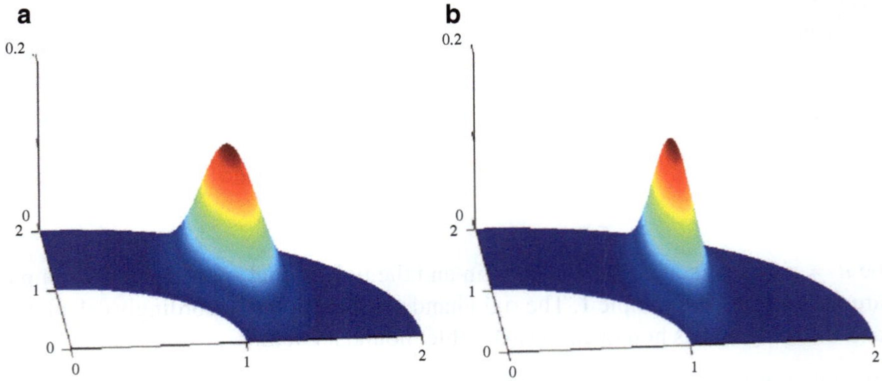

Fig. 1 Exact solutions u on Ω, Example 4. (**a**) Example 4.a ($\alpha = 20$). (**b**) Example 4.b ($\alpha = 50$)

Table 4 Efficiency indices for Example 4, $\hat{V}_h = \mathcal{S}_h^{2,2}$

	$\alpha = 20$		$\alpha = 50$	
$1/h$	$I_\ominus$	$I_\oplus$	$I_\ominus$	$I_\oplus$
16×8	0.9922	1.8614	0.9928	3.0624
32×16	0.9953	1.3357	0.9926	1.9711
64×32	0.9985	1.1596	0.9967	1.3923
128×64	0.9996	1.0792	0.9990	1.1891
256×128	0.9999	1.0396	0.9997	1.0930

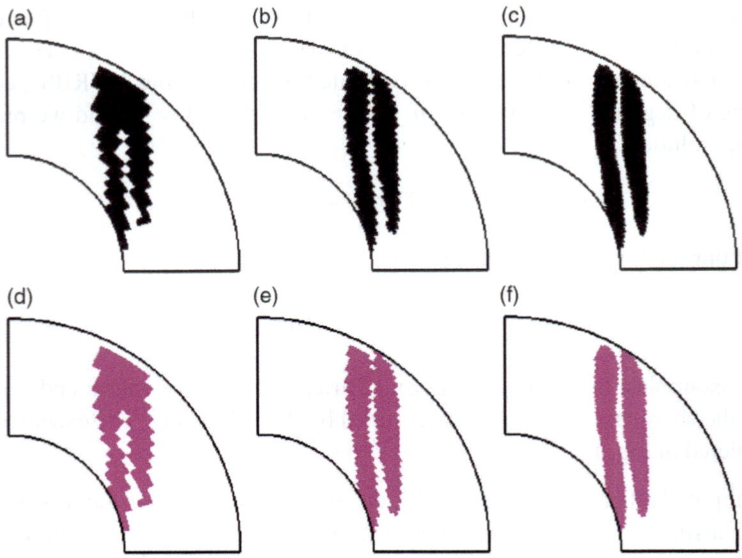

Fig. 2 Marked cells with $\psi = 20\,\%$ in Example 4.a ($\alpha = 20$), $\hat{V}_h = \mathcal{S}_h^{2,2}$. See [26] for details and the definition of Cases. (**a**) Exact, mesh 32×16. (**b**) Exact, mesh 64×32. (**c**) Exact, mesh 128×64. (**d**) Case 1, 32×16. (**e**) Case 2, 64×32. (**f**) Case 3, 128×64

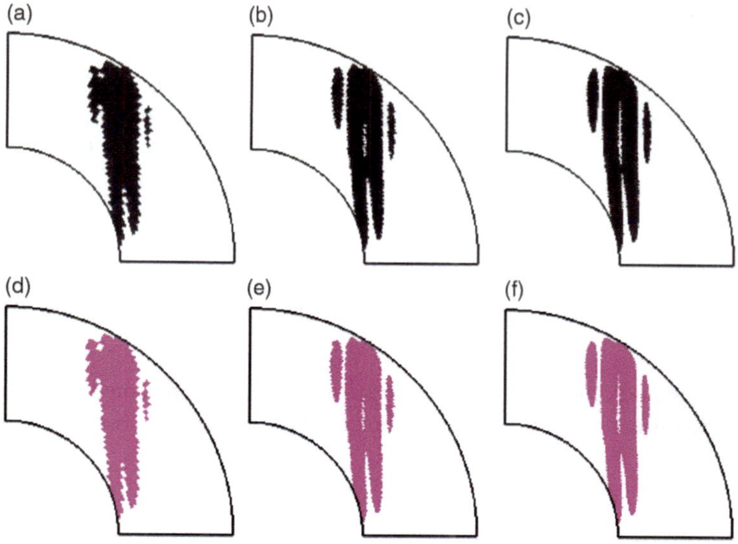

Fig. 3 Marked cells with $\psi = 20\,\%$ in Example 4.b ($\alpha = 50$), $\hat{V}_h = \mathcal{S}_h^{2,2}$. See [26] for details and the definition of Cases. (**a**) Exact, mesh 64 × 32. (**b**) Exact, mesh 128 × 64. (**c**) Exact, mesh 256 × 128. (**d**) Case 1, 64 × 32. (**e**) Case 2, 128 × 64. (**f**) Case 3, 256 × 128

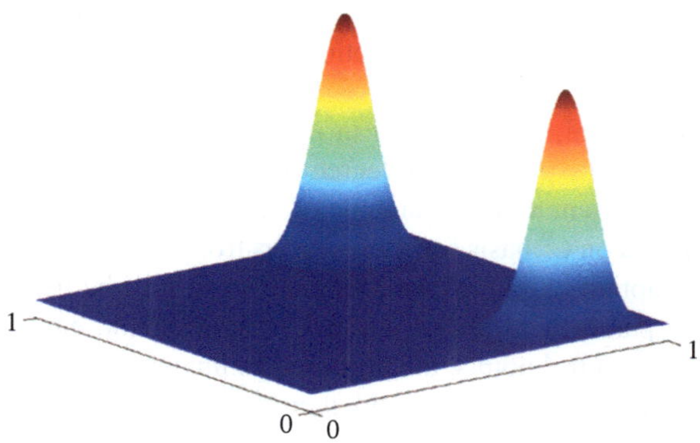

Fig. 4 Exact solution, Example 5

The next example is to test a basic adaptive refinement scheme based on the proposed upper bound.

Example 5 (Adaptive Refinement) The exact solution for this example is given by

$$u = (x^2 - x)(y^2 - y)e^{-100|(x,y)-(0.8,0.05)|^2 - 100|(x,y)-(0.8,0.95)|^2}.$$

The computational domain is again the unit square $\Omega = (0, 1)^2$, and is represented by B-splines of degree $p = q = 2$. The function u, which is illustrated in Fig. 4, has

Table 5 Efficiency indices for Example 5, $\hat{V}_h = \mathcal{S}_h^{2,2}$

$1/h$	$I_{\ominus}$	$I_{\oplus}$
16×16	0.9911	3.9972
25×26	0.9947	2.2161
38×44	0.9971	1.7824
64×74	0.9983	1.5383
92×136	0.9989	1.4599
184×256	0.9991	1.4058

zero Dirichlet boundary values and has two peaks at the coordinates $(0.8, 0.05)$ and $(0.8, 0.95)$.

The computed efficiency indices for this example are presented in Table 5. We see that the estimator is able to nicely direct the refinement process.

5 Conclusions

We have presented two-sided fully computable, guaranteed and sharp a posteriori error estimates in IGA. The computation of the presented estimates relies only on the use of NURBS basis functions, without the need for constructing complicated basis functions in $H(\Omega, \text{div})$. For the computation of these estimates within the context of IGA, two properties of NURBS basis functions are useful. Firstly, the basis functions are, in general, automatically in $H(\Omega, \text{div})$ due to their high continuity. Secondly, increasing the polynomial degree of NURBS basis functions adds only few DOFs. It is important to note that none of these properties are possible in FEM discretizations based on C^0 basis functions. Apart from the topical interest of quality assurance and adaptivity, the presented estimates can also be used in other topics in IGA, such as parametrization of computational domain, e.g., to optimize the placement of inner control points, the proposed estimator can be used to accurately detect the regions with large error and then use the optimization algorithm to reposition the control points. Finally, though in this paper we have only considered tensor-product NURBS discretizations, these estimates can be easily adapted to other type of basis functions as they are valid for *any* conforming approximation.

Acknowledgements The support from the Austrian Science Fund (FWF) through the project P21516-N18 is gratefully acknowledged.

References

1. M. Ainsworth, J.T. Oden, *A Posteriori Error Estimation in Finite Element Analysis* (Wiley Interscience, New York, 2000)

2. R.E. Bank, R.K. Smith A posteriori error estimates based on hierarchical bases. SIAM J. Numer. Anal. **30**(4), 921–935 (1993)
3. Y. Bazilevs, L. Beirão da Veiga, J.A. Cottrell, T.J.R. Hughes, G. Sangalli, Isogeometric analysis: approximation, stability and error estimates for h-refined meshes. Math. Models Methods Appl. Sci. **16**(7), 1031–1090 (2006)
4. Y. Bazilevs, V.M. Calo, J.A. Cottrell, T.J.R. Hughes, A. Reali, G. Scovazzi, Variational multiscale residual-based turbulence modeling for large eddy simulation of incompressible flows. Comput. Methods Appl. Mech. Eng. **197**(1–4), 173–201 (2007)
5. Y. Bazilevs, V.M. Calo, T.J.R. Hughes, Y. Zhang, Isogeometric fluid-structure interaction: theory, algorithms, and computations. Comput. Mech. **43**, 3–37 (2008)
6. Y. Bazilevs, V.M. Calo, J.A. Cottrell, J.A. Evans, T.J.R. Hughes, S. Lipton, M.A. Scott, T.W. Sederberg, Isogeometric analysis using T-splines. Comput. Methods Appl. Mech. Eng. **199**(5–8), 229–263 (2010)
7. L. Beirão da Veiga, A. Buffa, J. Rivas, G. Sangalli, Some estimates for h- p- k- refinement in isogeometric analysis. Numer. Math. **118**, 271–305 (2011)
8. D. Boffi, F. Brezzi, M. Fortin, *Mixed Finite Element Methods and Applications* (Springer, Berlin, 2013)
9. A. Buffa, G. Sangalli, R. Vázquez, Isogeometric analysis in electromagnetics: B-splines approximation. Comput. Methods Appl. Mech. Eng. **199**(17–20), 1143–1152 (2010)
10. A. Buffa, C. de Falco, G. Sangalli, IsoGeometric analysis: stable elements for the 2D Stokes equation. Int. J. Numer. Meth. Fluids **65**(11–12), 1407–1422 (2011)
11. A. Buffa, J. Rivas, G. Sangalli, R. Vázquez, Isogeometric discrete differential forms in three dimensions. SIAM J. Numer. Anal. **49**(2), 818–844 (2011)
12. J.A. Cottrell, T.J.R. Hughes, A. Reali, Studies of refinement and continuity in isogeometric structural analysis. Comput. Methods Appl. Mech. Eng. **196**, 4160–4183 (2007)
13. J. Cottrell, T.J.R. Hughes, Y. Bazilevs, *Isogeometric Analysis: Toward Integration of CAD and FEA* (Wiley, Chichester, 2009)
14. L. Dede, H.A.F.A. Santos, B-spline goal oriented error estimators for geometrically nonlinear rods. Comput. Mech. **49**, 35–52 (2012)
15. J. Deng, F. Chen, X. Li, C. Hu, W. Tong, Z. Yang, Y. Feng, Polynomial splines over hierarchical T-meshes. Graph. Models **70**, 76–86 (2008)
16. T. Dokken, T. Lyche, K.F. Pettersen, Polynomial splines over locally refined box-partitions Comput. Aided Geom. Des. **30**(3), 331–356 (2013)
17. M.R. Dörfel, B. Jüttler, B. Simeon, Adaptive isogeometric analysis by local h-refinement with T-splines. Comput. Methods Appl. Mech. Eng. **199**(5–8), 264–275 (2010)
18. T. Elguedj, Y. Bazilevs, V.M. Calo, T.J.R. Hughes, $\bar{B}$ and $\bar{F}$ projection methods for nearly incompressible linear and nonlinear elasticity and plasticity using higher-order NURBS elements. Comput. Methods Appl. Mech. Eng. **197**(33–40), 2732–2762 (2008)
19. J.A. Evans, T.J.R. Hughes, Isogeometric divergence-conforming B-splines for the Darcy-Stokes-Brinkman equations. Math. Models Methods Appl. Sci. **23**(04), 671–741 (2013)
20. C. Giannelli, B. Jüttler, H. Speleers, THB-splines: the truncated basis for hierarchical splines. Comput. Aided Geom. Des. **29**(7), 485–498 (2012)
21. C. Giannelli, B. Jüttler, H. Speleers, Strongly stable bases for adaptively refined multilevel splines spaces. Adv. Comput. Math. **40**, 459–490 (2014)
22. T.J.R. Hughes, J. Cottrell, Y. Bazilevs, Isogeometric analysis: CAD, finite elements, NURBS, exact geometry and mesh refinement. Comput. Methods Appl. Mech. Eng. **194**(39–41), 4135–4195 (2005)
23. T.J.R. Hughes, A. Reali, G. Sangalli, Efficient quadrature for NURBS-based isogeometric analysis. Comput. Methods Appl. Mech. Eng. **199**(5–8), 301–313 (2010)
24. K.A. Johannessen, An adaptive isogeometric finite element analysis. Master's thesis, Norwegian University of Science and Technology, 2009
25. K.A. Johannessen, T. Kvamsdal, T. Dokken, Isogeometric analysis using LR B-splines. Comput. Methods Appl. Mech. Eng. **269**, 471–514 (2014)

26. S.K. Kleiss, S.K. Tomar, Guaranteed and sharp a posteriori error estimates in isogeometric analysis. Comput. Math. Appl. **70**(3), 167–190 (2015)
27. G. Kuru, C.V. Verhoosel, K.G. van der Zee, E.H. van Brummelen, Goal-oriented isogeometric analysis with hierarchical splines. Comput. Methods Appl. Mech. Eng. **270**, 270–292 (2014)
28. R. Lazarov, S.I. Repin, S.K. Tomar, Functional a posteriori error estimates for discontinuous Galerkin approximations of elliptic problems. Numer. Methods Partial Differ. Equ. **25**(4), 952–971 (2009)
29. X. Li, J. Zheng, T.W. Sederberg, T.J.R. Hughes, M.A. Scott, On linear independence of T-spline blending functions. Comput. Aided Geom. Des. **29**(1), 63–76 (2012)
30. P.N. Nielsen, A.R. Gersborg, J. Gravesen, N.L. Pedersen, Discretizations in isogeometric analysis of Navier-Stokes flow. Comput. Methods Appl. Mech. Eng. **200**(45–46), 3242–3253 (2011)
31. L. Piegl, W. Tiller, *The NURBS Book*. Monographs in Visual Communications, 2nd edn. (Springer, Berlin/Heidelberg, 1997)
32. S.I. Repin, A posteriori error estimation for nonlinear variational problems by duality theory. J. Math. Sci. **99**, 927–935 (2000). See also: Zapiski Nauchnych Seminarov POMI **243**, 201–214 (1997)
33. S.I. Repin, A posteriori error estimation for variational problems with uniformly convex functionals. Math. Comput. **69**(230), 481–500 (2000)
34. S.I. Repin, Two-sided estimates of deviation from exact solutions of uniformly elliptic equations, in *Proceedings of the St. Petersburg Mathematical Society*, vol. IX, ed. by N. Uraltseva. Am. Math. Soc. Transl.: Ser. 2 **209**, 143–171 (2003)
35. S.I. Repin, *A Posteriori Estimates for Partial Differential Equations* (Walter de Gruyter, Berlin, 2008)
36. M.A. Scott, M.J. Borden, C.V. Verhoosel, T.W. Sederberg, T.J.R. Hughes, Isogeometric finite element data structures based on Bézier extraction of T-splines. Int. J. Numer. Meth. Eng. **88**, 126–156 (2011)
37. M.A. Scott, X. Li, T.W. Sederberg, T.J.R. Hughes, Local refinement of analysis-suitable T-splines. Comput. Methods Appl. Mech. Eng. **213–216**, 206–222 (2012)
38. T.W. Sederberg, D.L. Cardon, G.T. Finnigan, N.S. North, J. Zheng, T. Lyche, T-spline simplification and local refinement. ACM Trans. Graph. **23**(3), 276–283 (2004)
39. A.-V. Vuong, C. Giannelli, B. Jüttler, B. Simeon, A hierarchical approach to adaptive local refinement in isogeometric analysis. Comput. Methods Appl. Mech. Eng. **200**(49–52), 3554–3567 (2011)
40. P. Wang, J. Xu, J. Deng, F. Chen, Adaptive isogeometric analysis using rational PHT-splines. Comput.-Aided Des. **43**(11), 1438–1448 (2011)
41. G. Xu, B. Mourrain, R. Duvigneau, A. Galligo, Parameterization of computational domain in isogeometric analysis: methods and comparison. Comput. Methods Appl. Mech. Eng. **200**(23–24), 2021–2031 (2011)
42. G. Xu, B. Mourrain, R. Duvigneau, A. Galligo, Optimal analysis-aware parameterization of computational domain in 3D isogeometric analysis. Comput.-Aided Des. **45**, 812–821 (2013)
43. K.G. van der Zee, C.V. Verhoosel, Isogeometric analysis-based goal oriented error estimation for free boundary problems. Finite Elem. Anal. Des. **47**, 600–609 (2011)

Multilevel Preconditioning for Variational Problems

Angela Kunoth

Abstract For the numerical solution of linear variational problems involving elliptic partial differential operators in $n \geq 2$ space dimensions, iterative solution schemes are indispensable on account of their problem size. Our guiding principle is to devise iterative solvers which are optimal in the number of arithmetic operations, i.e., which are of linear complexity in the total number of unknowns. For these algorithms, asymptotically optimal preconditioners are required. The class of preconditioners for which this can be shown are of multilevel type, requiring nested approximation spaces to approximate the solution of the system on a fine user-specified grid. For smooth solutions of standard second and fourth order elliptic PDEs (partial differential equations) in variational form, approximations based on tensor products of higher-order B-splines yield high accuracy.

For such problem classes, this survey collects the main ingredients for multilevel preconditioners in terms of higher order B-splines. There are three types of multilevel preconditioners for which asymptotic optimality can be shown. Two of them, the so-called additive preconditioners, are specified for isogeometric analysis involving linear elliptic partial differential operators in terms of variants of the BPX (Bramble-Pasciak-Xu) preconditioner and wavelet preconditioners. The third class are the so-called multiplicative preconditioners, specifically, multigrid methods.

An essential ingredient for all these multilevel preconditioners are intergrid operators which transform vectors or matrices between grids of different grid spacing. For higher order B-splines, these intergrid operators can be derived from their refinement relations. In addition to a presentation of the theoretical ingredients, the performance of the different preconditioners will be demonstrated by some numerical examples.

A. Kunoth (✉)
Mathematisches Institut, Universität zu Köln, Weyertal 86–90, 50931 Köln, Germany
e-mail: kunoth@math.uni-koeln.de

B. Jüttler, B. Simeon (eds.), *Isogeometric Analysis and Applications 2014*,
Lecture Notes in Computational Science and Engineering 107,
DOI 10.1007/978-3-319-23315-4_11

1 Introduction

For variational systems involving linear elliptic partial differential equations (PDEs) in $n \geq 2$ space dimensions, a standard finite element or finite difference discretization on a uniform grid with grid spacing $0 < h < 1$ leads to the problem to numerically solve a large ill-conditioned system of linear equations. This is due to the fact that PDE operators have positive order $2r$, i.e., $r = 1$ for second order or $r = 2$ for fourth order operators. Ill-conditioned means that the system matrix $\mathbf{A}_h$ exhibits a spectral condition number $\kappa_2(\mathbf{A}_h)$ which is proportional to h^{-2r}, i.e., $\kappa_2(\mathbf{A}_h) \sim h^{-2r}$. Here and in the following, the relation $a \sim b$ stands for $a \lesssim b$ and $b \lesssim a$ where the latter inequality means that b can be bounded by some constant times a uniformly in all parameters on which a and b may depend. Since the convergence speed of any iterative solution scheme depends on the spectral condition number, the scheme will therefore become prohibitively slow. This effect becomes even worse when h is chosen smaller, in order to obtain more accurate approximations; the number of unknowns N increases like $N \sim h^{-n}$ and, thus, the system size also increases accordingly.

On the other hand, solutions of elliptic PDEs typically exhibit a multiscale behaviour. Enhancing iterative methods by multilevel ingredients, therefore, enables one to achieve much more efficient solution schemes. Ultimately, one strives for an 'optimally efficient scheme'. This means that one can solve the problem with fine grid accuracy with an amount of arithmetic operations that is proportional to the number of unknowns on this grid. The first such methods which were proven to provide an asymptotically optimal iterative scheme were geometric multigrid algorithms [9]. The basic idea of these schemes is to successively solve smaller approximations of the linear system on the finest grid. These can often be interpreted as discretizations with respect to coarser grids. Their iterative solutions can be seen as approximating the inverses of the system matrices applied to the right hand side on the different grids, thereby reducing the spectral condition number of the original system matrix. This idea has, therefore, suggested the term 'preconditioner'. We call a linear operator $\mathbf{C}_h$ an *(asymptotically) optimal preconditioner* if its set-up, storage and application is of optimal linear complexity in the number of unknowns N and if $\kappa_2(\mathbf{C}_h\mathbf{A}_h) \sim 1$ independent of h.

The search for such optimal preconditioners was a major topic for numerical elliptic PDE solvers in the 1980s. The goal was to better understand the ingredients which made a preconditioner optimal. Specifically, one aimed at finding directly applicable versions which could be interpreted as a change of basis. With the arrival of the hierarchical basis preconditioner [54], extending an idea of Babuška from the univariate case, a simple preconditioner became available. Although it is not optimal—one still has logarithmic growth in the grid spacing, $\kappa_2(\mathbf{C}_h\mathbf{A}_h) \sim \log h^{-2r}$, in the bivariate case, and exponential growth for $n = 3$—its simplicity still makes it popular up to now [42]. Another multilevel preconditioner was presented in [3, 4]. Within the last years, (tensor product) hierarchical B-spline discretizations were increasingly employed in the area of isogeometric analysis, mainly in the context of deriving discretizations which can be locally adapted to singularities, see, e.g., [36, 51], the survey [5] and the references therein.

At the end of the 1980s, another methodology to derive preconditioners via space decompositions and subspace corrections was developed, see the review [52]. The BPX (Bramble-Pasciak-Xu) preconditioner proposed first in [10] was numerically observed to be optimal; it is based on a hierarchical generator system with grid-dependent weights. With techniques from approximation theory, its optimality was theoretically established in [21, 45]. Since then, its range of application has been widened extensively. For second and fourth order elliptic problems on the sphere a BPX-type preconditioner was developed and its optimality proved for triangular finite elements in [43]. The survey article [53] records extensions of the BPX and of multigrid preconditioners to $H(\text{grad})$, $H(\text{curl})$, and $H(\text{div})$ systems on adaptive and unstructured grids; to name just two extensions.

Multigrid preconditioners for isogeometric analysis were presented in [27], whereas domain decomposition type preconditioners were proposed in [6]. Within the class of domain decomposition methods, also tearing and interconnecting methods (FETI-BETI) are important [34]. Note that not all of these preconditioners are or have been proven to be asymptotically optimal.

In this survey, we present the main ideas of the BPX preconditioner from [11] in the context of isogeometric analysis, employing tensor products of higher order B-splines in Sect. 2. We will see that the main theoretical tool to prove optimality of the BPX preconditioner are multilevel characterizations of the underlying energy space, so-called norm equivalences between Sobolev space norms and weighted sequence norms, describing their subspace contributions. We will also see that the main computational ingredients are linear intergrid operators which map vectors and matrices between grids of different grid spacings.

At about the same time, wavelets as a special example of a multiscale orthogonal basis of $L_2(\mathbb{R})$ with compact support were constructed [25]. While initially mainly developed and used for signal analysis and image compression, wavelets were soon discovered to provide optimal preconditioners in the above sense for second order elliptic boundary value problems [21, 33]. However, the fact that one cannot exploit L_2-orthogonality for elliptic boundary value problems together with the difficulty that the L_2-orthogonal Daubechies wavelets are only given implicitly led to the search for variants which are more practical for the numerical solution of PDEs. Soon, it was realized that biorthogonal spline-wavelets as developed in [17] are better suited since they allow one to work with piecewise polynomials for the discretization.

The principal and crucial property to prove optimality of a wavelet preconditioner are norm equivalences between Sobolev norms and sequence norms of weighted wavelet expansion coefficients. I mentioned before that multilevel characterizations for energy spaces are the crucial tool for proving optimality of the BPX preconditioner. Once a wavelet basis is available, one can represent the subspace contributions appearing there explicitly in terms of weighted wavelet coefficients. On this basis, optimal conditioning of the resulting linear system of equations can be achieved by applying the FWT (Fast Wavelet Transform) to a single-scale discretization on a uniform grid. While again multilevel characterizations of the underlying energy space play the crucial theoretical role for proving optimality, the intergrid operators which perform bases changes between different levels of resolution are the main practical ingredient for the efficiency of the FWT.

Nowadays, the terminology 'wavelets' is used in a more general sense that originally in [25]; we consider classes of multiscale bases with three main features: (R) they provide Riesz bases for the underlying function spaces, (L) the basis functions are local, and (CP) they exhibit cancellation properties. These will be detailed in Sect. 3.

After the initial results concerning optimal preconditioning with wavelets of local support in [21], research on employing wavelets for numerically solving elliptic PDEs went into different directions. One problem was that the original constructions in [17, 25] and many others were based on using the Fourier transform so that these constructions provide bases only for function spaces on all of $\mathbb{R}$, on the torus or, by tensorization, on $\mathbb{R}^n$. In contrast, PDEs naturally live on a bounded domain $\Omega \subset \mathbb{R}^n$. In order for wavelets to be employed for numerical PDEs, there arose the need for constructions of wavelets on bounded intervals and domains without, of course, loosing the crucial properties (R), (L) and (CP). The first such systematic construction of biorthogonal spline-wavelets on $[0, 1]$ of arbitrary order and, by tensor products, on $[0, 1]^n$, was provided in [22]. Different domain decomposition approaches yield constructions of biorthogonal wavelets on domains which can be represented as unions of parametric mappings of $[0, 1]^n$ [13, 23, 24, 40], see also [50] for details. Once such bases are available, the absolute value of the condition numbers of (systems of) elliptic PDEs can be ameliorated significantly by further inexpensive linear transformations taking into account a setup of the system matrices on the coarsest grid, a so-called operator-based preconditioning [12, 46]. A more recent survey on the results of wavelet-based preconditioning with extensions to PDE-constrained control problems can be found in [39].

Aside from optimal preconditioning, the built-in potential of local adaptivity for wavelets plays a prominent role when solving elliptic or parabolic PDEs with non-smooth solutions, on account of the fact that wavelets provide a locally supported Riesz basis for a whole function space. This issue is extensively addressed in the more recent survey paper [48]. In addition to the material in [26], there are at least four extensive surveys on wavelet and multiscale methods for more general PDEs addressing, among other things, the connection between adaptivity and nonlinear approximation and the evaluation of nonlinearities [15, 18–20].

In this article, I want to remain focused on multilevel preconditioning with higher-order discretizations for smooth solutions for which uniform grids provide a user-specified accuracy.

Isogeometric analysis is an the increasingly popular field in which higher order B-Splines are employed to reach this accuracy. Another such area is mathematical finance; specifically, option prizing problems. The fair pricing of an American option can in a standard model be formulated as a parabolic boundary value problem involving Black–Scholes' equation with a free boundary. The aim is to compute the free boundary, the optimal exercise price of the option, together with the solution of the PDE, the value of the option. For American put options, one generally does not have closed form solutions so that one has to resort to numerical schemes. In the simplest Black–Scholes' model, the volatility is assumed to be constant.

Numerical schemes for American option pricing are typically based on finite difference approaches, see, e.g., [1] and the references therein. In [14] and [44], corresponding multigrid methods are developed. However, here one is not only interested in the solution of the PDE, the option prize. For developing hedging strategies with options, it is of even more importance to accurately compute first and second derivatives of the solution, the so-called Greek letters. The availability of a smooth approximation enables to compute pointwise derivatives and, therefore, avoids additional numerical approximations of derivatives. Therefore, the idea in [30, 31] was to employ higher order B-splines for the solution of the variational inequality derived from Black-Scholes equation for American options. This approach enabled us to pointwise differentiate the option prize and, therefore, achieve a high accuracy approximation for the Greek letters. For the numerical solution, we extended the ideas about monotone multigrid methods with linear finite elements to solve variational inequalities from [35] to higher order B-splines. Here the intergrid operators mentioned before which stem from the refinement relations for B-splines had to be adapted in order not to violate the variational inequality; a task which is for higher order B-splines much more difficult than for linear finite elements since they are not interpolatory any more. Specific care has to be taken when deriving these intergrid operators in order not to violate the inequality constraints. The key feature was to replace function values by B-spline expansion coefficients which remain of the same sign because of the positivity of B-splines. Naturally, these ideas can be extended to obstacle problems which lead to variational inequalities and for which higher order approximations are sought.

While these results were very satisfying from a numerical point of view, the underlying Black-Scholes model has a severe deficiency in assuming that the volatility is constant. Particularly, an effect called volatility smile was observed in [32]. There are several approaches to estimate the volatility from observed stock data. Therefore, we adopted Heston's approach [28] to model the volatility to satisfy a stochastic differential equation. For Heston's model and one asset, the standard Itō approach yields a variational inequality in two space variables with a 'mildly' nonsymmetric parabolic differential operator. The ideas presented in [41] was as follows. A variational inequality for the American option pricing problem with Heston's model was discretized in terms of linear finite elements with respect to space. The resulting linear inequality system was solved in each time step with optimal linear computational complexity using a projective Gauss-Seidel scheme together with a monotone multigrid method. For this, again appropriate intergrid operators were constructed. Unfortunately, due to page limitations, I am not able to present these approaches here and have to refer to [30, 31, 41].

Once optimal preconditioners are available, for any of the problems described above, one can construct for a standard elliptic PDE optimal iterative schemes to achieve discretization error accuracy on the finest level with optimal linear complexity as follows. The idea, the so-called *nested iteration*, has been employed together with multigrid methods for a long time: starting on a coarse grid, one iteratively solves the preconditioned linear system of equations up to discretization error accuracy on this level. The resulting solution is transformed to the next finer grid, employing again the linear intergrid operator from the coarse to the finer grid. In this way, only a fixed number of iterations are necessary on each grid, independent of that grid spacing, see, e.g., [39]. One should note that, in contrast, iterating only on the finest grid introduces an additional $\log N$ factor, where N denotes the amount of unknowns on the finest grid.

The structure of this survey paper is as follows. Section 2 is devoted to the description of the BPX preconditioner for isogeometric analysis together with the proof of its optimality and corresponding numerical results. Section 3 is concerned with wavelet approximations of solutions of PDEs, and the derivation of Fast Wavelet Transforms for optimal preconditioning. We conclude in Sect. 4 with a short summary and some outlook.

2 BPX Preconditioning for Isogeometric Analysis

In this section, we consider linear elliptic PDEs in the framework of isogeometric analysis, combining modern techniques from computer aided design with higher order approximations of the solution. We treat the physical domain by means of a regular B-spline mapping from the parametric domain $\hat{\Omega} = (0,1)^n$, $n \geq 2$, to the physical domain Ω. The numerical solution of the PDE is computed by means of tensor product B-splines mapped onto the physical domain. We will construct additive BPX-type multilevel preconditioners and show that they are asymptotically optimal. This means that the spectral condition number of the resulting preconditioned stiffness matrix is independent of the grid spacing h. Together with a nested iteration scheme, this enables an iterative solution scheme of optimal linear complexity. The theoretical results are substantiated by numerical examples in two and three space dimensions. The results of this section are essentially contained in [11].

We consider linear elliptic partial differential operators of order $2r = 2, 4$ on the domain Ω in variational form: for given $f \in H^{-r}(\Omega)$, find $u \in H_0^r(\Omega)$ such that

$$a(u,v) = \langle f, v\rangle \qquad \text{for all } v \in H_0^r(\Omega) \tag{1}$$

holds. Here the energy space is $H_0^r(\Omega)$, a subset of the Sobolev space $H^r(\Omega)$, the space of square integrable functions with square integrable derivatives up to order r, containing homogeneous Dirichlet boundary conditions for $r = 1$ and homogeneous Dirichlet and Neumann derivatives for $r = 2$. The bilinear form $a(\cdot,\cdot)$ is derived from the linear elliptic PDE operator in a standard fashion, see, e.g., [8]. For example, the Laplacian is represented as $a(v,w) = \int_\Omega \nabla v \cdot \nabla w \, dx$. In order for the problem to be well-posed, we require the bilinear form $a(\cdot,\cdot) : H_0^r(\Omega) \times H_0^r(\Omega) \to \mathbb{R}$ to be symmetric, continuous and coercive on $H_0^r(\Omega)$. With $\langle\cdot,\cdot\rangle$, we denote on the right hand side of (1) the dual form between $H^{-r}(\Omega)$ and $H_0^r(\Omega)$. Our model problem (1) covers the second order Laplacian with homogeneous boundary conditions

$$-\Delta u = f \quad \text{on } \Omega, \qquad u|_{\partial\Omega} = 0, \tag{2}$$

as well as fourth order problems with corresponding homogeneous Dirichlet boundary conditions,

$$\Delta^2 u = f \quad \text{on } \Omega, \qquad u|_{\partial\Omega} = \mathbf{n} \cdot \nabla u|_{\partial\Omega} = 0 \tag{3}$$

where $\partial\Omega$ denotes the boundary of Ω and $\mathbf{n}$ the outward normal derivative at $\partial\Omega$. These PDEs serve as prototypes for more involved PDEs like Maxwell's equation or PDEs for linear and nonlinear elasticity. The reason we formulate these model problems of order $2r$ involving the parameter r is that this exhibits more clearly the order of the operator and the scaling in the subsequently used characterization of Sobolev spaces $H^r(\Omega)$. Thus, for the remainder of this section, the parameter $2r$ denoting the order of the PDE operator is fixed.

The assumptions on the bilinear form $a(\cdot,\cdot)$ entail that there exist constants $0 < c_A \leq C_A < \infty$ such that the induced self-adjoint operator $\langle Av, w\rangle := a(v,w)$ satisfies the isomorphism relation

$$c_A \|v\|_{H^r(\Omega)} \leq \|Av\|_{H^{-r}(\Omega)} \leq C_A \|v\|_{H^r(\Omega)}, \quad v \in H_0^r(\Omega). \tag{4}$$

If the precise format of the constants in (4) does not matter, we abbreviate this relation as $\|v\|_{H^r(\Omega)} \lesssim \|Av\|_{H^{-r}(\Omega)} \lesssim \|v\|_{H^r(\Omega)}$, or shortly as

$$\|Av\|_{H^{-r}(\Omega)} \sim \|v\|_{H^r(\Omega)}. \tag{5}$$

Under these conditions, Lax-Milgram's theorem guarantees that, for any given $f \in H^{-r}(\Omega)$, the operator equation derived from (1)

$$Au = f \quad \text{in } H^{-r}(\Omega) \tag{6}$$

has a unique solution $u \in H_0^r(\Omega)$, see, e.g., [8].

In order to approximate the solution of (1) or (6), we choose a finite-dimensional subspace $V_h \subset H_0^r(\Omega)$. We will construct these approximation spaces by using tensor products of B-splines as specified next.

2.1 B-Spline Discretizations

Our construction of optimal multilevel preconditioners will rely on tensor products so that principally any space dimension $n \in \mathbb{N}$ is permissible as long as storage permits; the examples cover the cases $n = 2, 3$. As discretization space, we choose in each spatial direction B-splines of the same degree p on uniform grids and with maximal smoothness. We begin with the univariate case and define B-splines on the interval [0, 1] recursively with respect to their degree p. Given this positive integer p and some $m \in \mathbb{N}$, we call $\Xi := \{\xi_1, \ldots, \xi_{m+p+1}\}$ a p-open knot vector if the knots are chosen such that

$$0 = \xi_1 = \ldots = \xi_{p+1} < \xi_{p+2} < \ldots < \xi_m < \xi_{m+1} = \ldots = \xi_{m+p+1} = 1, \tag{7}$$

i.e., the boundary knots 0 and 1 have multiplicity $p + 1$ and the interior knots are single. For Ξ, B-spline functions of degree p are defined following the well-known Cox-de Boor recursive formula, see [7]. Starting point are the piecewise constants for $p = 0$ (or characteristic functions)

$$N_{i,0}(\zeta) = \begin{cases} 1, & \text{if } 0 \le \xi_i \le \zeta < \xi_{i+1} < 1, \\ 0, & \text{otherwise}, \end{cases} \tag{8}$$

with the modification that the last B-spline $N_{m,0}$ is defined also for $\zeta = 1$. For $p \ge 1$ the B-splines are defined as

$$N_{i,p}(\zeta) = \frac{\zeta - \xi_i}{\xi_{i+p} - \xi_i} N_{i,p-1}(\zeta) + \frac{\xi_{i+p+1} - \zeta}{\xi_{i+p+1} - \xi_{i+1}} N_{i+1,p-1}(\zeta), \quad \zeta \in [0, 1], \tag{9}$$

with the same modification for $N_{m,p}$. Alternatively, one can define the B-splines explicitly by applying divided differences to truncated powers [7]. This gives a set of m B-splines that form a basis for the space of *splines*, that is, piecewise polynomials of degree p with $p - 1$ continuous derivatives at the internal knots ξ_ℓ for $\ell = p + 2, \ldots, m$. (Of course, one can also define B-splines on a knot sequence with multiple internal knots which entails that the spline space is not of maximal smoothness.) For $p = 1$, the B-splines are at least $C^0([0, 1])$ which suffices for the discretization of elliptic PDEs of order 2, and for $p = 2$ they are $C^1([0, 1])$ which suffices for $r = 2$. By construction, the B-spline $N_{i,p}$ is supported in the interval $[\xi_i, \xi_{i+p+1}]$.

These definitions are valid for an arbitrary spacing of knots in Ξ (7). Recall that smooth solutions of elliptic PDEs can be approximated best with discretizations on a uniform grid. Therefore, we assume from now on that the grid is *uniform*, i.e., $\xi_{\ell+1} - \xi_\ell = h$ for all $\ell = p + 1, \ldots, m$.

For n space dimensions, we employ tensor products of the one-dimensional B-splines. We take in each space dimension a p-open knot vector Ξ and define on the closure of the parametric domain $\hat{\Omega} = (0, 1)^n$ (which we also denote by $\hat{\Omega}$ for simplicity of presentation) the spline space

$$\begin{aligned} S_h(\hat{\Omega}) &:= \operatorname{span}\left\{ B_i(\mathbf{x}) := \prod_{\ell=1}^{n} N_{i_\ell,p}(x_\ell),\ i = 1, \ldots, N := mn,\ \mathbf{x} \in \hat{\Omega} \right\} \\ &=: \operatorname{span}\left\{ B_i(\mathbf{x}), i \in \mathbb{I},\ \mathbf{x} \in \hat{\Omega} \right\}. \end{aligned} \tag{10}$$

In the spirit of isogeometric analysis, we suppose that the computational domain Ω can also described in terms of B-splines. We assume that the computational domain Ω is the image of a mapping $\mathbf{F} : \hat{\Omega} \to \Omega$ with $\mathbf{F} := (F_1, \ldots, F_n)^T$ where each component F_i of $\mathbf{F}$ belongs to $S_{\bar{h}}(\hat{\Omega})$ for some given $\bar{h}$. In many applications, the geometry can be described in terms of a very coarse mesh, namely, $\bar{h} \gg h$. Moreover, we suppose that $\mathbf{F}$ is invertible and satisfies

$$\|D^\alpha \mathbf{F}\|_{L_\infty(\hat{\Omega})} \sim 1 \text{ for } |\alpha| \le r. \tag{11}$$

This assumption on the geometry can be weakened in that the mapping $\mathbf{F}$ can be a piecewise C^∞ function on the mesh with respect to $\bar{h}$, independent of h, or the domain Ω may have a multi-patch representation. This means that one can allow Ω also to be the union of domains Ω_k where each one parametrized by a spline mapping of the parametric domain $\hat{\Omega}$.

We now define the approximation space for (6) as

$$V_h^r := \{v_h \in H_0^r(\Omega) : v_h \circ \mathbf{F} \in S_h(\hat{\Omega})\}. \tag{12}$$

We will formulate three important properties of this approximation space which will play a crucial role later for the construction of the BPX-type preconditioners. The first one is that we suppose from now on that the B-spline basis is *normalized* with respect to L_2, i.e.,

$$\|B_i\|_{L_2(\hat{\Omega})} \sim 1, \text{ and, thus, also } \|B_i \circ \mathbf{F}^{-1}\|_{L_2(\Omega)} \sim 1 \text{ for all } i \in \mathbb{I}. \tag{13}$$

Then one can derive the following facts [11].

Theorem 1 *Let $\{B_i\}_{i\in\mathbb{I}}$ be the B-spline basis defined in (10) and normalized as in (13), $N = \#\mathbb{I}$ and V_h^r as in (12). Then we have*

(S) Uniform stability with respect to $L_2(\Omega)$
For any $\mathbf{c} \in \ell_2(\mathbb{I})$,

$$\left\| \sum_{i=1}^N c_i B_i \circ \mathbf{F}^{-1} \right\|_{L_2(\Omega)}^2 \sim \sum_{i=1}^N |c_i|^2 =: \|\mathbf{c}\|_{\ell_2}^2, \qquad \mathbf{c} := (c_i)_{i=1,\ldots,N}; \tag{14}$$

(J) Direct or Jackson estimates

$$\inf_{v_h \in V_h^r} \|v - v_h\|_{L_2(\Omega)} \lesssim h^s\, |v|_{H^s(\Omega)} \text{ for any } v \in H^s(\Omega),\ 0 \le s \le r, \tag{15}$$

where $|\cdot|_{H^s(\Omega)}$ denotes the Sobolev seminorm of highest weak derivatives s;

(B) Inverse or Bernstein estimates

$$\|v_h\|_{H^s(\Omega)} \lesssim h^{-s} \|v_h\|_{L_2(\Omega)} \text{ for any } v_h \in V_h^r \text{ and } 0 \le s \le r. \tag{16}$$

In all these estimates, the constants are independent of h but may depend on **F**, *i.e.,* Ω*, the polynomial degree p and the spatial dimension n.*

In the next section, we construct BPX-type preconditioners for (6) in terms of approximations with (12) and show their optimality.

2.2 Additive Multilevel Preconditioners

The construction of optimal preconditioners are based on a *multiresolution analysis* of the underlying energy function space $H_0^r(\Omega)$. As before, $2r \in \{2, 4\}$ stands for the order of the PDEs we are solving, and is always kept fixed.

We first describe the necessary ingredients within an abstract basis-free framework, see, e.g., [18]. Afterwards, we specify the realization for the parametrized tensor product spaces in (12).

Let $\mathcal{V}$ be a sequence of strictly nested spaces V_j, starting with some fixed coarsest index $j_0 > 0$, determined by the polynomial degree p which determines the support of the basis functions (which also depends on Ω), and terminating with a highest resolution level J,

$$V_{j_0} \subset V_{j_0+1} \subset \cdots \subset V_j \subset \cdots \subset V_J \subset H_0^r(\Omega). \tag{17}$$

The index j denotes the level of resolution defining approximations on a grid with dyadic grid spacing $h = 2^{-j}$, i.e., we use from now on the notation V_j instead of V_h to indicate different grid spacings. Then, V_J will be the space relative to the finest grid 2^{-J}. We associate with $\mathcal{V}$ a sequence of linear projectors $\mathcal{P} := \{P_j\}_{j \geq j_0}$ with the following properties.

Properties 2 *We assume that*

(P1) P_j *maps* $H_0^r(\Omega)$ *onto* V_j,
(P2) $P_j P_\ell = P_j$ *for* $j \leq \ell$,
(P3) $\mathcal{P}$ *is uniformly bounded on* $L_2(\Omega)$, *i.e.,* $\|P_j\|_{L_2(\Omega)} \lesssim 1$ *for any* $j \geq j_0$ *with a constant independent of j.*

These conditions are satisfied, for example, for $L_2(\Omega)$-orthogonal projectors, or, in the case of splines, for the quasi-interpolant proposed and analyzed in [47, Chapter 4]. The second condition (P2) ensures that the differences $P_j - P_{j-1}$ are also projectors for any $j > j_0$. We define next a sequence $\mathcal{W} := \{W_j\}_{j \geq j_0}$ of complement spaces

$$W_j := (P_{j+1} - P_j) V_{j+1} \tag{18}$$

which then yields the direct (but not necessarily orthogonal) decomposition

$$V_{j+1} = V_j \oplus W_j. \tag{19}$$

Thus, for the finest level J, we can express V_J in its *multilevel decomposition*

$$V_J = \bigoplus_{j=j_0-1}^{J-1} W_j \tag{20}$$

upon setting $W_{j_0-1} := V_{j_0}$. Setting also $P_{j_0-1} := 0$, the corresponding multilevel representation of any $v \in V_J$ is then

$$v = \sum_{j=j_0}^{J} (P_j - P_{j-1})v. \tag{21}$$

We now have the following result which will be used later for the proof of the optimality of the multilevel preconditioners.

Theorem 3 *Let $\mathcal{P}, \mathcal{V}$ be as above where, in addition, we require that for each V_j, $j_0 \leq j \leq J$, a Jackson and Bernstein estimate as in Theorem 1 (J) and (B) hold with $h = 2^{-j}$. Then one has the function space characterization*

$$\|v\|_{H^r(\Omega)} \sim \left(\sum_{j=j_0}^{J} 2^{2rj} \|(P_j - P_{j-1})v\|^2_{L_2(\Omega)} \right)^{1/2} \quad \textit{for any } v \in V_J. \tag{22}$$

Such a result holds for much larger classes of function spaces, Sobolev or even Besov spaces which are subsets of $L_q(\Omega)$ for general q, possibly different from 2 and for any function $v \in H^r(\Omega)$, then with an infinite sum on the right hand side, see, e.g., [18]. The proof of Theorem 3 for such cases heavily relies on tools from approximation theory and can be found in [21, 45].

Next we demonstrate how to exploit the norm equivalence (22) in the construction of an optimal multilevel preconditioner. Define for any $v, w \in V_J$ the linear self-adjoint positive-definite operator $C_J : V_J \to V_J$ given by

$$(C_J^{-1} v, w)_{L_2(\Omega)} := \sum_{j=j_0}^{J} 2^{2rj} \left((P_j - P_{j-1})v, (P_j - P_{j-1})w \right)_{L_2(\Omega)}, \tag{23}$$

which we call a multilevel *BPX-type preconditioner*. Let $A_J : V_J \to V_J$ be the finite-dimensional operator defined by $(A_J v, w)_{L_2(\Omega)} := a(v, w)$ for all $v, w \in V_J$, the approximation of A in (6) with respect to V_J.

Theorem 4 *With the same prerequisites as in Theorem 3, C_J is an asymptotically optimal symmetric preconditioner for A_J, i.e., $\kappa_2(C_J^{1/2} A_J C_J^{1/2}) \sim 1$ with constants independent of J.*

Proof For the parametric domain $\hat{\Omega}$, the result was proved independently in [21, 45] and is based on the combination of (22) together with the well-posedness of the continuous problem (6). The result on the physical domain follows then together with (11). □

Concrete realizations of the preconditioner defined in (23) based on B-splines lead to representations of the complement spaces W_j whose bases are called *wavelets*. For these, efficient implementations of optimal linear complexity involving the Fast Wavelet Transform can be derived explicitly, see Sect. 3.

However, since the order of the PDE operator r is positive, one can use here the argumentation from [10] which will allow to work with the same basis functions as for the spaces V_j. The first part of the argument relies on the assumption that the P_j are L_2-*orthogonal* projectors. For a clear distinction, we shall use the notation O_j for L_2-orthogonal projectors and reserve the notation P_j for the linear projectors with Properties 2. Then, the BPX-type preconditioner (23) (using the same symbol C_J for simplicity) reads as

$$C_J^{-1} := \sum_{j=j_0}^{J} 2^{2jr}(O_j - O_{j-1}), \tag{24}$$

which is by Theorem 4 a BPX-type preconditioner for the self-adjoint positive definite operator A_J. By the orthogonality of the projectors O_j, we can immediately derive from (24) that

$$C_J = \sum_{j=j_0}^{J} 2^{-2jr}(O_j - O_{j-1}). \tag{25}$$

Since $r > 0$, by rearranging the sum, the exponentially decaying scaling factors allow one to replace C_J by the spectrally equivalent operator

$$C_J = \sum_{j=j_0}^{J} 2^{-2jr} O_j. \tag{26}$$

Recall that two linear operators $\mathcal{A} : V_J \to V_J$ and $\mathcal{B} : V_J \to V_J$ are *spectrally equivalent* if they satisfy

$$(\mathcal{A}v, v)_{L_2(\Omega)} \sim (\mathcal{B}v, v)_{L_2(\Omega)}, \quad v \in V_J, \tag{27}$$

with constants independent of J. Thus, the realization of the preconditioner is reduced to a computation in terms of the bases of the spaces V_j instead of W_j. The orthogonal projector O_j can, in turn, be replaced by a simpler local operator which is spectrally equivalent to O_j, see [37] and the derivation below.

Up to this point, the introduction to multilevel preconditioners has been basis-free. We now show how this framework can be used to construct a BPX-preconditioner for the linear system (6). Based on the definition (12), we construct a sequence of spaces satisfying (17) such that $V_J = V_h^r$. In fact, we suppose that for each space dimension we have a sequence of p-open knot vectors $\Xi_{j_0,\ell}, \ldots, \Xi_{J,\ell}$, $\ell = 1, \ldots, n$, which provide a uniform partition of the interval $[0, 1]$ such that $\Xi_{j,\ell} \subset \Xi_{j+1,\ell}$ for $j = j_0, j_0 + 1, \ldots, J$. In particular, we assume that $\Xi_{j+1,\ell}$ is obtained from $\Xi_{j,\ell}$ by dyadic refinement, i.e., the grid spacing for $\Xi_{j,\ell}$ is proportional to 2^{-j} for each $\ell = 1, \ldots, n$. In view of the assumptions on the parametric mapping $\mathbf{F}$, we assume that $\bar{h} = 2^{-j_0}$, i.e., $\mathbf{F}$ can be represented in terms of B-splines on the coarsest level j_0. By construction, we have now achieved that

$$S_{j_0}(\hat{\Omega}) \subset S_{j_0+1}(\hat{\Omega}) \subset \ldots \subset S_J(\hat{\Omega}).$$

Setting $V_j^r := \{v \in H_0^r(\Omega) : \ v \circ \mathbf{F} \in S_j(\hat{\Omega})\}$, we arrive at a sequence of nested spaces

$$V_{j_0}^r \subset V_{j_0+1}^r \subset \ldots \subset V_J^r.$$

Setting $\mathbb{I}_j := \{1, \ldots, \dim S_j(\hat{\Omega})\}$, we denote by B_i^j, $i \in \mathbb{I}_j$, the set of L_2-normalized B-spline basis functions for the space $S_j(\hat{\Omega})$. Define now the positive definite operator $P_j : L_2(\Omega) \to V_j^r$ as

$$P_j := \sum_{i \in \mathbb{I}_j} (\,\cdot\,, B_i^j \circ \mathbf{F}^{-1})_{L_2(\Omega)}\, B_i^j \circ \mathbf{F}^{-1}. \tag{28}$$

Corollary 5 *For the basis $\{B_i^j \circ \mathbf{F}^{-1},\ i \in \mathbb{I}_j\}$, the operators P_j and the L_2-projectors O_j are spectrally equivalent for any j.*

Proof The assertion follows by combining (11), (14), with Remark 3.7.1 from [37], see [10] for the main ingredients. □

Finally, we obtain an explicit representation of the preconditioner C_J in terms of the mapped spline bases for $V_j^r, j = j_0, \ldots, J$,

$$C_J = \sum_{j=j_0}^{J} 2^{-2jr} \sum_{i \in \mathbb{I}_j} (\,\cdot\,, B_i^j \circ \mathbf{F}^{-1})_{L_2(\Omega)}\, B_i^j \circ \mathbf{F}^{-1}. \tag{29}$$

Note that this preconditioner involves *all* B-splines from all levels j with an appropriate scaling, i.e., in fact a properly scaled *generating system* for V_J^r.

Remark 6 The *hierarchical basis (HB) preconditioner* introduced for $n = 2$ in [54] for piecewise linear B-splines fits into this framework by choosing *Lagrangian interpolants* in place of the projectors P_j in (23). However, since these operators do not satisfy (P3) in Properties 2, they do not yield an asymptotically optimal preconditioner for $n \geq 2$. Specifically, for $n = 3$, this preconditioner does not have an effect at all.

So far we have not explicitly addressed the dependence of the preconditioned system on p. Since all estimates in Theorem 1 which enter the proof of optimality depend on p, it is to be expected that the absolute values of the condition numbers, i.e., the values of the constants, depend on and increase with p. Indeed, in the next section, we show some numerical results which also aim at studying this dependence.

2.3 *Realization of the BPX Preconditioner and Numerical Results*

Now we are in the position to describe the concrete implementation of the BPX preconditioner. Its main ingredient are linear *intergrid operators* which map vectors and matrices between different grids. Specifically, we need to define prolongation and restriction operators.

Since $V_j^r \subset V_{j+1}^r$, each B-spline B_i^j on level j can be represented by a linear combination of B-splines B_k^{j+1} on level $j+1$. Arranging the B-splines in the set $\{B_i^j, i \in \mathbb{I}_j\}$ into a vector $\mathbf{B}^j$ in a fixed order, this relation denoted as *refinement relation* can be written as

$$\mathbf{B}^j = \mathbf{I}_j^{j+1}\mathbf{B}^{j+1} \tag{30}$$

with *prolongation operator* $\mathbf{I}_j^{j+1}$ from the trial space V_j^r to the trial space V_{j+1}^r. The restriction $\mathbf{I}_{j+1}^j$ is then simply defined as the transposed operator, i.e., $\mathbf{I}_{j+1}^j = (\mathbf{I}_j^{j+1})^T$. In case of piecewise linear B-splines, this definition coincides with the well known prolongation and restriction operators from finite element textbooks obtained by interpolation, see, e.g., [8].

We will exemplify the construction in case of quadratic and cubic B-splines on the interval, see, e.g., [7], as follows. We equidistantly subdivide the interval $[0, 1]$ into 2^j subintervals and obtain 2^j and $2^j + 1$, respectively, B-splines for $p = 2, 3$ and the corresponding quadratic and cubic spline space V_j^r which is given on this partition, respectively, see Fig. 1 for an illustration. Note that the two boundary functions which do not vanish at the boundary were removed in order to guarantee that $V_j^r \subset H_0^r(\Omega)$. Moreover, recall that the B-splines are L_2 normalized according to (13) which means that B_i^j is of the form $B_i^j(\zeta) = 2^{j/2}B(2^j\zeta - i)$ if B_i^j is an interior function, and correspondingly for the boundary functions.

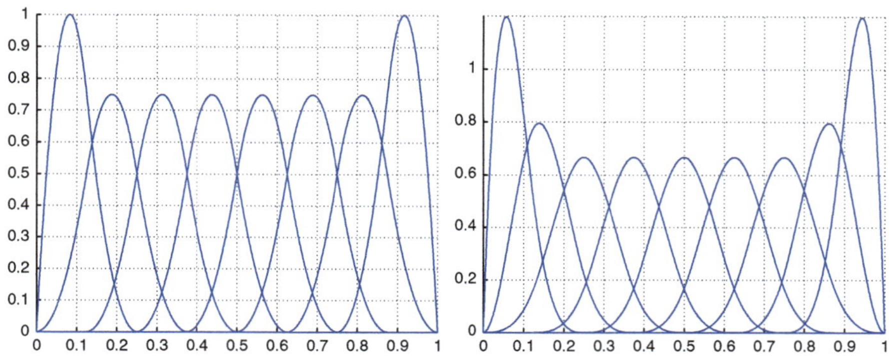

Fig. 1 Quadratic ($p = 2$) (*left*) and cubic ($p = 3$) (*right*) L_2-normalized B-splines (see (13)) on level $j = 3$ on the interval $[0, 1]$, yielding basis functions for $V_j^r \subset H_0^r(\Omega)$

In case of quadratic B-splines ($p = 2$), the restriction operator $\mathbf{I}_{j+1}^j$ reads

$$\mathbf{I}_{j+1}^j = 2^{-1/2} \begin{bmatrix} \frac{1}{2} & \frac{9}{8} & \frac{3}{8} & & & & & & \\ & \frac{1}{4} & \frac{3}{4} & \frac{3}{4} & \frac{1}{4} & & & & \\ & & & \frac{1}{4} & \frac{3}{4} & \frac{3}{4} & \frac{1}{4} & & \\ & & & & \ddots & \ddots & & & \\ & & & & \frac{1}{4} & \frac{3}{4} & \frac{3}{4} & \frac{1}{4} & \\ & & & & & & \frac{3}{8} & \frac{9}{8} & \frac{1}{2} \end{bmatrix} \in \mathbb{R}^{2^j \times 2^{j+1}}.$$

For cubic B-splines ($p = 3$), it has the form

$$\mathbf{I}_{j+1}^j = 2^{-1/2} \begin{bmatrix} \frac{1}{2} & \frac{9}{8} & \frac{3}{8} & & & & & & & \\ & \frac{1}{4} & \frac{11}{12} & \frac{2}{3} & \frac{1}{6} & & & & & \\ & & \frac{1}{8} & \frac{1}{2} & \frac{3}{4} & \frac{1}{2} & \frac{1}{8} & & & \\ & & & & \ddots & \ddots & \ddots & & & \\ & & & & \frac{1}{8} & \frac{1}{2} & \frac{3}{4} & \frac{1}{2} & \frac{1}{8} & \\ & & & & & \frac{1}{6} & \frac{2}{3} & \frac{11}{12} & \frac{1}{4} & \\ & & & & & & & \frac{3}{8} & \frac{9}{8} & \frac{1}{2} \end{bmatrix} \in \mathbb{R}^{(2^j+1) \times (2^{j+1}+1)}.$$

The normalization factor $2^{-1/2}$ stems from the L_2-normalization (13). The matrix entries are scaled in the usual fashion such that their rows sum to two. From these restriction operators for one dimensions, one obtains the related restriction operators on arbitrary unit cubes $[0, 1]^n$ via tensor products. Finally, we set $\mathbf{I}_j^J := \mathbf{I}_{J-1}^J \mathbf{I}_{J-2}^{J-1} \cdots \mathbf{I}_j^{j+1}$ and $\mathbf{I}_J^j := \mathbf{I}_j^{j+1} \mathbf{I}_{j+1}^{j+2} \cdots \mathbf{I}_{J-1}^J$ to define prolongations and restrictions between arbitrary levels j and J.

In order to derive the explicit form of the discretized BPX-preconditioner, for given functions $u_J, v_J \in V_J$ with expansion coefficients $u_{J,k}$ and $v_{J,\ell}$, respectively, we conclude from (29) that

$$(C_J u_J, v_J)_{L_2(\Omega)} = \sum_{k,\ell\in\mathbb{I}_J} u_{J,k} v_{J,\ell} (C_J(B_k^J \circ \mathbf{F}^{-1}), B_\ell^J \circ \mathbf{F}^{-1})_{L_2(\Omega)}$$

$$= \sum_{k,\ell\in\mathbb{I}_J} u_{J,k} v_{J,\ell} \sum_{j=j_0}^{J} 2^{-2jr} \sum_{i\in\mathbb{I}_j} (B_k^J \circ \mathbf{F}^{-1}, B_i^j \circ \mathbf{F}^{-1})_{L_2(\Omega)} \times (B_i^j \circ \mathbf{F}^{-1}, B_\ell^J \circ \mathbf{F}^{-1})_{L_2(\Omega)}.$$

Next, one can introduce the mass matrix $\mathbf{M}_J = [(B_k^J \circ \mathbf{F}^{-1}, B_\ell^J \circ \mathbf{F}^{-1})_{L_2(\Omega)}]_{k,\ell}$ and obtains by the use of restrictions and prolongations

$$(C_J u_J, v_J)_{L_2(\Omega)} = \sum_{j=j_0}^{J} 2^{-2jr} \mathbf{u}_J^T \mathbf{M}_J \mathbf{I}_j^J \mathbf{I}_J^j \mathbf{M}_J \mathbf{v}_J.$$

The mass matrices which appear in this expression can be further suppressed since $\mathbf{M}_J$ is spectrally equivalent to the identity matrix. Finally, the *discretized BPX-preconditioner* to be implemented is of the simple form

$$\mathbf{C}_J = \sum_{j=j_0}^{J} 2^{-2jr} \mathbf{I}_j^J \mathbf{I}_J^j, \tag{31}$$

involving only restrictions and prolongations. A further simple improvement can be obtained by replacing the scaling factor 2^{-2jr} by $\operatorname{diag}(\mathbf{A}_j)^{-1}$, where $\operatorname{diag}(\mathbf{A}_j)$ denotes the diagonal matrix built from the diagonal entries of the stiffness matrix $\mathbf{A}_j$. This diagonal scaling has the same effect as the levelwise scaling by 2^{-2jr} but improves the condition numbers considerably, particularly if parametric mappings are involved. Thus, the discretized BPX-preconditioner takes on the form

$$\mathbf{C}_J = \sum_{j=j_0}^{J} \mathbf{I}_j^J \operatorname{diag}(\mathbf{A}_j)^{-1} \mathbf{I}_J^j \tag{32}$$

which we will use in the subsequent computations presented in Tables 1 and 2. If the condition number $\kappa(\mathbf{A}_{j_0})$ is already high in absolute numbers on the coarsest level j_0, it is worth to use its exact inverse on the coarse grid, i.e., to apply instead of (32) the operator

$$\mathbf{C}_J = \mathbf{I}_{j_0}^J \mathbf{A}_{j_0}^{-1} \mathbf{I}_J^{j_0} + \sum_{j=j_0+1}^{J} \mathbf{I}_j^J \operatorname{diag}(\mathbf{A}_j)^{-1} \mathbf{I}_J^j,$$

Table 1 Condition numbers of the BPX-preconditioned Laplacian on $\hat{\Omega} = (0,1)^n$ for $n = 1, 2, 3$

Level	Interval ($n = 1$)				Square ($n = 2$)				Cube ($n = 3$)			
	$p = 1$	$p = 2$	$p = 3$	$p = 4$	$p = 1$	$p = 2$	$p = 3$	$p = 4$	$p = 1$	$p = 2$	$p = 3$	$p = 4$
3	7.43	3.81	7.03	5.93	5.93	7.31	22.8	133	3.49	39.5	356	5957
4	8.87	4.40	9.47	7.81	5.00	9.03	40.2	225	4.85	50.8	624	9478
5	10.2	4.67	11.0	9.36	5.70	9.72	51.8	293	5.75	56.6	795	11,887
6	11.3	4.87	12.1	10.7	6.27	10.1	58.7	340	6.40	59.7	895	13,185
7	12.2	5.00	12.7	11.5	6.74	10.4	63.1	371	6.91	61.3	961	13,211
8	13.0	5.10	13.0	11.9	7.14	10.5	66.0	391	7.34	62.2	990	13,234
9	13.7	5.17	13.2	12.1	7.48	10.6	68.0	403	7.70	62.6	1016	13,255
10	14.2	5.22	13.4	12.2	7.77	10.6	69.3	411	7.99	62.9	1040	–

Table 2 Condition numbers of the BPX-preconditioned Laplacian on the analytic arc seen on the right hand side. The bracketed numbers are the related condition numbers without preconditioning

Level	$p = 1$	$p = 2$	$p = 3$	$p = 4$
3	5.04 (21.8)	12.4 (8.64)	31.8 (31.8)	184 (184)
4	11.1 (90.2)	16.3 (34.3)	54.7 (32.9)	291 (173)
5	25.3 (368)	19.0 (139)	70.1 (98.9)	376 (171)
6	31.9 (1492)	21.4 (560)	79.2 (401)	436 (322)
7	37.4 (6015)	23.1 (2255)	84.4 (1620)	471 (1297)
8	42.1 (241,721)	24.3 (9062)	87.3 (6506)	490 (5217)
9	45.7 (969,301)	25.2 (36,353)	89.0 (26,121)	500 (20,945)
10	48.8 (388,690)	25.9 (145,774)	90.1 (104,745)	505 (83,975)

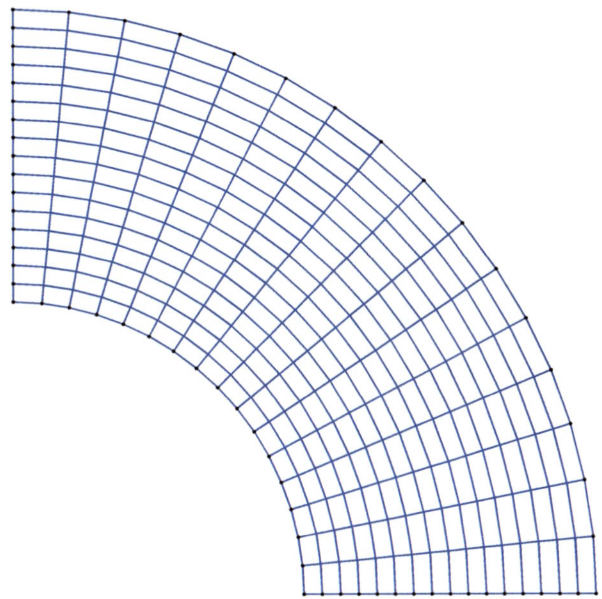

see [12, 46]. Another substantial improvement of the BPX-preconditioner can be achieved by replacing the diagonal scaling on each level by, e.g., a SSOR preconditioning as follows. We decompose the system matrix as $\mathbf{A}_j = \mathbf{L}_j + \mathbf{D}_j + \mathbf{L}_j^T$ with the diagonal matrix $\mathbf{D}_j$, the lower triangular part $\mathbf{L}_j$, and the upper triangular part $\mathbf{L}_j^T$. Then we replace the diagonal scaling on each level of the BPX-preconditioner (32) by the SSOR preconditioner, i.e., instead of (32) we apply the preconditioner

$$\mathbf{C}_J = \sum_{j=j_0}^{J} \mathbf{I}_j^J (\mathbf{D}_j + \mathbf{L}_j)^{-T} \mathbf{D}_j (\mathbf{D}_j + \mathbf{L}_j)^{-1} \mathbf{I}_J^j. \tag{33}$$

In doing so, the condition numbers can be improved impressively. In Table 3, we list the ℓ_2-condition numbers for the BPX-preconditioned Laplacian in case of cubic B-splines in two spatial dimensions. By comparing the numbers with those found in Tables 1 and 2 one can infer that the related condition numbers are all reduced by a factor about five. Note that the setup, storage and application of the operator defined in (33) is still of optimal linear complexity.

Finally, we provide numerical results in order to demonstrate the preconditioning and to specify the dependence on the spatial dimension n and the spline degree p. We consider an approximation of the homogeneous Dirichlet problem for the Poisson equation on the n-dimensional unit cube $\hat{\Omega} = (0, 1)^n$ for $n = 1, 2, 3$. The mesh on level j is obtained by subdividing the cube j-times dyadically into 2^n subcubes of mesh size $h_j = 2^{-j}$. On this subdivision, we consider the B-splines of degree $p = 1, 2, 3, 4$ as defined in Sect. 2.1. The ℓ_2-condition numbers of the related stiffness matrices, preconditioned by the BPX-preconditioner (32), are shown in Table 1. The condition numbers seem to be independent of the level j, but they depend on the spline degree p and the space dimension n for $n > 1$. For fourth order problems on the sphere, corresponding results for the bi-Laplacian with and without BPX preconditioning were presented in [43].

We study next the dependence of the condition numbers on the parametric mapping $\mathbf{F}$. We consider the case $n = 2$ in case of a smooth mapping (see the plot on the right hand side of Table 2 for an illustration of the mapping). As one can see from Table 2, the condition numbers are at most about a factor of five higher than the related values in Table 1. Nearly the same observation holds if we replace the parametric mapping by a C^0-parametrization which maps the unit square onto an L-shaped domain, see [11].

If we consider a singular map $\mathbf{F}$, that is, a mapping that does not satisfy (11), the condition numbers grow considerably as expected, see [11]. But even in this case, the BPX-preconditioner with SSOR acceleration (33) is able to drastically reduce the condition numbers of the system matrix in all examples, see Table 3.

Table 3 Condition numbers of the BPX-preconditioned Laplacian for cubic B-splines on different geometries in case of using a BPX-SSOR preconditioning on each level

Level	Square		$\mathcal{C}^0$-map of the L-shape	Singular $\mathcal{C}^1$-map of the L-shape
3	3.61	3.65	3.67	3.80
4	6.58	6.97	7.01	7.05
5	8.47	10.2	10.2	14.8
6	9.73	13.1	13.2	32.2
7	10.5	14.9	15.2	77.7
8	11.0	15.9	16.3	180
9	11.2	16.5	17.0	411
10	11.4	16.9	17.7	933

3 Wavelets and the FWT (Fast Wavelet Transform)

Returning to the abstract setting at the beginning of Sect. 2.2, it will now be specified how to construct and realize the FWT (fast wavelet transform) for preconditioning. I would like to emphasize at this point that wavelets are, more importantly than for preconditioning, an appropriate theoretical tool to derive adaptive discretizations in case of singularities in the domain or data. In fact, they enable one to not only prove convergence of a corresponding adaptive scheme but also optimal complexity of these schemes, see [16] and several subsequent papers or [48] for a more recent survey.

3.1 Some Basic Notions

In view of the problem formulation (6), we need to have a wavelet basis for the space $H_0^r(\Omega)$ and its dual. We formulate these basic properties for a general Hilbert space H, following [18] or [38]. A *wavelet basis* for H, shortly *wavelets*, is understood here as a collection of functions

$$\Psi_H := \{\psi_{H,\lambda} : \lambda \in \mathbb{I}_H\} \subset H \tag{34}$$

which are indexed by elements λ from an infinite index set $\mathbb{I}_H$. Each of the indices λ comprises different information $\lambda = (j, \mathbf{k}, \mathbf{e})$ such as the *refinement scale* or *level of resolution* j and a spatial location $\mathbf{k} = \mathbf{k}(\lambda) \in \mathbb{Z}^n$. In more than one space dimension, the basis functions are built from taking tensor products of certain univariate functions, and in this case the third index $\mathbf{e}$ contains information on the *type* of wavelet. We will frequently use the symbol $|\lambda| := j$ to access the resolution level j. In the univariate case on all of $\mathbb{R}$, $\psi_{H,\lambda}$ is typically generated by means of shifts and dilates of a single function ψ, i.e., $\psi_\lambda = \psi_{j,k} = 2^{j/2}\psi(2^j \cdot -k), j, k \in \mathbb{Z}$, normalized with respect to $\|\cdot\|_{L_2(\mathbb{R})}$. On bounded domains which can be represented as unions of mappings of tensor product domains, the structure of the functions is essentially the same up to modifications near the boundary.

We assume that the wavelets satisfy the following crucial properties.

Riesz basis property (R): Every $v \in H$ has a unique expansion in terms of Ψ_H,

$$v = \sum_{\lambda \in \mathbb{I}_H} v_\lambda \, \psi_{H,\lambda} =: \mathbf{v}^T \Psi_H, \quad \mathbf{v} := (v_\lambda)_{\lambda \in \mathbb{I}_H}, \tag{35}$$

and its expansion coefficients satisfy the *norm equivalence*

$$\|\mathbf{v}\| \;\sim\; \|\mathbf{v}^T \Psi_H\|_H, \quad \mathbf{v} \in \ell_2(\mathbb{I}_H), \tag{36}$$

where $\|\cdot\| := \|\cdot\|_{\ell_2(\mathbb{I}_H)}$.

Locality (L): The functions $\psi_{H,\lambda}$ have compact support which decreases with increasing level $j = |\lambda|$, i.e.,

$$\operatorname{diam}(\operatorname{supp} \psi_{H,\lambda}) \;\sim\; 2^{-|\lambda|}. \tag{37}$$

Cancellation property (CP): There exists an integer $\tilde{d} = \tilde{d}_H$ such that

$$\langle v, \psi_{H,\lambda} \rangle \;\lesssim\; 2^{-|\lambda|(n/2 - n/p + \tilde{d})} |v|_{W_p^{\tilde{d}}(\operatorname{supp} \psi_{H,\lambda})}. \tag{38}$$

This means that integrating against a wavelet has the effect of taking an $\tilde{m}$th order difference which annihilates the smooth part of v. This property is for wavelets defined on Euclidean domains typically realized by constructing Ψ_H in such a way that it possesses a *dual* or *biorthogonal* basis $\tilde{\Psi}_H \subset H'$ such that the multiresolution spaces $\tilde{S}_j := \operatorname{span}\{\tilde{\psi}_{H,\lambda} : |\lambda| < j\}$ contain all polynomials of order $\tilde{m}$. Here *dual basis* means that $\langle \psi_{H,\lambda}, \tilde{\psi}_{H,\nu} \rangle = \delta_{\lambda,\nu}$, $\lambda, \nu \in \mathbb{I}_H$.

A few remarks on these properties should be made. In (R), the norm equivalence (36) is crucial since it means control over a function measured in $\|\cdot\|_H$ from above and below by its expansion coefficients: small changes in the coefficients only cause small changes in the function. Together with the locality (L), this also means that local changes stay local. This stability is an important feature which is used for deriving optimal preconditioners. Finally, the cancellation property (CP) entails that smooth functions have small wavelet coefficients which, on account of (36) may be neglected in a controllable way. Moreover, (CP) can be used to derive quasi-sparse representations of a wide class of operators which is important for compression purposes, see, e.g., [48].

By duality arguments one can show that (36) is equivalent to the existence of a biorthogonal collection which is *dual* or *biorthogonal* to Ψ_H,

$$\tilde{\Psi}_H := \{\tilde{\psi}_{H,\lambda} : \lambda \in \mathbb{I}_H\} \subset H', \quad \langle \psi_{H,\lambda}, \tilde{\psi}_{H,\mu} \rangle = \delta_{\lambda,\mu}, \qquad \lambda, \mu \in \mathbb{I}_H, \tag{39}$$

which is a Riesz basis for H', that is, for any $\tilde{v} = \tilde{\mathbf{v}}^T \tilde{\Psi}_H \in H'$ one has

$$\|\tilde{\mathbf{v}}\| \sim \|\tilde{\mathbf{v}}^T \tilde{\Psi}_H\|_{H'} \tag{40}$$

see [19]. Here the tilde expresses that the collection $\tilde{\Psi}_H$ is a dual basis to a primal one for the space identified by the subscript, so that $\tilde{\Psi}_H = \Psi_{H'}$. Here and in the following, we will view Ψ_H both as in (34) as a *collection* of functions as well as a (possibly infinite) column *vector* containing all functions always assembled in some fixed unspecified order. For a countable collection of functions Θ and some single function σ, the term $\langle \Theta, \sigma \rangle$ is then the column vector with entries $\langle \theta, \sigma \rangle$, $\theta \in \Theta$, and correspondingly $\langle \sigma, \Theta \rangle$ the row vector. For two collections Θ, Σ, the quantity $\langle \Theta, \Sigma \rangle$ is then a (possibly infinite) matrix with entries $(\langle \theta, \sigma \rangle)_{\theta \in \Theta,\, \sigma \in \Sigma}$ for which $\langle \Theta, \Sigma \rangle = \langle \Sigma, \Theta \rangle^T$. This also implies for a (possibly infinite) matrix $\mathbf{C}$ that $\langle \mathbf{C}\Theta, \Sigma \rangle = \mathbf{C}\langle \Theta, \Sigma \rangle$ and $\langle \Theta, \mathbf{C}\Sigma \rangle = \langle \Theta, \Sigma \rangle \mathbf{C}^T$.

In this notation, the *biorthogonality* or *duality conditions* (39) can be expressed shortly as

$$\langle \Psi, \tilde{\Psi} \rangle = \mathbf{I} \tag{41}$$

with the infinite identity matrix $\mathbf{I}$.

Wavelets with the above properties can be obtained in the following way. We start from an *anchor basis* $\Psi = \{\psi_\lambda : \lambda \in \mathbb{I} = \mathbb{I}_H\}$ which is a Riesz basis for $L_2(\Omega)$, meaning that Ψ is scaled such that $\|\psi_\lambda\|_{L_2(\Omega)} \sim 1$. Moreover, its dual basis $\tilde{\Psi}$ is also a Riesz basis for $L_2(\Omega)$. Ψ and $\tilde{\Psi}$ are constructed in such a way that rescaled versions of *both bases* $\Psi, \tilde{\Psi}$ form Riesz bases for a whole range of (closed subspaces of) Sobolev spaces $H^s := H^s(\Omega)$, for $0 < s < \gamma, \tilde{\gamma}$, respectively, or subspaces of these. Consequently, one can derive that for each $s \in (-\tilde{\gamma}, \gamma)$ the collection

$$\Psi_s := \{2^{-s|\lambda|}\psi_\lambda : \lambda \in \mathbb{I}\} =: \mathbf{D}^{-s}\Psi \tag{42}$$

is a Riesz basis for H^s [18], i.e., it holds

$$\|\mathbf{v}\| \sim \|\mathbf{v}^T \Psi_s\|_{H^s}, \quad \mathbf{v} \in \ell_2(\mathbb{I}), \tag{43}$$

for each $s \in (-\tilde{\gamma}, \gamma)$. Such a scaling represented by a diagonal matrix $\mathbf{D}^s$ introduced in (42) will play an important role later on. Concrete constructions of wavelet bases with the above properties for parameters $\gamma, \tilde{\gamma} \leq 5/2$ can be found in [22] which cover the situation in (6) for $r = 1, 2$.

3.2 Multiscale Decomposition of Function Spaces

In this section, the basic construction principles of the biorthogonal wavelets with properties (R), (L) and (CP) are summarized, see, e.g., [18]. Their cornerstones are *multiresolution analyses* of the function spaces under consideration as in Sect. 2.2 and the concept of *stable completions*. These concepts are free of Fourier techniques and can therefore be applied to derive constructions of wavelets on domains or manifolds which are subsets of $\mathbb{R}^n$.

Multiresolution of L_2 (univariate case). Practical constructions of wavelets typically start out with multiresolution analyses of function spaces. Consider a *multiresolution* $\mathcal{S}$ of L_2 which consists of closed subspaces S_j of L_2, called *trial spaces*, such that they are nested and their union is dense in L_2,

$$S_{j_0} \subset S_{j_0+1} \subset \ldots \subset S_j \subset S_{j+1} \subset \ldots L_2, \qquad \operatorname{clos}_{L_2}\Big(\bigcup_{j=j_0}^{\infty} S_j\Big) = L_2. \tag{44}$$

The index j is the refinement level which appeared already in the elements of the index set $\mathbb{I}$ in (34), starting with some coarsest level $j_0 \in \mathbb{N}_0$. We abbreviate for a finite subset $\Theta \subset L_2$ the linear span of Θ as

$$S(\Theta) = \operatorname{span}\{\Theta\}.$$

Typically the multiresolution spaces S_j have the form

$$S_j = S(\Phi_j), \qquad \Phi_j = \{\phi_{j,k} : k \in \Delta_j\}, \tag{45}$$

for some finite index set Δ_j, where the set $\{\Phi_j\}_{j=j_0}^{\infty}$ is *uniformly stable* in the sense that

$$\|\mathbf{c}\|_{\ell_2(\Delta_j)} \sim \|\mathbf{c}^T\Phi_j\|_{L_2}, \qquad \mathbf{c} = \{c_k\}_{k\in\Delta_j} \in \ell_2(\Delta_j), \tag{46}$$

holds uniformly in j. Here we have used again the shorthand notation

$$\mathbf{c}^T\Phi_j = \sum_{k\in\Delta_j} c_k\phi_{j,k}$$

and Φ_j denotes both the (column) vector containing the functions $\phi_{j,k}$ as well as the set of functions (45).

The collection Φ_j is called *single scale basis* since all its elements live only on one scale j. In the present context of multiresolution analysis, Φ_j is also called *generator basis* or shortly *generators* of the multiresolution. We assume that the $\phi_{j,k}$ are compactly supported with

$$\operatorname{diam}(\operatorname{supp}\phi_{j,k}) \sim 2^{-j}. \tag{47}$$

It follows from (46) that they are scaled such that

$$\|\phi_{j,k}\|_{L_2} \sim 1 \tag{48}$$

holds. It is known that nestedness (44) together with stability (46) implies the existence of matrices $\mathbf{M}_{j,0}$ such that the two-scale relation

$$\Phi_j = \mathbf{M}_{j,0}^T \Phi_{j+1} \tag{49}$$

holds. Any set of functions satisfying an equation of this form, the *refinement* or *two-scale relation*, will be called *refinable*. The matrices $\mathbf{M}_{j,0}$ are up to boundary modifications exactly the intergrid operators occurring in (30).

Denoting by $[X, Y]$ the space of bounded linear operators from a normed linear space X into the normed linear space Y, one has that

$$\mathbf{M}_{j,0} \in [\ell_2(\Delta_j), \ell_2(\Delta_{j+1})]$$

is *uniformly sparse* which means that the number of entries in each row or column is uniformly bounded. Furthermore, one infers from (46) that

$$\|\mathbf{M}_{j,0}\| = \mathcal{O}(1), \quad j \geq j_0, \tag{50}$$

where the corresponding operator norm is defined as

$$\|\mathbf{M}_{j,0}\| := \sup_{\mathbf{c} \in \ell_2(\Delta_j),\ \|\mathbf{c}\|_{\ell_2(\Delta_j)}=1} \|\mathbf{M}_{j,0}\mathbf{c}\|_{\ell_2(\Delta_{j+1})}.$$

Since the union of $\mathcal{S}$ is dense in L_2, a basis for L_2 can be assembled from functions which span any complement between two successive spaces S_j and S_{j+1}, i.e.,

$$S(\Phi_{j+1}) = S(\Phi_j) \oplus S(\Psi_j) \tag{51}$$

where

$$\Psi_j = \{\psi_{j,k} : k \in \nabla_j\}, \qquad \nabla_j := \Delta_{j+1} \setminus \Delta_j. \tag{52}$$

The functions Ψ_j are called *wavelet functions* or shortly *wavelets* if, among other conditions detailed below, the union $\{\Phi_j \cup \Psi_j\}$ is still uniformly stable in the sense of (46). Since (51) implies $S(\Psi_j) \subset S(\Phi_{j+1})$, the functions in Ψ_j must also satisfy a matrix-vector relation of the form

$$\Psi_j = \mathbf{M}_{j,1}^T \Phi_{j+1} \tag{53}$$

with a matrix $\mathbf{M}_{j,1}$ of size $(\#\Delta_{j+1}) \times (\#\nabla_j)$. Furthermore, (51) is equivalent to the fact that the linear operator composed of $\mathbf{M}_{j,0}$ and $\mathbf{M}_{j,1}$,

$$\mathbf{M}_j = (\mathbf{M}_{j,0}, \mathbf{M}_{j,1}), \tag{54}$$

is *invertible* as a mapping from $\ell_2(\Delta_j \cup \nabla_j)$ onto $\ell_2(\Delta_{j+1})$. One can also show that the set $\{\Phi_j \cup \Psi_j\}$ is uniformly stable if and only if

$$\|\mathbf{M}_j\|, \|\mathbf{M}_j^{-1}\| = \mathcal{O}(1), \quad j \to \infty. \tag{55}$$

The particular cases that will be important for practical purposes are when not only $\mathbf{M}_{j,0}$ and $\mathbf{M}_{j,1}$ are uniformly sparse but also the inverse of $\mathbf{M}_j$. We denote this inverse by $\mathbf{G}_j$ and assume that it is split into

$$\mathbf{G}_j = \mathbf{M}_j^{-1} = \begin{pmatrix} \mathbf{G}_{j,0} \\ \mathbf{G}_{j,1} \end{pmatrix}. \tag{56}$$

A special situation occurs when $\mathbf{M}_j$ is an orthogonal matrix,

$$\mathbf{G}_j = \mathbf{M}_j^{-1} = \mathbf{M}_j^T$$

which corresponds to the case of L_2 *orthogonal wavelets* [25]. A systematic construction of more general $\mathbf{M}_j$, $\mathbf{G}_j$ for spline-wavelets can be found in [22], see also [18] for more examples, including the hierarchical basis.

Thus, the identification of the functions Ψ_j which span the complement of $S(\Phi_j)$ in $S(\Phi_{j+1})$ is equivalent to completing a given refinement matrix $\mathbf{M}_{j,0}$ to an invertible matrix $\mathbf{M}_j$ in such a way that (55) is satisfied. Any such completion $\mathbf{M}_{j,1}$ is called *stable completion* of $\mathbf{M}_{j,0}$. In other words, the problem of the construction of compactly supported wavelets can equivalently be formulated as an algebraic problem of finding the (uniformly) sparse completion of a (uniformly) sparse matrix $\mathbf{M}_{j,0}$ in such a way that its inverse is also (uniformly) sparse. The fact that inverses of sparse matrices are usually dense elucidates the difficulties in the constructions. Constructions that yield compactly supported wavelets are particularly suited for computations in numerical analysis.

Combining the two-scale relations (49) and (53), one can see that $\mathbf{M}_j$ performs a change of bases in the space S_{j+1},

$$\begin{pmatrix} \Phi_j \\ \Psi_j \end{pmatrix} = \begin{pmatrix} \mathbf{M}_{j,0}^T \\ \mathbf{M}_{j,1}^T \end{pmatrix} \Phi_{j+1} = \mathbf{M}_j^T \Phi_{j+1}. \tag{57}$$

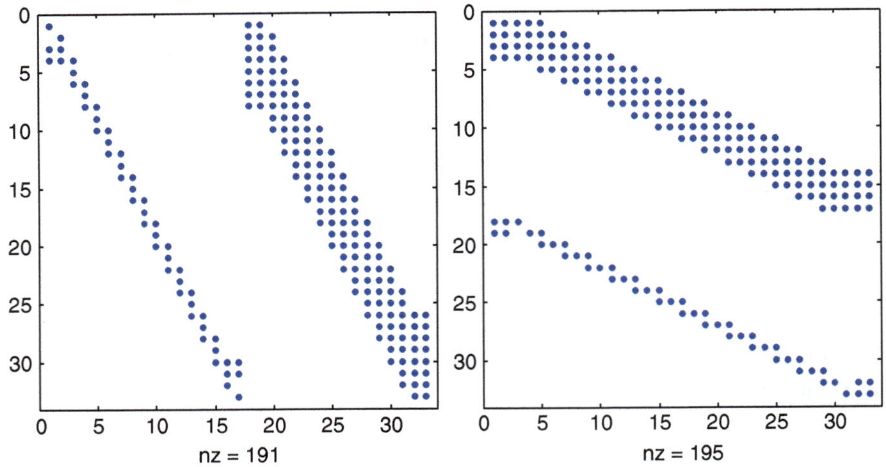

Fig. 2 Nonzero pattern of matrices $\mathbf{M}_j$ (*left*) and $\mathbf{G}_j$ (*right*) for boundary-adapted B-splines of order $d = 2$ (degree $p = 1$) as generators and duals of order $\tilde{d} = 4$

Conversely, applying the inverse of $\mathbf{M}_j$ to both sides of (57) results in the *reconstruction identity*

$$\Phi_{j+1} = \mathbf{G}_j^T \begin{pmatrix} \Phi_j \\ \Psi_j \end{pmatrix} = \mathbf{G}_{j,0}^T \Phi_j + \mathbf{G}_{j,1}^T \Psi_j. \tag{58}$$

An example of the structure of the matrices $\mathbf{M}_j$ and $\mathbf{G}_j$ is given in Fig. 2.

Fixing a *finest resolution level* J, one can repeat the decomposition (51) so that $S_J = S(\Phi_J)$ can be written in terms of the functions from the coarsest space supplied with the complement functions from all intermediate levels,

$$S(\Phi_J) = S(\Phi_{j_0}) \oplus \bigoplus_{j=j_0}^{J-1} S(\Psi_j). \tag{59}$$

Thus, every function $v \in S(\Phi_J)$ can be written in its *single-scale representation*

$$v = (\mathbf{c}_J)^T \Phi_J = \sum_{k \in \Delta_J} c_{J,k} \phi_{J,k} \tag{60}$$

as well as in its *multi-scale form*

$$v = (\mathbf{c}_{j_0})^T \Phi_{j_0} + (\mathbf{d}_{j_0})^T \Psi_{j_0} + \cdots + (\mathbf{d}_{J-1})^T \Psi_{J-1} \tag{61}$$

with respect to the *multiscale* or *wavelet basis*

$$\Psi^J := \Phi_{j_0} \cup \bigcup_{j=j_0}^{J-1} \Psi_j =: \bigcup_{j=j_0-1}^{J-1} \Psi_j \tag{62}$$

Often the single-scale representation of a function may be easier to compute and evaluate while the multi-scale representation allows one to separate features of the underlying function characterized by different length scales. Since therefore both representations are advantageous, it is useful to determine the transformation between the two representations, commonly referred to as the *Wavelet Transform*,

$$\mathbf{T}_J : \ell_2(\Delta_j) \to \ell_2(\Delta_j), \qquad \mathbf{d}^J \mapsto \mathbf{c}_J, \tag{63}$$

where

$$\mathbf{d}^J := (\mathbf{c}_{j_0}, \mathbf{d}_{j_0}, \ldots, \mathbf{d}_{J-1})^T.$$

The previous relations (57) and (58) indicate that this will involve the matrices $\mathbf{M}_j$ and $\mathbf{G}_j$. In fact, $\mathbf{T}_J$ has the representation

$$\mathbf{T}_J = \mathbf{T}_{J,J-1} \cdots \mathbf{T}_{J,j_0}, \tag{64}$$

where each factor has the form

$$\mathbf{T}_{J,j} := \begin{pmatrix} \mathbf{M}_j & \mathbf{0} \\ \mathbf{0} & \mathbf{I}^{(\#\Delta_J - \#\Delta_{j+1})} \end{pmatrix} \in \mathbb{R}^{(\#\Delta_J)\times(\#\Delta_J)}. \tag{65}$$

Schematically $\mathbf{T}_J$ can be visualized as a pyramid scheme

$$\begin{array}{ccccccccc}
\mathbf{c}_{j_0} & \xrightarrow{\mathbf{M}_{j_0,0}} & \mathbf{c}_{j_0+1} & \xrightarrow{\mathbf{M}_{j_0+1,0}} & \mathbf{c}_{j_0+2} & \longrightarrow \cdots & \mathbf{c}_{J-1} & \xrightarrow{\mathbf{M}_{J-1,0}} & \mathbf{c}_J \\
 & \mathbf{M}_{j_0,1} & & \mathbf{M}_{j_0+1,1} & & & & \mathbf{M}_{J-1,1} & \\
 & \nearrow & & \nearrow & & \nearrow \cdots & & \nearrow & \\
\mathbf{d}_{j_0} & & \mathbf{d}_{j_0+1} & & \mathbf{d}_{j_0+2} & & \mathbf{d}_{J-1} & &
\end{array} \tag{66}$$

Accordingly, the inverse transform $\mathbf{T}_J^{-1}$ can be written also in product structure (64) in reverse order involving the matrices $\mathbf{G}_j$ as follows:

$$\mathbf{T}_J^{-1} = \mathbf{T}_{J,j_0}^{-1} \cdots \mathbf{T}_{J,J-1}^{-1}, \tag{67}$$

where each factor has the form

$$\mathbf{T}_{J,j}^{-1} := \begin{pmatrix} \mathbf{G}_j & \mathbf{0} \\ \mathbf{0} & \mathbf{I}^{(\#\Delta_J - \#\Delta_{j+1})} \end{pmatrix} \in \mathbb{R}^{(\#\Delta_J)\times(\#\Delta_J)}. \tag{68}$$

The corresponding pyramid scheme is then

$$\begin{array}{ccccccccc} & \mathbf{G}_{J-1,0} & & \mathbf{G}_{J-2,0} & & & & \mathbf{G}_{j_0,0} & \\ \mathbf{c}_J & \longrightarrow & \mathbf{c}_{J-1} & \longrightarrow & \mathbf{c}_{J-2} & \longrightarrow & \cdots & \longrightarrow & \mathbf{c}_{j_0} \\ & \mathbf{G}_{J-1,1} & & \mathbf{G}_{J-2,1} & & & & \mathbf{G}_{j_0,1} & \\ & \searrow & & \searrow & & \searrow & \cdots & \searrow & \\ & & \mathbf{d}_{J-1} & & \mathbf{d}_{J-2} & & \mathbf{d}_{J-1} & & \mathbf{d}_{j_0} \end{array} \tag{69}$$

Property (55) and the fact that $\mathbf{M}_j$ and $\mathbf{G}_j$ can be applied in $(\#\Delta_{j+1})$ operations uniformly in j entails that the complexity of applying $\mathbf{T}_J$ or $\mathbf{T}_J^{-1}$ using the pyramid scheme is of order $\mathcal{O}(\#\Delta_J) = \mathcal{O}(\dim S_J)$ uniformly in J. For this reason, $\mathbf{T}_J$ is called the *Fast Wavelet Transform* (FWT). Note that one should *not* explicitly assemble $\mathbf{T}_J$ or $\mathbf{T}_J^{-1}$. In fact, due to the particular band structure of $\mathbf{M}_j$ and $\mathbf{G}_j$, this would result in matrices with $\mathcal{O}(J\,\#\Delta_J)$ entries. In Table 4 at the end of this section, spectral condition numbers for the FWT for different constructions of biorthogonal wavelets on the interval computed in [46] are displayed.

Since $\cup_{j\geq j_0} S_j$ is dense in L_2, a basis for the whole space L_2 is obtained when letting $J \to \infty$ in (62),

$$\begin{aligned} \Psi &:= \bigcup_{j=j_0-1}^{\infty} \Psi_j = \{\psi_{j,k} : (j,k) \in \mathbb{I}\}, \qquad \Psi_{j_0-1} := \Phi_{j_0} \\ \mathbb{I} &:= \{\{j_0\} \times \Delta_{j_0}\} \cup \bigcup_{j=j_0}^{\infty} \{\{j\} \times \nabla_j\}. \end{aligned} \tag{70}$$

Theorem 7 ([18]) *The multiscale transformations $\mathbf{T}_J$ are well-conditioned in the sense*

$$\|\mathbf{T}_J\|, \|\mathbf{T}_J^{-1}\| = \mathcal{O}(1), \quad J \geq j_0, \tag{71}$$

if and only if the collection Ψ defined by (70) is a Riesz basis for L_2.

A detailed construction of the *dual wavelets* $\tilde{\Psi}$, can be found in [18, 39].

Multiresolution of Sobolev spaces. Let now $\mathcal{S}$ be a multiresolution sequence consisting of closed subspaces of H^s with the property (44) whose union is dense in H^s. The following result from [18] ensures under which conditions norm equivalences hold for the H^s-norm.

Theorem 8 *Let* $\{\Phi_j\}_{j=j_0}^{\infty}$ *and* $\{\tilde{\Phi}_j\}_{j=j_0}^{\infty}$ *be uniformly stable, refinable, biorthogonal collections. If the* Jackson-type estimate

$$\inf_{v_j \in S_j} \|v - v_j\|_{L_2} \lesssim 2^{-sj} \|v\|_{H^s}, \quad v \in H^s,\ 0 < s \leq \bar{d}, \tag{72}$$

and the Bernstein inequality

$$\|v_j\|_{H^s} \lesssim 2^{sj} \|v_j\|_{L_2}, \quad v_j \in S_j,\ s < \bar{t}, \tag{73}$$

hold for

$$S_j = \begin{Bmatrix} S(\Phi_j) \\ S(\tilde{\Phi}_j) \end{Bmatrix} \text{ with order } \bar{d} = \begin{Bmatrix} d \\ \tilde{d} \end{Bmatrix} \text{ and } \bar{t} = \begin{Bmatrix} t \\ \tilde{t} \end{Bmatrix}, \tag{74}$$

then for

$$0 < \sigma := \min\{d, t\}, \qquad 0 < \tilde{\sigma} := \min\{\tilde{d}, \tilde{t}\}, \tag{75}$$

we have the norm equivalence

$$\|v\|_{H^s}^2 \sim \sum_{j=j_0-1}^{\infty} 2^{2sj} \|\langle \tilde{\Psi}_j, v\rangle\|_{\ell_2(\nabla_j)}^2, \quad s \in (-\tilde{\sigma}, \sigma). \tag{76}$$

For many applications it suffices to have established (76) only for certain $s > 0$ for which one only needs to require (72) and (73) for $\{\Phi_j\}_{j=j_0}^{\infty}$. The Jackson estimates (72) of order $\tilde{d}$ for $S(\tilde{\Phi}_j)$ imply the cancellation properties (CP) (38), see, e.g., [20].

Remark 9 When the wavelets live on $\Omega \subset \mathbb{R}^n$, (72) means that all polynomials up to order $\tilde{d}$ are contained in $S(\tilde{\Phi}_j)$. One also says that $S(\tilde{\Phi}_j)$ is *exact* of order $\tilde{d}$. Because of the biorthogonality conditions, this implies that the wavelets $\psi_{j,k}$ are orthogonal to polynomials up to order $\tilde{d}$ or have $\tilde{d}$th order *vanishing moments*. By Taylor expansion, this in turn yields (38).

For a summary of different constructions of biorthogonal wavelets on $\mathbb{R}$ and bounded domains in $\mathbb{R}^n$, see, e.g., [39]. We display spectral condition numbers for the FWT for two different constructions of biorthogonal wavelets on the interval in Table 4. The first column denotes the finest level on which the spectral condition numbers of the FWT are computed. The next column contains the numbers for the construction of biorthogonal spline-wavelets on the interval from [22] for the case $d = 2, \tilde{d} = 4$ while the last column displays the condition numbers for a scaled version derived in [12]. We observe that the absolute numbers stay constant and low even for high levels j. We will see later in Sect. 3.3 how the transformation $\mathbf{T}_J$ is used for preconditioning.

Table 4 Computed spectral condition numbers for the FWT on [0, 1] for different constructions of biorthogonal wavelets on the interval [46]

j	$\kappa_2(\mathbf{T}_{\mathrm{DKU}})$	$\kappa_2(\mathbf{T}_{\mathrm{B}})$	j	$\kappa_2(\mathbf{T}_{\mathrm{DKU}})$	$\kappa_2(\mathbf{T}_{\mathrm{B}})$
4	4.743e+00	4.640e+00	11	1.097e+01	8.011e+00
5	6.221e+00	6.024e+00	12	1.103e+01	8.034e+00
6	8.154e+00	6.860e+00	13	1.106e+01	8.046e+00
7	9.473e+00	7.396e+00	14	1.107e+01	8.051e+00
8	1.023e+01	7.707e+00	15	1.108e+01	8.054e+00
9	1.064e+01	7.876e+00	16	1.108e+01	8.056e+00
10	1.086e+01	7.965e+00			

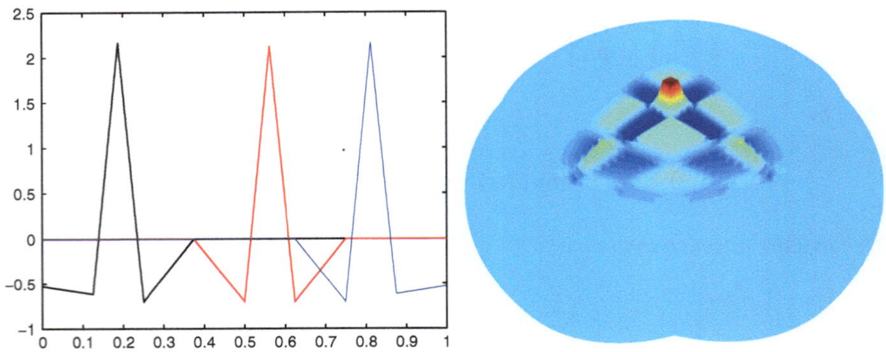

Fig. 3 Primal wavelets for $d = 2$ on [0, 1] (*left*) and on a sphere as constructed in [40] (*right*)

We see in Fig. 3 some primal biorthogonal wavelets of order $d = 2$ which consist of piecewise linear B-splines. These can be employed in the construction in wavelets on manifolds [24] which were optimized and implemented to construct biorthogonal wavelet bases on the sphere in [40], see the right graphic in Fig. 3.

3.3 Elliptic Boundary Value Problems in Wavelet Coordinates

We now derive a representation of the elliptic boundary value problem (6) in terms of (initially infinite) wavelet coordinates.

Let for $H = H_0^r(\Omega)$ the collection Ψ_H be a wavelet basis with corresponding dual $\tilde{\Psi}_H$ which satisfies the properties (R), (L) and (CP) from Sect. 3.1. Expanding the solution $u = \mathbf{u}^T\Psi_H$, the right hand side $f = \mathbf{f}^T\tilde{\Psi}_H$ and recalling the definition of A in (6), the wavelet representation of the elliptic boundary value problem (6) is given by

$$\mathbf{A}\mathbf{u} = \mathbf{f} \tag{77}$$

where

$$\mathbf{A} := a(\Psi_H, \Psi_H), \qquad \mathbf{f} := \langle \Psi_H, f \rangle. \tag{78}$$

Then the mapping property (5) and the Riesz basis property (R) yield the following fact.

Proposition 10 *The infinite matrix* $\mathbf{A}$ *is a boundedly invertible mapping from* $\ell_2 = \ell_2(\mathbb{I}_H)$ *into itself, and one has*

$$\|\mathbf{v}\| \sim \|\mathbf{A}\mathbf{v}\|, \qquad \mathbf{v} \in \ell_2. \tag{79}$$

The proof follows by combining the isomorphism relation (5) for the elliptic operator with the Riesz basis property (43). This entails the following consequence with respect to *preconditioning*. Let for $\mathbb{I} = \mathbb{I}_H$ the symbol Λ denote *any* finite subset of the index set $\mathbb{I}$. For the corresponding set of wavelets $\Psi_\Lambda := \{\psi_\lambda : \lambda \in \Lambda\}$, denote by $S_\Lambda := \operatorname{span} \Psi_\Lambda$ the respective finite-dimensional subspace of H. For the wavelet representation of A in terms of Ψ_Λ, $\mathbf{A}_\Lambda := a(\Psi_\Lambda, \Psi_\Lambda)$, we then have that $\kappa_2(\mathbf{A}_\Lambda) \sim 1$ independent of Λ, on account of the ellipticity of the operator. In other words, representations of A with respect to properly scaled wavelet bases for H entail well-conditioned system matrices $\mathbf{A}_\Lambda$ independent of Λ.

Fast wavelet transform. We briefly explain now in the situation of uniform refinements, i.e., when $S(\Phi_J) = S(\Psi^J)$, how the FWT $\mathbf{T}_J$ from (64) can be used for preconditioning linear elliptic operators, together with a diagonal scaling induced by the norm equivalence (76) [21]. We recall the notation from Sect. 3.2 where the wavelet basis is in fact the (unscaled) anchor basis from Sect. 3.1. Thus, the norm equivalence (36) using the scaled wavelet basis Ψ_H is the same as (76) in the anchor basis. Recall that the norm equivalence (76) implies that every $v \in H^s$ can be expanded uniquely in terms of the Ψ and its expansion coefficients $\mathbf{v}$ satisfy

$$\|v\|_{H^s} \sim \|\mathbf{D}^s \mathbf{v}\|_{\ell_2}$$

where $\mathbf{D}^s$ is a diagonal matrix with entries $\mathbf{D}^s_{(j,k),(j',k')} = 2^{sj}\delta_{j,j'}\delta_{k,k'}$. Depending on the order of the elliptic operator, we have $H \subset H^r(\Omega)$ for $r = 1$ or $r = 2$. We have, therefore, already identified the diagonal (scaling) matrix $\mathbf{D}_J$ consisting of the finite portion of the matrix $\mathbf{D} = \mathbf{D}^1$ for which $j_0 - 1 \leq j \leq J - 1$. The representation of A with respect to the (unscaled) wavelet basis Ψ^J can be expressed in terms of the FWT $\mathbf{T}_J$, that is,

$$\langle \Psi^J, A\Psi^J \rangle = \mathbf{T}_J^T \langle \Phi_J, A\Phi_J \rangle \mathbf{T}_J, \tag{80}$$

Table 5 Optimized spectral condition numbers of the operator **A** using tensor products of biorthogonal wavelets on the interval for space dimensions $n = 1, 2, 3$ [12]

j	$n=1$	$n=2$	$n=3$
3	22.3	9.6	18.3
4	23.9	11.8	37.1
5	25.0	14.3	39.8
6	25.7	16.0	40.9
8	26.6	18.4	
10	27.1		
12	27.3		

where Φ_J is the single-scale basis for $S(\Psi^J)$. Thus, we first set up the operator equation as in finite element settings in terms of the single-scale basis Φ_J. Applying the FWT $\mathbf{T}_J$ together with $\mathbf{D}_J$ yields that the operator

$$\mathbf{A}_J := \mathbf{D}_J^{-1} \mathbf{T}_J^T \langle \Phi_J, A\Phi_J \rangle \mathbf{T}_J \mathbf{D}_J^{-1} \tag{81}$$

has uniformly bounded condition numbers independent of J. This can be seen by combining the properties of A according to (5) with the norm equivalences (36) and (40).

It is known that the boundary adaptations of the generators and wavelets aggravate the absolute values of the condition numbers. Nevertheless, these constants can be substantially reduced by an operator-adapted transformation which takes into account only the coarsest discretization level and, thus, is inexpensive [12]. Numerical tests confirm that the absolute constants can further be improved by taking instead of $\mathbf{D}_J^{-1}$ the inverse of the diagonal of $\langle \Psi^J, A\Psi^J \rangle$ for the scaling in (81) [12, 46].

In Table 5 we display the condition numbers for discretizations using the weak form of the elliptic operator $-\Delta + I$ on $(0,1)^n$ in up to three dimensions using boundary adapted biorthogonal spline-wavelets in the case $d = 2$, $\tilde{d} = 4$ with such a scaling and additional shifts of small eigenvalues which is an inexpensive operation [12].

4 Conclusion and Outlook

The central theme of this paper was to present optimal multilevel preconditioners which enable us to reduce the spectral condition number of the system matrix to be independent of the grid spacing h. Specifically in the context of isogeometric analysis in Sect. 2, the issue arises how the spectral condition numbers depend on the polynomial degree p; we have seen a corresponding behaviour already in Table 1. From a theoretical point of view, it is not clear yet how to estimate the dependence on p and how to remedy its influence.

Some recent results in this directions may be promising. In [29], a full geometric multigrid method was proposed for problems in isogeometric analysis with an increased number of smoothing steps to account for the dependence of the spectral condition number on the polynomial degree p. A different approach is presented with different multigrid approaches involving high-and low-order variants in [49].

I expect that the intergrid operators used in Sect. 2 will also be efficient for isogeometric collocation methods as presented in [2] although it is not clear how to prove optimality of the corresponding preconditioner.

This paper dealt with preconditioners for discretizations on uniform grids which provide best approximations for PDEs with smooth solutions. In case when the solution is not smooth, adaptivity may capture the optimal degrees of freedom when compared to a best N-term approximation. Also in an adaptive method one has to iteratively solve linear systems of equations which again requires a preconditioner. For a BPX-type or FWT preconditioner, the same principles can immediately be applied as long as the spaces generated in the adaptive process are nested as in (17). Of course, then the intergrid operators in Sect. 2.3 have to be adapted accordingly.

References

1. Y. Achdou, O. Pironneau, *Computational Methods for Option Pricing* (SIAM, Philadelphia, 2005)
2. F. Auriccio, L. Beirão da Veiga, T.J.R. Hughes, A. Reali, C. Sangalli, Isogeometric collocation methods. Math. Models Methods Appl. Sci. **20**(11), 2075–2107 (2010)
3. O. Axelsson, P.S. Vassilevski, Algebraic multilevel preconditioning methods I. Numer. Math. **56**, 157–177 (1989)
4. O. Axelsson, P.S. Vassilevski, Algebraic multilevel preconditioning methods II. SIAM J. Numer. Anal. **27**, 1569–1590 (1990)
5. L. Beirão da Veiga, A. Buffa, C. Sangalli, R. Vázquez, Mathematical analysis of variational isogeometric methods. Acta Numer. **23**, 157–287 (2014)
6. L. Beirão da Veiga, D. Cho, L.F. Pavarino, S. Scacchi, Overlapping Schwarz methods for isogeometric analysis. SIAM J. Numer. Anal. **50**(3), 1394–1416 (2012)
7. C. de Boor, *A Practical Guide to Splines*, rev. edn. (Springer, New York, 2001)
8. D. Braess, *Finite Elements: Theory, Fast Solvers and Applications in Solid Mechanics*, 2nd edn. (Cambridge University Press, Cambridge, 2001)
9. D. Braess, W. Hackbusch, A new convergence proof for the multigrid method including the V-cycle. SIAM J. Numer. Anal. **20**, 967–975 (1983)
10. J.H. Bramble, J.E. Pasciak, J. Xu, Parallel multilevel preconditioners. Math. Comput. **55**, 1–22 (1990)
11. A. Buffa, H. Harbrecht, A. Kunoth, G. Sangalli, BPX preconditioning for isogeometric analysis. Comput. Methods Appl. Mech. Eng. **265**, 63–70 (2013)
12. C. Burstedde, Fast optimised wavelet methods for control problems constrained by Elliptic PDEs, PhD Dissertation, Mathematisch-Naturwissenschaftliche Fakultät, Universität Bonn, Bonn, 2005, http://hss.ulb.uni-bonn.de/2005/0682/0682.htm
13. C. Canuto, A. Tabacco, K. Urban, The wavelet element method, part I: construction and analysis. Appl. Comput. Harm. Anal. **6**, 1–52 (1999)
14. N. Clarke, K. Parrot, Multigrid for American option pricing with stochastic volatility. Appl. Math. Financ. **6**, 177–195 (1999)

15. A. Cohen, *Numerical Analysis of Wavelet Methods*. Studies in Mathematics and its Applications, vol. 32 (Elsevier, Amsterdam/Boston, 2003)
16. A. Cohen, W. Dahmen, R. DeVore, Adaptive wavelet methods for elliptic operator equations—convergence rates. Math. Comput. **70**, 27–75 (2001)
17. A. Cohen, I. Daubechies, J.-C. Feauveau, Biorthogonal bases of compactly supported wavelets. Commun. Pure Appl. Math. **45**, 485–560 (1992)
18. W. Dahmen, Wavelet and multiscale methods for operator equations. Acta Numer. **6**, 55–228 (1997)
19. W. Dahmen, Wavelet methods for PDEs—some recent developments. J. Comput. Appl. Math. **128**, 133–185 (2001)
20. W. Dahmen, Multiscale and wavelet methods for operator equations, in *Multiscale Problems and Methods in Numerical Simulation*, ed. by C. Canuto. C.I.M.E. Lecture Notes in Mathematics, vol. 1825 (Springer, Heidelberg, 2003), pp. 31–96
21. W. Dahmen, A. Kunoth, Multilevel preconditioning. Numer. Math. **63**, 315–344 (1992)
22. W. Dahmen, A. Kunoth, K. Urban, Biorthogonal spline wavelets on the interval—stability and moment conditions. Appl. Comput. Harm. Anal. **6**, 132–196 (1999)
23. W. Dahmen, R. Schneider, Composite wavelet bases for operator equations. Math. Comput. **68**, 1533–1567 (1999)
24. W. Dahmen, R. Schneider, Wavelets on manifolds I: construction and domain decomposition. SIAM J. Math. Anal. **31**, 184–230 (1999)
25. I. Daubechies, Orthonormal bases of compactly supported wavelets. Commun. Pure Appl. Math. **41**, 909–996 (1988)
26. R.A. DeVore, A. Kunoth, *Multiscale, Nonlinear and Adaptive Approximations* (Springer, Heidelberg/New York, 2009)
27. K.P.S. Gahalaut, J.K. Kraus, S.K. Tomar, Multigrid methods for isogeometric discretization. Comput. Methods Appl. Mech. Eng. **253**(1), 413–425 (2013)
28. S.L. Heston, A closed-form solution for options with stochastic volatility with applications to bond and currency options. Rev. Financ. Stud. **6**(2), 327–343 (1993)
29. C. Hofreither, W. Zulehner, On full multigrid schemes for isogeometric analysis, NuMa-Report No. 2014–03, Johannes-Kepler-Universität Linz, May 2014, http://www.numa.uni-linz.ac.at/Publications/
30. M. Holtz, A. Kunoth, B-spline based monotone multigrid methods with an application to the pricing of American options, in *Multigrid, Multilevel and Multiscale Methods*, ed. by P. Wesseling, C.W. Oosterlee, P. Hemker. (Electr.) Proc. EMG, 2005
31. M. Holtz, A. Kunoth, B-spline based monotone multigrid methods. SIAM J. Numer. Anal. **45**(1), 1175–1199 (2007)
32. J.C. Hull, *Options, Futures and Other Derivatives* (Prentice Hall International Editions, Upper Saddle River, 2000)
33. S. Jaffard, Wavelet methods for fast resolution of elliptic problems. SIAM J. Numer. Anal. **29**, 965–986 (1992)
34. S.K. Kleiss, C. Pechstein, B. Jüttler, S. Tomar, IETI—isogeometric tearing and interconnecting. Comput. Methods Appl. Mech. Eng. **247–248**, 201–215 (2012)
35. R. Kornhuber, *Adaptive Monotone Multigrid Methods for Nonlinear Variational Problems* (Teubner, Stuttgart, 1997)
36. R. Kraft, Adaptive and linearly independent multilevel B-splines, in *Surface Fitting and Multiresolution Methods*, ed. by A. Le Méhauté, C. Rabut, L.L. Schumaker (Vanderbilt University Press, Nashville, 1997), pp. 209–218
37. A. Kunoth, *Multilevel Preconditioning* (Shaker-Verlag, Aachen, 1994)
38. A. Kunoth, *Wavelet Methods—Elliptic Boundary Value Problems and Control Problems*. Advances in Numerical Mathematics (Teubner, Stuttgart, 2001)
39. A. Kunoth, Optimized wavelet preconditioning, in *Multiscale, Nonlinear and Adaptive Approximations*, ed. by R.A. DeVore, A. Kunoth (Springer, Heidelberg/New York, 2009), pp. 325–378
40. A. Kunoth, J. Sahner, Wavelets on manifolds: an optimized construction. Math. Comput. **75**, 1319–1349 (2006)

41. A. Kunoth, Chr. Schneider, K. Wiechers, Multiscale methods for the valuation of American options with stochastic volatility. Int. J. Comput. Math. **89**(9), 1145–1163 (2012)
42. J. Maes, A. Bultheel, A hierarchical basis preconditioner for the biharmonic equation on the sphere. IMA J. Numer. Anal. **26**(3), 563–583 (2006)
43. J. Maes, A. Kunoth, A. Bultheel, BPX-type preconditioners for 2nd and 4th order elliptic problems on the sphere. SIAM J. Numer. Anal. **45**(1), 206–222 (2007)
44. C.W. Oosterlee, On multigrid for linear complementary problems with application to American-style options. Electron. Trans. Numer. Anal. **15**, 165–185 (2003)
45. P. Oswald, On discrete norm estimates related to multilevel preconditioners in the finite element method, in *Constructive Theory of Functions*, ed. by K.G. Ivanov, P. Petrushev, B. Sendov. Proceedings of the International Conference Varna 1991, Bulgarian Academy of Sciences, Sofia, 1992, pp. 203–214
46. R. Pabel, Wavelet methods for PDE constrained elliptic control problems with dirichlet boundary control, Diploma Thesis, Institut für Angewandte Mathematik, Universität Bonn (2006), Shaker-Verlag Aachen (2007). doi:10.2370/236_232
47. L.L. Schumaker, *Spline Functions: Basic Theory*. Cambridge Mathematical Library, 3rd edn. (Cambridge University Press, Cambridge, 2007)
48. R. Stevenson, Adaptive wavelet methods for solving operator equations: an overview, in *Multiscale, Nonlinear and Adaptive Approximations*, ed. by R.A. DeVore, A. Kunoth (Springer, Heidelberg/New York, 2009), pp. 543–598
49. H. Sundar, G. Stadler, G. Biros, Comparison of multigrid algorithms for high-order continuous finite element discretizations, submitted, 2014. [arXiv:1402.5938]
50. K. Urban, *Wavelet Methods for Elliptic Partial Differential Equations* (Oxford University Press, Oxford, 2009)
51. A.-V. Vuong, C. Giannelli, B. Jüttler, B. Simeon, A hierarchical approach to adaptive local refinement in isogeometric analysis. Comput. Methods Appl. Mech. Eng. **200**(49–52), 3554–3567 (2011)
52. J. Xu, Iterative methods by space decomposition and subspace correction. SIAM Rev. **34**(4), 581–613 (1992)
53. J. Xu, L. Chen, R.H. Nochetto, Optimal multilevel methods for H(grad), H(curl), and H(div) systems on graded and unstructured grids, in *Multiscale, Nonlinear and Adaptive Approximations*, ed. by R.A. DeVore, A. Kunoth (Springer, Heidelberg/New York, 2009), pp. 599–659.
54. H. Yserentant, On the multilevel splitting of finite element spaces. Numer. Math. **49**, 379–412 (1986)

Editorial Policy

1. Volumes in the following three categories will be published in LNCSE:

i) Research monographs
ii) Tutorials
iii) Conference proceedings

Those considering a book which might be suitable for the series are strongly advised to contact the publisher or the series editors at an early stage.

2. Categories i) and ii). Tutorials are lecture notes typically arising via summer schools or similar events, which are used to teach graduate students. These categories will be emphasized by Lecture Notes in Computational Science and Engineering. **Submissions by interdisciplinary teams of authors are encouraged.** The goal is to report new developments – quickly, informally, and in a way that will make them accessible to non-specialists. In the evaluation of submissions timeliness of the work is an important criterion. Texts should be well-rounded, well-written and reasonably self-contained. In most cases the work will contain results of others as well as those of the author(s). In each case the author(s) should provide sufficient motivation, examples, and applications. In this respect, Ph.D. theses will usually be deemed unsuitable for the Lecture Notes series. Proposals for volumes in these categories should be submitted either to one of the series editors or to Springer-Verlag, Heidelberg, and will be refereed. A provisional judgement on the acceptability of a project can be based on partial information about the work: a detailed outline describing the contents of each chapter, the estimated length, a bibliography, and one or two sample chapters – or a first draft. A final decision whether to accept will rest on an evaluation of the completed work which should include

- at least 100 pages of text;
- a table of contents;
- an informative introduction perhaps with some historical remarks which should be accessible to readers unfamiliar with the topic treated;
- a subject index.

3. Category iii). Conference proceedings will be considered for publication provided that they are both of exceptional interest and devoted to a single topic. One (or more) expert participants will act as the scientific editor(s) of the volume. They select the papers which are suitable for inclusion and have them individually refereed as for a journal. Papers not closely related to the central topic are to be excluded. Organizers should contact the Editor for CSE at Springer at the planning stage, see *Addresses* below.

In exceptional cases some other multi-author-volumes may be considered in this category.

4. Only works in English will be considered. For evaluation purposes, manuscripts may be submitted in print or electronic form, in the latter case, preferably as pdf- or zipped ps-files. Authors are requested to use the LaTeX style files available from Springer at http://www.springer.com/gp/authors-editors/book-authors-editors/manuscript-preparation/5636 (Click on LaTeX Template → monographs or contributed books).

For categories ii) and iii) we strongly recommend that all contributions in a volume be written in the same LaTeX version, preferably LaTeX2e. Electronic material can be included if appropriate. Please contact the publisher.

Careful preparation of the manuscripts will help keep production time short besides ensuring satisfactory appearance of the finished book in print and online.

5. The following terms and conditions hold. Categories i), ii) and iii):

Authors receive 50 free copies of their book. No royalty is paid.
Volume editors receive a total of 50 free copies of their volume to be shared with authors, but no royalties.

Authors and volume editors are entitled to a discount of 33.3 % on the price of Springer books purchased for their personal use, if ordering directly from Springer.

6. Springer secures the copyright for each volume.

Addresses:

Timothy J. Barth
NASA Ames Research Center
NAS Division
Moffett Field, CA 94035, USA
barth@nas.nasa.gov

Michael Griebel
Institut für Numerische Simulation
der Universität Bonn
Wegelerstr. 6
53115 Bonn, Germany
griebel@ins.uni-bonn.de

David E. Keyes
Mathematical and Computer Sciences
and Engineering
King Abdullah University of Science
and Technology
P.O. Box 55455
Jeddah 21534, Saudi Arabia
david.keyes@kaust.edu.sa

and

Department of Applied Physics
and Applied Mathematics
Columbia University
500 W. 120 th Street
New York, NY 10027, USA
kd2112@columbia.edu

Risto M. Nieminen
Department of Applied Physics
Aalto University School of Science
and Technology
00076 Aalto, Finland
risto.nieminen@aalto.fi

Dirk Roose
Department of Computer Science
Katholieke Universiteit Leuven
Celestijnenlaan 200A
3001 Leuven-Heverlee, Belgium
dirk.roose@cs.kuleuven.be

Tamar Schlick
Department of Chemistry
and Courant Institute
of Mathematical Sciences
New York University
251 Mercer Street
New York, NY 10012, USA
schlick@nyu.edu

Editor for Computational Science
and Engineering at Springer:
Martin Peters
Springer-Verlag
Mathematics Editorial IV
Tiergartenstrasse 17
69121 Heidelberg, Germany
martin.peters@springer.com

Lecture Notes in Computational Science and Engineering

1. D. Funaro, *Spectral Elements for Transport-Dominated Equations.*
2. H.P. Langtangen, *Computational Partial Differential Equations.* Numerical Methods and Diffpack Programming.
3. W. Hackbusch, G. Wittum (eds.), *Multigrid Methods V.*
4. P. Deuflhard, J. Hermans, B. Leimkuhler, A.E. Mark, S. Reich, R.D. Skeel (eds.), *Computational Molecular Dynamics: Challenges, Methods, Ideas.*
5. D. Kröner, M. Ohlberger, C. Rohde (eds.), *An Introduction to Recent Developments in Theory and Numerics for Conservation Laws.*
6. S. Turek, *Efficient Solvers for Incompressible Flow Problems.* An Algorithmic and Computational Approach.
7. R. von Schwerin, ***Multi Body System SIMulation.*** Numerical Methods, Algorithms, and Software.
8. H.-J. Bungartz, F. Durst, C. Zenger (eds.), *High Performance Scientific and Engineering Computing.*
9. T.J. Barth, H. Deconinck (eds.), *High-Order Methods for Computational Physics.*
10. H.P. Langtangen, A.M. Bruaset, E. Quak (eds.), *Advances in Software Tools for Scientific Computing.*
11. B. Cockburn, G.E. Karniadakis, C.-W. Shu (eds.), *Discontinuous Galerkin Methods.* Theory, Computation and Applications.
12. U. van Rienen, *Numerical Methods in Computational Electrodynamics.* Linear Systems in Practical Applications.
13. B. Engquist, L. Johnsson, M. Hammill, F. Short (eds.), *Simulation and Visualization on the Grid.*
14. E. Dick, K. Riemslagh, J. Vierendeels (eds.), *Multigrid Methods VI.*
15. A. Frommer, T. Lippert, B. Medeke, K. Schilling (eds.), *Numerical Challenges in Lattice Quantum Chromodynamics.*
16. J. Lang, *Adaptive Multilevel Solution of Nonlinear Parabolic PDE Systems.* Theory, Algorithm, and Applications.
17. B.I. Wohlmuth, *Discretization Methods and Iterative Solvers Based on Domain Decomposition.*
18. U. van Rienen, M. Günther, D. Hecht (eds.), *Scientific Computing in Electrical Engineering.*
19. I. Babuška, P.G. Ciarlet, T. Miyoshi (eds.), *Mathematical Modeling and Numerical Simulation in Continuum Mechanics.*
20. T.J. Barth, T. Chan, R. Haimes (eds.), *Multiscale and Multiresolution Methods.* Theory and Applications.
21. M. Breuer, F. Durst, C. Zenger (eds.), *High Performance Scientific and Engineering Computing.*
22. K. Urban, *Wavelets in Numerical Simulation.* Problem Adapted Construction and Applications.
23. L.F. Pavarino, A. Toselli (eds.), *Recent Developments in Domain Decomposition Methods.*

24. T. Schlick, H.H. Gan (eds.), *Computational Methods for Macromolecules: Challenges and Applications.*

25. T.J. Barth, H. Deconinck (eds.), *Error Estimation and Adaptive Discretization Methods in Computational Fluid Dynamics.*

26. M. Griebel, M.A. Schweitzer (eds.), *Meshfree Methods for Partial Differential Equations.*

27. S. Müller, *Adaptive Multiscale Schemes for Conservation Laws.*

28. C. Carstensen, S. Funken, W. Hackbusch, R.H.W. Hoppe, P. Monk (eds.), *Computational Electromagnetics.*

29. M.A. Schweitzer, *A Parallel Multilevel Partition of Unity Method for Elliptic Partial Differential Equations.*

30. T. Biegler, O. Ghattas, M. Heinkenschloss, B. van Bloemen Waanders (eds.), *Large-Scale PDE-Constrained Optimization.*

31. M. Ainsworth, P. Davies, D. Duncan, P. Martin, B. Rynne (eds.), *Topics in Computational Wave Propagation.* Direct and Inverse Problems.

32. H. Emmerich, B. Nestler, M. Schreckenberg (eds.), *Interface and Transport Dynamics.* Computational Modelling.

33. H.P. Langtangen, A. Tveito (eds.), *Advanced Topics in Computational Partial Differential Equations.* Numerical Methods and Diffpack Programming.

34. V. John, *Large Eddy Simulation of Turbulent Incompressible Flows.* Analytical and Numerical Results for a Class of LES Models.

35. E. Bänsch (ed.), *Challenges in Scientific Computing - CISC 2002.*

36. B.N. Khoromskij, G. Wittum, *Numerical Solution of Elliptic Differential Equations by Reduction to the Interface.*

37. A. Iske, *Multiresolution Methods in Scattered Data Modelling.*

38. S.-I. Niculescu, K. Gu (eds.), *Advances in Time-Delay Systems.*

39. S. Attinger, P. Koumoutsakos (eds.), *Multiscale Modelling and Simulation.*

40. R. Kornhuber, R. Hoppe, J. Périaux, O. Pironneau, O. Wildlund, J. Xu (eds.), *Domain Decomposition Methods in Science and Engineering.*

41. T. Plewa, T. Linde, V.G. Weirs (eds.), *Adaptive Mesh Refinement – Theory and Applications.*

42. A. Schmidt, K.G. Siebert, *Design of Adaptive Finite Element Software.* The Finite Element Toolbox ALBERTA.

43. M. Griebel, M.A. Schweitzer (eds.), *Meshfree Methods for Partial Differential Equations II.*

44. B. Engquist, P. Lötstedt, O. Runborg (eds.), *Multiscale Methods in Science and Engineering.*

45. P. Benner, V. Mehrmann, D.C. Sorensen (eds.), *Dimension Reduction of Large-Scale Systems.*

46. D. Kressner, *Numerical Methods for General and Structured Eigenvalue Problems.*

47. A. Boriçi, A. Frommer, B. Joó, A. Kennedy, B. Pendleton (eds.), *QCD and Numerical Analysis III.*

48. F. Graziani (ed.), *Computational Methods in Transport.*

49. B. Leimkuhler, C. Chipot, R. Elber, A. Laaksonen, A. Mark, T. Schlick, C. Schütte, R. Skeel (eds.), *New Algorithms for Macromolecular Simulation.*

50. M. Bücker, G. Corliss, P. Hovland, U. Naumann, B. Norris (eds.), *Automatic Differentiation: Applications, Theory, and Implementations.*

51. A.M. Bruaset, A. Tveito (eds.), *Numerical Solution of Partial Differential Equations on Parallel Computers.*

52. K.H. Hoffmann, A. Meyer (eds.), *Parallel Algorithms and Cluster Computing.*

53. H.-J. Bungartz, M. Schäfer (eds.), *Fluid-Structure Interaction.*

54. J. Behrens, *Adaptive Atmospheric Modeling.*

55. O. Widlund, D. Keyes (eds.), *Domain Decomposition Methods in Science and Engineering XVI.*

56. S. Kassinos, C. Langer, G. Iaccarino, P. Moin (eds.), *Complex Effects in Large Eddy Simulations.*

57. M. Griebel, M.A Schweitzer (eds.), *Meshfree Methods for Partial Differential Equations III.*

58. A.N. Gorban, B. Kégl, D.C. Wunsch, A. Zinovyev (eds.), *Principal Manifolds for Data Visualization and Dimension Reduction.*

59. H. Ammari (ed.), *Modeling and Computations in Electromagnetics: A Volume Dedicated to Jean-Claude Nédélec.*

60. U. Langer, M. Discacciati, D. Keyes, O. Widlund, W. Zulehner (eds.), *Domain Decomposition Methods in Science and Engineering XVII.*

61. T. Mathew, *Domain Decomposition Methods for the Numerical Solution of Partial Differential Equations.*

62. F. Graziani (ed.), *Computational Methods in Transport: Verification and Validation.*

63. M. Bebendorf, *Hierarchical Matrices.* A Means to Efficiently Solve Elliptic Boundary Value Problems.

64. C.H. Bischof, H.M. Bücker, P. Hovland, U. Naumann, J. Utke (eds.), *Advances in Automatic Differentiation.*

65. M. Griebel, M.A. Schweitzer (eds.), *Meshfree Methods for Partial Differential Equations IV.*

66. B. Engquist, P. Lötstedt, O. Runborg (eds.), *Multiscale Modeling and Simulation in Science.*

67. I.H. Tuncer, Ü. Gülcat, D.R. Emerson, K. Matsuno (eds.), *Parallel Computational Fluid Dynamics 2007.*

68. S. Yip, T. Diaz de la Rubia (eds.), *Scientific Modeling and Simulations.*

69. A. Hegarty, N. Kopteva, E. O'Riordan, M. Stynes (eds.), *BAIL* 2008 – *Boundary and Interior Layers.*

70. M. Bercovier, M.J. Gander, R. Kornhuber, O. Widlund (eds.), *Domain Decomposition Methods in Science and Engineering XVIII.*

71. B. Koren, C. Vuik (eds.), *Advanced Computational Methods in Science and Engineering.*

72. M. Peters (ed.), *Computational Fluid Dynamics for Sport Simulation.*

73. H.-J. Bungartz, M. Mehl, M. Schäfer (eds.), *Fluid Structure Interaction II - Modelling, Simulation, Optimization.*

74. D. Tromeur-Dervout, G. Brenner, D.R. Emerson, J. Erhel (eds.), *Parallel Computational Fluid Dynamics 2008.*

75. A.N. Gorban, D. Roose (eds.), *Coping with Complexity: Model Reduction and Data Analysis.*

76. J.S. Hesthaven, E.M. Rønquist (eds.), *Spectral and High Order Methods for Partial Differential Equations.*

77. M. Holtz, *Sparse Grid Quadrature in High Dimensions with Applications in Finance and Insurance.*

78. Y. Huang, R. Kornhuber, O.Widlund, J. Xu (eds.), *Domain Decomposition Methods in Science and Engineering XIX.*

79. M. Griebel, M.A. Schweitzer (eds.), *Meshfree Methods for Partial Differential Equations V.*

80. P.H. Lauritzen, C. Jablonowski, M.A. Taylor, R.D. Nair (eds.), *Numerical Techniques for Global Atmospheric Models.*

81. C. Clavero, J.L. Gracia, F.J. Lisbona (eds.), *BAIL 2010 – Boundary and Interior Layers, Computational and Asymptotic Methods.*

82. B. Engquist, O. Runborg, Y.R. Tsai (eds.), *Numerical Analysis and Multiscale Computations.*

83. I.G. Graham, T.Y. Hou, O. Lakkis, R. Scheichl (eds.), *Numerical Analysis of Multiscale Problems.*

84. A. Logg, K.-A. Mardal, G. Wells (eds.), *Automated Solution of Differential Equations by the Finite Element Method.*

85. J. Blowey, M. Jensen (eds.), *Frontiers in Numerical Analysis - Durham 2010.*

86. O. Kolditz, U.-J. Gorke, H. Shao, W. Wang (eds.), *Thermo-Hydro-Mechanical-Chemical Processes in Fractured Porous Media - Benchmarks and Examples.*

87. S. Forth, P. Hovland, E. Phipps, J. Utke, A. Walther (eds.), *Recent Advances in Algorithmic Differentiation.*

88. J. Garcke, M. Griebel (eds.), *Sparse Grids and Applications.*

89. M. Griebel, M.A. Schweitzer (eds.), *Meshfree Methods for Partial Differential Equations VI.*

90. C. Pechstein, *Finite and Boundary Element Tearing and Interconnecting Solvers for Multiscale Problems.*

91. R. Bank, M. Holst, O. Widlund, J. Xu (eds.), *Domain Decomposition Methods in Science and Engineering XX.*

92. H. Bijl, D. Lucor, S. Mishra, C. Schwab (eds.), *Uncertainty Quantification in Computational Fluid Dynamics.*

93. M. Bader, H.-J. Bungartz, T. Weinzierl (eds.), *Advanced Computing.*

94. M. Ehrhardt, T. Koprucki (eds.), *Advanced Mathematical Models and Numerical Techniques for Multi-Band Effective Mass Approximations.*

95. M. Azaïez, H. El Fekih, J.S. Hesthaven (eds.), *Spectral and High Order Methods for Partial Differential Equations ICOSAHOM 2012.*

96. F. Graziani, M.P. Desjarlais, R. Redmer, S.B. Trickey (eds.), *Frontiers and Challenges in Warm Dense Matter.*

97. J. Garcke, D. Pflüger (eds.), *Sparse Grids and Applications – Munich 2012.*

98. J. Erhel, M. Gander, L. Halpern, G. Pichot, T. Sassi, O. Widlund (eds.), *Domain Decomposition Methods in Science and Engineering XXI.*

99. R. Abgrall, H. Beaugendre, P.M. Congedo, C. Dobrzynski, V. Perrier, M. Ricchiuto (eds.), *High Order Nonlinear Numerical Methods for Evolutionary PDEs - HONOM 2013.*

100. M. Griebel, M.A. Schweitzer (eds.), *Meshfree Methods for Partial Differential Equations VII.*

101. R. Hoppe (ed.), *Optimization with PDE Constraints - OPTPDE 2014.*

102. S. Dahlke, W. Dahmen, M. Griebel, W. Hackbusch, K. Ritter, R. Schneider, C. Schwab, H. Yserentant (eds.), *Extraction of Quantifiable Information from Complex Systems.*

103. A. Abdulle, S. Deparis, D. Kressner, F. Nobile, M. Picasso (eds.), *Numerical Mathematics and Advanced Applications - ENUMATH 2013.*

104. T. Dickopf, M.J. Gander, L. Halpern, R. Krause, L.F. Pavarino (eds.), *Domain Decomposition Methods in Science and Engineering XXII.*

105. M. Mehl, M. Bischoff, M. Schäfer (eds.), *Recent Trends in Computational Engineering - CE2014.* Optimization, Uncertainty, Parallel Algorithms, Coupled and Complex Problems.

106. R.M. Kirby, M. Berzins, J.S. Hesthaven (eds.), *Spectral and High Order Methods for Partial Differential Equations - ICOSAHOM '14.*

107. B. Simeon, B. Jüttler (eds.), *Isogeometric Analysis and Applications 2014.*

For further information on these books please have a look at our mathematics catalogue at the following URL: `www.springer.com/series/3527`

Monographs in Computational Science and Engineering

1. J. Sundnes, G.T. Lines, X. Cai, B.F. Nielsen, K.-A. Mardal, A. Tveito, *Computing the Electrical Activity in the Heart.*

For further information on this book, please have a look at our mathematics catalogue at the following URL: www.springer.com/series/7417

Texts in Computational Science and Engineering

1. H. P. Langtangen, *Computational Partial Differential Equations.* Numerical Methods and Diffpack Programming. 2nd Edition
2. A. Quarteroni, F. Saleri, P. Gervasio, *Scientific Computing with MATLAB and Octave.* 4th Edition
3. H. P. Langtangen, *Python Scripting for Computational Science.* 3rd Edition
4. H. Gardner, G. Manduchi, *Design Patterns for e-Science.*
5. M. Griebel, S. Knapek, G. Zumbusch, *Numerical Simulation in Molecular Dynamics.*
6. H. P. Langtangen, *A Primer on Scientific Programming with Python.* 4th Edition
7. A. Tveito, H. P. Langtangen, B. F. Nielsen, X. Cai, *Elements of Scientific Computing.*
8. B. Gustafsson, *Fundamentals of Scientific Computing.*
9. M. Bader, *Space-Filling Curves.*
10. M. Larson, F. Bengzon, *The Finite Element Method: Theory, Implementation and Applications.*
11. W. Gander, M. Gander, F. Kwok, *Scientific Computing: An Introduction using Maple and MATLAB.*
12. P. Deuflhard, S. Röblitz, *A Guide to Numerical Modelling in Systems Biology.*

For further information on these books please have a look at our mathematics catalogue at the following URL: www.springer.com/series/5151

Mangrove Ecosystem Restoration

Edited by Sahadev Sharma

Published in London, United Kingdom

IntechOpen

Supporting open minds since 2005

Mangrove Ecosystem Restoration
http://dx.doi.org/10.5772/intechopen.77881
Edited by Sahadev Sharma

Contributors
Aroloye O. Numbere, Claude Gasper Mung'Ong'O, V. Shiva Shankar, Neelam Purti, Ravi Pratap Singh, Faiyaz A. Khudsar, Takayoshi Koike, Ruili Li, Minwei Chai, Xiooxue Shen, Guoyu Qiu, Cong Shi, Djumanto, Christoph Zöckler, Dominic Wodehouse, Matthias Markolf, Delianis Pringgenies, Ali Ridlo, Ali Djunaedi, Lutfianna Fatma Dewi, Ícaro Thiago Andrade Moreira, Olívia Maria Cordeiro de Oliveira, Gisele Mara Hadlich, Antônio Fernando de Souza Queiroz, Célia Karina Maia Cardos, Isadora Machado Marques, Evelin Daiane Serafim Santos Franco, Ana Katerine de Carvalho Lima Lobato

First published in London, United Kingdom, 2021 by IntechOpen
IntechOpen is the global imprint of INTECHOPEN LIMITED, registered in England and Wales, registration number: 11086078, 5 Princes Gate Court, London, SW7 2QJ, United Kingdom
Printed in Croatia

British Library Cataloguing-in-Publication Data
A catalogue record for this book is available from the British Library

Additional hard and PDF copies can be obtained from orders@intechopen.com

Mangrove Ecosystem Restoration
Edited by Sahadev Sharma
p. cm.
Print ISBN 978-1-83962-799-6
Online ISBN 978-1-83962-800-9
eBook (PDF) ISBN 978-1-83962-801-6

For EU product safety concerns:
IN TECH d.o.o., Prolaz Marije Krucifikse Kozulić 3, 51000 Rijeka, Croatia, info@intechopen.com or visit our website at intechopen.com.

Meet the editor

Dr. Sahadev Sharma obtained a DSc in Plant Ecology and Physiology in 2012. He is a mangrove ecologist and physiologist with research interests in mangrove forest ecology, physiology, and blue carbon dynamics from local site to the landscape level, using a wide spectrum of methods and technologies such as remote sensing and field-based monitoring and sampling, and integrating ecological, chemical and hydrological data. He is currently focusing on the Asia Pacific region mangrove ecosystem. He is a senior lecturer at the Institute of Ocean and Earth Sciences, Universiti Malaya, Malaysia. Formerly he was a research faculty at the Department of Natural Resources and Environmental Management, the University of Hawaii at Manoa, and a JSPS Postdoctoral Fellow at Tokyo Institute of Technology, Japan. He is currently involved in many national and international research projects for the Universiti Malaya. He is the author of more than sixty papers in international peer-reviewed journals and books, and a referee for about ten journals.

Contents

Preface

The mangrove ecosystem provides several ecosystem services such as timber, charcoal, wave attenuation, nursery and spawning area for fishes, food, tourism, and climate change mitigation. Although they provide these important ecosystem services, we have lost a significant number of mangroves over time. Mangrove degradation and deforestation are caused by both natural disasters (tsunami, cyclones, storms, disease, and erosion) and anthropogenic activities (illegal logging, aquaculture pond, agriculture, urban development and other exploitation). Climate change is happening, therefore we need to protect the remaining mangroves and restore them to enhance the ecosystem services they provide to achieve sustainable development goals. Recently, mangroves have been recognized as one of the nature-based solutions for coastal communities. We are now almost at the tipping point where we can restore mangroves ecologically to mitigate climate change and other important ecosystem services under the United Nations Decade on Ecosystem Restoration. This book is comprised of three sections containing eight chapters that focus on mangrove ecosystem restoration, the ecosystem services they provide, and how to manage and conserve mangroves.

The first section on restoration and natural regeneration includes three chapters. Chapter 1 examines the possibility of utilizing natural means of forest recovery through seedling recruitment and regeneration of mangroves. Chapter 2 presents a guide to rapid evaluation of mangrove degradation where later mangrove restoration can be done. Chapter 3 explains how mangroves regenerate naturally after natural disasters such as tsunamis.

The second section on ecosystem services includes three chapters that discuss how planted mangroves provide different ecosystem services. Chapter 4 discusses the perspective of pollutant deposition in mangrove wetlands, physiological ecology of mangrove species on the impact of heavy metal pollution, and seeking ecosystem restoration in terms of environmental education. Chapter 5 discusses mangrove-based natural dye for batik fabric from *Rhizopora mucronata* waste as a promising product to increase people's income. Chapter 6 talks about the use of advanced molecular biology in biotechnologies as a promising path to faster, more economically viable and ecologically correct mangrove restoration.

The third and final section includes two chapters related to the conservation and management of restored mangroves. Chapter 7 documents the various strategies and approaches used in mangrove restoration in the world generally, and then specifically in the Rufiji Delta, Tanzania. The chapter also presents related policies to protect mangroves. Chapter 8 explains how conservation management of restored mangroves can be done through evaluating ecosystem services such as tourism in Indonesia.

We hope this book helps in the restoration, conservation and management of mangroves to enhance the ecosystem services they provide and help in mitigating climate change.

Sahadev Sharma
Institute of Ocean and Earth Sciences,
University of Malaya,
Kuala Lumpur, Malaysia

Section 1

Restoration and Natural Regeneration

Chapter 1

Mangrove Restoration under Different Disturbances Regime in the Niger Delta, Nigeria

Aroloye O. Numbere

Abstract

Mangroves of the Niger Delta are the largest in Africa and are the source of numerous ecosystem services such as firewood, seafood, building materials and medicinal herbs. Their sustainable use and protection are important for future generations. However, anthropogenic activities such as oil and gas exploration, urbanization, industrialization, dredging, overexploitation and sand mining are the major disturbances that have pushed the mangroves to the brink of extinction. Therefore, in other to restore lost areas of the mangroves natural and artificial means can be adopted to bring them to a restored state. More often than not emphasis of recovery had been placed on artificial remediation and restoration, where polluted sites are cleaned with chemicals and nursery seedlings transplanted to remediated such sites. Nevertheless, this chapter discusses the possibility of utilizing natural means of forest recovery through seedling recruitment and regeneration. This can be achieved by establishing the right environmental conditions such as setting up of a hydro-channel to ensure smooth inflow and out flow of river water carrying seeds, availability of parent mangrove trees to supply the seeds, and the availability of the right soil condition to enable seedling germination and growth. The use of dried and ground mangrove parts as a new way for restoring polluted soil is discussed; in addition, the unconventional proposition of using low key pollution to manage and increase forest resilience is highlighted in this work even though further studies are recommended. Future direction of mangrove restoration should be tilted towards the application of the force of nature, which has the potentials of reversing the adverse effect of anthropogenic activities in well managed and protected sites.

Keywords: ecology, hydrocarbon pollution, remediation, recruitment, succession, urbanization

1. Introduction

Mangroves of the Niger Delta are the most abundant and most productive forest in Africa [1]. It is also the third largest in the world. The significance of mangroves unlike other rain forest ecosystem e.g. Mahogany (*Khaya ivorensis*), obeche (*Triplochiton scleroxylon*) and iroko (*Malicia excelsa* is the kind of ecosystem services they provide [2]. This is because in addition to purifying the air, stabilizing the soil and being used as timber, mangroves play special role in the environment by serving as one of the biggest carbon sink in the world [3] based on the kind of terrain they occupy. They are the only tree species that grow within the swamps and at the fringes of the sea in highly saline environment [4]. They are adapted to one of the most gruesome environments for any tree to survive. For instance, apart from their salty environment, they grow in soft and muddy soil and are constantly bashed by violent tidal currents. In spite of all these environmental difficulties the mangroves had come out unscathed. Mangroves tend to survive very difficult environmental conditions. To deal with high salt, their system shuts off, sweats out or pumps out excessive salt to survive their environment. Their adventitious root system grows not only from the bottom of the tree but also grow out from the branches in an octopus-like manner to be rooted in the swamp, which provide additional support from the ground-based roots. These roots system if not carefully identified can easily be mistaken for mature stems. Excess salt that will easily kill off other trees act as nutrients for their rapid growth. The survival of mangrove in its difficult environment can be a lesson on resiliency for humans.

1.1 Natural mangrove recruitment

Seedling recruitment in mangroves occurs when juvenile organisms survive to be added to a population, by birth or immigration, usually a stage whereby the organisms are settled and are able to be detected by an observer in natural mangroves forest [5]. The Nigerian landscape has significantly changed over the last few decades and anthropogenic activities by man such as sand mining practices is one of the most important causes of this change. Rapid re-establishment of native vegetation, particularly after a large-scale disturbance, can be critical in preventing soil erosion, invasion by exotics, and other unwanted species such as *Nypa fruticans*. Re-colonization of disturbed sites may be slow and unpredictable, especially if seed sources are remote. Ecological restoration may involve not only artificial reintroduction of the original community dominants, but also nurse species that improve seed trapping and establishment [6], attract seed carriers, enhance soil conditions through organic matter or nutrient accumulation [7, 8], or provide protection of sensitive seedlings [9]. Ecological restoration approaches, however, must be based on a thorough understanding of the natural successional dynamics of the system as well as the growth requirements of the dominant plant species. The current challenge in ecological restoration is to manipulate development so that recovery of the entire suite of structural and functional features is achieved as quickly as possible [10]. Few studies have experimentally examined facilitation in the context of restoration [11, 12]. Facilitation may not only involve amelioration of environmental conditions that promote growth of a beneficiary species, but can also arise from effects of dispersal and establishment, e.g., trapping of seeds. Facilitation has been studied in extreme environments such as salt marshes [13, 14] where plants must cope with stresses such as salinity, flooding, and variable sediment and nutrient supplies such as mangroves. Mangroves are the tropical equivalent of temperate salt marshes, but in contrast to marsh grasses that can propagate vegetatively, these tidal forests are dominated by tree species dependent upon seedling recruitment for regeneration. Mangroves are frequently disturbed by hurricanes and human activities, which severely damage or eliminate the forest community [15]. Mangrove plant communities often contain herbaceous species, which are common components of the tropical beach habitat, salt marshes, or other wet coastal communities [16]. Although factors influencing mangrove recruitment such as seed and seedling predators [17, 18], flooding and salinity [19, 20], and sedimentation [21] had been studied in neotropical forests. Mangroves may be extremely slow to recolonize and grow, especially in harsh (e.g., arid, hypersaline) environments [22]. Mangrove ecosystems thus constitute not only a critical habitat with important ecological and societal benefits, but are systems in which facilitative interactions might be applied to improve restoration techniques.

1.2 Ecological restoration

Ecological restoration is defined as the process of repairing damage caused by humans to the diversity and dynamics of indigenous ecosystem [23]. Ecological restoration includes a wide scope of projects including erosion control, reforestation, removal of non-native species (e.g. *N fruticans*) and weeds, revegetation of disturbed areas, day lighting streams, reintroduction of native species, and habitat and range improvement for targeted species. Hydrological connections to natural restoration site are also very important to allow in inflow of saline water and mangrove seeds [24]. Inflow of water will also clean the site from pollutants such as oil spillage [25].

Factors that affect ecological restoration include the following:

1. **Disturbance:** is a change in environmental conditions that disrupts the functioning of an ecosystem [26]. Disturbance can occur at a variety of spatial and temporal scales, and is a natural component of many communities. For instance, sand mining and oiling activities urban development are disturbances. Differentiating between human-caused and naturally occurring disturbances is important in restoration and minimization of anthropogenic impacts.

   Human disturbance: This is a kind of disturbance caused by humans e.g. urbanization and industrialization. Humans build bridges, shopping malls, roads, schools, hospitals etc. in cleared mangrove forests. These activities eliminate the natural wetland system and destroy numerous biodiversity that inhabit this environment (barnacles, mussel, periwinkles, crabs etc). During oil and gas exploration humans deliberately bulldoze and clear large acre of forest to make way for the establishment of booth camps, oil wells and crude oil pipelines. The pipelines conveys petroleum products from oil wells to the refinery while finished products are transported back via pipelines to tankers evacuating products at the port (**Figure 1A**). These pipelines are established by creating right of way passage (ROW) through deforestation (**Figure 1B**), furthermore, crude oil spills occur from these pipelines due to sabotage or mechanical failure leading to the destruction of vast amount of mangrove forest (**Figure 1C**).

   Natural disturbances: This is a disturbance that is caused by force of nature. This includes flood, erosion, hurricanes, tsunami and earthquake [27]. These disturbances are controlled by weather conditions and cause massive damage to mangrove forest, which changes the forest structure and composition (**Figure 2**).

2. **Ecological Succession**: is the process by which a community changes over time, especially following a disturbance [28]. In many instances, an ecosystem will change from a simple level of organization with a few dominant pioneer species to an increasingly complex community with many interdependent species. Restoration often consists of initiating, assisting, or accelerating ecological successional processes, depending on the severity of the disturbance. Following mild to moderate natural and anthropogenic disturbances, restoration in these systems involves hastening natural successional path.

3. **Habitat Fragmentation**: describes spatial discontinuities in a biological system, where ecosystems are broken up into smaller parts through land use changes (e.g. agriculture) and natural disturbance [29]. This reduces the size of the populations and increases the degree of isolation. Thus, the smaller and isolated populations are more vulnerable to extinction whereas fragmenting ecosystems decreases quality of the habitat.

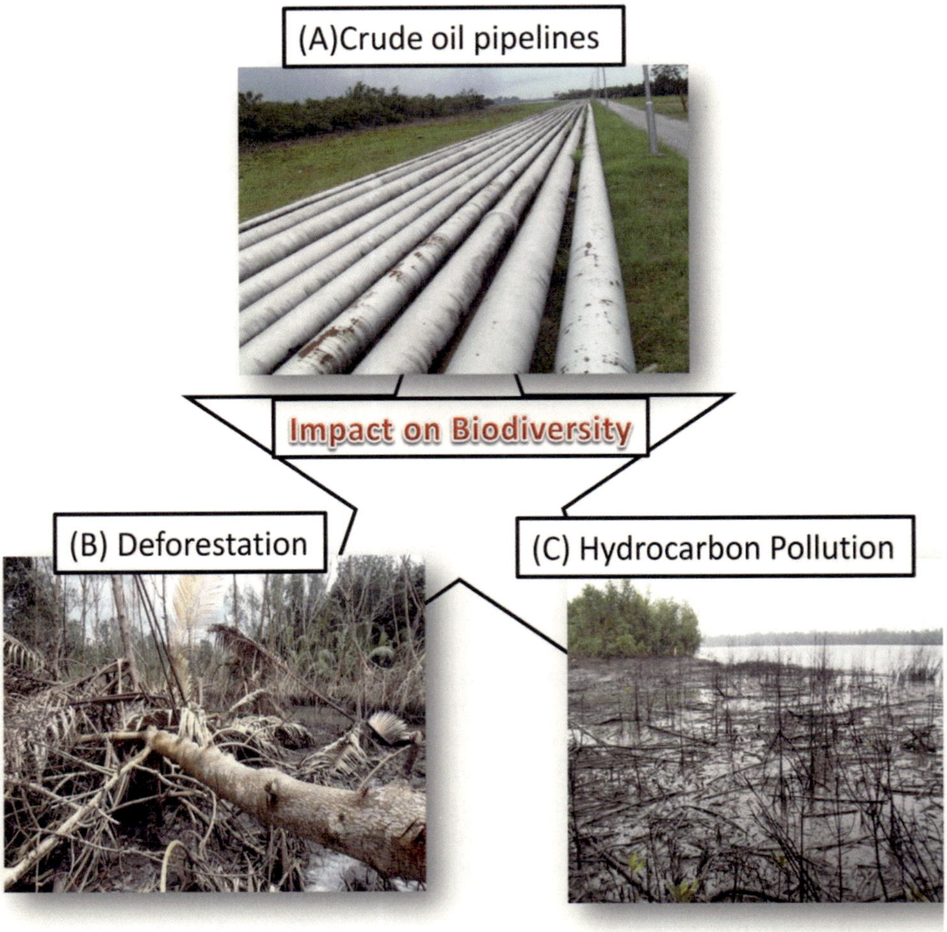

Figure 1.
Picture a: Pipelines' leading from Port Harcourt refinery to Okrika Jetty was taken in 2010. Spillages do occur, which affects neighboring mangrove forest. Picture B: Deforestation of mangrove forest at Okrika refinery Jetty to make way for pipelines right of way (ROW) taken in 2015. Picture C: Massive death of mangrove forest at Abbi Ama, Buguma Asari-Toru local government area of Rivers state following a major oil spillage. The picture was taken in 2010.

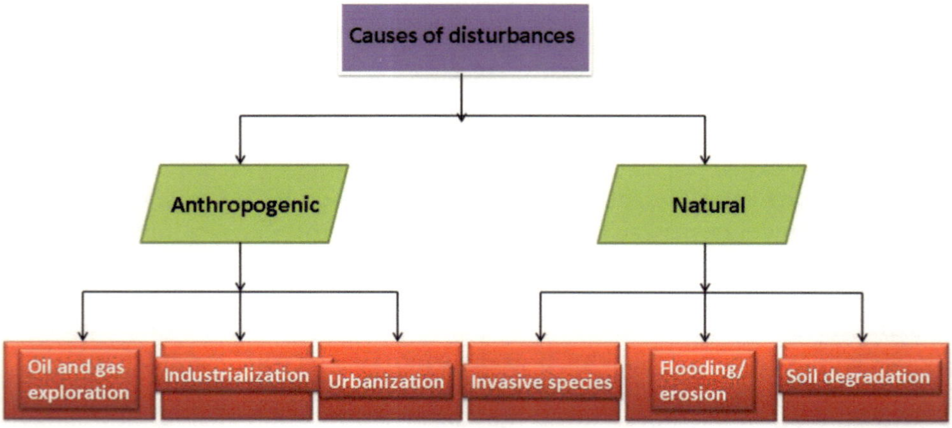

Figure 2.
Causes of disturbances in mangrove forest in the Niger Delta, Nigeria.

4. **Ecosystem function**: describes the most basic and essential foundational processes of any natural systems, including nutrient cycles and energy fluxes [30]. An understanding of ecosystem functions is necessary to address any ecological processes that may be degraded. Ecosystem functions are emergent properties of the system as a whole, thus monitoring and management are crucial for the long-term stability of ecosystems. Mangrove ecosystem functions include three major aspects namely: (1) good and services e.g. timber, fuel, food, medicine and dyes; (2) Environmental and ecological services such as (i) regulatory services e.g. coastal protection, climate regulation and (ii) supporting services e.g. nursery, biodiversity, nutrient cycling and soil formation; (3) Cultural services e.g. spiritual, esthetic, recreational and educational.

5. **Community assembly**: is a framework that can unify virtually all of community ecology under a single conceptual umbrella. Community assembly theory attempts to explain the existence of environmentally similar sites with differing assemblages of species [31]. It assumes that species have similar niche requirements, so that community formation is a product of random fluctuations from a common species pool.

6. **Population genetics**: Genetic diversity has shown to be as important as species diversity for restoring ecosystem processes [32]. Hence ecological restorations are increasingly factoring genetic processes into management practices. Such processes can predict whether or not a species successfully establishes at a restoration site.

7. **Pollution:** is the emission of toxic substances into the environment. Mangroves are impacted by hydrocarbon pollution [33], and this occurs during crude oil spillage from punctured pipelines at both offshore and onshore sites. The crude oil spilled into the water coat the roots of mangroves and suffocates them to death. Oil pollution in mangrove forest lead to the increase in heavy metal concentration, which creates toxic condition and lead to the death of immature mangroves [34].

1.3 Impact of pollution on mangrove restoration and ecosystem services

Pollution prevents propagule germination and growth, and causes mutation of mangrove which results to stunted growth and eventually death [35, 36]. Hydrocarbon pollution increases litter fall via defoliation, which increases the rate of productivity [37]. Pollution also slows, but do not stop the rate of decomposition [38]. Based on the action of pollution on mangrove structure it is as an impairing agent of its ecosystem functions. This is because it impedes the air purification role of mangrove trees because of increase in defoliation. Death of immature trees prevents their use as a source of firewood production, a major source of cooking energy by poor rural people already wallowing in poverty. The role of mangrove as a biodiversity hotspot is affected because of oil spill that kills other organisms that live on, within and around the mangrove forest.

1.4 Mangrove as a bio-remediation agent and pollution inhibitor

The ability of mangroves to survive in difficult terrain goes beyond the natural perturbations (sodium chloride, heat, waves etc) to anthropogenic activities (oil spillage, organic waste, pollutants etc.). This is because mangrove survive polluted environment by using similar means to survive a highly saline environment. This is because studies had revealed that mangroves growing in highly polluted sites have higher productivity than mangroves growing in lowly polluted sites [37]. The growth of mangroves in a polluted soil reduces the pollutant load by accelerating microbial action in the soil through decomposition of litter materials. Soil pollutants absorbed into mangrove parts are locked up and deactivated in guard cells which prevent the pollutants from becoming harmful to the internal organs of the plant just the same way salinity is controlled and eliminated in mangrove cells [39].

1.5 Mangrove resistance to pollution

Mangrove cope with pollution through the following means: (1) by acting as a sink for pollutants, this is because mangroves absorb heavy metals and prevent them from circulating in the ecosystem, (2) through defoliations of leaves that have absorbed pollutants. Here the leaves accumulate pollutants and later fall off to prevent the contamination of the tree [40]; (3) Tough giant root system, the adventitious root system of mangroves grow to a maximum height of above 5 meters from the ground and have only 25% of the root embedded in the swamp and 75% hanging in the atmosphere. This therefore makes oil spill or any pollutant to have limited focus of attack and thus less effect on the tree because of the low root-soil contact. (4) Mangrove root is tough and coated with algal growth, which further provides a layer of protection against external pollutants from the watery environment. This prevents the diffusion of crude oil into the root of mangroves and (5) Tidal flushing, is a process where the tides wipe away oil spills from the forest floor.

Nypa palm (*N fruticans*) is invasive in the Niger Delta because it is a foreign species brought in from Indonesia [41], and over the years they have driven away the mangroves and colonized their territory. *Nypa fruticans* are from the family palmae and have different bio-physical properties from the mangroves, which makes them to have an antagonistic relationship with the mangroves. Currently, nypa palms have limited ecosystem services as compared to the mangroves in the Niger Delta. This is the reason why they are removed from most location in favor of the mangrove forest. Although, there are ongoing research to manufacture manure and life buoy from the palms.

1.6 Mangroves (*Rhizophora* spp) parts as tools for bio-remediation of degraded forest

Mangrove parts (leaf, stem, root and seed) can be used in attenuating pollutants load in the soil. A recent study using ground mangrove parts on polluted soil shows a drastic reduction in oil pollution level (**Table 1**). The preliminary results shows that roots of mangrove and nypa palm performed better than other parts whereas the stem of mangrove had the least remediating effect (Numbere unpublished). Mangrove of the Niger Delta has one of the highest productivity levels in the world [29]. High litter fall causes high microbial activities, which in turn leads to high decomposition rate [38]. This has made the mangrove of the Niger Delta to survive a 50 year period of constant pollution from oil spillages. During major oil spillages (**Table 1**) hydrocarbon pollution suffocates the trees causing death and fragmentation of the forest. In addition mangrove response to stress includes the following (adapted from [42]):

1. Tree mortality
2. Defoliation of canopy
3. Root mortality
4. Bark fissuring/epithelial scarring
5. Development of abnormal adventitious root pneumatophores
6. Leaf deformities and chlorosis
7. Propagule shrinking
8. Alterations in the numbers of lenticels
9. Reduction in tree snail and crab mortality
10. Changes in in-faunal density

Mangrove survives pollution by shutting down pollutants from being absorbed into the root. It also survives by concentrating pollutants in the leaves which are later expelled from the tree via defoliation.

In addition, to pollution, construction or industrial activities carried out by government and private agencies lead to increased deforestation. It causes difficulty in the restoration of the mangrove forests. However, to recover such areas those structures have to be removed by bulldozing the buildings, excavating the soil and replacing them with mangrove swamp soil. Mangrove propagules should then be transferred from the nursery to the restoration sites after two years. The removal of invasive *N. fruticans*, which thrives in disturbed environment, is also important because they are the second most significant threat to mangrove forest after hydrocarbon pollution [33]. The palms grow mainly in fresh water but have adapted to salt water conditions having lived several years in this environment, where they compete effectively with mangroves [41].

Types of oil spills	Land	Swamp	Offshore	Total
Minor spills (1–249)				
Number of spills	457	446	130	1, 033
Quantity of spills (barrels)	7, 565	14, 317	21, 297	43, 179
Medium oil spills (250–2499 barrels)				
Number of spills	596	91	31	712
Quantity of spills (barrels)	17, 203	33, 139	49, 359	99, 701
Major oil spills (over 2500 barrels)				
Number of spills	206	32	16	256
Quantity of spills (barrels)	76, 996	44, 775	1, 379, 2423	1, 921, 013

Source: [43].

Table 1.
Size of spills with respect to ecological zones in Nigeria 1976–1985.

2. Methodology

2.1 Sample collection

1. **Plant parts**: Mangrove *Rhizophotra* (branch, leave, root, seed and stem) and nypa palm (leave, root and seed) parts were retrieved from the forest at Eagle Island (4°43'N and 7 °58′E). These parts were put in polyethylene bags and sent to the laboratory. They were oven-dried at 70 °C for 48 hours and then ground into fine powder by a hand grinding machine. The powdered form of the leaves were bagged and labeled (**Figure 3**).

2. **Soil**: soil samples were collected randomly at ten points with a soil augur 5 cm below the soil surface from a polluted site at Okrika. Some samples of the collected soil were bagged and sent to the laboratory for physicochemical analysis. The soil is then put in 27 (9 plant parts × 3 replicates) seedling containers for the remediation experiment.

Physicochemical analysis of the soil and ground parts were analyzed for Cadmium (Cd), Iron (Fe), Lead (Pb), total hydrocarbon content (THC) and Zinc (Zn).

Remediation experiment: the ground plant parts are applied to the soil surface and monitored for six months with monthly soil samples sent to the lab for physicochemical analysis.

Figure 3.
Remediation experiment using ground parts of mangrove (Rhizophora spp.) and Nypa palm (N. fruticans).

3. Results

The result (**Table 2**) shows that there was no significant difference in heavy metal concentration in the soil samples treated with the ground plant parts ($P > 0.05$).

Metal					Mangrove parts			Nypa palm parts	
	Control	M.branch	M.leaf	M.root	M.seed	M.stem	P.leaf	P.root	P.seed
Cd	0.36 ± 0.01	0.94 ± 0.01	0.04 ± 0.01	0.06 ± 0.01	0.001 ± 0.00	0.4 ± 0.01	0.2 ± 0.01	0.001 ± 0.00	0.001 ± 0.00
Fe	2562.2 ± 0.25	454.2 ± 1.29	419.4 ± 0.46	431.9 ± 0.18	453.2 ± 0.61	436.8 ± 0.48	434.1 ± 0.48	457.6 ± 0.59	421.1 ± 0.45
Pb	12.02 ± 0.24	0.002 ± 0.001	0.001 ± 0.00	0.001 ± 0.00	0.001 ± 0.00	0.002 ± 0.001	0.001 ± 0.00	0.001 ± 0.00	0.001 ± 0.00
THC	607.9 ± 0.6	218.2 ± 2.15	118.3 ± 0.4	6.3 ± 0.11	224.1 ± 0.25	1137.9 ± 0.6	67.8 ± 0.65	15.3 ± 0.35	91.2 ± 0.39
Zn	14.18 ± 0.4	6.79 ± 0.02	6.9 ± 0.22	3.5 ± 0.01	12.2 ± 0.51	6.9 ± 0.15	8.0 ± 0.06	11.8 ± 0.46	5.3 ± 0.01

Where M refers to mangrove and P refers to Nypa palm species.

Table 2.
Concentration of total hydrocarbon content (THC) and heavy metals in soil samples after treatment with ground mangrove and nypa palm parts.

4. Discussion

Heavy metal concentration in mangrove versus nypa palm parts in soil.

The result reveals that there was significant difference in the heavy metal concentration between mangrove and nypa palm parts ($P < 0.05$). Soils treated with ground *N. fruticans* parts have lower concentration of THC and heavy metals as compared to *Rhizophora* species parts. This result revealed that *N. fruiticans* parts remediated the soil better than the *Rhizophora* species parts. *N. fruiticans* parts can therefore be used as a biological remediating agent to clean crude oil contaminated soil. This will be an advantage and a reason for protecting the palms in the long run because they have been seen as having no usefulness in this region.

4.1 Post-remediation of mangrove forest

After the remediation of polluted mangrove forest soils intensive re-planting can be done to recover the devastated forest [44]. This can come through natural or artificial process depending on the nature of the terrain. If it is an area that has the natural setting for recruitment, this strategy can be used. The set up that facilitates natural recruitment are an enclosed coastal channel, connection to an active river with good tidal pressure (i.e. fluctuation of high and low tides) or hydrology, swampy soil that contains soil nutrients such as Nitrates and Iron, nearby parent plants that supplies viable seeds, high litter fall, and high microbial activities. If these conditions are not already set up it can be deliberately established to accelerate the natural process of seedling recruitment as long as it is close to a river. Natural remediation can be facilitated by practically changing coastal structures to create a barrier to trap mangrove seedlings once they are brought in by tidal current to enable the seeds to settle down and grow. An example is the research at Eagle Island, which facilitated mangrove seedling recruitment and growth. But for a non-coastal area such as upland mangroves that are far away from the river natural recruitment will not be possible except artificial recruitment is done where seedlings are grown in nursery and transplanted to the field to facilitate growth. However if the disturbance type was sand mining activity the remediation, recruitment and regeneration methods will vary as given below.

Restoration of Mangrove Forest For a sand dump the sand on the surface has to be scooped away and replaced with swamp soil collected from nearby mangrove forest. After piling the area with mangrove swamp, it should be left for some weeks to settle and consolidate. River connection should be established if it does not exist already so that tidal water will flow in and out to deposit seeds and seedlings from parent trees in the catchment area. In flow of estuarine water will also change the soil chemistry through increase in salinity level. In addition, seedlings can be brought in from the nursery to supplement the ones recruited naturally. Details of the different regeneration methods that can be adopted are given below.

4.2 Mangrove forest regeneration

Forest regeneration is a process by which forest is renewed, and mangrove forest can be regenerated in two ways: (1) Natural and (2) Artificial (**Figure 4**).

1. **Natural regeneration**: This occurs when seedlings of mangroves sprout naturally without human intervention. It involves the provision of suitable environment for the growth and development of volunteer tree species, which are growing in the area. Mangrove forests require suitable environmental condition for them to grow such as high temperature to enhance productivity, precipitation and saline water. Natural regeneration usually occur after

disturbance such as deforestation, dredging, urbanization, clearing of forest to create right of way (ROW) passage for oil and gas pipelines [45]. The success of natural regeneration in mangroves is dependent on some factors, which are:

1. Presence of sufficient numbers of parent trees that would supply enough seeds that would be carried by tides to the regeneration sites.

2. Connecting hydrology that would bring in the seeds

3. Presence of an enclosure containing wire gauze to filter debris, but allow water to flow. Similarly, litter materials in the enclosure trap seeds and prevent them from being flushed away into the open river by tidal currents.

4. The right soil type (swamp locally called "chikoko") and chemistry that would accelerate the growth of the seeds

5. Production of viable seeds by the parent plant that would germinate fast within short period of time.

6. Low population of fiddler crab (*Uca tangeri*) that feed on the seeds

7. Reduction in anthropogenic disturbances in mangrove wetland can lead to the proliferation of fiddler crabs. The action of *U. tangeri* consuming seeds do not entirely affect seed growth negatively, but also helps to redistribute and bury mangrove seeds around the forest, which causes zonation and rapid growth of seeds [46].

Figure 4.
The growth of mangrove seedling can occur through (a) artificial and (b) natural means of regeneration. The natural regeneration is occurring at Eagle Island, Niger Delta, Nigeria.

Advantages of natural regeneration: (i) The forest is established naturally and is not expensive when compared with artificial regeneration that involves finances; (ii) Natural regeneration of the forest do not require site recreation, a forest which naturally recreates will have much forest species; (iii) It brings about the establishment of natural ecosystem in an area of its choice. Natural regeneration does not permit the outbreak of pest and disease epidemic; (iv) It does not require management skill, it is rather based on experience. Mangrove forest restoration is mainly carried out using *Rhizophora* species (i.e. monospecific restoration). This is because of its ability to grow speedily in both nurseries and natural environment. However, in the natural environment many species (e.g. red, white and black mangroves and nypa palm) (See **Table 3**) are carried into restoration sites and grow at different pace depending on their ability to adapt to soil physico-chemical conditions, for instance natural mangrove recruitment site at Eagle Island, Niger Delta (**Figure 4**).

Disadvantage of natural regeneration: (i) It results in lack of uniformity of trees because there are differences in size classes; (ii) Lack of uniformity of stands, so that they cannot be used for suitable purpose such as logging that provides same size stems; (iii) It lacks uniform management e.g. rate of growth, and maturity of each tree is slow.

The development of forest via natural regeneration takes 10 to 20 years to grow to maturity and start fruiting. For the forest to develop quickly there need to be salt water that will facilitate growth, since most mangrove species are halophilic. Furthermore, there need to be adequate soil nutrients such as nitrate, phosphorus, calcium and zinc to aid growth. Mangrove forest is also facilitated by litter decomposition through microbial action that converts organic materials (leaves, seeds, and branches) to soil nutrients.

Artificial regeneration: This is the total replacement of old stand that has been cut down or affected by any form of disturbance with new seedlings, which are deliberately planted in nurseries and later transferred to the field. It involves deliberate establishment of forest trees in remediated polluted site. Direct planting of seedlings on the remediated site can also be done, especially if there is barrier created by swamp embankments to slow down tidal pressure to prevent erosion from carrying away the seeds. Artificial regeneration is used because there is adequate nurturing of seeds for 1 to 2 years to enable them to develop root system so that when planted in the field they will start to grow immediately to withstand environmental changes such as erosion, climate change and pollution. Similarly, during growth in nursery pest can easily be controlled and diseases prevented through the administration of chemicals to mitigate against future attack by

Species	Common name	Abundance	Proportion (p_i)	Ln (p_i)	P_i*Ln (p_i)	H	Rank
R. racemosa	Red mangrove	63	0.0334	−3.399	−0.114	−0.114	3rd
L. racemosa	Black mangrove	1079	0.5721	−0.558	−0.319	−0.319	2nd
A. germinans	White mangrove	709	0.3759	−0.978	−0.368	−0.368	1st
N. fruticans	Nypa palm	35	0.0186	−3.985	−0.074	−0.074	4th
Total		1886					

Table 3.
Abundance and diversity of different species of mangroves and nypa palms in a natural regeneration site at Eagle Island, Niger Delta.

disease-causing insects. Operations of artificial regeneration involve the following: (a) Site preparation such as clearing and removing of thorns, tree stems and other dirt (b) Seedling collection, which is very important in the establishment of a forest (c) nursery practice, i.e. raising the seeds in the nursery or seeds pots or bags. Nurseries are places where seedlings are raised before they are taken to the planting site. Its success lies in the production of adequate number of seedlings of the right quality and fast growing ability.

Two types of nurseries are permanent and temporary nurseries, permanent nursery is meant for large scale, continuous and sustained production of seedlings while temporary nursery is for a short term period of seedling growth.

Advantage of artificial regeneration: (i) There is a high rate of uniformity of the growth of the trees; (ii) Uniformity results in the production of trees very suitable for specific purposes e.g. red mangroves (*Rhizophora* spp) for firewood; (iii) The trees grow and mature faster than in natural regeneration.

Disadvantage of artificial regeneration: (i) It is very expensive and requires a lot of skills; (ii) Environmentally, it brings about change in existing ecosystem particularly in the area of its establishment; (iii) It may result into an outbreak of pest and diseases.

4.3 Species diversity

Species diversity describes the diversity of important ecological entities that span multiple spatial scales from genes to communities. This has to do with species richness and evenness in a specific area. In a second study of species abundance and diversity done at Eagle Island, which measures about 3900 m^2, it was found that black mangroves (*Laguncularia racemosa*) were the most abundant species (n = 1079) followed by white mangroves (*Avicennia germinans*) (n = 709), red mangroves (*Rhizophora racemosa*) (63) and nypa palm (*Nypa fruticans*) (35) (**Table 3**). *A. germinans* had the highest species diversity while *N. fruticans* had the least species diversity.

4.4 Mangrove reserve management

There are three aspects of mangrove management, the ecological, human and the ecosystem.

1. **Ecological management**: This occurs when there is a disturbance like hydrocarbon pollution, deforestation to establish urban areas and sand mining and shortly after these events take their courses through ecological succession. It involves a progressive change of plant growth through the replacement of destroyed mangrove with new mangrove community. An example is successional process of different species of mangrove seedlings including nypa palm colonization of an abandoned sand mining site at Eagle Island, Niger Delta (**Figure 5**).

The four stages of succession to be encountered in the above kind of environment include:

1. Pioneer species (P): This is the first species after a disturbance. It is common 1–5 years after a major disturbance had occurred such as earth quake, flood or volcano. An example of this species is annual plant such as weeds and grasses. Everything is killed leaving behind bare soil, but after a period of time there is a dramatic increase in weed,

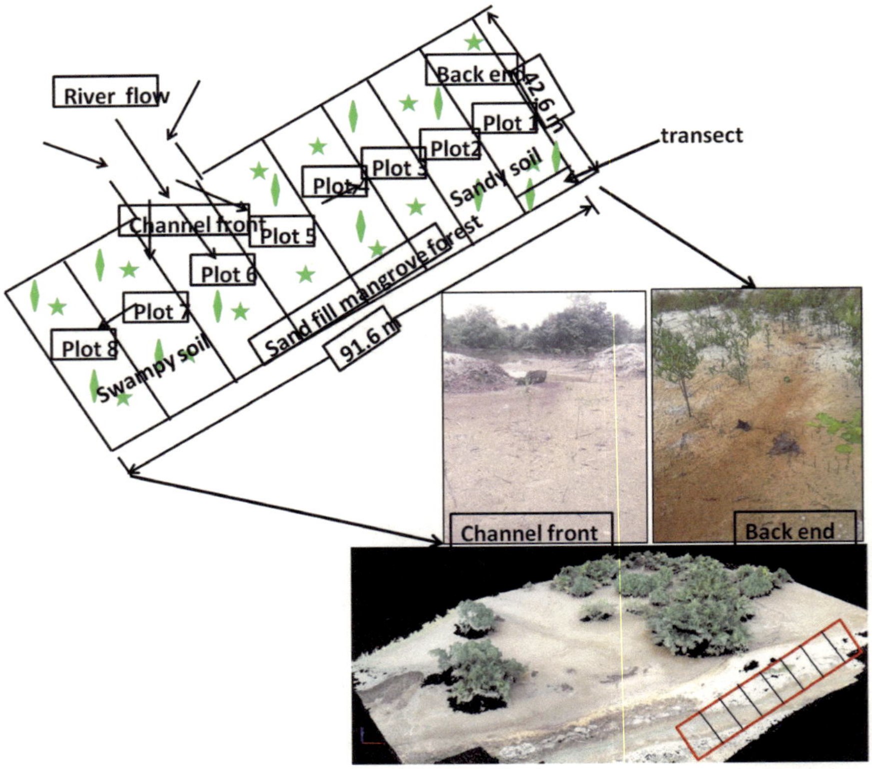

Figure 5.
Plant succession in deforested mangrove forest at Eagle Island, Niger Delta, Nigeria.

2. Early successional species (E): They are a group of species that replace pioneer species. They are dominant for 5—10 years e.g. woody shrub

3. Mid successional species (M): They are large shrubs or small trees that are dominant for 10—30 years. They basically replace their early successional species.

4. Late successional species (L): They are also called climax community. They are tall tree communities that exist in the absence of disturbance.

The cause of the regular progression of change in the mangrove community is seed dispersal, which is facilitated by water. Mangrove takes a long time to mature between i.e. 10—20 years. But the annuals like grasses grow faster and colonize the disturbed site few weeks after disturbance. The mangroves are the top species in the intertidal marine environment because of their ability to withstand the tough environmental conditions created by nature and humans.

4.5 Theoretical basis of succession as a means of restoration of mangrove forests

There is evidence to support the above successional pattern in marine intertidal system where progressive change in disturbed area lead to increase in diversity. One of such hypothesis is the intermediate disturbance hypothesis (IDH). It is an empirical relationship between the time a disturbance occurs and the time species diversity increases in a given location. This is because the climax no longer has the highest diversity. It explains why in some areas we have high diversity while in others we have low diversity. The cause of decline in diversity is competition. Low diversity is also caused by the suppression of early species. The implication for mangrove management is that classical pattern minimizes disturbance to maximize diversity whereas IDH pattern manages disturbance at an intermediate level. Therefore, to identify an IDH system the following should be noted: (a) there will be complete replacement during succession; (b) climax species need to competitively suppress all other successional species, (c) The climax species does not create a new species.

4.6 Regional versus local forest management

In mangrove forest restoration, regional management is better than local site management. This is because disturbance is managed to maximize diversity but make sure all successional stages exist within the management area. Secondly, at the regional level everything is reproducing. Moreover, disturbance does occur at different time and at different places so each one of four successional processes are often dominant. This is done by introducing disturbance in the system in other to maximize diversity. For example, after the Ozark forest is burnt down different species begin to sprout rapidly that were not originally present. This means humans can play a role in natural disturbance to maximize natural biodiversity. In the Niger Delta low level of hydrocarbon pollution i.e. minor crude oil spill (See **Table 1**) does not impact mangrove growth and development, even when deliberately added it will have little or no effect on plant growth. However, in major oil spill it may be harmful to growth by deforming the seedlings or killing them outright. Years of study mangrove forest in Niger Delta has shown that low level introduction of crude oil in mangrove forest could help facilitate the growth of seedling that has resilient qualities, and the elimination of weak species. However, further studies are needed to validate this field observation. This will ensure a long term positive feedback of rejuvenated growth, as recorded in previous findings, where it was discovered that highly polluted sites had higher productivity, species diversity and mangrove tree structure than less polluted site [29]. Although, this revelation has not been thoroughly studies to make a conclusive statement, there are some observations that point to the fact that crude oil pollution in low amount can facilitate mangrove growth in some areas in the Niger Delta. However, the implication is that there might be bioaccumulation of pollutant up the food chain. This suggestion is made because the mangroves of the Niger Delta had been growing in polluted environment for over fifty years without dying. Rather the major killer of mangroves is deforestation through urbanization and fire wood production. It is suspected but not proven that the DNA of this set of mangroves might have been imprinted with "pollution resisting genes" that has made them less vulnerable in the face of high pollution. The complete removal of a disturbance regime is a form of disturbance in itself because we are basically altering the order [47].

Human management: Humans are the ultimate problem of biodiversity especially the mangroves. This is because of their affinity for the mangroves due to the ecosystem services they render. All aspects of the mangroves are useful to humans such as the leaves, stem, root, leaves and seeds, which are used for producing medicinal herbs, fire wood, food etc. Hazardous climatic effect such as earthquake, tsunami, hurricane, cyclone and flood do not affect the mangroves of the Niger Delta rather humans are the greatest threat to the mangroves in this region. The aim of human management is to prevent negative anthropogenic effect on the mangroves by keeping people away from the forest. This is because human activities such as deforestation, logging, and oil and gas exploration are the major threats of mangroves. Since people cause problems for mangroves it is necessary to create zones of use, buffer and transition zones to protect forest resources. The purpose of this method is to allow plants and animals to be protected [32].

Ecosystem management: It is a new way of managing reserve to benefit biodiversity and people. It is a strategy for protecting or restoring the function, structure and species composition of an ecosystem while providing for its sustainable socio-economic use. The tenets of ecosystem management of mangrove forest are as follows:

1. Ecosystems are dynamic- This means they change during succession because they are not static.

2. Ecosystems are subject to unpredictable events or disturbances such as fire, pest and insect attack, crab and animal herbivory, hydrocarbon pollution, earthquake and cyclone, so management need to be flexible in a process called adaptive management.

3. Humans are integral part of the ecosystem

4. Ecosystem requires constant monitoring of populations.

4.7 Sustainable management of mangrove forest

Sustainability means the ability to preserve an environmental resource to last for future generations. Therefore, sustainable management of mangrove forest is the process whereby mangrove forest is managed to last long for the benefit of incoming future generations. Studies have already shown that 5% of mangrove forest has already been lost in the Niger Delta due to oil and gas exploration, urbanization and invasive nypa palm species [48]. To manage mangrove sustainably, those aforementioned key factors that lead to their decimation need to be eliminated. Both onshore and offshore oil and gas exploration results to many cases of oil spillages recorded in the Niger Delta which has devastated large amount of mangrove forest for a period of over a sixty years (1956 to 2020). The rate of oil spillage has to be reduced drastically by the oil companies through the constant maintenance of old pipelines. In the same vein, sabotage of pipelines by local vandals has to be checked to prevent incessant oil spillages. Urbanization is a necessity to modernize the city, but the mangroves areas can be avoided or put into the city plan through urban ecology, where city and forest would exist side-by-side with each other, this will guarantee the survival of the forest. As for the *N. fruticans* they can be removed within the mangrove forest through the use of bulldozers so as to provide breathing space for the mangroves. When all these suggested changes are executed mangrove forest can last for centuries and become beneficial to future generations.

4.8 Mangrove restoration ecology

It is a process of intentionally altering a site to establish a given indigenous historic ecosystem. In this method we try to bring degraded location to what it was originally. There are four ways of accomplishing restoration; three are active method while one is passive method.

1. **Passive method**; It involves the stopping of degradation and allowing succession to occur. This method is good because it is predictable and attains climax. It involves natural succession. Here degraded land can be sealed off

and allowed to recuperate through natural seedling recruitment and regeneration. A classical example of this method is the natural mangrove seedling recruitment at an abandoned sand fill site at Eagle Island (**Figure 4**). Three year (2016—2019) monitoring of this area shows that seedlings that were carried into the site by tidal water had been growing naturally for the past 3 years without human intervention. This method holds great potentials in re-populating many polluted and destroyed sites in the Niger Delta. To achieve this method we need to create the right environmental condition such as establishing water channels, creating an enclosure to trap the seeds from escaping, improving soil chemistry and increasing microbial population of the soil. An example is the natural regeneration of mangroves at Sungei Api-Api, a man-made estuarine channel in Singapore [49]. Factors considered in this study include: the establishment of mangroves include: slope gradients, salinity and tidal inundation levels substrate type, tidal currents and propagule establishment.

2. **Active method**: This is required for areas that have been severely degraded. There are 3 active methods, and these include:

    a. **Replacement** – instead of going back to the original forest, which is impossible a new set of forest or plant community is established as a replacement. For instance, changing mangrove forest to another type of tropical forest (i.e. inter-species replacement) will not be the best option, this is because mangroves are habitat-specific and can only occupy swampy areas. Therefore to restore mangroves, the right soil condition need to be established. This method will work if nypa palm forest is replaced with mangrove forest and soil conditioning carried out. Why it would work out is because both species occupy nearly similar environment. Another kind of replacement that can be done is intra-species replacement where red mangroves (*Rhizophora* spp) are replaced by white (*Avicennia germinans*) or black mangroves (*Laguncularia racemosa*). This one is better than bringing in a completely different species. Mangrove fern (*Acrostichum aureum*) is a species that can grow in disturbed area. It can be planted as a pioneer species in remediated site.

    b. Rehabilitation- In this method an attempt is made to restore the original ecosystem, but it cannot be fully restored because most of the species had gone extinct. Mangrove forest can be rehabilitated after damage by carrying out artificial seedling regeneration or direct planting.

    c. Restoration- It is the attempt to fully restore the original ecosystem. Here the degraded mangrove forest can be restored through the provision of more species to enhance ecosystem structure and function. For instance, mangrove forest in Armacao dos Buzios Brazil was managed by establishing environmental protection unit, education and enlightenment campaigns to support active regeneration [50].

5. Conclusion and recommendation

Mangroves are unique species of plants that are useful to the environment therefore, their protection is important to prevent their extinction as a result of harmful practices such as sand mining, oiling activities, dredging, urbanization etc.

The resilience of mangroves in the face of pollution should not be overestimated because just like other species they have a threshold of resistance against environmental perturbation. And once this limit is surpassed they will become vulnerable to the slightest environmental change. There is therefore a need for repopulating lost mangroves to recover lost stands. Increase in population through natural and artificial means will ensure that mangroves become dominant again in their coastal habitat so as to withstand invasion by foreign species such as *N. fruticans*, which are the most dominant invasive species wreaking havoc to the mangroves. It is recommended that more emphasis should be placed on natural recovery of mangrove forest by deliberately facilitating this process. This can be achieved by the removal of foreign species, the establishment of connecting water channels, soil conditioning and seed transportation to sites of restoration. Finally to accelerate recovery process there can be a combination of natural and artificial seedling recruitment and regeneration methods.

Definition of terms

Replacement: This is to process of bringing in a new set of species in place of a destroyed system.

Rehabilitation: This is the process of fixing the damage caused by disturbance in other to take the system to look like the former system.

Restoration: This is to bring back a disturbed area to almost 100% similar to the previous system.

Author details

Aroloye O. Numbere
Department of Animal and Environmental Biology, University of Port Harcourt, Nigeria

*Address all correspondence to: aroloyene@yahoo.com

References

[1] Diop ES, Gordon C, Semesi AK, Soumaré A, Diallo N, Guissé A, Diouf M, Ayivor JS. Mangroves of Africa. In Mangrove ecosystems. Springer, Berlin, Heidelberg. 2002; 63-121.

[2] Numbere AO. Mangrove Species Distribution and Composition, Adaptive Strategies and Ecosystem Services in the Niger River Delta, Nigeria. Mangrove Ecosystem Ecology and Function. 2018; 7:17.

[3] Donato DC, Kauffmann JB, Murdiyarso D, Kurnianto S, Stidham M, Kanninen M. Mangrove among the most carbon-rich forests in the tropics. Nature Geoscience. 2011; 4(5): 293-297.

[4] Parida AK, Jha B. Salt tolerance mechanisms in mangroves: a review. Trees. 2010; 1; 24(2):199-217.

[5] Rahmania R, Kepel TL, Arifin T. Evaluating the effectiveness of mangroves rehabilitation efforts by comparing the beta diversity of rehabilitated and natural mangroves. In IOP Conference Series. Earth and Environmental Science. 2020; 404(1): 012070.

[6] Day TA, Wright RG. Positive plant spatial association with *Eriogonum ovalifolium* in primary succession on cinder cones: seed-trapping nurse plants. Vegetatio. 1989; 1; 80(1):37-45.

[7] Pugnaire FI, Haase P, Puigdefábregas J, Cueto M, Clark SC, Incoll LD. Facilitation and succession under the canopy of a leguminous shrub, Retama sphaerocarpa, in a semi-arid environment in south-east Spain. Oikos. 1996; 1:455-64.

[8] Maestre FT, Bautista S, Cortina J, Bellot J. Potential for using facilitation by grasses to establish shrubs on a semiarid degraded steppe. Ecological Applications. 2001; 11(6):1641-55.

[9] Gómez-Aparicio L, Valladares F, Zamora R, Luis Quero J. Response of tree seedlings to the abiotic heterogeneity generated by nurse shrubs: an experimental approach at different scales. Ecography. 2005; 28(6):757-68.

[10] Walker LR, Walker J, Del Moral R. Forging a new alliance between succession and restoration. In Linking restoration and ecological succession. Springer, New York, NY. 2007; 1-18).

[11] Brooker RW, Maestre FT, Callaway RM, Lortie CL, Cavieres LA, Kunstler G, Liancourt P, Tielbörger K, Travis JM, Anthelme F, Armas C. Facilitation in plant communities: the past, the present, and the future. Journal of Ecology. 2008; 96(1):18-34.

[12] Gómez-Aparicio L, Zamora R, Gómez JM, Hódar JA, Castro J, Baraza E. Applying plant facilitation to forest restoration: a meta-analysis of the use of shrubs as nurse plants. Ecological Applications. 2004; 14 (4):1128-38.

[13] Ludewig K, Wanner A, Jensen K. Recolonization and facilitation in Baltic salt marsh vegetation. InAnnales Botanici Fennici. Finnish Zoological and Botanical Publishing Board. 2015; 1: 181-191.

[14] Peterson JM, Bell SS. Tidal events and salt-marsh structure influence black mangrove (*Avicennia germinans*) recruitment across an ecotone. Ecology. 2012; 93 (7):1648-58.

[15] Alongi DM.. Mangrove forests: resilience, protection from tsunamis, and responses to global climate change. Estuarine, Coastal and Shelf Science, 2008; 76(1), pp.1-13.

[16] Tomlinson PB. The botany of mangroves. Cambridge University Press; 2016 Oct 27.

[17] McKee KL. Seedling recruitment patterns in a Belizean mangrove forest: effects of establishment ability and physico-chemical factors. Oecologia. 1995; 101(4):448-60.

[18] Lindquist ES, Carroll CR. Differential seed and seedling predation by crabs: impacts on tropical coastal forest composition. Oecologia. 2004; 141(4):661-71.

[19] Elster C, Perdomo L, Schnetter ML. Impact of ecological factors on the regeneration of mangroves in the Ciénaga Grande de Santa Marta, Colombia. In Diversity and Function in Mangrove Ecosystems. Springer, Dordrecht. 1999; 35-46.

[20] Sherman RE, Fahey TJ, Battles JJ. Small-scale disturbance and regeneration dynamics in a neotropical mangrove forest. Journal of Ecology. 2000; 88(1):165-78.

[21] Elster C. Reasons for reforestation success and failure with three mangrove species in Colombia. Forest Ecology and Management. 2000; 131(1-3):201-14.

[22] Imbert D, Rousteau A, Scherrer P. Ecology of mangrove growth and recovery in the Lesser Antilles: state of knowledge and basis for restoration projects. Restoration Ecology. 2000; 8 (3):230-6.

[23] Jackson LL, Lopoukhine N, Hillyard D. Ecological restoration: a definition and comments. Restoration Ecology. 1995; 3(2): 71-75.

[24] Turner RE, Lewis RR. Hydrologic restoration of coastal wetlands. Wetlands Ecology and Management, 1996; 4(2), 65-72.

[25] Lewis III RR. Ecological engineering for successful management and restoration of mangrove forests. Ecological engineering, 2015; 24(4), 403-418.

[26] Wang W, Fu H, Lee SY, Fan H, Wang M. Can Strict Protection Stop the Decline of Mangrove Ecosystems in China? From Rapid Destruction to Rampant Degradation. Forests. 2020; 11(1):55.

[27] Cochard R. Scaling the costs of natural ecosystem degradation and biodiversity losses in Aceh Province, Sumatra. In Redefining Diversity & Dynamics of Natural Resources Management in Asia, Elsevier, 2017; 1: 231-271.

[28] Mitra A. Mangroves: A Barrier Against Erosion. In Mangrove Forests in India. Springer, Cham. 2020; 59-86.

[29] Faridah-Hanum I, Salleh MN. Tertiary Forestry Education Beyond 2020: The Case for Malaysia. Journal of Tropical Forest Science. 2018; 30(5):439-45.

[30] Luque S, Pettorelli N, Vihervaara P, Wegmann M, Vamosi J. Improving biodiversity monitoring using satellite remote sensing to provide solutions towards the 2020 conservation targets. Methods Ecol Evol. 2018; 111:07030-5774.

[31] Mazzei V, Gaiser E. Periphyton, hydrological and environmental data in a coastal freshwater wetland (FCE), Florida Everglades National Park, USA (2014-2015), 2018.

[32] Iuit C, Landy R, Machkour-M'Rabet S, Espinoza-Ávalos J, Hernández-Arana HA, López-Adame H, Hénaut Y. Genetic Structure and Connectivity of the Red Mangrove at Different Geographic Scales through a Complex Transverse Hydrological System from Freshwater to Marine Ecosystems. Diversity. 2020; 12(2):48.

[33] Numbere AO. The impact of oil and gas exploration: invasive nypa palm species and urbanization on mangroves in the Niger River Delta, Nigeria. In *Threats to Mangrove Forests*, 2018; (pp. 247-266). Springer, Cham.

[34] Analuddin K, Sharma S, Septiana A, Sahidin I, Rianse U, Nadaoka K. Heavy metal bioaccumulation in mangrove ecosystem at the coral triangle ecoregion, Southeast Sulawesi, Indonesia. Marine pollution bulletin, 2017; 125(1-2), 472-480.

[35] Veldkornet D, Rajkaran A, Paul S, Naidoo G. Oil induces chlorophyll deficient propagules in mangroves. Marine pollution bulletin, 2020; 150: 110667.

[36] Ellison AM, Farnsworth EJ. Anthropogenic disturbance of Caribbean mangrove ecosystems: past impacts, present trends, and future predictions. Biotropica, 1996; pp.549-565.

[37] Numbere AO, Camilo GR. Structural characteristics, above-ground biomass and productivity of mangrove forest situated in areas with different levels of pollution in the Niger Delta, Nigeria. African journal of ecology, 2018, 56(4): 917-927.

[38] Numbere AO, Camilo GR. Mangrove leaf litter decomposition under mangrove forest stands with different levels of pollution in the Niger River Delta, Nigeria. *African journal of ecology*, 2017, *55*(2): 162-167.

[39] Rabinowitz D. Dispersal properties of mangrove propagules. Biotropica. 1978; 1:47-57.

[40] Connolly RM, Connolly FN, Hayes MA. Oil spill from the Era: Mangroves taking eons to recover. Marine Pollution Bulletin, 2020; 153: 110965.

[41] Numbere AO. The impact of oil and gas exploration: invasive nypa palm species and urbanization on mangroves in the Niger River Delta, Nigeria. In Threats to Mangrove Forests, 2018; (pp. 247-266). Springer, Cham.

[42] Getter CD, Scott GI, Michel J. The effects of oil spills on mangrove forests: A comparison of five oil spill sites in the Gulf of Mexico and the Caribbean Sea. In International Oil Spill Conference, American Petroleum Institute. 1981; 1: 535-540.

[43] Ifeadi CN, Nwankwo JN. 'A critical analysis of oil spill incidents in Nigerian Petroleum Industry,' Proceedings of international seminar on the Petroleum industry and Nigerian Environment.1987; 104-114.

[44] Little DI. Mangrove Restoration and Mitigation After Oil Spills and Development Projects in East Africa and the Middle East. In Threats to Mangrove Forests, 2018; (pp. 637-698). Springer, Cham.

[45] Numbere AO. Impact of invasive nypa palm (*Nypa fruticans*) on mangroves in coastal areas of the Niger Delta Region, Nigeria. In Impacts of Invasive Species on Coastal Environments 2019; (pp. 425-454). Springer, Cham.

[46] Sreelekshmi S, Nandan SB, Kaimal SV, Radhakrishnan CK, Suresh VR. Mangrove species diversity, stand structure and zonation pattern in relation to environmental factors—A case study at Sundarban delta, east coast of India. Regional Studies in Marine Science, 2020; 35, p.101111.

[47] Scatena FN, Blanco JF, Beard KH, Waide RB, Lugo AE, Brokaw N, Silver WL, Haines BL, Zimmerman JK. Disturbance regime. A Caribbean forest tapestry: The multidimensional nature of disturbance and response. Oxford University Press, New York. NY, 2012; pp.164-200.

[48] Wang P, Numbere AO, Camilo GR. Long term changes in mangrove landscape of the Niger River Delta, Nigeria. Am J Environ Sci, 2016; 12, pp.248-259.

[49] Ramos S, Hesp PA. Mangrove Establishment in an Artificially Constructed Estuarine Channel, Sungei Api-Api, Singapore. In Threats to Mangrove Forests, 2018; (pp. 419-427). Springer, Cham.

[50] Obraczka M, Mansur KL, da Silva GC. Towards Sustainability and Protection of Threatened Coastal Ecosystems: Management Strategies for a Rare Stone Mangrove in Gorda Beach, Armação dos Búzios, Brazil. In Threats to Mangrove Forests, 2018; (pp. 377-400). Springer, Cham.

Chapter 2

A Visual Assessment Scale for Rapid Evaluation of Mangrove Degradation, Using Examples from Myanmar and Madagascar

Christoph Zöckler, Dominic Wodehouse and Matthias Markolf

Abstract

Mangroves are globally threatened, disappearing and degraded. They are lost due to land use changes, mostly agricultural expansion and aquaculture, but also degraded by cutting by villagers and logging and timber extraction for domestic and economic purposes. Extent and conversion of mangroves can usually be estimated by applying remote sensing and modern drone technology, but the scale of degradation of mangrove habitats is not easily detected by such methods. In this paper we propose an assessment tool for a rapid evaluation on the degradation, using examples from different regions in Myanmar and Madagascar. We propose a visual and practical guide listing a range of 1–6 to identify and quantify the level of degradation. We demonstrate the application by displaying various examples from Myanmar and Madagascar and how this tool can be used for wider applications, discussing advantages scope, and limitations.

Keywords: Mangroves, Myanmar, Madagascar, degradation, scale, restoration

1. Introduction

Globally mangroves are one of the most threatened ecosystems. In 1980 there were globally 198,000 km^2 of mangroves [1], but by 2003 this had reduced to 154,000km^2 [2]. By 2010, 38% of the global mangrove cover had been lost and for Asia the figure is over 50% [3] and the trend is still continuing [4]. The main drivers are agricultural expansion and aquaculture, while a growing rural population increasingly encroach remaining areas [5]. Moreover, the remaining mangroves are widely subjected to degradation, threatened by legal and illegal logging for domestic and commercial use, consequently reducing the ecosystem services that they provide as summarised for Myanmar [6].

In 2000, Myanmar still had the seventh highest mangrove forest cover in the world, but between 2000 and 2012 had lost mangroves at a much faster rate than almost any other country [5–7]. Myanmar continues to have a relatively high rate of loss of 0.8% per annum (p.a.) in the 21st century [7]. Specifically, 1924–1999, 83% of the mangroves in the Ayeyarwady Delta in Myanmar were cleared [8, 9]. While this central delta area has suffered most of the losses, the southern region of Taninthary still holds vast swathes of pristine mangrove.

Madagascar still holds large areas of mangrove forests, but many of them are also subjected to pressures from a growing local population. In 2013, the total area of mangroves for the country, situated almost exclusively on the West coast, was estimated at 303,000 ha. From 1990 to 2010 Madagascar experienced a net loss of about 21% of its mangroves, a total of 2,868 ha per year [10, 11]. These losses are mainly due to the massive exploitation of mangroves for firewood, charcoal and timber (housing and fencing), the development of aquaculture, cyclones and other causes [11, 12].

However, the rate of loss declined in recent years and globally mangroves have become prime conservation targets [13]. While in the period from 1990 to 2000 there was a net loss of almost 12% (or 34,418 ha) of Madagascar's mangroves, the net loss in the period of 2000–2010 was estimated at 22,941 ha or 8.6%), the most significant of which is in the Tsiribihina Delta (4,177 ha/25.5%) [11]. The mangroves of the area, however, are still one of the largest remaining dense mangroves in Madagascar [10].

Restoration and rehabilitation efforts have largely focused on areas previously covered by mangroves (e.g. Lewis et al., [14]), but little attention has been paid to rehabilitating degraded mangrove areas. It is important to be able to describe degraded mangrove areas that would benefit from improvement activity such as the enhancement of hydrological connectivity and protection measures. Rehabilitation will increase their ability to provide the full range of ecosystem services as well as preserve the whole ecosystem integrity. Therefore, the proposed degradation scale can also provide a reliable and cost-effective methodology to accurately describe mangrove conditions, also in recently restored mangroves.

Despite constantly improving technologies, remote sensing and more recently drone-based surveys, have struggled to depict accurately the condition of mangroves [7, 15]. Although mangrove conversion and deforestation can be reliably monitored using such techniques, mangrove loss is only one indicator of mangrove status. The importance of mangrove degradation has gained considerably less attention [16]. Modern technologies still fail to reveal the scale and the extent of forest degradation and hence poorly describe the state of the remaining forest [10, 17]. While it is acknowledged that there have been great strides in the development of remote sensing and drone/LIDAR capability, this technology will not be available to local NGOs, government mangrove agency field offices and village conservation groups until it becomes much cheaper and simpler.

Therefore, we propose here a rapid assessment tool that is ground- or boat-based, which uses visible features of the mangrove forest structure. This is a simple tool to describe and categorise mangrove forest degradation for Indo-West Pacific non-arid areas, using photographic examples from Myanmar and Madagascar. Comments and suggestions from the mangrove community are welcome to improve this degradation scale.

There is an increasing need to identify the real status of a mangrove, its ecosystem health and the scale of degradation. Degraded mangroves can give a false impression of being superficially healthy but might no longer fully provide the full range of expected ecosystem services, such as the buffering of storm surges, benthic biomass production and others [18].

2. Methodology

The authors visited several different sites between 2013 and 2019 in SE Asia and Africa to assess their conservation status and degree of degradation. The mangroves of Taninthary in southern Myanmar were visited eight times between Dec 2013 and Nov 2019. Mangroves further north on the west coast of Myanmar in the Ayeyarwady Division were surveyed in January and February 2016, [19, 20].

DOI: http://dx.doi.org/10.5772/intechopen.95340

Scale			
1	**Shape**: Very low / very few mangroves. **Height**: <1 m due to possible presence of ferns, herb layer or climbers (e.g. *Acrostichum*, *Acanthus*, *Finlaysonia, photos on right side, which can block natural regeneration.)* **DBH of remaining mangroves**: NA **Logging**: Everything cut. Visible stumps, clearly cut by blade. **Light to Floor**: 100% **Notes**: Reliant on external seed / propagule sources. Ensure area was originally mangrove. Substrate might be eroding, revealing dead mangrove roots. In extreme cases the area might erode so much as to be below an elevation suitable for mangrove growth.		
2	**Shape:** Low. Bushes, establishing seeds and propagules, saplings, regenerating stumps, dead stumps and roots possibly still visible. **Height**: <2 m **DBH of remaining mangroves**: Could be large if species like *Xylocarpus* present, otherwise small (<5 cm DBH) **Logging**: Mangrove has been clear cut of everything but low bushes. Some stumps might be growing back, if species 'coppice', e.g. *Sonneratia*, *Heritiera*, *Avicennia*, *Laguncularia* and *Xylocarpus*, sometimes creating a 'bonsai' effect, lowest left photo. (N.B. *Rhizophora* and *Ceriops* will not grow back as they have no reserve meristem.) **Light to Floor**: 95% **Notes**: Ensure area used to be mangrove. Few or no seeding trees. Reliant on external supply of seeds / propagules.		 

Scale		
3	**Shape**: Low forest. Dense, bushy vegetation. Young trees, saplings. Often a heterogeneous mix of gaps and a few young trees. **Height:** 2-5 m. Very few trees taller than 5 m / 20 cm DBH. **DBH of Remaining Mangroves**: <15 cm DBH. **Logging:** Larger trees (>15 cm DBH) were removed, stumps of which might be visible. Gaps might also have been produced by logged trees damaging neighbouring trees / saplings when felled. **Light to floor**: 25–75% **Notes**: Forest will have a lot of gaps, but is likely to have enough seeding trees to regenerate. (Difficult to depict as similar to level 4, but overall tree height lower and less homogeneous in structure.)	  

Scale			
4	**Shape**: Mangrove clearly defined in the shape of forest with gaps in between few larger trees, but more even canopy than in Level 3. **Height**: > 6-12 m **DBH of Remaining Mangroves**: 20 cm and larger unless a densely stocked plantation. **Logging**: All the very big trees have been logged as well as some of the mid-sized trees. **Light to Floor**: 25–75% **Notes**: In a plantation forest, the majority of trees are in place, but some stems have been removed (<30%). If the planting density was high, there will be little light to the forest floor.	 	
5	**Shape**: High forest with large trees. Plantations will have an even height canopy. Limited understory. Canopy undisrupted. **Height**: > 12 m **DBH of Remaining Mangroves**: Up to 1 m DBH **Logging**: Only a few trees extracted and few cut stumps evident. **Light to Floor**: <25%. Canopy largely closed. **Notes**: Mature mangroves, particularly *Sonneratia* / *Avicennia*-dominated can be naturally quite open forest.		

Scale			
6	**Shape**: Tall forest. Limited to no understory. Continuous cover except for natural disturbance or gaps. **Height**: >12 m. **DBH of Remaining Mangroves**: Up to and over a 1 m. **Logging**: N/A. Trees intact. Very limited extraction. **Light to Floor**: <25%. Canopy largely closed. **Notes**: Likely to have limited understory where canopy is closed. As in 5, areas at the front low zone and back can be naturally quite open, with significant spaces between trees, and tree form very open, e.g. *Avicennia*, *Sonneratia*.		

Table 1.
Mangrove degradation scale 1–6, based on mangrove forest structural features such as shape, height, visible logging, light reaching the mangrove floor and stem diameter of the remaining trees. This scale is not applicable in northern latitudes where cryptic mangrove stands are reaching their limits of range, such as in southern China, North Vietnam, the Red Sea and North Africa. This scale is also not relevant within arid mangrove zones.

Most of the surveys were conducted using small boats, but many mostly non-estuarine mangroves were surveyed on foot and some even accessed by motorbike. These ground surveys were essential to access the interior of mangrove areas [21].

Georeferenced point assessments were conducted using a specifically designed KOBO smart phone app that uses our proposed mangrove degradation scale from 1 (very poor) - 6 (excellent), see **Table 1**. Inevitably the GPS point taken with the smart phone app is likely to be several meters up to 200 m distant from the actual observed mangrove stand providing inaccuracies that can be ignored as they give a rough first assessment of the mangrove nearby. However the GPS points do not allow accurate analysis using remote sensing tools. Where possible, visible additional information on the causes of mangrove loss were noted. The app is designed to be simple and user friendly.

3. Selected examples of application

Myanmar has suffered large losses of mangroves and the remaining forest has been subjected to many pressures. While huge areas have been lost or been

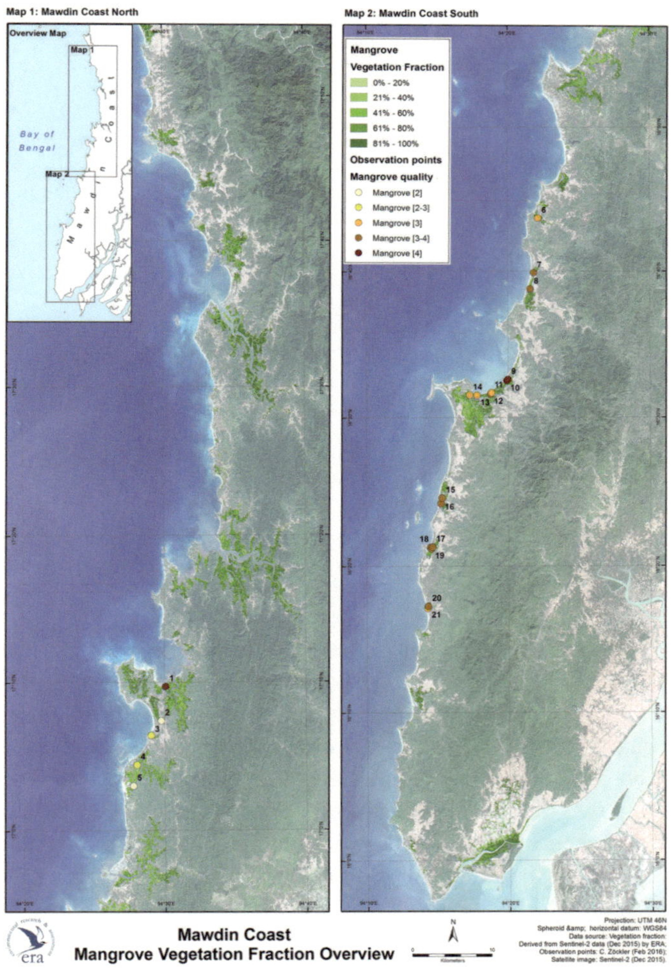

Figure 1.
Mangrove status of non-estuarine mangrove stands on the Mawdin coast in the Ayeyarwady Region on the west coast of Myanmar [22]. Each symbol represents an assessment point (21).

Figure 2.
Mangrove distribution (pink) and status of the mangroves within the Myeik archipelago, Taninthary, Myanmar in 2016, based on our scale with symbols from 2.5 (pale pink) to 5.5 (dark green) and based on 282 assessment points (100 in the northern part and 182 in the southern part, [19]). See also ***Table 2****.*

converted to agricultural land or aquaculture, many of the remaining areas have been heavily degraded by local logging and timber harvesting for building materials. Recently and with increasing severity, mangroves have been extensively harvested for charcoal production [6]. Our degradation scale has been applied to several mangrove sites in Myanmar in 2016 and 2017. **Figures 1** and **2** show the results mapped at two distinct coastal areas. **Figure 1** shows the Mawdin shoreline on the west coast in the Ayeyarwady Division which has only marginal and often small coastal mangrove areas. This region also includes minor areas that have been recently selected for small-scale mangrove restoration. **Figure 2** depicts the mangrove rich region of southern Taninthary, south of Myeik town. These large estuarine mangroves contain mature mangrove stands of well over 150,000 ha. Although most of the mangroves are still in good condition, recent increased usage and harvesting by local communities have left signs of degradation which this rapid assessment tool has depicted.

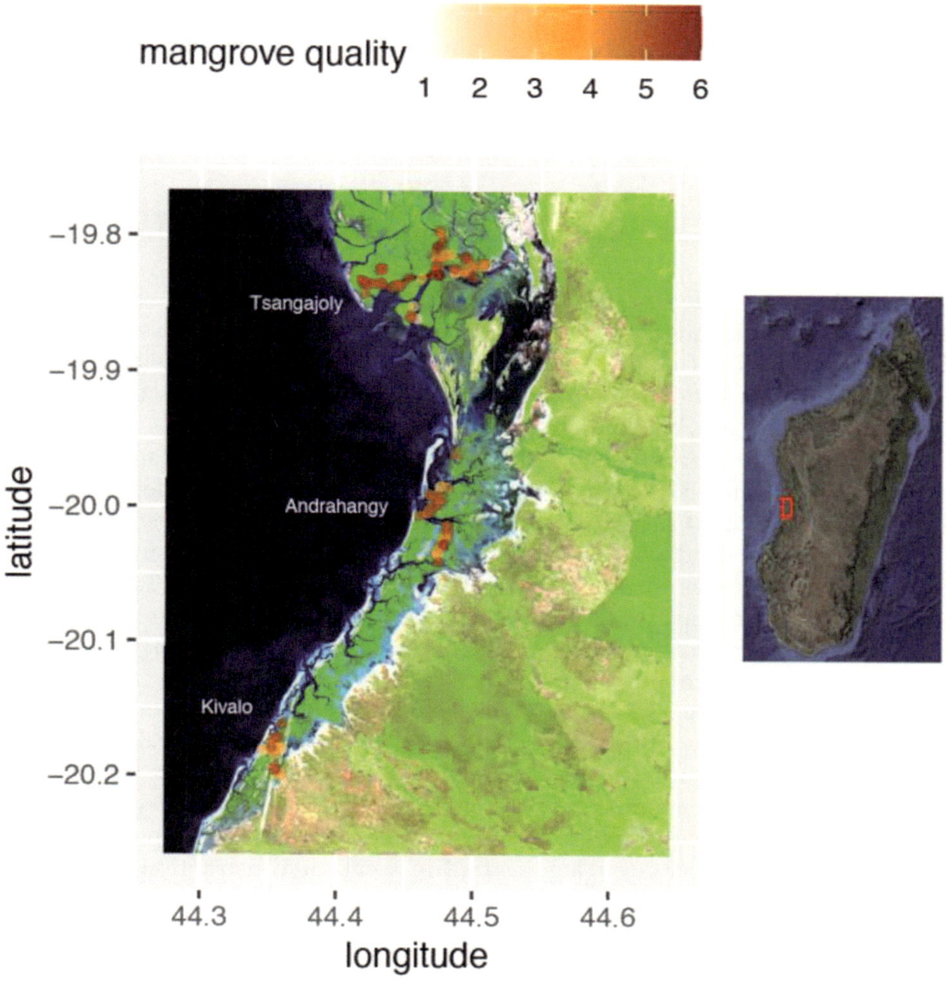

Figure 3.
Left: Landsat 8 image with remaining vegetation and the mangrove degradation status in three sub-regions in the Menabe Antimena protected area, Western Madagascar. A total of 114 assessment points were taken across all three areas. Right: Location (red square) of surveyed area in Madagascar.

No	Site	Average mangrove quality and range of assessments	No of mangrove assessment points
	Myanmar (see Figures 1, 2)		
1	Mawdin Coast, Ayeyarwady Division	3.2 (2.0–4.0)	21
2	North of Myeik, Taninthary	3.4 (3.0–5.5)	100
3	South of Myeik, Taninthary	4.5 (3.5–5.5)	182
	Madagascar (Menabe) (see Figure 3)		
1	Tsangajoly/ Baie de Borongeny	4.0 (3.5–5.0)	58
2	Andrahangy	3.8 (3.0–4.5)	29
3	Kivalo	3.3 (3.0–4.5)	27

Table 2.
Average mangrove degradation (range from 1 = much degraded to 6 = intact, high-quality) at selected sites in Myanmar and Madagascar between 2016 and 2019 [6, 19, 20].

Examples from Madagascar, show similar patterns and demonstrate the value of a scale that can be widely applied across the Indian Ocean. **Figure 3** shows the results of the application of this degradation scale in 2019 on the western coast within the Menabe Antimena Protected Area. **Table 2** shows the average assessment scale as a measure of the overall status of the mangrove quality in each region or sub-region.

4. Summary assessment of selected mangrove areas

Table 2 shows the average degradation levels observed in different coastal regions in Myanmar and Madagascar. The first area on the Mawdin coast was based on a small sample size (n = 21). It suggests a relatively low average of just over 3, reflecting the wide-scale destruction and degradation of mangroves in the region as well as early stages of rehabilitation efforts.

In the Taninthary region, the northern side, closer to the business capital Myeik, appears to have suffered more mangrove losses and disturbances, the degradation is lower with a score of 3.4 than the southern more remote mangroves around Whale Bay and Kan Maw island which averaged over 4.5 (see **Figure 3**). This suggests that the southern mangroves are healthier than the northern mangroves of Taninthary.

Madagascar also displayed differences in mangrove status in the three selected sub-regions (see **Table 2** and **Figure 3**). Kivalo, followed by Andrahangy and Tsangajoly/Baie de Borongeny showed the highest overall degradation. Although this was not specifically tested, it might well be due to higher population densities in the southern areas, which are closer to the biggest regional city of Morondava. All three areas show significant signs of degradation of which most are rather unlikely to be detected using remote sensing methods. Most signs of degradation were spatially associated with local communities depicting increased pressure on the mangroves mainly due to logging for fire wood and construction material. Over-exploitation of mangrove wood in the region by local fisherman for cooking, treatment of fishery products, and construction of boats and houses was already described by Rasolofo [23]. In some surveyed areas, grazing of zebu or goats also present increasing threats to mangroves.

5. Discussion

This simple, rapid degradation assessment tool allows the assessment of the present status and degree of degradation of a mangrove forest, but it also demonstrates the state of forest succession and rate of restoration after intervention and restoration activities have taken place. The tool is applicable over at least the Indo-West Pacific and West-Indian Ocean regions in non-arid situations, where high salinity is not the limiting factor. In the northern margins of the mangrove belt, mangroves develop much smaller 'dwarf' versions, which do not allow the application of the full range of the degradation scale, particularly the assessment of height. We hope that beyond these areas, where similar species at genus level provide comparable forest structures, this assessment tool will also allow comparisons across regions and possibly also for mangroves across the Pacific, Caribbean and South America.

Like any tool this degradation scale approach has its limitations. It only provides a restricted window from the sea front or from a boat, at best within navigable channels or small access roads, excluding large areas of the inner part of the mangroves, which are often, especially in levels 3 and 4, inaccessible on foot. While this is certainly a restriction, this rapid assessment tool is only meant to provide an initial, qualitative assessment of damage by logging and cutting or other degrading activities. We are encouraging assessors to get out of survey boats as much as possible to provide additional survey points on foot. In addition, assessments might be hampered by observer bias or difficulty in allocating local degradation to the appropriate class.

This degradation scale has not been tested and verified, but initial comparisons by different observers using the same locations did not indicate a significant difference in the assessment results. This first draft would benefit from further testing in other mangrove systems including non-deltaic mangroves to develop a more robust scale of degradation. Later on, a combination of this rapid assessment tool together with drone surveys would provide a more accurate scale of degradation and present status of any chosen mangrove forest. Repeated surveys are encouraged as they could reveal changes in the status of a mangrove stand over time. This would be particularly valuable to assess the effectiveness of in-situ protection measures, community forest agreements and active restoration schemes if baseline data is collected before, and then at intervals afterwards. Additionally, it is hoped that the scale can be tested and used on its own by community groups and government mangrove agency field officers to assess and rank their mangroves in order to prioritise rehabilitation and protection measures. Being simple and cheap the proposed rapid assessment tool has major advantages in comparison with remote sensing and LIDAR approaches and could provide substantial benefits to community-based mangrove conservation projects.

The tool offers the identification of degraded areas that have not appeared to be in need for restoration based on superficial consideration or often remote sensing. In addition, the tool can also be applied in recently restored mangroves and plantations and could also provide a good measure for success of restoration projects and activities, whereby the age of the restoration activities needs to be taken into consideration. It also allows comparisons and can point to errors and failures of the restoration efforts and highlights mitigation measures required.

In comparison to deforested mangroves, areas with reversible mangrove degradation represent opportunities for rapid and effective conservation interventions, and thus can substantially facilitate mangrove restoration initiatives [24]. The tool provides rapid and effective identification of sites most suitable for mangrove rehabilitation.

We would welcome input, comments and improvements, including extra photos, particularly from groups that have tried to use this scale. Eventually it will be available for download and printing as well as a smart phone app. It is suggested that a version of it is laminated for use in the mangrove while conducting surveys.

Acknowledgements

We are very grateful for Fauna Flora International (FFI) and the Manfred-Hermsen Foundation who have supported the mangrove surveys financially and logistically and encouraged the development of a rapid assessment tool. In particular we like to thank Frank Momberg (FFI) for his vision and support for coastal mangrove surveys and Mark Grindley (FFI) for his continued support. Patrick Oswald and Milan Fanck helped with the GIS mapping. Surveys in Madagascar were supported by Chances for Nature and the Stiftung Artenschutz.

Author details

Christoph Zöckler[1]*, Dominic Wodehouse[2] and Matthias Markolf[3]

1 Manfred-Hermsen Foundation, Bremen, Germany

2 Mangrove Action Project (MAP), Thailand

3 German Primate Center (DPZ), University of Göttingen, Germany

*Address all correspondence to: christoph.zoeckler@m-h-s.org

References

[1] FAO, 2003. Status and trends in mangrove area extent worldwide, in: Wilkie, M.L., Fortuna, S. (Eds.), Forest Resources Assessment Working Paper No. 63. FAO, Rome, Italy.

[2] FAO, 2007. The world's mangroves 1980-2005. FAO For. Pap. 153, 89. https://doi.org/978-92-5-105856-5

[3] Thomas N, Lucas R, Bunting P, Hardy A, Rosenqvist A, Simard M (2017) Distribution and drivers of global mangrove forest change, 1996-2010. PLoS ONE 12(6): e0179302. https://doi.org/10.1371/journal.pone.0179302

[4] Bryan-Brown, D. N., Connolly, R. M., Richards, D. R., Adame, F., Friess, D. A., & Brown, C. J. (2020). Global trends in mangrove forest fragmentation. Scientific Reports, *10*(1), 1-8.

[5] Richards, D.R. & Friess, D.A. 2016. Rates and drivers of mangrove deforestation in Southeast Asia, 200-2012. PNAS **113** (2): 344-349

[6] Zöckler C., Aung C. (2019) The Mangroves of Myanmar. In: Gul B., Böer B., Khan M., Clüsener-Godt M., Hameed A. (eds) Sabkha Ecosystems. Tasks for Vegetation Science, vol 49. Springer, Cham

[7] Hamilton, S.E., Casey, D., 2016. Creation of a high spatio-temporal resolution global database of continuous mangrove forest cover for the 21st century (CGMFC-21). Glob. Ecol. Biogeogr. 25, 729-738. https://doi.org/10.1111/geb.12449

[8] Ohn, U., n.d. Coastal Resource Management with Special Reference to Mangroves of Myanmar. FREDA.

[9] Webb, E.L, Jachowski, N.R.A, Phelps, J., Fries, D. A., Than ,M.M. & Ziegler, A.D. 2014. Deforestation in the Ayeyarwady Delta and the conservation implications of an internationally-engaged Myanmar. *Global Environmental Change* **24:** 321-333. doi:10.1016/j.gloenvcha.2013.10.007

[10] Jones, T., Glass, L., Gandhi, S., Ravaoarinorotsihoarana, L., Carro, A., Benson, L., Ratsimba, H., Giri, C., Randriamanatena, D., Cripps, G., 2016. Madagascar's Mangroves: Quantifying Nation-Wide and Ecosystem Specific Dynamics, and Detailed Contemporary Mapping of Distinct Ecosystems. Remote Sensing 8, 106. https://doi.org/10.3390/rs8020106

[11] Razakanirina, H. & E. Roger (2013). Mangrove status and management in the Western Indian Ocean Region: Madagascar. WIOMSA. 29p.

[12] Scales, I.R., Friess, D.A., 2019. Patterns of mangrove forest disturbance and biomass removal due to small-scale harvesting in southwestern Madagascar. Wetlands Ecology and Management 27, 609-625. https://doi.org/10.1007/s11273-019-09680-5

[13] Friess, D. A., Yando, E. S., Abuchahla, G. M., Adams, J. B., Cannicci, S., Canty, S. W. & Diele, K. (2020). Mangroves give cause for conservation optimism, for now. Current Biology, 30(4), R153-R154.

[14] Ellison, A. M., Felson, A. J., & Friess, D. A. (2020). Mangrove Rehabilitation and Restoration as Experimental Adaptive Management. Frontiers in Marine Science.

[15] Yong, J., W., H., 2016. An Ecological and Plant Biodiversity assessment of the Meinmahla Kyun Wildlife Sanctuary (MKWS) in relation to biodiversity conservation and restoration, and human livelihood. FFI report. 37 pp.

[16] Worthington, T. A., Andradi-Brown, D. A., Bhargava, R., Buelow, C., Bunting, P., Duncan, C., ... & Lagomasino, D. (2020). Harnessing Big Data to Support the Conservation and Rehabilitation of Mangrove Forests Globally. One Earth, 2(5), 429-443.

[17] Giri, C., Ochieng, E., Tieszen, L.L., Zhu, Z., Singh, A., Loveland, T., Masek, J., Duke, N.C., 2011. Status and distribution of mangrove forests of the world using earth observation satellite data. Glob. Ecol. Biogeogr. 20, 154-159. https://doi.org/10.1111/j.1466-8238.2010.00584.x

[18] Carugati, L., Gatto, B., Rastelli, E., Lo Martire, M., Coral, C., Greco, S., Danovaro, R., 2018. Impact of mangrove forests degradation on biodiversity and ecosystem functioning. Scientific Reports 8. https://doi.org/10.1038/s41598-018-31683-0

[19] Zöckler, C. & Saw Moses. 2016. Bird survey report Eastern Delta and Mawdin Coast, Ayeyarwaddy Region, Myanmar, 18-28 February 2016. Unpubl. Report for FFI

[20] Zöckler, C., 2016. Bird Fauna of the Southern Myeik Archipelago: Report on Historic and New Surveys in the Tanintharyi Coast of Southern Myanmar. Report No. 32 of the Tanintharyi Conservation Programme. Yangon, Myanmar.

[21] Lewis, R. R., Brown, B. M., and Flynn, L. L. (2019). "Methods and criteria for successful mangrove forest rehabilitation," in Coastal Wetlands: An Integrated Ecosystem Approach, 2nd Edn, eds G. M. E. Perillo, E. Wolanski, D. R. Cahoon, and C. S. Hopkinson (Amsterdam: Elsevier), 863-887. doi: 10.1016/B978-0-444-63893-9.00024-1

[22] Harris C., K. Lorenz & C. Zöckler. (2016). Land cover classification, Mawdin Coast, Ayeyarwady Division, Myanmar. Unpubl. Report for Fauna Flora International. 27p

[23] Rasolofo, MV (1997) Use of mangroves by traditional fishermen in Madagascar. *Mangroves and Salt Marshes* 1:243-253.

[24] Worthington, T., and Spalding, M. (2018). Mangrove Restoration Potential: A Global Map Highlighting a Critical Opportunity. https://doi.org/ 10.17863/CAM.39153.

Chapter 3

Secondary Ecological Succession of Mangrove in the 2004 Tsunami Created Wetlands of South Andaman, India

V. Shiva Shankar, Neelam Purti, Ravi Pratap Singh and Faiyaz A. Khudsar

Abstract

Andaman and Nicobar Islands (ANI's) being situated in the Tropical zone is the cradle of multi-disasters viz., cyclones, floods, droughts, land degradation, runoff, soil erosion, shallow landslides, epidemics, earthquakes, volcanism, tsunami and storm surges. Mangroves are one of the first visible reciprocators above land and sea surface to cyclonic storms, storm surges, and tsunamis among the coastal wetlands. The Indian Ocean 2004 tsunami was denoted as one of the most catastrophic ever recorded in humankind's recent history. A mega-earthquake of Magnitude (9.3) near Indonesia ruptured the Andaman-Sunda plate triggered this tsunami. Physical fury, subsidence, upliftment, and prolonged water logging resulted in the massive loss of mangrove vegetation. A decade and half years after the 2004 tsunami, a study was initiated to assess the secondary ecological succession of mangrove in Tsunami Created Wetlands (TCWs) of south Andaman using Landsat satellite data products. Since natural ecological succession is a rather slow process and demands isotope techniques to establish a sequence of events succession. However, secondary ecological succession occurs in a short frame of time after any catastrophic event like a tsunami exemplifying nature's resilience. Band-5 (before tsunami, 2003) and Band-6 (after tsunami, 2018) of Landsat 7 and Landsat-8 satellite respectively were harnessed to delineate mangrove patches and TCWs in the focus area using ArcMap 10.5, Geographic Information Systems (GIS) software. From the study, it was understood that *Fimbrisstylis littoralis* is the pioneering key-stone plant followed by *Acrostichum aureum* and *Acanthus ilicifolius* facilitating *Avicennia spp/Rhizophora spp* for ecological succession in the TCWs.

Keywords: natural disasters, Landsat (7 & 8), satellite image, Short Wave Infra-Red, GIS (Geographic Information Systems), fluvial influx, mangrove biodiversity

1. Introduction

A befitting example of the interaction of Sea, land, and air is the 'coastal frontier'. This Coastal frontier comprises of fragile, sensitive, dynamic, and diverse ecosystems like forests, estuaries, coral reefs, tidal mudflats, salt marshes, seagrass, and mangroves [1, 2]. Mangroves are circum-tropical halophytes representing an ecotone between terrestrial and marine habitats which are adapted to wet and saline conditions having a vital ecological and economic relevance at global, regional, and local scales [3]. These mangrove forests comprise of 65 true mangrove species and 6 hybrids [4], housed in one hundred and twenty-three countries between 32°N and 38°S covering an area of 1.5 million sq. Km [5]. The highest concentration (60%) of global mangrove species (44) are reported from southeast Asia [5, 6]. The mangroves of Andaman and Nicobar Islands (ANI's) represent the third-largest cover on the Indian subcontinent next to Gujarat and Sunderbans respectively [5]. ANI's comprise 38 true mangrove species belonging to 19 genera, and 13 families. Thus, ANI's houses 50% of the global mangrove species [7, 8].

Globally mangrove forests are known as among one of the most productive and biologically important ecosystems because they deliver a variety of vital and distinctive ecosystem goods and services to humankind and other coastal marine ecosystems like the mudflats, coral reefs, seagrass, etc [9]. Since time immemorial mangrove is been conventionally used for firewood, charcoal, alcohol, folk-lore therapeutics, roof thatching [10, 11]. They act as nursery and breeding ground for the juveniles of many commercial fish, crustaceans, including avifauna and reptiles [12–15]. Also, they reduce coastal erosion, stabilize the shoreline, provide sediment and nutrient retention, improve water quality, and provide both flood and flow control as well as protection against storms, hurricanes, and tsunamis [16–21]. Carbon sequestration is presently recognized as the most important service of the mangrove owing to the growing appreciation of the efficacy of these habitats in climate regulation through fixing carbon from the atmosphere [22–24].

The mangrove forests of the world are dwindling at a rate of 1–2% annually and if this trend continues the mangrove and its ecosystem shall be erased from the face of the earth by the 21st century [25–27]. The deterioration of mangrove is more alarming than any other ecosystem like the coral reef and marine forests. At this rate of destruction, the world would be deprived of mangrove and its ecosystem services by the end of this century [28]. The loss of mangrove forests can be attributed to anthropogenic and natural factors. Anthropogenic factors such as dumping of wet and solid wastes generated by the urban population, deforestation, conversion for aquaculture, agriculture, industrial discharge, petroleum spills, the combustion of fossil fuels, automobile exhaust are responsible for the loss of mangrove forests [25, 27, 29–33]. Although the mangrove forest act as a bio-shield against natural disasters such as climate change, cyclones, hurricanes, typhoons, storm surges, and tsunamis [3, 16–21]. On the contrary, these natural factors are also partly responsible for the loss of mangrove forests [34]. However, Mangroves demonstrates the ability to be resilient to natural eventualities [18, 35–40] by following the fluvial influx [39, 41].

The resilience of mangrove is naturally ensured by ecological succession. It is rather a slow process of development and adjustment of species compositions of the mangrove communities over time and space. Further, the ecological succession is dependent on the vital driving factors such as growth potential of the mangrove species, dispersal, settlement, competition, and external or biogenic changes in abiotic conditions [42]. The fluvial influx in the landmass subsided zones due to the 2004 tsunami created a conducive environment for mangrove colonization (ecological succession). Hence, the present study aims at understanding the secondary ecological succession of mangrove in Tsunami Created Wetlands (TCWs) of South Andaman so that it would help in initiating anthropogenically induced massive restoration and rehabilitation of it in the future [6, 28, 43–47].

2. Study area

ANI's is a union territory of India in the Bay of Bengal between peninsular India and Myanmar, trending in a north-south direction. Bounded by the coordinates (92° to 94° East and 6° to 14° North), it is an archipelago with > 500 islands/islets, stretching over 700 km [39]. They are closer to the Indonesian landmass than to mainland India (1200 km), with the southernmost island only 150 km from Sumatra and the northernmost landfall, 190 km south of West Myanmar. ANI's being the cradle of multi-disasters like cyclones, storm surges, earthquakes, and tsunami, the mangroves of this region are vulnerable to disaster. However, nature has its own plans for resilience after any disaster. The present study illustrates the ecological succession of mangrove in south Andaman after the 2004 devastating tsunami. Subsidence and Upliftment of landmass were observed in ANI's due to the 2004 tsunami [48]. Subsidence of landmass around the coastal frontiers rendered it to be permanently waterlogged thus creating wetlands that are very conducive for the mangroves to colonize [37–39]. The area under focus is bounded by the coordinates 11°27′00″and 11°45′00"N and 92°30′00″and 92°46′47″E (**Figure 1**) covering a land area of 333.18 km^2 that encountered destruction from the 2004 tsunami and subsidence as well [48, 49].

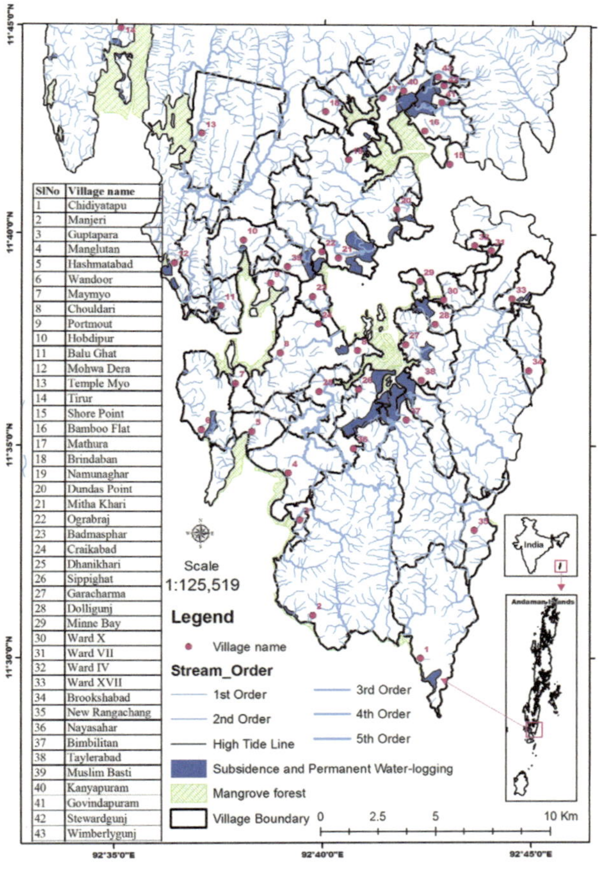

Figure 1.
Study area map showing TCWs with mangrove forest.

2.1 Geology, soils, geomorphology, and drainage

The origin of the Andaman-Nicobar islands is approximately dated as late Pliocene to Pleistocene [50]. The subsidence of landmass is defined by the rock type. Two types of rocks are encountered in the study area viz., (1) Sedimentary rock (Andaman flysch), and (2) Ophiolite suite of volcanic origin [51, 52]. Sedimentary rock comprises of greywacke, siltstone, chalk, limestone are soft and more susceptible to subsidence due to tectonic activity when compared to the Ophiolite suite (**Figure 2a**).

Geomorphogically the study area is dominated by the structural hill, valley trending N-S direction followed by pediments and coastal plains (**Figure 2b**). The coastal plains are dominated by alluvium and colluvium.

The soils of the study area have developed under the dominant influence of vegetation and climate and over diverse parent material. The soil is either present on the hill tops or deposited in the valleys or along the coast as escorted soil. Along the coast, the soil is sandy and contains shingles and old corals, etc. It is extremely porous. In the valley and in the lower slopes of hills, the soil is clayey loam. On the hills, it is rigid clay and dark red loam. There are three orders of soil Entisols, Inceptisols, and Alfisols [53] in six soil texture class viz., Clay, Clay loam, Loamy sand, Sandy, Sandy Clay, Sandy Clay loam. Clay loam is the dominant textural class of soil well distributed throughout the study area followed by clay. Sandy texture was seen along the coastal fringes (**Figure 2c**).

The drainage in the area under investigation exhibits dendritic and trellis patterns a typical structurally controlled drainage pattern of volcanic origin. In general, almost all the drainages are very young and terminate their first or second-order

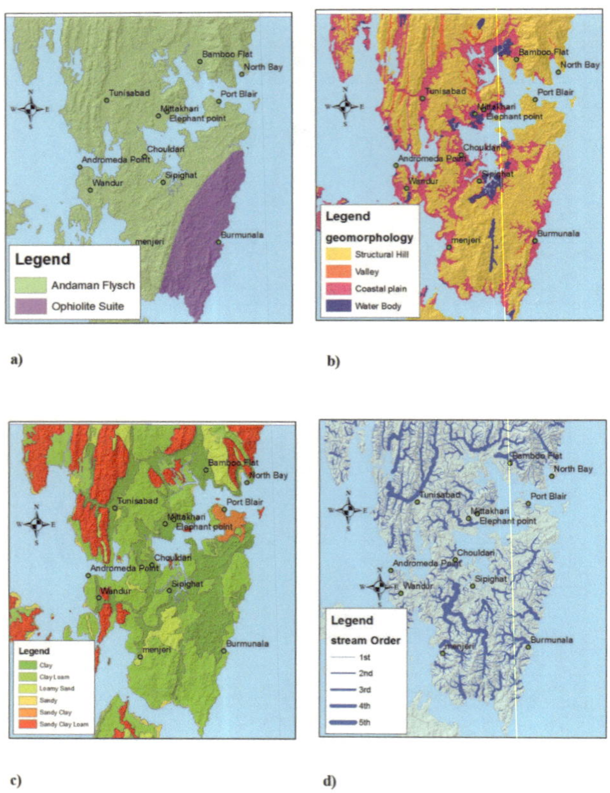

Figure 2.
Maps of (a) Geology, (b) Geomorphology, (c) Soil texture, and (d) Drainage.

stream within a short distance. There are no landlocked watersheds hence all the streams empties into the adjacent sea (**Figure 2d**).

2.2 Meteorology

The study area is situated south of Tropic of Cancer and the region is surrounded by warm seas. The climate of this region is categorized as Warm and Humid. The recorded average temperature ranges from 25°C to 30.5°C. The prevalent temperature along with relatively high humidity gives rise to perceptible and sultry weather. However, this type of weather is moderated with pleasant sea breezes. The relative humidity is high throughout the year reaching > 90 % during the northeast monsoons. The maximum temperature recorded at Port Blair is 32°C. The average annual rainfall is around 3000 to 3500 mm. May to August is the rainiest months and April is the warmest month in this region. It is observed that the South-West monsoon brings in most of the rainfall. During May-June, the onset of the monsoon occurs and in September-October withdrawal of monsoon is observed. The North-East monsoons beginning in November and persists till the end of February. This transitional period is nonetheless disturbed by cyclonic storms which may be accompanied by thundershowers. Most of the storms experienced by the mainland and the area under investigation originate in the Bay of Bengal [54].

3. Conducive environment for mangrove ecosystem

The prevalent geology, soil, geomorphology, drainage system, and climatic conditions in the study area favour the tall and gregarious growth of mangrove flora. The rocks of sedimentary origin are more susceptible to weathering than volcanic rocks. Tropical rains weather the rock material and escort them to the coastal front through the natural drainage system along with abundant freshwater. The climate of any tropical intertidal zone acts as a vital and requisite factor for the natural growth, development, and succession of mangroves. Among these necessary climatic factors are (i) the temperature fluctuation-ranges between 20°C and 30°C [55, 56], (ii) the humidity is of a higher range [57], (iii) the total annual rainfall is above 1000 mm [58], (iv) there is regular wind flow, (v) the area is frost free [59], (vi) radiation and (vii) sedimentation along with upstream water supply plays a very dominant role for the growth and viability of mangrove in a holistic manner [60].

4. Materials and methodology

Landsat (7 & 8) satellite data products before (2003) and after (2018) tsunami respectively, for the study were downloaded from the website (www.earthexplorer.usgs.gov/). The study area is covered by the scene with path (134) and row (52). Mangrove patches and water bodies decipherably picked up very well by band-5 and band-6 by the short-wave infrared (SWIR) sensor of Landsat 7 and 8 satellites respectively from other features like the forest, human settlements, etc. Using ArcGIS Desktop 10.5 software mangrove patches and TCWs were demarcated.

Apart from the demarcation of TCWs, stream networks were delineated from the 1979 Survey of India (SOI) toposheet. An overlay analysis of stream network was comprehended over (1) satellite imageries, (2) geology map, (3) geomorphology map, (4) soil texture map, and (5) village administrative boundary map to understand the source of fluvial Influx dynamics and ecological succession.

Village-wise mangrove stand and TCWs (subsided landmass and permanent waterlogging thereafter) were inferred from before and after tsunami satellite image interpretation. A fishnet grid of 1 km^2 covering mangroves and TCWs was generated with unique ID's and the same was converted into Global Positioning System (GPS) compatible format (*.gpx). These grids were loaded in the handheld Garmin 62CSX, GPS for field investigation. Enumeration of mangrove species was carried out through a 150 m line transect technique [61] with a 50 m interval between each transect within the 1 km^2 grid during the dry season (January-May, 2019 and March-April, 2020). These line transects were laid orthogonal to the coast either ways (land to sea and sea to land). A subplot of 4 m^2 dimension was laid for enumerating individual plants [8]. Mangrove phenology and habitat description were carried out as per Debnath 2004 [62].

5. Results and discussion

Through field survey, a total of twenty-eight mangrove species around existing mangrove and TCWs in forty-three village locations were enumerated and presented in **Table 1**. Also, village-wise pre-tsunami landuse with soil type and the maximum distance from the existing mangrove patch (km) were tabulated in **Table 2**.

Tsunami is rather a rare disaster in the Indian Ocean [63]. A mega-earthquake of magnitude 9.3 on the Richter scale struck near Indonesia On December 26th, 2004 at 07:58:53 local time [64, 65]. The epicenter was located 80km west of the coast of Northern Sumatra (at approximately 95°51' W and 3°25'N). The earthquake advanced thereafter approximately northward rupturing 1200 km to 1300 km (with an average rupture speed of 2.5 to 3 km/s) of the Andaman-Sunda plate in about 8 to 10 minutes [66–68] causing up to ~6 m of bottom subsidence and ~10 m of uplift parallel to the rupture and about100-150 km wide across the subduction area [69]. Upliftment and subsidence of landmass [38] were generated as a consequence of earthquake elastic rebound, offshore of Banda Aceh, the northern tip of Sumatra [70]. Rupture of the plate and coseismic activities spontaneously triggered a tsunami catastrophic devastation ever witnessed in the modern history of humankind [70–73]. All the above sequential events just occurred in a short span of few hours resulting in unprecedented destruction and mangroves were one of the first visible responders of the tsunami [3, 74–77].

Voluminous literature speaks about mangrove demonstrating resilience after a disaster like hurricane, cyclone, and tsunami [18, 35–40, 78, 79]. However, very few studies were conducted to understand the dynamics of the ecological succession of mangroves after natural disasters like hurricanes and tsunamis [80]. The mangroves of the study area faced the double impact of mortality due to 26th December 2004 tsunami viz., (1) physical fury, and (2) prolonged submergence due to subsidence of land mass [38–48]. Zones of subsided landmass were waterlogged permanently resulting in (TCWs). Nudation of mangrove (**Figure 3**) occurred due to a catastrophic event [81].

Overlay analysis of geology geomorphology and stream network of pre-post–tsunami satellite imageries suggest that subsidence of landmass (TCWs) has occurred in the regions of sedimentary rock and on the coastal plains. Sedimentary rocks (Andaman flysh) being soft are more susceptible to deformation due to tectonic activity when compared to volcanic rock (Ophiolite suite). Also, the streams once which were emptying itself in the shallow depths of the coastal frontiers started depositing in the TCWs (**Figures 1, 3**, and **4**). Mineral-rich fine sediments and abundant freshwater were deposited into TCWs through the

Species Name	1*	2*	3	4	5	6*	7	8*	9*	10*	11*	12*	13	14*	15	16	17	18	19*	20*	21*	22*	23	24	25	26*	27*	28*	29	30	31	32	33*	34	35	36**	37**	38**	39**	40**	41**	42**	43**
Acanthus ebracteatus	–	–	–	–	–	–	–	–	–	–	–	–	–	–	–	–	–	–	–	–	–	–	+	+	+	*+	+	+	+	+	–	–	–	–	–	+	+	+	–	–	–	–	–
Acanthus ilicifolius	* +	* +	+	+	+	*+	+	*+	*+	*+	*+	*+	+	*+	+	+	+	+	*+	*+	+	*+	+	+	+	*+	*+	+	+	+	–	–	*+	+	+	+	+	+	+	+	+	+	+
Acrostichum aureum	* +	* +	+	+	+	*+	+	*+	*+	*+	*+	*+	+	*+	+	+	+	+	*+	*+	+	*+	+	+	+	*+	*+	*+	+	+	+	+	*+	+	+	+	+	+	+	+	+	+	+
Aegiceras corniculatum	–	–	–	–	–	–	–	–	–	–	–	–	–	–	–	–	–	–	–	–	–	–	–	–	–	–	–	–	–	–	–	–	+	+	+	–	–	–	–	–	–	–	–
Avicennia marina	* +	* +	+	+	+	*+	+	*+	*+	*+	–	–	+	*+	+	+	+	+	*+	*+	*+	*+	+	+	+	*+	*+	*+	+	+	+	+	*+	+	+	–	+	+	+	+	+	+	+
Avicennia officinalis	* +	* +	+	+	+	*+	+	*+	*+	*+	–	–	+	*+	+	+	+	+	*+	*+	*+	–	+	+	+	*+	*+	*+	+	+	+	+	*+	+	+	+	+	+	+	+	+	+	+
Bruguiera cylindrica	+	+	+	+	+	+	+	+	+	+	+	+	+	+	+	+	+	+	+	+	+	+	+	+	+	+	+	+	+	+	–	–	–	–	–	–	–	–	–	–	–	–	–
Bruguiera gymnorhiza	+	+	+	+	+	+	+	+	+	+	+	+	+	+	+	+	+	+	+	+	+	+	+	+	+	+	+	+	+	+	–	–	–	+	+	–	–	–	–	–	–	–	–
Bruguiera parviflora	+	–	–	–	–	+	+	+	+	+	–	–	+	+	+	+	+	+	+	+	+	+	+	+	+	+	+	+	+	+	–	–	+	–	–	–	–	–	–	–	–	–	–
Ceriops tagal	+	+	+	+	+	+	+	+	+	+	+	+	+	+	+	+	+	+	+	+	+	+	+	+	+	+	+	+	+	+	–	–	+	–	–	–	–	–	–	–	–	–	–
Cynometra iripa	+	–	–	–	–	–	–	–	–	–	–	–	–	–	–	–	–	–	–	–	–	–	–	–	–	–	–	–	–	–	–	–	–	+	+	–	–	–	–	–	–	–	–
Dolichandrone spathacea	+	+	+	+	+	+	+	+	+	+	+	+	+	+	+	+	+	+	+	+	+	+	+	+	+	+	+	+	+	+	–	–	+	+	+	–	–	–	–	–	–	–	–
Excoecaria agallocha	+	+	+	+	+	+	+	+	+	+	+	+	+	+	+	+	+	+	+	+	+	+	+	+	+	+	+	+	+	+	–	–	+	+	+	–	–	–	–	–	–	–	–
Heritiera littoralis	+	+	+	+	+	+	+	+	+	+	+	+	+	+	+	+	+	+	+	+	+	+	+	+	+	+	+	+	+	+	–	–	+	+	+	–	–	–	–	–	–	–	–
Lumnitzera littorea	+	+	+	+	+	–	–	–	–	–	+	+	–	–	–	–	–	–	–	–	+	+	–	–	–	–	–	–	–	–	–	–	–	–	–	–	–	–	–	–	–	–	–
Lumnitzera racemosa	–	–	–	–	–	–	–	–	–	–	–	–	–	–	–	–	–	–	–	–	–	–	–	–	–	–	–	–	–	–	–	–	–	+	+	–	–	–	–	–	–	–	–

Species Name	1*	2*	3	4	5	6*	7	8*	9*	10*	11*	12*	13	14*	15	16	17	18	19*	20*	21*	22*	23	24	25	26*	27*	28*	29	30	31	32	33*	34	35	36**	37**	38**	39**	40**	41**	42**	43**
Nypa fruticans	–	–	–	–	–	–	–	–	–	–	–	–	+	+	+	+	+	+	+	+	+	+	+	+	+	+	+	+	+	+	–	–	+	+	+	–	–	–	–	–	–	–	–
Pemphis acidula	–	–	–	–	–	–	–	–	–	–	–	–	–	–	–	–	–	–	–	–	–	–	–	–	–	–	–	–	–	–	–	–	+	–	–	–	–	–	–	–	–	–	–
Phoenix paludosa	–	–	–	–	–	–	–	–	–	–	+	+	+	+	–	–	–	–	–	–	–	–	–	–	–	–	–	–	–	–	–	–	–	–	–	–	–	–	–	–	–	–	–
Pandanus tectorius	–	–	–	–	–	–	–	–	–	–	–	–	–	–	–	–	–	–	–	–	+	–	–	–	–	–	–	–	–	–	–	–	–	–	–	–	–	–	–	–	–	–	–
Rhizophora apiculata	* +	* +	+	+	+	*+	+	*+	*+	*+	*+	*+	+	*+	+	+	+	+	*+	*+	*+	*+	+	+	+	*+	*+	*+	+	+	+	+	*+	+	+	+	+	+	+	+	+	+	+
Rhizophora mucronata	* +	* +	+	+	+	*+	+	*+	*+	*+	*+	*+	+	*+	+	+	+	+	*+	*+	*+	*+	+	+	+	*+	*+	*+	+	+	+	+	*+	+	+	+	+	+	+	+	+	+	+
Rhizophora stylosa	+	+	–	+	+	+	+	+	+	+	–	–	–	–	–	–	–	–	–	–	–	–	–	–	–	–	–	–	–	–	–	–	+	+	+	–	–	–	–	–	–	–	–
Scyphiphora hydrophylacea	–	+	+	+	+	+	+	+	+	+	+	+	+	+	+	+	+	+	+	+	+	+	+	+	+	+	+	+	+	+	–	–	–	–	–	–	–	–	–	–	–	–	–
Sonneratia alba	+	+	–	+	+	+	+	+	+	+	+	+	–	–	+	+	+	+	+	+	+	+	+	+	+	+	+	+	+	+	–	–	+	+	+	–	–	–	–	–	–	–	–
Sonneratia ovata	+	–	–	+	+	+	+	+	+	+	–	–	–	–	+	+	+	+	+	+	–	–	–	–	–	–	–	–	–	–	–	–	–	+	+	–	–	–	–	–	–	–	–
Xylocarpus granatum	+	+	+	+	+	+	+	+	+	+	+	+	+	+	+	+	+	+	+	+	+	+	+	+	+	+	+	+	+	+	–	–	+	+	+	–	–	–	–	–	–	–	–
Fimbrisstylis littoralis	* +	* +	+	+	+	*+	+	*+	*+	*+	*+	*+	+	*+	+	+	+	+	*+	*+	*+	*+	+	+	+	*+	*+	*+	+	+	+	+	*+	+	+	+	+	+	+	+	+	+	+

"" -Village-wise distribution of Mangroves, "**" -Village-wise distribution of Mangrove in new venues of TCWs, "–" - Indicates absence of Mangrove species, "+" - Indicates presence of Mangrove Species, "*+"- Indicates presence of Mangrove species in TCWs*

Table 1.
Village-wise distribution of mangrove and its associated species.

Sl no	Village name	Soil texture type	Pre-tsunami land use land cover	Max distance from the existing mangrove patch (km)
1*	Chidiyatapu	Clay loam	Agricultural Land & Settlement	1.2
2*	Manjeri	Clay loam	Agricultural Land & Settlement	0.12
3	Guptapara	Clay and loamy sand	—	—
4	Manglutan	Clay loam	—	—
5	Hashmatabad	Sandy clay loam	—	—
6*	Wandoor	Clay	Agricultural Land	1.14
7	Maymyo	Clay	—	—
8*	Chouldari	Loamy sand	Agricultural Land & Settlement	0.55
9*	Portmout	Clay loam	Agricultural Land & Settlement	0.29
10*	Hobdipur	Clay	Agricultural Land & Settlement	0.35
11*	Balu Ghat	Clay loam and sandy clay loam	Open jungle	0.15
12*	Mohwa Dera	Sandy and sandy clay loam	Open jungle	1.5
13	Temple Myo	Clay loam	—	—
14*	Tirur	Clay loam, Loamy sand, clay, sandy clay loam	Agricultural Land & Settlement	0.25
15	Shore Point	Clay loam	—	—
16	Bamboo Flat	Clay loam and clay	—	—
17	Mathura	Clay and clay loam	—	—
18	Brindaban	Clay and clay loam	—	—
19*	Namunaghar	Clay and clay loam	Agricultural Land & Settlement	0.27
20*	Dundas Point	Clay loam	OpenJungle	0.65
21*	Mitha Khari	Clay	Plantation/ Agricultural land	1.5
22*	Ograbraj	Clay and clay loam	Agricultural Land & Settlement	0.22
23	Badmasphar	Clay and clay loam	—	—
24	Craikabad	Sandy and clay loam	—	—
25	Dhanikhari	Clay and clay loam	—	—
26*	Sippighat	Clay and clay loam	Agricultural Land & Settlement	1.01
27*	Garacharma	Clay and clay loam	Agricultural Land & Settlement	0.45
28*	Dolligunj	Clay and clay loam	OpenJungle	1.05
29	Minne Bay	Clay and clay loam	—	—
30	Ward X	Clay loam	—	—

Sl no	Village name	Soil texture type	Pre-tsunami land use land cover	Max distance from the existing mangrove patch (km)
31	Ward VII	Sandy clay	—	—
32	Ward IV	Sandy clay	—	—
33*	Ward XVII	Clay	Agricultural Land	1.01
34	Brookshabad	Clay	—	—
35	New Rangachang	Clay loam	—	—
36**	Nayasahar	Clay and clay loam	Agricultural Land & Settlement	2.5
37**	Bimblitian	Clay and clay loam	Agricultural Land & Settlement	1.2
38**	Taylerabad	Clay and clay loam	Agricultural Land & Settlement	0.87
39**	Muslim Basti	Clay and clay loam	Agricultural Land & Settlement	1.32
40**	Kanyapuram	Clay loam and clay	Agricultural Land & Settlement	0.95
41**	Govindapuram	Clay loam and clay	Agricultural Land & Settlement	1.47
42**	Stewardgunj	Clay loam and clay	Agricultural Land & Settlement	1.91
43**	Wimberlygunj	Clay loam and clay	Agricultural Land	2.02

**Indicates TCWs in the vicinity of pre-existing mangrove.*
***Indicates new venues of TCWs.*

Table 2.
Village-wise soil texture, pre-tsunami landuse pattern and maximum distance from the existing mangrove patch.

conduits of natural streams network (**Figure 4**). Freshly deposited fine sediments are barren and are called as mud banks.

These mud banks in the TCWs were wet, saline, and poorly aerated proves unfavourable for higher plants [82] so, microbes and algae prepare the mud banks for the utilization of higher plants by aerating them [83, 84]. Also, sediments are counteracted by compaction and consolidation of both mud and peat [82]. Coaction of non-woody key-stone species like *Fimbrisstylis littoralis, Acrostichum aureum,* and *Acanthus ilicifolius* subsequently colonized the TCWs (**Figure 5**, **Tables 1–3**). The aforementioned key-stone species were the pioneer plants to colonize the landmass subsided zones thus trapping the sediments and nutrients resulting in the invasion of novel mangrove species [85, 86]. Basically, key-stone species for the initial succession perform the role of nurse plants which start on the bare aerated soil, modifying its conditions like decreasing interstitial salinity and increasing nutrient, enabling the succession of mangroves and can thus be called facilitator species [87, 88]. key-stone species like *Fimbrisstylis littoralis and Acrostichum aureum* were invariably found in all the forty-three sites. Similarly, mangrove species like *Rhizophora* and *Avicennia spp* were also encountered in all the stations. *Pandanus tectorius* and *Pemphis acidula* were found in Mitha Khari and Ward XVII respectively (**Table 1**). Basic soil textures like clay, sand, and loam in different

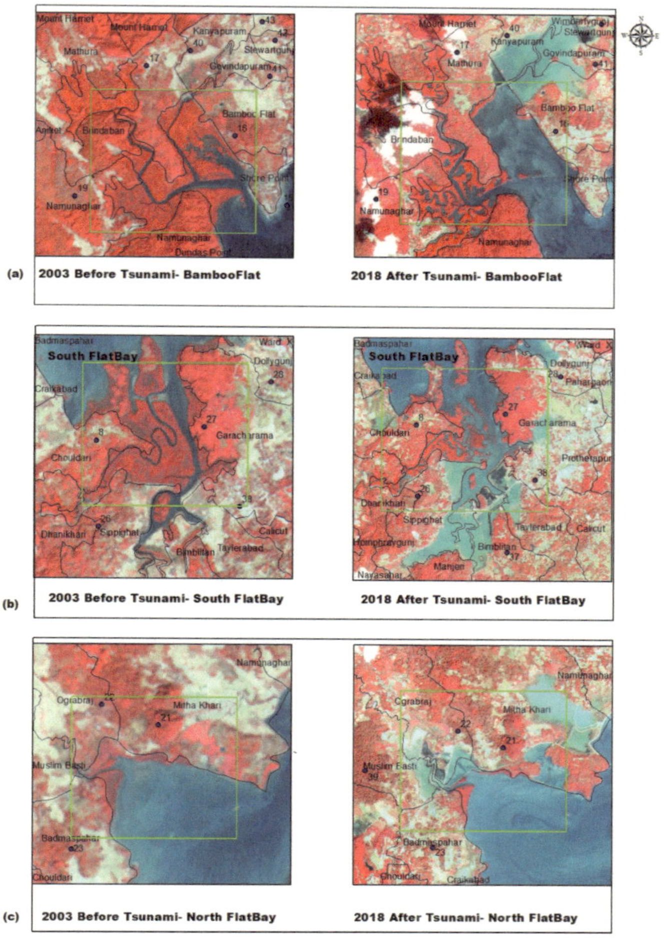

Figure 3.
Satellite image showing before and nudation of mangrove after tsunami (a) Bambooflat, (b) South Flat Bay and (c) North Flat Bay.

combinations were found in the focus area (**Table 2**). The flowering and fruiting phenology along with the habitat descriptions are presented in **Table 3**.

The mangrove seedlings were transported to the TCWs through the tidal influx from pre-existing mangrove (**Tables 1** and **2**) and thus had a stable environment. TCWs being situated in the shallow sheltered bays with low tidal amplitude favours the rooting of propagules [82, 89]. Mangroves follow the existing patterns of fluvial influx and their distribution is determined by the formation of banks, deltas, channels, levees, lagoons, and bays [55, 90–92]. Mangroves respond to geomorphic changes [93, 94] and attain a steady-state system in low energy tropical saline environments [95]. Mangrove succession is a continuous process, where the species recruitment and replacement is systematic and anticipated [96]. It has to be noted here that the likelihood of this phenomenon is of enormous benefit in assessing the evolution towards the climax species complex. The ecological succession from land towards the sea in TCWS in south Andaman is as follows: *Fimbrisstylis littoralis* is the pioneering key-stone plant followed by *Acrostichum aureum* and *Acanthus ilicifolius*. *Avicennia spp/Rhizopara spp* are the prime mangroves to colonize. The ecological succession of mangrove in TCWs are

Figure 4.
Siltation and freshwater influx by natural stream network in TCWs.

considered as secondary ecological succession, which is caused by a natural disaster like the tsunami, subsidence of landmass followed by permanent waterlogging. This type of succession was studied worldwide [35, 37, 38, 80].

6. Conclusion

From the present study, it is understood that secondary ecological succession has occurred in Andaman after the catastrophic 2004 tsunami. Key-stone species like *Fimbrisstylis littoralis, Acrostichum aureum* and *Acanthus ilicifolius* acting as a facilitator species were first to colonize the TCWs and followed by mangrove species like *Avicennia spp/Rhizopara spp.* Infact the key-stone species were the pioneer lower plants to colonize the landmass subsided zones of the 2004 tsunami. The nutrient-rich upstream sediments trapped amongst the roots of the key-stone species provides a conducive environment for the mangrove to colonize. The present study provides a window for anthropogenically induced rehabilitation and restoration of mangrove forests. For any rehabilitation and restoration endeavor of mangrove firstly, the area should be seeded with key-stone species after couple of years mangrove species like *Avicennia spp and Rhizopara spp* has to planted. Thereafter it may take 15–20 years for dense patch of mangrove. A broad avenues for future

Figure 5.
Field photos of key-stone species and mangroves TCWs.

Sl no	Species name	Phenology		Habitait
		Flowering	Fruiting	
1	*Acanthus ebracteatus*	Mar-Jun	Jun-Aug	Common along tidal streams, inland borders of Mangrove swamps under the influence of salt or Brackish water
2	*Acanthus ilicifolius*	Apr-Jun	Jun-Aug	Gregarious in brachish swamps along the seashore and tidal streams
3	*Acrostichum aureum*	NA	NA	Landward side of mangrove, survives in TCWs completely cut off from sea
4	*Aegiceras corniculatum*	Throughout the year		Often found in inner mangroves along with *Bruguiera* spp., *Ceriops* spp., and *Xylocarpus* spp. Also present at landward margin of mangroves inundated during normal high tides and fringing the banks at upstream region.
5	*Avicennia marina*	Apr-Jun	Jun-Aug	Often found in high intertidal and intermediate estuarine position also present in downstream and low intertidal areas. It is a dominant species in highly polluted areas

Sl no	Species name	Phenology		Habitait
		Flowering	Fruiting	
6	*Avicennia officinalis*	Jun-Aug	Aug-Oct	Often found in low and high intertidal position and also occur in mid and upper estuarine position along the banks of the creek.
7	*Bruguiera cylindrica*	Mar-Jun	Jun-Aug	Gregrious on stiff clay behind Avicennia, sometimes in association with Bruguiera gymnorrhiza
8	*Bruguiera gymnorhiza*	Throughout the year		Commonly occur in intertidal zone, along creeks, usually associated with Rhizophora apiculata and R. mucronata
9	*Bruguiera parviflora*	Apr-Jul	Jul-Sep	Occur in intertidal zones of esturaine swamps in association with Bruguriera gymnorrhiza,Rhizophora apiculata and R. mucronata
10	*Ceriops tagal*	Mar-Jul	Jul-Oct	Occur in intertidal banks of mangrove, also in areas nearer to and under esturaine influnce
11	*Cynometra iripa*	Sep-Nov	Dec-Feb	Frequent to back mangrove in the Heritiera littoralis zones
12	*Dolichandrone spathacea*	May-Jun	Jun-Aug	Sporadic occurrence, found around the inner edge ofmangrove swamps in association with Sonneratia caseolaris and Heritiera littoralis
13	*Excoecaria agallocha*	Mar-Jun	Jun-Aug	Occurs in muddy or sandy shores, commonly found in intertidal
14	*Heritiera littoralis*	Mar-Jun	Jun-Aug	Commonly found in intertial zone, frequently extending into muddy or sandy shores
15	*Lumnitzera littorea*	Jan-Apr	Apr-Oct	Occurs in middle zone of mangrove forest, where soulble salts are more
16	*Lumnitzera racemosa*	Jan-Apr	Apr-Jul	Occurs in muddy or sandy elevated zones of esturaine and backwater
17	*Nypa fruticans*	Feb-Jun	Jun-Sep	Sheltered intertidal creeks of mangrove swamps, preferably low saline regions
18	*Pemphis acidula*	Aug-Dec	Dec-Apr	Occur along the edge of mangrove forests
19	*Phoenix paludosa*	Jan-May	May-Aug	Elevated muddy swamps, esturaine banks, can tolerate higher percentage of salt, even found on the sea coast too
20	*Pandanus tectorius*	Sep-Nov	Nov-Marp	Littoral shrub,often found in the tidal forests
21	*Rhizophora apiculata*	May-Jul	Jul-Sep	Occur in intertidal regions of the creek in the in sheltered parts of mangrove
22	*Rhizophora mucronata*	Jul-Sep	Sep-Nov	Occur in intertidal banks of creeks or in estuaries
23	*Rhizophora stylosa*	Jul-Sep	Sep-Nov	Often found in mid to low intertidal and downstream tidal creeks; grows in a variety of habitats and disrupted mangrove areas. One distinctive niche is its ability to grow on edges of small coral islands, establishing on the coral substrate.
24	*Scyphiphora hydrophylacea*	Mar-Aug	Mar-Aug	Along mangrove creeks
25	*Sonneratia alba*	Mar-Jun	Jun-Sep	Occurs along the mouth of tidal creeks, grows on sandy or rocky soil

DOI: http://dx.doi.org/10.5772/intechopen.94113

Sl no	Species name	Phenology		Habitait
		Flowering	Fruiting	
26	*Sonneratia ovata*	Mar-Jun	Jun-Sep	Often occurs on the landward edge of mangrove swamps in brackish water and muddy soil.
27	*Xylocarpus granatum*	Throughout the year		Occur in the sheltered banks in association with Kandelia candel, Rhizophora sp and Sonneratia alba

Pehonology and habitat description is as per Ref. [62].

Table 3.
Phenology of mangrove and it associated species.

research are generated like role of benthic community, avian population, physico-chemical and biological parametric studies, etc., in TCWs.

Acknowledgements

The author would like to acknowledge Dr. G. Narshimulu, guest faculty, Dept. Geography, JNRM, Port Blair, Andaman for the capturing field photos. Also, gratitude's is due to Dr. Manoharan, PG- teacher, Rangat, Middle Andaman for inspiring to write this book chapter.

Field visits and other incidentals were spent from our own coffer and no external funding's were received from either national or international agencies.

Conflict of interest

None.

Author details

V. Shiva Shankar[1]*, Neelam Purti[2], Ravi Pratap Singh[3] and Faiyaz A. Khudsar[4]

1 Department of Coastal Disaster Management, Pondicherry University, Brookashabad Campus, Port Blair, Andaman, India

2 Department of Environment and Forest, Manglutan Range, South Andaman Forest Division, Andaman, India

3 Department of Ocean Studies and Marine Biology, Pondicherry University, Brookashabad Campus, Port Blair, Andaman, India

4 Yamuna Biodiversity Park, CEMDE, University of Delhi, Delhi, India

*Address all correspondence to: shivashankarvj@gmail.com

References

[1] Pereira FRDS, Kampel M, Cunha-Lignon M. Mapping of mangrove forests on the southern coast of São Paulo, Brazil, using synthetic aperture radar data from ALOS/PALSAR. Remote Sensing Letters. 2011;**3**(7):567-576

[2] Gnanappazham L, Selvam V. The dynamics in the distribution of mangrove forests in Pichavaram, South India – perception by user community and remote sensing. Geocarto International. 2011;**26**(6):475-490

[3] Kathiresan K, Rajendran N. Coastal mangrove forests mitigated tsunami. Estuarine, Coastal and Shelf Science. 2006;**65**(3):601-606 https://doi.org/10.1016/j.ecss.2005.06.022

[4] FAO. The world's mangroves 1980-2005. FAO forestry paper. 2007:153

[5] Spalding, M., Kainuma, M., and Collins, F., (2010). World mangrove atlas. Earthscan, London, 319pp. ISBN: 978-1-84407-657-4.

[6] Sharma S, MacKenzie RA, Tieng T, Soben K, Tulyasuwan N, Resanond A, et al. The impacts of degradation, deforestation and restoration on mangrove ecosystem carbon stocks across Cambodia. Science of the Total Environment. 2019. DOI: https://doi.org/10.1016/j.scitotenv.2019.135416

[7] Goutham Bharathi MP, Roy SD, Krishnan P, Kaliyamoorthy M, Immanuel T. Species diversity and distribution of mangroves in Andaman and Nicobar Islands, India. Botanica Marina. 2014;**57**(6):421-432 https://doi.org/10.1515/bot-2014-0033

[8] Ragavan P, Saxena A, Mohan PM, Ravichandran K, Jayaraj RSC, Saravanan S. Diversity, distribution and vegetative structure of mangroves of the Andaman and Nicobar Islands, India. Journal of Coastal Conservation. 2015;**19**(4):417-443 https://doi.org/10.1007/s11852-015-0398-4

[9] Ragavan P, Mohan PM, Saxena A, Jayaraj RSC, Ravichandran K, Saxena M. Mangrove floristics of the Andaman and Nicobar Islands: critical review and current scenario. Marine Biodiversity. 2016 https://doi.org/10.1007/s12526-016-0581-3

[10] Lignon, C.M., Kampel, M., Menghini, R.P., Novelli, S.Y., Cintrón., Guebas, D.F., (2011). Mangrove Forests Submitted to Depositional Processes and Salinity Variation Investigated using satellite images and vegetation structure surveys. Journal of Coastal Research. SI 64, 344-348.

[11] Dodd RS, Ong JE. Future of Mangrove Ecosystems to 2025. In: Polunin NV, editor. Aquatic Ecosystems: Trends and Global Prospects. New York, NY, USA: Cambridge University Press; 2008. pp. 172-287

[12] Robertson AI, Duke NC. Mangroves as nursery sites, comparisons of the abundance of fish and crustaceans in mangroves and other nearshore habitats in tropical Australia. Marine. Biology. 1987;**96**(2):193-205 https://doi.org/10.1007/BF00427019

[13] Twilley RR. The exchange of organic carbon in basin mangrove forests in a southwest Florida estuary. Estuarine, Coastal and Shelf Science. 1985;**20**(5):543-557 https://doi.org/10.1016/0272-7714(85)90106-4

[14] Moran MA, Wicks RJ, Hodson RE. Export of dissolved organic matter from a mangrove swamp ecosystem – evidence from natural fluorescence, dissolved lignin phenols and bacterial secondary production. Marine Ecology Progress Series. 1991;**76**:175-184

[15] Kamal M, Phinn S. Hyper-spectral Data for Mangrove Species Mapping: A Comparison of Pixel-Based and Object-Based Approach. Remote Sensing. 2011; *3*(10):2222-2242 https://doi.org/10.3390/rs3102222

[16] Ewel KC, Twilley RR, Ong JE. Different kinds of mangrove forests provide different goods and services. Global Ecology. Biogeography Letters. 1998;**7**:83-94 https://doi.org/10.1111/j.1466-8238.1998.00275.x

[17] Mumby P, Edwards A, Arlas-Gonzalez J, Lindeman K, Blackwell P, Gall A, et al. Mangroves enhance the biomass of coral reef fish communities in the Caribbean. Nature. 2004;**427**:533-536 https://doi.org/10.1038/nature02286

[18] Alongi DM. Mangrove forests: Resilience, protection from tsunamis, and responses to global climate change. Estuarine, Coastal and Shelf Science. 2008;**76**:1-13 https://doi.org/10.1016/j.ecss.2007.08.024

[19] Nagelkerken I, Blaber SJM, Bouillon S, Green P, Haywood M, Kirton LG, et al. The habitat function of mangroves for terrestrial and marine fauna: A review. Aquatic Botany. 2008; **89**(2):155-185 https://doi.org/10.1016/j.aquabot.2007.12.007

[20] Walters BB, Rönnbäck P, Kovacs JM, Crona B, Hussain SA, Badola R, et al. Ethnobiology, socio-economics and management of mangrove forests: A review. *Aquatic* Botany. 2008;**89**(2):220-236 https://doi.org/10.1016/j.aquabot.2008.02.009

[21] Giri C, Zhu Z, Tieszen LL, Singh A, Gillette S, Kelmelis JA. Mangrove forest distributions and dynamics (1975-2005) of the tsunami-affected region of Asia. Journal of Biogeography. 2008;**35**(3): 519-528 https://doi.org/10.1111/j.1365-2699.2007.01806.x

[22] Murray BC, Pendleton L, Jenkins WA, Sifleet S. Green Payments for Blue Carbon Economic Incentives for Protecting Threatened Coastal Habitats. Durham: Duke University; 2011

[23] Pendleton L, Donato DC, Murray BC, Crooks S, Jenkins WA, Sifleet S, et al. Estimating global "blue carbon" emissions from conversion and degradation of vegetated coastal ecosystems. PloS One. 2012;**7**:e43542 https://doi.org/10.1371/journal.pone.0043542

[24] Cornell, H.A., Susan O. Grose, Pendleton. L., 2018. Mangrove Ecosystem Service Values and Methodological Approaches to Valuation: Where Do We Stand? Frontiers in Marine Sciences https://doi.org/10.3389/fmars.2018.00376.

[25] Valiela I, Bowen JL, York JK. Mangrove forests: One of the world's threatened major tropical environment. Bioscience. 2001;**51**:807-815

[26] Ellison AM. Macroecology of mangroves: Large scale patterns and processes in tropical coastal forests. Trees Struct. Funct. 2002;**16**:181-194

[27] Duke NC, Meynecke J-O, Dittmann S, Ellison AM, Anger AM, Berger U, et al. A world without mangroves. Science. 2007;**317**:41-42

[28] Friess DA, Yando ES, Abuchahla GMO, Adams JB, Cannicci S, Canty SWJ, et al. Mangroves give cause for conservation optimism. *Current Biology*. 2020;***30***(4):R153-R154

[29] IUCN. The impact of climatic change and sea-level rise on ecosystems. London: Report for the Commonwealth Secretariat; 1989

[30] Primavera J. Socio-economic impacts of shrimp culture. Aquaculture Research. 1997;**28**:815-827

[31] Alongi DM. Present state and future of the world's mangrove forests. Environmental Conservation. 2002;**29**: 331-349

[32] Kruitwagen G, Pratap HB, Covaci A, Wendelaar Bonga SE. Status of pollution in mangrove ecosystems along the coast of Tanzania. Marine Pollution Bulletin. 2008;**56**:1022-1031

[33] Cavalcante, M.R., Sousa,W.F., Nascimento, F.R., Nascimento, R.E., Freire, S.S.G., (2009). The impact of urbanization on tropical mangroves (Fortaleza, Brazil): Evidence from PAH distribution in sediments. Journal of Environmental Management 91, 328–335. https://doi.org/10.1016/j.jenvman.2009.08.020.

[34] Gilman E, Ellison J, Duke NC, Field C. Threats to mangroves from climate change and adaptation options: a review. Aquatic Botany. 2008;**89**(2): 237-250

[35] Roth LC. Hurricanes and mangrove regeneration: effects of Hurricane Joan, October 1988, on the vegetation of Isla del Venado, Bluefields, Nicaragua. Biotropica. 1992;**24**:375-384

[36] Sherman RE, Fahey TJ, Battles JJ. Small-scale disturbance and regeneration dynamics in a neo-tropical mangrove forest. Journal of Ecology. 2000;**88**(1):165-178 https://doi.org/10.1046/j.1365-2745.2000.00439.x

[37] Nehru P, Balasubramanian P. Re-colonizing mangrove species in tsunami devastated habitats at Nicobar Islands, India. Check List. 2016; 7(3):253-256 https://doi.org/10.15560/7.3.253

[38] Nehru P, Balasubramanian P. Mangrove species diversity and composition in the successional habitats of Nicobar Islands, India: A post-tsunami and subsidence scenario. Forest Ecology and Management. 2018; **427**:70-77 https://doi.org/10.1016/j.foreco.2018.05.063

[39] ShivaShankar, V., Narshimulu, G., Kaviarasan, T., Narayani, S., Dharanirajan, K., James, R.A., Singh, R. P., (2019). 2004 Post Tsunami Resilience and Recolonization of Mangroves in South Andaman, India. Wetlands. https://doi.org/10.1007/s13157-019-01211-5.

[40] Sopher MF, MacKenzie RA, Sharma S, Cole TG, Litton CM, Sparks JP. Non-native mangroves support carbon storage, sediment carbon burial, and accretion of coastal ecosystems. In: Global Change Biology. 2019 https://doi.org/10.1111/gcb.14813

[41] Xiao H, Su F, Fu D, Qi W, Huang C. Coastal Mangrove Response to Marine Erosion: Evaluating the Impacts of Spatial Distribution and Vegetation Growth in Bangkok Bay from 1987 to 2017. Remote Sensing. 2020;**12**:220 https://doi.org/10.3390/rs12020220

[42] Berger U, Adams M, Grimm V, Hildenbrandt H. Modeling secondary succession of neotropical mangroves: causes and consequences of growth reduction in pioneer species. Perspectives in Plant Ecology, Evolution and Systematics. 2006;7(4):243-252 https://doi.org/10.1016/j.ppees.2005.08.001

[43] Primavera, J.H,, Esteban, J., (2008). A review of mangrove rehabilitation in the Philippines: successes, failures and future prospects. Wetlands Ecology and Management 2008; 16: 345-358.

[44] Lewis RR, Milbrandt EC, Brown B, Krauss KW, Rovai AS, Beever JW, et al. Stress in mangrove forests: Early detection and preemptive rehabilitation are essential for future successful worldwide mangrove forest management. Marine Pollution Bulletin. 2016;**09**:764-771

[45] Feller IC, Friess DA, Ken W, Krauss RR, Lewis III. The state of the

world's mangroves in the 21st century under climate change. Hydrobiologia. 2017 https://doi.org/10.1007/s10750-017-3331-z

[46] Onrizal, Ahmad, A.G., Mansor, M., (2017) Assessment of natural regeneration of mangrove species at tsunami affected areas in Indonesia and Malaysia. IOP Conference. Series: Materials Science and Engineering 180 01204. https://doi.org/10.1088/1757-899X/180/1/012045.

[47] Teutli-Hernández C, Herrera-Silveira JA, Comín FA, López MM. Nurse species could facilitate the recruitment of mangrove seedlings after hydrological rehabilitation. Ecological Engineering. 2017;**130**:263-270 https://doi.org/10.1016/j.ecoleng.2017.07.03

[48] Roy SD, Krishnan P. Mangrove stands of Andamans: Vis-à-Vis tsunami. Current Science. 2005;**89**(11):1800-1804 https://www.jstor.org/stable/24111111

[49] Dharanirajan K, Kasinatha Pandian P, Gurugnanam B, Narayanan RM, Ramachandran S. An integrated study for the assessment of tsunami impacts: a case study of South Andaman Island, India using remote sensing and GIS. Coastal Engineering Journal. 2007;**49**(3):229-266 https://doi.org/10.1142/S0578563407001617

[50] Chibber HL. Geology of Burma. London: Macmillan; 1934

[51] Karunakaran, C., Ray, K.K., Saha, S. S., (1964). Geology of South Andaman Island. Proceedings in 22nd International Geological Congress, India. 11, 79-97.

[52] Karunakaran C, Ray KK, Saha SS. Tertiary sedimentation in the Andaman and Nicobar geosyncline. Journal Geological Society of India. 1968;**9**:32-39

[53] Ganeshamurthy AN, Dinesh R, Ravisankar N, Nair AK, Ahlawat SPS. Land resources of Andaman and Nicobar islands. Andaman and Nicobar, India: CARI publication; 2000

[54] Meteorological Statistics of Andaman & Nicobar Islands, 2019. Directorate of economics and statistics Andaman and Nicobar Administration Port Blair.

[55] West RC. Mangrove Swamps of the pacific coast of Columbia. Annals of the Association of American Geographers banner. 1956;**46**(1):98-121 https://doi.org/10.1111/j.1467-8306.1956.tb01498.x

[56] Blasco, F., (1984). Climate factors and the biology of mangrove plants. In: The Mangrove Ecosystem: Research Method. Snedaker SC, Snedaker JG, (Eds) UNESCO. 18-35.

[57] Davis JH. The Ecology and Geologic role of mangroves in Florida. Carnegie Institution of Washington publication. 1940;**517**:303-412

[58] Galloway, R.W., (1982). Distribution and physiographic patterns of Australian mangroves In: Clough BF (ed): Mangrove Ecosystems in Australia: Structure, Function and Management, Canberra B.F. Australian National University Press 31-54.

[59] Clough BF, Andrews TJ, Cowan IR. Physiological process in Mangrove. In: Clough BF, editor. Mangrove Ecosystems in Australia: Structure, Function and Management, Canberra B.F. Australian National University Press; 1982. pp. 193-210

[60] Venkatesan T. The Mangroves of Madras State. Indian Forester. 1966; **92**(1):27-34

[61] Altamirano, J.P., Primavera, J.H., Banaticla, M.R.N., Kurokura, H., (2010). Practical techniques for mapping small patches of mangroves. Wetlands Ecology and Management 18 (6):707–715, https://doi. org/10.1007/s11273-010-9190-2.

[62] Debnath HS. Mangroves of Andaman and Nicobar Islands: Taxonomy and Ecology (A community profile). Bishen singh Mahendra Pal Singh publication, DehraDun, India. 2004;**248001** 81-211-0420-3

[63] O'Neil, T., (2005). Tsunamis: Where Next? With danger waiting in every ocean, detection is key. Geographica, National Geographic. Available online at: www.nationalgeographic.com/ngm/0504/resources_geo.html (accessed 8 Jan 2016).

[64] Stein S, Okal EA. Size and speed of the Sumatra earthquake. Nature. 2005; **434**:581-582 https://doi.org/10.1038/434581a

[65] Micheal, J, G., David, J, D., (2007). Multi-angle Imaging Spectro Radiometer (MISR) time- lapse imagery of tsunami waves from the 26 December 2004 Sumatra-Andaman earthquake. Remote Sensing of Environment. 107, 256-263.

[66] Ammon CJ, Ji C, Thio H-K, Robinson D, Ni S, Hjorleifsdottir V, et al. Rupture process of the 2004 Sumatra-Andaman earthquake. Science. 2005;**308**:1133-1139 https://doi.org/10.1126/science.1112260

[67] Bilham R, Engdahl R, Feldl N, Satyabala SP. Partial and complete rupture of the indo-Andaman plate boundary 1847-2004. Seismological Research Letters. 2005;**76**(3):299-311 https://doi.org/10.1785/gssrl.76.3.299

[68] Lay, T., Kanamori, H., Ammon, C. J., Nettles, M., Ward, S.N., Aster, R.C., Beck, S.L., Bilek, S.L., Brudzinski, M.R., Butler, R., DeShon, H.R., Ekström, G., Satake, K., Sipkin, S., (2005). The great Sumatra-Andaman earthquakeof 26 December 2004. Science 308:1127–1133. https://doi.org/10. 1126/science.1112250.

[69] Ioualalen M, Asavanant J, Kaewbanjak N, Grilli ST, Kirby JT, Watts P. Modeling the 26 December 2004 Indian Ocean tsunami: case study of impact in Thailand. Journal of Geophysical Research, Oceans. 2007;**112** (C7) https://doi.org/10.1029/2006JC003850

[70] Bilham R. A flying start, then a slow slip. Science. 2005;**308**:1126-1127 https://doi.org/10.1126/science.1113363

[71] Nagarajan B, Suresh I, Sundar D, Sharma R, Lal AK, Neetu S, et al. The great tsunami of 26 December 2004: a description based on tide- gauge data from the Indian subcontinent and surrounding areas. Earth, Planets and Space. 2006;**58**:211-215 https://doi.org/10.1186/BF03353380

[72] Subarya C, Chlieh M, Prawirodirdjo L, Avouac JP, Bock Y, Sieh K, et al. Plate boundary deformation associated with the great Sumatra–Andaman earthquake. Nature. 2006;**440**:46-51 https://doi.org/10.1038/nature04522

[73] Piatanesi A, Lorito S. Rupture process of the 2004 Sumatra–Andaman earthquake from tsunami waveform inversion. Bulletin of the Seismological Society of America. 2007;**97**(1A):S223-S231 https://doi.org/10.1785/0120050627

[74] Danielsen, F., Sørensen, M.K., Olwig, M.F., Selvam, V., Parish, F., Burgess, N.D., Hiraishi, T., Karunagaran, V.M., Rasmussen, M.S/, Hansen, L.B., (2005). The Asian tsunami: a protective role for coastal vegetation. Science310:643. https://doi.org/10.1126/science.1118387.

[75] Dahdouh-Guebas F, Jayatissa LP, Di ND, Bosire JO, Lo SD, Koedam N. How effective were mangroves as a defence against the recent tsunami? Current Biology. 2005a;**15**(12):R443-R447 https://doi.org/10.1016/j.cub.2005.06.008

[76] Dahdouh-Guebas F, Hettiarachchi S, Lo SD, Batelaan O,

Sooriyarachchi S, Jayatissa LP, et al. Transitions in ancient inland freshwater resource management in Sri Lanka Affect Biota and Human Populations in and around coastal lagoons. Current Biology. 2005b;**15**:579-586 https://doi.org/10.1016/j.cub.2005.01.053

[77] Dahdouh-Guebas F, Koedam N, Danielsen F, Sørensen MK, Olwig MF, Selvam V, et al. Coastal vegetation and the Asian tsunami. Science. 2006;**311**: 37-38 https://doi.org/10.1126/science.311.5757.37

[78] Mateo DM, Ruiz Bruce Taylor BR, Rangel-Salazar LJ, Hernández CB. Resilience in a Mexican Pacific Mangrove after Hurricanes: Implications for Conservation-Restoration. Journal of Environmental Protection. 2013;**4**:1383-1391 http://dx.doi.org/10.4236/jep.2013.412159

[79] Das AK, Jha DK, Devi MP, Sahu BK, Vinithkumar NV, Kirubagaran R. Post tsunami mangrove evaluation in coastal vicinity of Andaman Islands, India. Journal of Coastal Conservation. 2014; **18**(3):249-255 https://doi.org/10.1007/s11852-014-0312-5

[80] Ross, M.S., Meeder, J.F., Sah, J.P., Ruiz, P.L., Telesnicki, G.J., (2000) The Southeast Saline Everglades revisited: 50 years of coastal vegetation change. Journal of Vegetation Science 11(1):101–112. https://doi.org/10.2307/3236781.

[81] Bullock EL, Fagherazzi S, Nardin W, Vo-Luong P, Nguyen P, Woodcock CE. Temporal patterns in species zonation in a mangrove forest in the Mekong Delta, Vietnam, using a time series of Landsat imagery. Continental Shelf Research. 2017;**147**: 144-154 https://doi.org/10.1016/j.csr.2017.07.007

[82] Doeke Eisma and Dijkema, K.S., (1997). The influence of salt marsh vegetation on sedimentation. In Doeke Eisma and Dijkema, K. (eds): Intertidal Deposits: River mouths, Tidal flats and Coastal lagoons. 403-435.

[83] Schuster, W.H., (1952). Fish Culture in brackish water ponds of Java. Spec. Publ. Indo Pac Fish Counc., no.1.

[84] Emerit M. Etude granulométrique de la mangrove de Joal, Senegal. Ann. Fac. Sci. University Dakar. 1960;**5**:107-115

[85] Milbrandt EC, Tinsley MN. The role of saltwort (Batis maritime L.) in regeneration of degraded mangrove forests. Hydrobiologia. 2006;**568**(1): 369-377 https://doi.org/10.1007/s10750-006-0203-3

[86] Lewis RR. Successful mangrove Forest restoration informs the process of successful general wetland restoration. National Wetlands News. 2011;**33**:23-25

[87] Lewis, R.R., (1982). Creation and restoration of coastal plant communities. In: Lewis RR (ed) Mangrove forests. CRC Press, Boca Raton, pp 153–172.

[88] Teutli-Hernández C, Herrera-Silveira JA, Comín FA, López MM. Nurse species could facilitate the recruitment of mangrove seedlings after hydrological rehabilitation. Ecological Engineering. 2019;**130**: 263-270 https://doi.org/10.1016/j.ecoleng.2017.07.030

[89] Oliver, J., (1982). The geographic and environmental aspect of mangrove communities: Climate. In: B.F. Clough (ed.). Mangrove Ecosystem of Australia. *Austr. Inst.Mar.Sci.* with A.N.U. Press. 19-30.

[90] Thom BG. Mangrove ecology and deltaic geomorphology: Tabasco, Mexico. J. ecol. 1967;**55**:301-343

[91] Egler F. Southeast Saline Everglades Vegetation, Florida, and its management. Vegetatio. 1950;**3**:213-265

[92] Vann J. Landform-vegetation relationship in the Atrato Delta. Annals of the Association of American Geographer. 1959;**49**:345-360

[93] Odum EP. Fundamentals of ecology, pp xiv, 574. Philadelphia, London, Toronto: W.B. Saunders Company; 1971

[94] Smith, W. G., (1968). Sedimentary Environments and Environmental Change in the Peat forming Area of South Florida. Pennsylvania State University, 254pp.(Ph.D . thesis).

[95] Lugo EA. Mangrove Ecosystem: Successional or Steady State? Biotopica. 1980;**12**(2):65-72

[96] Zedler, J.D., (2000) Handbook of restoring tidal wetlands. CRC, London.

Section 2

Ecosystem Services

Chapter 4

Environmental and Education Trials for Mangrove Ecosystem Rehabilitation in China

Ruili Li, Minwei Chai, Xiaoxue Shen, Cong Shi, Guoyu Qiu and Takayoshi Koike

Abstract

Based on Chinese ecological policy, we have been studying mangrove ecosystems in southern China, especially from the perspective of pollutants deposition in mangrove wetlands, physiological ecology of mangrove species on the impact of heavy metal pollution and seeking ecosystem restoration. For these, we explored in three aspects: 1) pollutants distribution and ecological risk in main distribution of mangrove, China, 2) eco-statistics and microbial analyses of mangrove ecosystems (including shellfish) in representative locations where mangrove plants are well developed, especially in Shenzhen, a rapid developing economic city in Guangdong Province, 3) ecophysiological experiments on a representative species of mangrove for evaluating combination effects of major nutrient elements and heavy metal pollution on growth and physiological responses of the seedlings. Based on the results, we proposed how to rehabilitate mangrove ecosystem in China under rapidly changing environmental conditions, with a view to our future survival and to provide nature-based solution as well as the public with more ecosystem services.

Keywords: ecosystem, wetland, pollution, restoration, subtropical mangroves, nature based solution

1. Introduction

Well-developed mangrove forest in southern China has increased their values of environment, eco-tourism resources, and conservation of biodiversity, etc. [1–3]. Mangrove ecosystems are also expected to provide many ecological services: (1) provisioning, (2) regulation, (3) culture, and (4) basic service [4, 5]. Basic service, i.e., means primary productivity of plants, soil formation and nutrient cycling, etc. The rest services are depending on basic services. Therefore, ecophysiology of mangrove is the most fundamental and essential information for this chapter.

In terms of ecological functions, mangroves can provide (1) many foods including fish, shellfish via offering their habitats, dye materials, wood and materials for high quality charcoal, etc., (2) maintain marsh ecosystems: soil conservation, reduction of storm disasters, wave attenuation, acceleration of reclamation, contaminant degradation, clean the atmosphere and marine environment, (3) eco-tourism, culture, scientific resources, etc., (4) CO_2 fixation and O_2 evolution, biomass production, nutrient circulation [6]. However, an increase of anthropogenic activities in coastal areas reduced mangrove cover and functions, with environment deterioration to be important factor, such as pollution caused by heavy metals. Recent degradation of mangrove functions, such as offering habitat for many living organisms is also reduced by persistent organic pollutants (POPs) including micro-plastics (MPs), etc. [7–9].

In China, large area of mangroves mainly distributed in six provinces (Zhejiang, Taiwan, Fujian, Hainan, Guangxi, and Guangdong), and two special administrative regions (Hong Kong and Macao). There were 37 mangrove species, representing 20 families and 25 genera, with thermophilic eurytopic species being the dominant components [10]. Mangrove could accumulate various pollutants derived from rivers and tidal waters due to its unique properties, such as high productivity, organic-rich matter scrap, fine grains of wetland soil, and anoxic environment [11, 12].

Pollutants like heavy metals and organic contaminants are generally toxic and persistent in mangrove ecosystems. In estuarine mangroves of New Zealand, the soils were characterized with lower Eh and currents upstream trapped more macro-nutrients and heavy metals compared to downstream [13]. In Southeast Sulawesi of Indonesia, mangrove species significantly bioaccumulated heavy metals (such as Cu, Hg, Cd, Zn, and Pb), with different partitioning and uptake capacity of heavy metals to be detected in tissues of mangrove species [14]. Marchand et al. (2011) explored the relationships between heavy metals and organic matter cycling in mangrove sediments of Conception Bay, New Caledonia [15]. Currently, various pollutions caused by anthropogenic activities are well recognized in mangrove in Shenzhen, the most rapidly developed city in China [16, 17]. Especially, Futian mangrove in Shenzhen has been recognized as one of the most typical urban mangroves located in a big city (**Figure 1**), which has been receiving more and more attention. Shenzhen municipal government has decided to keep their environment in order to achieve the sustainable development goals (SDGs) and adopt mangrove as the iconic plant.

Figure 1.
Futian mangrove located in the center of Shenzhen, China (http: //mcf.org.cn/mobile/).

In this chapter, first, we briefly summarized the history of studies in East Asia; then, we take China as an example to explore current progress in mangrove management and research: i.e. heavy metal distribution and ecological risk in mangrove sediment, as well as POPs (such as polybrominated diphenyl ethers). POPs could affect photosynthesis via belowground root vigor and function, and transfer along food chain/web to bio-accumulate [18]. Thirdly, the status of education effect of mangroves in China, especially for Shenzhen was stated. At last, the future perspective of mangrove research in China was provided. These mangrove functions under current and rapid environmental changes in China may provide a hint of conservation and restoration of mangrove ecosystems in the rest of the world.

2. Brief history of mangrove studies in East Asia

In recent years, intensively research have been conducted on the productivities of mangrove forests in tropical and sub-tropical areas of southeast Asian countries, including Indonesia, Philippines, Thailand, and southern Japan, etc. [19–22]. As a contribution to the IBP (International Biological Program), we globally estimated the biomass productivity of different types of vegetation and ecosystem. This IBP is dedicated to human survival from the perspective of biomass production as well as conservation of biodiversity including local people's lives [23–25]. In this project, general empirical models for estimating above-ground biomass, especially universal equation for root biomass was developed (**Figure 2**) [26].

Mangrove plants are characterized by their unique growth characteristics [28]. They can grow and develop along with brackish region of estuaries, wetlands and sea-shores, and are well-known as "marine forest." Furthermore, the growth of mangrove plant would be affected by other climate factors. For example, from dendrometer monitoring of the diameter growth of *Avicennia alba* growing at brackish region, a pioneer species in east Thailand, diameter growth increased with flooding of fresh water during rainy season [26]. In fact, "annual" ring was found in this species even though *A. alba* grows in western Mahachai bay, Thailand (tropical region without any clear dry season) [29] . This species may use intensively fresh

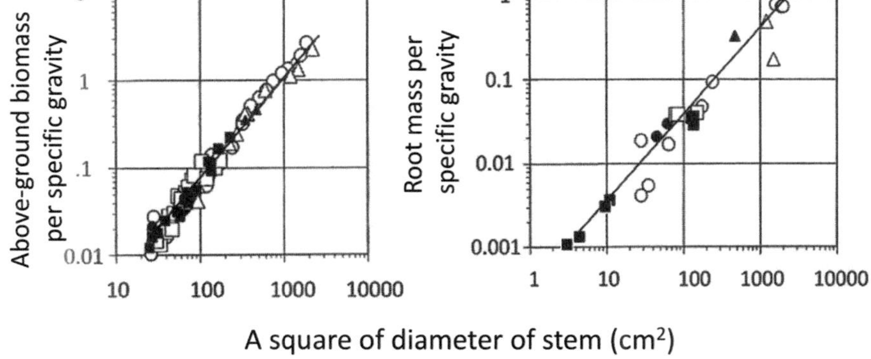

Figure 2.
Universal allometry equation of above- and below-ground biomass of mangrove forests in Asia (Adopted from Komiyama 2017 [26] modified Komiyama et al. [23] and Komiyama et al. [27]. Sampling; ○: Rhizophora *sp. at south Thailand ● n of above- and below-ground bioma △:*Rhizophora *sp. at east Indonesia, ▲: The other species at east Indonesia, □:* Rhizophora *sp. at east Thailand, ■:* Rhizophora *sp. in east Thailand.* $W_{top} = 0.251\ sD^{4.46}$; $W_{root} = 0.199\ sD^{2.22}$.

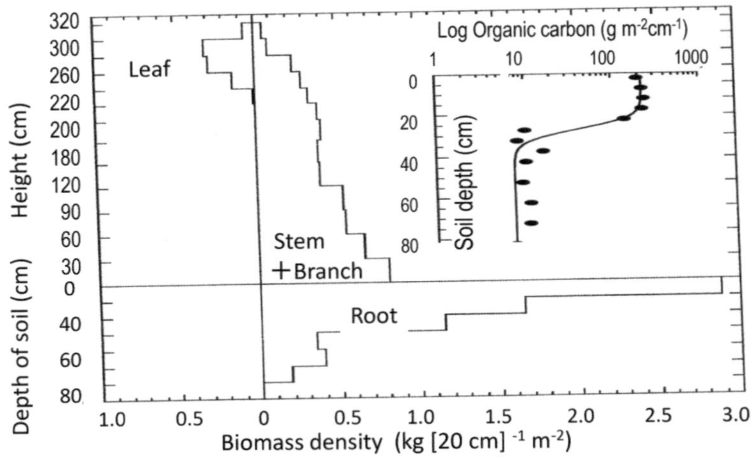

Figure 3.
The stratification diagram of a mangrove stand at the northern boundary in the Ryukyu islands, south Japan (Adopted from Hagihara [35]).

water for its growth. These areas are usually zoned by different species of mangrove depending on their adaptive traits in salt tolerance [30–33] and light utilization characteristics of each species [34]. Mangrove species grow in the front of seashore are light demanding type while a component of well-developed stands is low-light utilization type based on chlorophyll fluorescence.

Hagihara (2006) reviewed production ecology and carbon cycling of mangrove stand (*Kandelia obovata*) in the Ryukyu Islands (**Figure 3**): i.e. northern limit for the distribution of mangrove plants in Japan [35]. This species is very popular in China for ecological study [26, 32]. Hoque et al. (2011) found that Top/Root (T/R) ratio of *K. obovata* in Manko wetland, Okinawa, Japan was around 1.87 [36].

Generally, T/R ratio of mangrove stands was around 1.0 [26, 37] whereas it is ranged between 3 ~ 4 for most forests. From the stratification diagram, estimated value of the extinction coefficient (K_F) including branches for *K. obovata* stand was 0.50. The K_F was 0.56 ~ 0.69 for *Bruguiera gymnorhiza* stands in Ryukyu Islands, 0.54 ~ 0.57 for *Rhizophora apiculate* in Thailand. K_F values of deciduous conifer larch (Larix kaempferi), evergreen oak (*Quercus phillyraeoides*) and Hinoki cypress (*Chamaecyparis obtusa*) were 0.31, 0.36, 0.67 and 0.37, respectively. Thus, compared with other plants, *K. obovata* stand can make good use of incident sunlight in their canopy with their photosynthetic capacity [37].

People recently have again recognized that mangroves have played an important role in preventing Tsunami tide wave after earthquakes in Indonesia, which lead to tsunami in Southeast Asia and cause huge losses and casualties to Southeast Asian Countries, particularly Indonesia [38, 39]. In fact, the mangrove stands along with seacoast of Indonesia protected and lessened destructive power of tide wave at the time [4]. In China, though very few typical examples were available for weaken effect of mangrove on tsunami, the storm prevention of mangrove is one important aspect of mangrove ecosystem services, which have been evaluated to be 10473.3×10^4 RMB in terms of energy value [40].

Recently, due to our concern that global warming will cause rapid rise in sea level, new aspects of salt resistance (tolerance and avoidance) of mangrove species has been studied intensively [33, 41–43]. Sea level rise would increase tidal inundation period and make mangrove species beyond the specific thresholds of flooding tolerance [44, 45]. With intense environmental change, the related knowledge about mangroves have also been systematically summarized through publish of revised editions of botanical books on mangroves as well as ethnobotanical books [31, 32, 46, 47]. On the other hand, researchers recognized the impact of polluted water caused by anthropogenic activities on growth of mangrove plants, without paying much attention to negative impact of heavy metal pollution on mangrove ecosystems [26, 31, 46, 48].

The exploration on heavy metal pollution in mangrove wetlands would understand the source, history, and status of heavy metals, and obtain the relationship between heavy metals and mangrove ecosystem, which is important for coordination between economic development and environment protection. With positive leadership of Chinese ecological policy [49], we have been studying on heavy metal pollution, such as mercury (Hg), cadmium (Cd), copper (Cu), and its counter effects on physiology and growth of the representative mangrove plants dominated around southern China. A typical example of pollution is an intensive study on mangrove ecosystems in Shenzhen city where is located north of Hong Kong, one of the most dramatically developed cities in China. Shenzhen City has decided to adopt mangrove plants as the symbolic trees and well organized ecological and environmental education by establishing mangrove museums and field education parks [7]. Environmental education trail will also be briefly discussed in latter part of this review.

3. Mangrove management and research in China

3.1 Mangrove management

Mangrove species in China belong to the Indo-Malaysia Northeast subgroup of East group and covered >50,000 ha in 1950s [47]. Before 1990s, mangroves in China had been degraded and the areas greatly reduced, with only 22,752 ha remained [47]. Furthermore, mangrove ecological exploitation in China existed many problems, including imbalance between protection and utilization, simple ecological development mode, low economic benefit of ecological development, planning management and related policies and insufficient regional cooperation. Since then, increased government investments have greatly improved the research on mangroves in China.

In 1995, China's Biodiversity Conservation Action Plan included the action plans, which called for "Increasing mangrove conservation areas". As a result, majority of the national mangroves have been protected as a part of the national wide mangrove nature reserves (**Figure 4**). On the announcement of the leader Mr. Xi Jinping, one of the Chinese ecological policies orients us how to conserve mangrove forest as an ecological unit [49]. Based on this statement, conservation of mangrove ecosystem is one of the national key projects, especially at Fujian and Guangdong Province, especially Shenzhen city government, the most rapid developing economical city in China.

Over the past decades, large number of studies have significantly improved our understanding of structure and function of the mangrove ecosystems, however, there are still many areas needed to be strengthened: (1) The construction of sea walls plus many skyscrapers behind natural mangrove wetlands may prevent migration landward into areas of higher elevations in response to sea level rise; (2) Biological invasions such as those of *Spartina alterniflora* may compromise habitats. Their invasive mechanisms and efficient measures for controlling such invasion is still unclear; (3) There is still a lack of universal standard system for evaluating the efforts and achievements of mangrove afforestation and restoration in China; (4) Cooperation among related mangrove research institutions should be strengthened to ensure successful conservation, restoration of mangroves in China.

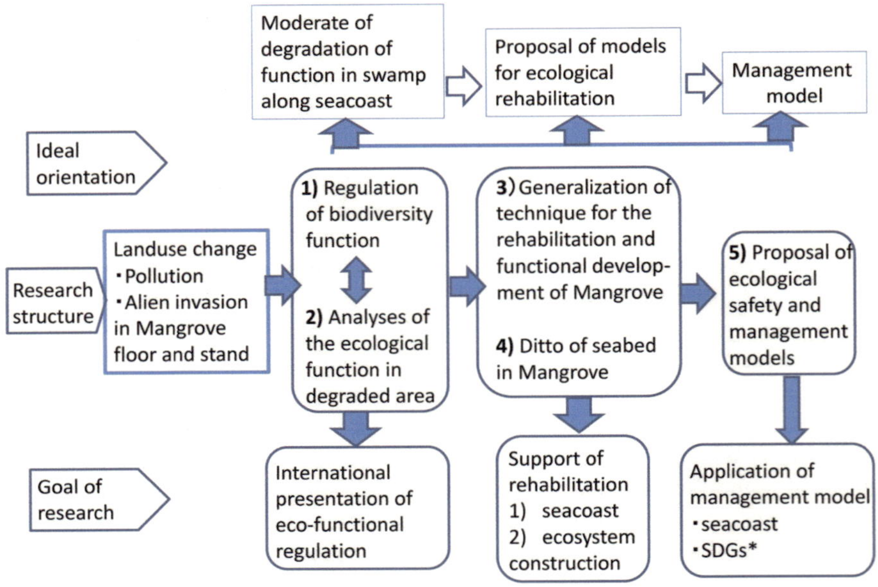

Figure 4.
A Flame work of conservation strategy of mangrove ecosystem in China proposed by Dr. Hailei Zheng at Ningde, China in 2019 [49].

3.2 Mangrove researches

Since 1970s, mangrove researches in China have mainly focused on taxonomy and ethno-botanical view point, medical use and practical use of mangrove plants [6, 50, 51]. According to uncompleted statistic, China researchers have published 24 monographs or proceedings related with mangrove, including comprehensive basic research, ecological restoration, macro-benthos, birds, pest control, and ecological location, remote sensing monitoring and evaluation, resource management, and popular science [52]. The Shenzhen City Office strongly supports researches and education in mangrove ecosystem (see education section).

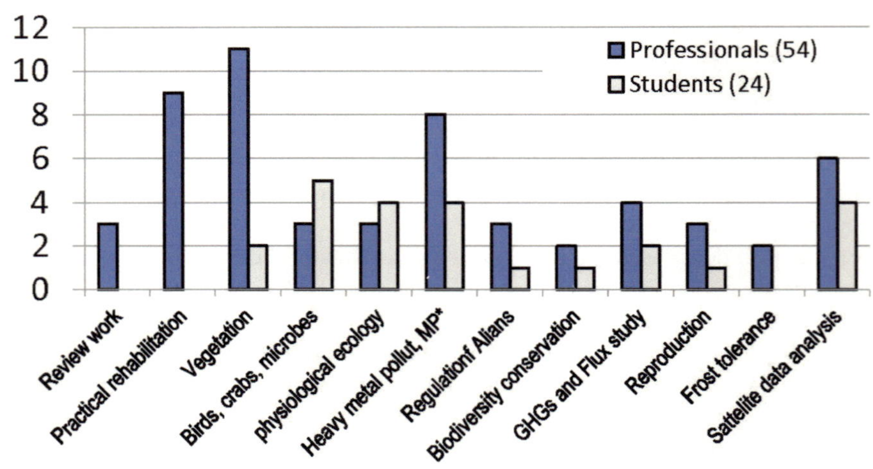

Figure 5.
Research topics on mangrove forests in Chinese Ecological Society (CES) in 2019 (Adopted from program leaflet of the meeting of CES [49, 53]).

Chinese mangrove ecologists hold workshop every two years to make further progress in mangrove researches and practical works. For most researchers, they are engaged in the topics such as remote sensing, biodiversity in mangrove ecosystems, physiological ecology, heavy metal pollutions, etc. These topics are relatively regarded as short-term target compared to those topics more professional (**Figure 5**).

At the 9th workshop in Ningde 2019 (in Fujian Province) [53], the recent ecological efforts on mangrove conservation in accordance with the SDGs (Sustainable developmental goals) was summarized as follows (**Figure 6**) [49].

The contents were: i.e. policy making, scientific review, practical forestry, and frost resistant researches, which were mainly reported by national and regional research institutes. Among them, one unique research is how to increase mangrove's frost tolerance and freezing avoidance research (**Figure 7**) [56, 57]. In 2010, sudden snow-fall caused death of newly planted mangroves because they were originally grown in sub-tropical and tropical, where there would not be a severe

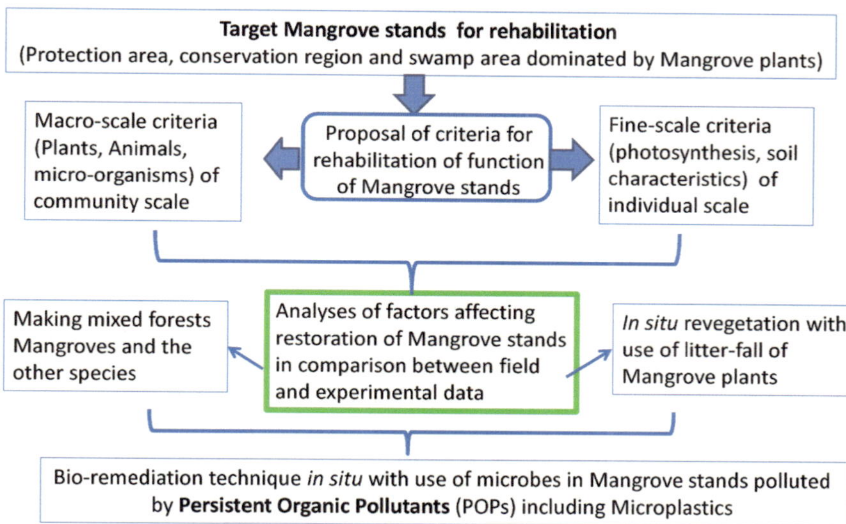

Figure 6.
Rehabilitation strategy of mangrove forests under pollution, (Adopted from the statement by Dr. Hailei Zheng, 2019) [52–54].

Figure 7.
Photos of snow on mangrove seedlings planted in Zhejiang Province, China in 2010. Left photo was offered by Dr. Jianbiao Qiu of Zhejiang Mariculture Research Institute. And right photo at nursery was offered by Dr. QiuXia Chen of Zhejiang Sub-tropical research institute [55].

low temperature environment. Glycine-Betaine was employed as a target chemical for species selection and breeding, as this compound is also common in relation to increase of desiccation and frozen tolerance [57]. Clearly, more work is needed to perform practical applications in the future.

4. Ecological statistics and diagnoses of heavy metal pollution in China

The current situation of heavy metal pollution of the southern part of China from ecological statistics was reported by Shi et al. [58]. As the first integrated analysis of heavy metal pollution in mangrove sediments across China, this study covered whole mangroves in China by selecting 6 sites including Hainan Island and near the border of Vietnam. If we focus on common mangrove species of these regions, *Avicennia marina* (characteristics: this species has pneumatophores [looks like young shoot of bamboo] for respiration in few centimeters from muddy soil) [59], the heavy metal pollution of this species shows very strong correlations in polluted condition of tideland soil (**Figure 8**).

From data of heavy metal pollution at the 6 sites, the pollution level in Futian district of Shenzhen city has higher pollution level. We detected representative heavy metal pollutants cause by high concentration of zinc (Zn), chromium (Cr), lead (Pb), nickel (Ni), arsenic (As) and relatively low concentration of molybdenum (Mo), cobalt (Co) and cadmium (Cd). Although proportion of Cd in the total heavy metal concentrations was low (based on evaluation of geo-accumulation index, contamination factor, potential ecological risk coefficient, pollution load index, and potential ecological risk index), Cd is a cause of the "Itai-itai" disease in Japan [60] and was detected as 0.66 ~ 3.30 μg/g at Futian district of Shenzhen city [61].

The factor affecting capture of heavy metal is particle size of soil with different specific surface areas and adsorption capacities [60, 61]. In fact, the heavy metal concentration was lower at Fangchenggang in Guangxi province where sand is dominant, while it was higher at Dong fang, Yunxiao and Futian district of Shenzhen where silt (between sand and clay) is dominant [62, 63]. We regarded mangrove stands, coral-reef and seaweed fauna as three major inshore marine ecosystems [1, 64, 65]. Furthermore, we explored the heavy metal pollution in Futian mangroves of Shenzhen, China. Futian mangrove is a mangrove forest area of 304 ha located in the Guangdong Province, and is the only mangrove forests located in the middle of Shenzhen, China. Futian mangrove was adjacent to the Mai

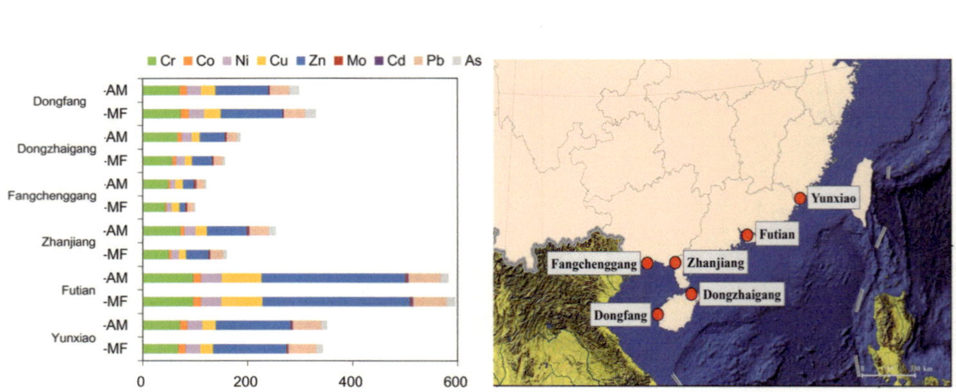

Figure 8.
Ecological statistics of heavy metals pollution in south China (AM, Avicennia marina a kind of mangrove; MF, mudflat) (Adopted from Shi et al. [58]).

Po Nature Reserve, Hong Kong, and has suffered serious heavy metal pollution since the early 1990s [62].

In order to systematically explore ecological risk of heavy metal contamination in Futian mangrove forest, being important for designing management and conservation policy, we quantify the concentrations of heavy metals (Cd, Cr, Cu, Pb and Zn) in mangrove sediments, assess the potential ecological risk and sources of heavy metals, and identify the speciation of heavy metals [66]. The results showed that heavy metal concentrations in surface sediments (0–20 cm depth) varied greatly along the coastline, demonstrating the heterogeneity of sediment to some extent. As for different heavy metal species, the concentrations reduced in the order of Zn > Cr > Pb > Cu > Cd [66]. Furthermore, the combination of studied metals had a 21% probability of being toxic, based on analysis of mean probable effects level quotient. Similarly, high heavy metal contamination was also revealed in term of potential ecological risk index and geo-accumulation index. Among all heavy metals, Cd has higher potential for adverse biological effects, being of primary concern. Take into account the sediment characteristics, clay and silt were important in raising deposition/accumulation of Cr, Cu, and Zn. As for different speciation of heavy metals, the percentage of mobile heavy metals was relatively higher than other fractions; while, no considerable ecological risk to the biota was detected in terms of the risk assessment code. The mobile heavy metals referred to the sum of acid-soluble, reducible, and the oxidizable fractions in terms of heavy metal speciation [67].

5. Ecophysiology of heavy metal resistance for mangrove plants

Mangrove plants have specific nutrient balance for growth and survival because they grow in very special environment (i.e. high NaCl, flooding environment, etc.).

Therefore, response of mangrove plants to various environmental stresses is a key information of rehabilitation of degraded regions. Cadmium (Cd), a non-essential element, can easily be taken up by plants and cause chlorosis [69], wilting [70] and cell death [71]. Heavy metals and large amounts of nutrients including nitrogen from domestic sewage also accumulate in mangrove sediment, and change its oligotrophic state [72] and pH [73]. We quantified the effects of ammonium nitrogen on the accumulation, subcellular distribution, and chemical forms of cadmium (Cd) in *K. obovata* [68]. The concentration and total amounts of Cd in leaves, stems and roots increased with NH_4^+-N supply (**Figure 9**).

In terms of subcellular distribution, Cd in roots of *K. obovata* was mainly deposited on cell wall, and the largest chemical forms of Cd was pectate and protein integrated form in all treatment. Under Cd treatments of 1 and 5 mg/L, 50 mg/L NH_4^+-N enhanced the transfer of root cell wall-combined Cd into cell, improved bioaccumulation of pectate and protein integrated Cd in cell to reduce toxicity caused by Cd. Under 10 mg/L Cd treatment, NH_4^+-N addition improved bioaccumulation of Cd on root cell wall, and limited enter of Cd into cell, which were also verified by decreased bioaccumulation of pectate and protein integrated Cd in root cell to some extent. Thus, under Cd stress, NH_4^+-N supply improved Cd immobilization in roots of *K. obovata*; the results of subcellular distribution and chemical forms showed that root cell wall combination and integration with pectate and protein acted as the mainly detoxification strategies of *K. obovata*. Plant transpiration transfer coefficient performed well in indicating the water and high temperature stress conditions plants experienced [74].

The use of chlorophyll florescence for diagnose of plant health status under stress, especially for mangrove plants is important in the rapid non-destructive

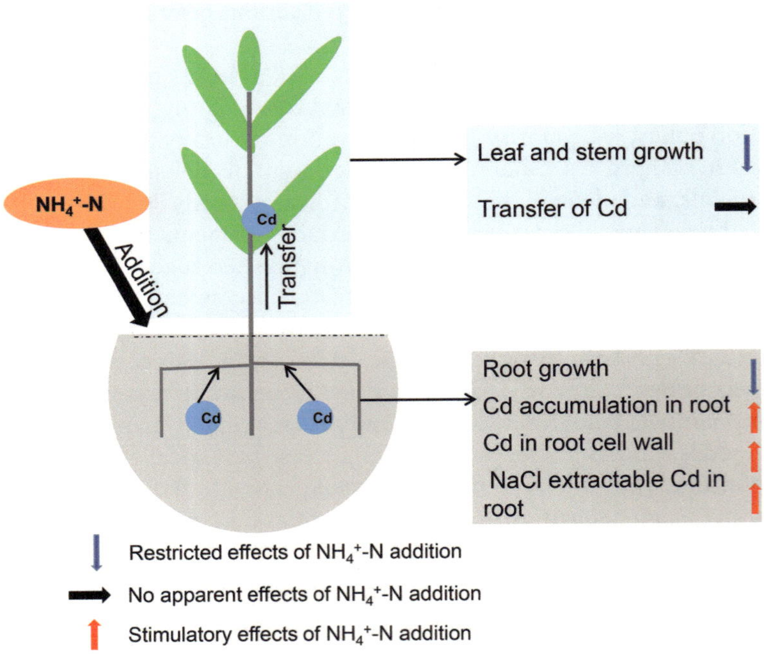

Figure 9.
The distribution of cadmium (Cd) in Kandelia obovata under nitrogen addition (Adopted from Chai et al.) [68].

assessment [34]. Physiological activity of coastal mangrove species is evaluated based on a three-temperature (3 T) model using high-resolution thermal infrared remote sensing. This evaluation method is based on growth evaluation of the representative mangrove species, *K. obovata* seedlings treated with different concentrations of Cd and nitrogen input as a simulation of environment at estuary of Shenzhen, China [75].

The cultured *K. obovata* seedlings were promoted for their photosynthesis due to the application of nitrogen, which brought a decrease leaf temperature detected by remote thermometer. In contrast, leaf temperature of *K. obovata* seedlings treated Cd increased due to stomatal closure with Cd toxicity [75]. The toxicity of Cd may be moderated by the existence of nitrogen addition, which can be clearly detected by the images from thermometer. The high-resolution thermal infrared remote sensing +3 T model is practicable for diagnosing plant health status. We should always consider synergy and antagonism in each element including heavy metals and nutritional elements [76]. Generally, the information in single metal contamination research does not reflect the biological toxicity when multiple metals are present together, with their combination toxicity having complicated mutual interactions on plants [77].

Furthermore, the interactive effects of multiple heavy metals (Cu, Pb, and Zn) on growth of *K. obovata* were studied, including plant biomass, photosynthetic parameters, lipid peroxidation and compatible osmolytes [78]. The results showed no significant reduction of biomass under heavy metal stresses, except for decreased root biomass under higher Pb + Cu treatment. This evidence indicates that *K. obovata* is highly tolerant to heavy metal stress. With increasing heavy metal stress (except for Pb + Cu and Pb + Zn + Cu), the photosynthetic parameters detected by a potable porometer (LI-6400; NE, USA) (net photosynthetic rate [Pn], transpiration rate [Tr], and stomatal conductance [Gs]) decreased at ambient CO_2 concentration (about 400 ppm) [75].

Compared to external binary metal treatment, trinary treatments (Pb + Zn + Cu) improved plant biomass and the photosynthetic capacity. As for root of *K. obovata*, binary treatment reduced biomass and soluble sugar content compared with ternary treatment. These results showed that in Pb combined treatments, the combination of Zn and Cu improved alleviating toxicity than each of them alone. Malondialdehyde (MDA) can reflect the degree of cell membrane lipid peroxidation and plant response to stress conditions [79, 80].

In leaves of *K. obovata*, Zn-containing combined treatments significantly reduced MDA, soluble sugar, and proline content under low concentration, demonstrating antagonistic effects; while, Pb + Cu treatments significantly promoted these parameters, indicated synergistic effects. Furthermore, there were negative correlations between leaf MDA and proline content with Zn concentration ($P < 0.05$). Leaf MDA content was positively correlated with the osmotic parameters, indicating co-existence of osmotic stress and lipid membranes oxidation under multiple heavy metal stresses. Thus, as for *K. obovata*, the toxicity caused by multiple heavy metals could be indicated by responses of leaf biomass, Tr, leaf MDA, leaf proline and soluble sugar [75, 78].

6. Evaluation of mangrove ecosystem: microbes and shellfish

6.1 Microbe in mangrove sediments

At the ecosystem level, mangrove plants, microbes in the soil, any other living organisms, and their natural environment cooperated with each other [3, 81, 82]. There are many places in mangrove stands producing sulfate compounds (e.g. hydrogen sulfide, H_2S). We can identify the smell of H_2S, implying importance of sulfate producing microbe activities. In the field conditions, heavy metal accumulation is important environmental factor regulating bacterial communities [83, 84]. Sulfate-reducing bacteria (SRB) could utilize sulfate as an electron acceptor in the dissimilatory reduction of sulfate [85]. How about the impacts of heavy metals pollutant on SRB? The effects of heavy metal contamination on sulfate-reducing bacteria (SRB) with both field survey and experimental approaches have been revealed [3]. SRB communities were investigated in mangrove sediments (0–30 cm depth) from 3 districts of mangrove wetlands in Shenzhen with different heavy metal contamination levels.

The results revealed that SRB community abundance was correlated with depth of mangrove sediments, especially significant correlation was found in soil concentration of Cd and Ni concentrations. From 1980 to 1990s, almost no analysis was done from the view point of biodiversity [81]. The α-diversity index of SRB community was significantly correlated with Cd level in mangrove sediments. Dominant 3 SRB groups (Desulfo-bacteraceae, Desulfobulbaceae, Syntrophobacteraceae) were isolated in the mangrove sediments of Shenzhen mangrove, China [3]. Among these families, Syntrophobacter-aceae was most sensitive to heavy metal contamination. The Unifrace clustering analysis revealed that SRB community structure was influenced by heavy metal stress. Moreover, redundancy analysis (RDA) indicated that Cd and total phosphorus were the major element affecting the SRB structure in the mangrove sediments [3].

Generally, the structure of mangrove sediment bacterial community could be affected by various factors, including plantation species [86], sediment depths [87], physico-chemical properties of sediment [88], and anthropogenic activities [87]. Different mangroves might reveal the specific biogeographic distribution pattern of bacterial community [89]. We explored the biogeographic distribution of sediment bacterial community in six mangroves across China, including two mangroves in Hainan Province, two in Guangdong Province, one in Guangxi Province and one in Fujian Province [90]. Among all six mangroves, the sediment bacterial demonstrated different characteristics in terms of bacterial abundance, bacterial richness and diversity, and bacterial community structure. Compared with intertidal mudflat, *A. marina* planted zone improved sediment bacterial abundance, richness, and diversity. Furthermore, *Proteobacteria* acted as the largest bacterial phylum in sediments in both intertidal mudflat and *A. marina* planted zone. Therefore, the biogeographic distribution of bacterial community across six mangroves in China was driven by variable wetland tropic status and other physicochemical factors (such as salinity) [90].

In mangrove sediment, *Archaea* sp. played important role in biogeochemical processes, such as ammonia oxidation [91] and methanogenesis [79]. Understanding biogeographic distribution pattern can be helpful to increase the knowledge of microbial function and to predict ecosystem responses to environmental variability [89]. We explored the effect of geographic location on mangrove archaeal community in six mangroves of China [92].

In different geographic location, mangrove archaeal have different community characteristics, which might be related to various environmental factors, including pH, carbon, and nitrogen contents in sediment. Furthermore, the main archaeal communities in mangrove sediments were genus *Thaumarchaeota* and *Euryarchaeota*, with their percentages to be 54.7–85.2% and 11.8–43.9%, respectively. This work would be useful for understanding the characteristics of archaeal community in mangrove ecosystem, which may provide new insight for exploring microbial function and ecosystem responses to variable environment.

6.2 Shellfish as food resources

Shellfish is an important component in mangrove ecosystems similar to well-known crabs [32]. Recently, people have eaten more shellfishes as healthy food than before; however, the bio-accumulation of heavy metals in the shellfish can endanger the health of consumer [93]. Shenzhen is a fast-developing city in south China, and has been developed from a small fishermen village to a modern metropolis with about 12 million populations since the reform and opening policy in 1978 [94].

A case study on 3 markets of Shenzhen has received increasing attention. Arsenic (As), Cd, Cu, Hg, and Pb in 10 popular shellfish species and associated health risks were analyzed for Shenzhen's consumers by evaluation of estimated weekly intake (EWI), non-carcinogenic and carcinogenic health risks to the 3 stages in a human life-cycle (children, adolescents, adults) [95]. Based on 50 shellfish samples in each site of market there, they found that the levels of inorganic arsenic (iAs) in *Babylonia areolata* exceeded the maximum permissible-limit decided by the food safety guidelines (0.5 mg/kg), while other elements were below the limit of the guidelines as shown in Ministry of Health of the PR. China: GB 2762–2012. EWI values of the 3 stages of human development were lower than provisional tolerable weekly intakes (PTWIs) of all shellfish species. Analysis of the total target hazard quotients (TTHQ) showed that the consumption of *B. areolata* in all stages of people would cause non-carcinogenic risks; as for children (< 10 years old), the ingested *Argopecten irradians* and *Chlamys farreri* were at non-carcinogenic risks. As for children, adolescents, and adults, the bioaccumulation of Cd caused by shellfish consumption (*A. irradians, B. areolata, C. farreri*, and *Crassostrea ariakensis*) would lead to cancer risk during life-time, with Pb and iAs to be toxicity acceptable or negligible [96].

From the perspective of species, the concentration of Zn and Cu in the *C. ariakensis* was the highest, the concentration of Cd in *C. farreri* was the highest, and the concentration in the *Sinonovacula constricta* was relatively low. From the regional differences, Cd content in *C. farreri* of Dandong was the highest; the overall concentration of heavy metals in Qingdao was relatively low; Cu and Zn concentration in *C. ariakensis* and Cd concentration in *C. farreri* of Zhoushan were relatively high; Zn and Cu content in *C. ariakensis* of Shenzhen was the highest. The calculated daily intake-limit-results show that the feeding rate in the *C. farreri* only is 0.04 ~ 0.15 kg/d to reach the limit of Cd, and the feeding rate of Zn, Cu and Cd in the *C. ariakensis* is less than 1 kg/d can reach the limit value.

The results of the risk assessment showed that the weekly intake of heavy metals by eating shellfish did not exceed the provisional tolerable weekly intake set by Joint Expert Committee on Food Additives (JECFA*, *JECFA, 2010. Joint FAO/WHO Expert Committee on Food Additives, Summary and Conclusions by the JECFA.), and there was no non-carcinogenic risk to the human body. However, in terms of the long overdose of all sampling points of the *C. farreri* and *C. ariakensis* in Dandong, the accumulated Cd has a potential carcinogenic risk to the human body.

7. Persistent organic pollutants (POPs) in mangroves of China

Overuse of POPs like plastic bags pollutes mangrove stands and POPs are hardly decomposed in mangrove ecosystems and consequently degrade habitat for most living organisms in a mangrove stand. The value of mangrove environment is getting worse by accumulation of the POPs, therefore, we should know the effect of POPs on mangrove plants as well as ecosystem for improvement of rehabilitation strategy of degraded mangrove stands.

7.1 Microplastics

Microplastics (MPs) researches have been mainly investigating on food-web and bioconcentration in Japan [97]. For example, in Tokyo Bay, the bowel of several kinds of seabird, sardine, etc. contains huge amount of MP. In detail study on a kind of seabird (*Puffinus tenuirostris*), this bird cannot eat foods due to full with MPs in the bowel. Much worse, the accumulated MPs in their stomach gradually release toxic substances [98], such as PCB. Now, if we eat such polluted sardine, our health condition would not be injured. However, we cannot deny the harmful effects of PCB and related toxic substances in MPs.

MPs are now worldwide serious problem [97, 99]. MPs have a long life-span in ecosystem and become smaller in size of less than 5 mm over time, and deposit toward deep sea [9]. The existence form of MPs in south China was classified as fiber, film, and fine particle with several colors of original products [99]. At the 6 mangrove stands located along the coast of Southern China, the top 3 MPs were detected as health safety substance, including polypropylene (PP), polyethylene (PE), and polystyrene (PS). In terms of shape, color and size, MPs were mainly fibrous, white-transparent and 500 μm-5000 μm, respectively. MPs pollution in mangroves was significantly linked to surrounding socio-economic development. The TOC and silt content of mangrove sediments also affect the deposition of MPs [99]. Based on a comprehensive evaluation using the potential ecological risk factor, potential ecological risk, polymer risk index and pollution load index, MPs showed highest ecological risk in Futian mangrove of Shenzhen, China. These fundamental data on MPs occurring in mangroves of Southern China could support further studies of the ecological consequences of MPs on mangrove macro-fauna, shrimp, fish and even human.

7.2 Polybrominated diphenyl ethers (PBDEs)

Unfortunately, polluted condition with MPs of mangrove forest is getting worse in Shenzhen. As harmful MPs, PBDEs (polybrominated diphenyl ethers) have recently been detected. PBDEs are structurally similar to PCBs and other polyhalogenated compounds. Exposure to PBDEs would cause problems in the hormone system, liver and kidney morphology, neuro-behavioral and sexual development [100, 101]. The health risk of PBDEs and PCBs are increasing, and these chemical compounds have been shown to reduce human fertility to some extent [97]. The amount of PBDEs in a mangrove ecosystem increases with increasing amount of organic matters in soil [95, 102]. Concentration [unit: ng/g-dw] of PBDE-209 in mangrove stands of Hong Kong was 0.5 ~ 5.4 except for Mai Po mangrove (47.2–112.0) opposite to Futian district of Shenzhen [103]. These values were lower than those of Shenzhen (2.1–1987.6) [102].

In Shenzhen mangroves, the levels of PBDE-209 [unit: ng/g-dw] were 2.1 ~ 110.0 in mangrove sediment and 180 ~ 600 in the leaves. The highest value was detected as 3600 in bark of the avenue trees in Beijing [104]. We found that PBDE-209 was the dominant PBDE congener in all six mangroves in China, including Yunxiao, Futian, Zhanjiang, Fangchenggang, Dongzhaigang and Dongfang [5]. Futian mangrove in Shenzhen was seriously polluted by PBDEs (in particular PBDE-209), compared to the other 5 mangrove wetlands. Total organic matter acted as an effective factor in affecting spatial distribution and ecological risk of PBDEs in sediment of mangroves. In 6 mangroves, sediments may pose low/moderate risk of exposure to penta- and deca-BDE congeners for sediment-dwelling organisms, with penta- and deca-BDE congeners to be major drivers of ecological risk. Furthermore, we explored PBDEs contamination in 4 urban mangroves of Shenzhen, including Shajing mangrove (SJM), Xixiang mangrove (SJM), Futian mangrove (FTM) and Baguang mangrove (BGM) [102]. Regarding urban functional zoning, urban mangroves were featured with industry district (SJM and XXM), central business district (CBD) (FTM), and ecological preserve (BGM) [105–107]. Our result showed that the ranking order of PBDEs contamination in urban mangroves was BGM ≈ FTM < XXM < SJM.

Compositions of PBDEs were complex in SJM, XXM, and FTM, with surface runoff to be the main source apportionment of PBDEs. Thus, in urban mangroves with different urban functional zonings, PBDEs accumulation in mangrove sediment and their bioaccumulation in mangrove plants were different. In the future, much work should be done to decrease e the input of PBDEs into the urban mangrove, such as the inspection of the illegal waste recycling sites and promotion of sewage treatment capacity of PBDEs-related enterprises.

7.3 Other POPs

Levels of polycyclic aromatic hydrocarbons (PAHs: unit: [ng/g-dw]) in mangrove sediments of China ranged from 15 to 11,098 and decreased in the order of Hong Kong (56–11,098) > Fujian (171–1074) > Guangdong (15–726) > Hainan (31–63) > Guangxi (24) [11, 12, 108]. Higher levels of PAH s in Hong Kong mangrove might be attributed to the intense anthropogenic activities. Levels of PAHs in mangrove sediments of China (24–11,098) were far below effects range mean (ERM) (44,792), with levels of PAHs in Hong Kong to be lower than effects range low (ERL) (4022), indicating that PAHs in Hong Kong may pose little risk to biota in mangrove ecosystems [63, 108]. Li et al. (2014) explored PAH pollution in sediments of three mangrove swamps of Shenzhen, China, namely Futian, Baguang and Water-lands, and found that the mean concentrations of PAHs in Futian (4480) was higher than that in Baguang (1262) and Watersheds (2711) [109]. The higher levels of PAHs in Futian mangrove may be related to various anthropogenic activities, such as continuously discharges of domestic sewage from households and Fengtang River, effluents from industrial processes, construction of highways and heavy traffic [48].

In China including Hong Kong, Guangdong, Guangxi, Hainan and Fujian, levels of polychlorinated biphenyls (PCBs) in mangrove sediments ranged from 0.1 to 47 ng/g dry weight [110–113]. In general, PCB concentrations in mangrove sediments from Hong Kong, Shenzhen, and Zhuhai were relatively higher, indicating heavily PCBs-polluted mangrove sediments in the Pearl River Estuary to some extent [114]. The higher PCB levels in sediments from Pearl River Estuary could be linked to high density of electronic/electrical industries and electronic waste recycling activities [110].

Other organic pollutants in mangrove sediment were limited. Total petroleum hydrocarbons (TPHs) in sediments were reported to be 32–579 mg/kg-dw and a higher level was observed in the Pearl River Estuary, China [82, 115], which mainly derived from vehicle exhausts and incomplete combustion [82, 88]. Tam et al. (2008) reported that the levels of dichlorodiphenyltrichloroethanes (DDTs) and hexachlorcyclohexanes (HCHs) were 28 and 0.07 ng/g-dw in Leizhou Peninsula [116].

8. Education effort

8.1 General effort

Although Chinese ecological policy was proposed, ecological and environmental education should be made for conservation and increasing ecological services of mangrove forests [31, 98, 117]. In China, the environmental education mainly focused on the popular science and propaganda of mangrove reserve and park [97, 118–120]. Since there are multiple education stations in China, we show an example of Shenzhen because mangrove species are applied as the symbolic woody plants for the city. The City Hall shows SDGs (sustainable developmental goals) to the public with both ordinal (indoor type) museum as well as field museum (outdoor type). The former exhibits basic information of mangrove ecosystem by indoor exhibition. The latter mainly shows 2 parts; one is a practical method of how to rehabilitate mangrove stand at water front and the other is to conduct ecological research, including pollination biology, vegetation, etc.

8.2 Indoor and outdoor exhibition

Shenzhen shows all aspects of mangrove conservation and the latest advances in natural education at the city museums. The museum is unique in interactive exhibits, such as many educational quizzes about how to develop mangrove forests as well as to conserve the ecosystem. There are two field museums on the coast of Shenzhen: (1) one is a kind of park to walk in mangrove stand along with the interior experience of different species of forest stands, and (2) the other is to show the ecological rehabilitation of mangrove stands (**Figure 10**).

Figure 10A showed the rehabilitation of mangrove stands in Futian mangrove in Shenzhen. In order to improve ecological function of mangroves [121], the rehabilitation of Geiwai pond in mangrove ecosystem was conducted (**Figure 10B**). Among

Figure 10.
Example of mangrove ecosystem rehabilitation in south China. (A) Rehabilitation of mangrove stand; (B) Rehabilitation of Geiwai pond of mangrove ecosystem; (C) Typical example of mangrove ecosystem plantations by Prof. Changyi Lu of Xianmen University.

these trails, a distinguish exhibition for both tourists and education of mangrove park was established and popular trees were provided to the public near Xiamen, Fujian province, southeast China (**Figure 10C**). The star shaped mangrove restoration indicated that suitable mangrove species selection and planting design would create a beautiful landscape. This is a park showing a symbol of restoration success with mangrove.

8.3 Non-Government Organization in education effect of mangrove

The protection of mangrove in Shenzhen was closely related with large amounts of non-governmental organization (NGO) and volunteer groups. In 2017, the number of professional volunteers registered in Mangrove Conservation Foundation (MCF) exceeded 300, with 198 newly trained volunteer in one year, and 100,000 people participated in relevant activities in three years. In Shenzhen, China, the above-mentioned citizen acted as the volunteer labor group for Futian mangrove ecological park. These volunteers mainly come from enterprise, residence community and school, and took part in various environmental improvement activities,

Figure 11.
Activities carried out by nongovernment organizations. (A) Marine protection activity; (B) vein painting of popular science education; (C) display board of natural education. All photos were offered by RL Li and MW Chai.

including cleaning of invasive plants, collection of marine garbage, replantation of plant, and construction of ecological floating island, etc.

The implication of these activities increased volunteer number of mangrove protection in Shenzhen. Furthermore, there were also a certain number of volunteers in Overseas Chinese Town (OCT) wetland, Shenzhen Green Fund Association, Shenzhen Spring Environmental Protection Volunteer Association. Some activities related with mangrove education was implicated to improve protection awareness of mangrove protection (**Figure 11**).

9. Future perspective

Anthropogenic pollution in mangrove ecosystems has been intensively studied (including heavy metals and POPs, etc.) and consequently we can obtain many phenomena of current situation, such as sea level-rise [27, 122–124]. Based on the data of mangrove ecosystems in China, we should make further progress in increasing ecological services and forest rehabilitation. In the future, we believe that several aspects should be further explored to improve mangrove afforestation and restoration: (1) Ecological adaptation mechanism of mangrove species under various environmental stresses; (2) The development of mangrove plant breeding and colonization techniques; (3) The remediation of degraded mangrove ecosystem; (4) Digital technology research and development in mangrove ecological engineering.

Ecological exploitation of mangrove would create better ecological environment, economic and social benefits, being important for sustainable development of mangrove resource in the future. Several countermeasures are included to engage ecological exploitation of mangroves: coordinating development and alleviating the contradiction between protection and development; promoting diversified ecological development based on local conditions; improving the economic benefits of ecological development by scientific evaluation; improving relevant policies and plans while promoting regional cooperation and scientific research. We hope our data may contribute for improving restoration practices in the rapid economic development regions.

Acknowledgements

This work was supported by the Program of Science and Technology of Shenzhen (JCYJ20160330095549229, JSGG20170413103811649), Special Fund of State Key Joint Laboratory of Environment Simulation and Pollution Control (18K05ESPCP), and Shenzhen Municipal Development and Reform Commission (Discipline construction of watershed ecological engineering). We thank Emeritus Prof. A. Hagihara of The Univ. of Ryukyu for his kind permission to modify his original figures to cite in the text.

Author details

Ruili Li[1*], Minwei Chai[1], Xiaoxue Shen[1], Cong Shi[1], Guoyu Qiu[1] and Takayoshi Koike[1,2*]

1 School of Environment and Energy, Peking University Shenzhen Graduate School, Guangdong, China

2 Research Faculty of Agriculture, Hokkaido University, Japan

*Address all correspondence to: liruili@pkusz.edu.cn
and tkoike@for.agr.hokudai.ac.jp

References

[1] Rivera-Monroy VH, Lee S-Y, Kristensen E, Twilley RR. Mangrove Ecosystems: A lobal Biogeographic Perspective -Structure, Function, and Services-, 2017; pp.381. Springer Verlag.

[2] Sharma S. Introductory Chapter: Mangrove Ecosystem Research Trends - Where has the Focus been So Far, *In*: Mangrove Ecosystem Ecology and Function, 2018; **DOI:** 10.5772/intechopen.80962.

[3] Wu SJ, Li RL, Xie SG, Shi C. Depth-related change of sulfate-reducing bacteria community in mangrove sediments: The influence of heavy metal contamination. Marine Pollution Bulletin, 2019; **140**: 443-450.

[4] Faridah-Hanum I, Latiff A, Hakeem KR, Őzturk M. Mangrove Ecosystems of Asia- Status, Challenges and Management Strategies-, 2014; Springer-Verlag, pp. 453.

[5] MA (Millennium ecosystem assessment). Ecosystems and human well-being: Current state and trends: Chapter **22** dryland systems. Island Press, 2005; pp. 623-662.

[6] Wang YN, Fu XM, Shao CL, Wang CY, Li GQ, Liu GX, Sun SC, Zeng XQ, Ye ZJ, Guan HS. Investigation on the status of mangrove resources and medicinal research in China I. Ecological functions and values. Periodical of Ocean University of China, 2009; **39**: 699-704 (in Chinese with English abstract).

[7] Koike T. Research activities of mangrove ecosystems in south China with special references to heavy metal pollution. Japanese Journal of International Forest and Forestry, 2019; **106**:8-13 (in Japanese).

[8] Mato Y, Isobe T, Takada H, Kanehiro H, Ohtake C, Kaminuma T. Plastic resin pellets as a transport medium for toxic chemicals in the marine environment. Environmental Science & Technology, 2001; **35**: 318-324.

[9] Takada H. Chemical substances pollution and micro plastic problems from the view point of matter cycling. Environ Information Science, 2019; **46**: 12-16. (in Japanese).

[10] Li MS, Lee SY. Mangroves of China; a brief review. Forest Ecology and Management, 1997; **96**: 241-259.

[11] Tam NFY, Fang Z, Yu S. Heavy metals, polycyclic aromatic hydrocarbons and organochlorine pesticides in the surface sediments of mangrove swamps from coastal sites along the Leizhou Pennsula, South China. Acta Oceanologica Sinica, 2008; **27:** 42-53.

[12] Zhang ZW, Xu XR, Sun YX, Yu S, Chen YS, Peng JX. Heavy metal and organic contaminants in mangrove ecosystems of China: a review. Environmental Science and Pollution Research, 2014; **21:** 11938-11950.

[13] Bourgeois C, Alfaro AC, Brown AD, Duprey JL, Denudes A, Marchand C. Stocks and soil-plant transfer of macro-nutrients and trace metals in temperature New Zealand estuarine mangroves. Plant and Soil, 2019; **436**: 565-586.

[14] Analuddin K, Sharma S, Jamili, Septiana A, Sahidin I, Rianse U, Nadaoka K. Heavy metal bioaccumulation in mangrove ecosystem at the coral triangle ecoregion, Southeast Sulawesi, Indonesia. Marine Pollution Bulletin, 2017; **125**: 472-480.

[15] Marchand C, Allenbach M, Vergès EL. Relationships between heavy metals distribution and organic matter cycling in mangrove sediments (Conception Bay, New Caledonia). Geoderma, 2011; **160**: 444-456.

[16] Chai MW, Li RL, Shi C, Shen XX, Li RY, Zan QJ. Contamination of polybrominated diphenyl ethers (PBDEs) in urban mangroves of Southern China. Science of the Total Environment, 2019a; **646:** 390-399.

[17] Chai MW, Li RL, Ding H, Zan QJ. Occurrence and contamination of heavy metals in urban mangroves: A case study in Shenzhen, China. Chemosphere, 2019b; **219**:165-173.

[18] Ding H, Chai MW, Gong Y, Li RL. Reviews on organic pollution in mangrove wetlands. Marine Environmental Science, 2019; **38**:153-160 (in Chinese with English abstract).

[19] Poungparn S, Komiyama A, Sangteian T, Maknual C, Patanaponpaiboon P, Suchewaboripont V. High primary productivity under submerged soil raises the net ecosystem productivity of a secondary mangrove forest in eastern Thailand. Journal of Tropical Ecology, 2012; **28**: 303-306.

[20] Abino AC, Castillo JAA, Lee YJ. Assessment of species diversity, biomass and carbon sequestration potential of a natural mangrove stand in Samar, the Philippines. Forest Science and Technology, 2014; **10**: 2-8.

[21] Kamruzzaman M, Mouctar K, Sharma S, Osawa A. Comparison of biomass and net primary productivity among three species in a subtropical mangrove forest at Manko Wetland, Okinawa, Japan. Regional Studies in Marine Science, 2019; **25**: 100475.

[22] Nordhaus I, Yoben M, Fauziyah A. Impact of deforestation on mangrove tree diversity, biomass and community dynamics in the Segara Anakan lagoon, Java, Indonesia: A ten-year perspective. Estuarine, Coastal and Shelf Science, 2019; **227:** 106300.

[23] Komiyama A, Ong JE, Pounngparn S. Allometry, biomass, and productivity of Mangrove forest: A review. Aquatic Botany, 2008; **89**: 128-137.

[24] Nakamura T, Nakasuka T. An introduction to Mangrove - Green forest stand in a sea, 1998; pp. 243, Mekon Publisher, Okinawa, (in Japanese).

[25] Ogino K. Biological processes and regulation mechanism of Mangrove ecosystem in Asian Pacific Ocean. 1990; pp. 116, Final report of KAKEN (Scientific aid by JSPS) No. 63041095.

[26] Komiyama A. Mangrove forests: changing forest ecosystem along seacoast, 2017; pp. 273, Kyoto University Publisher, pp. 273. (in Japanese).

[27] Schuerch M, Spencer T, Temmerman S, Kirwan ML, Wolff C, Lincke D, McOwen CJ. Future response of global wetlands to sea-level rise. Nature, 2018; **561**: 231-234.

[28] Wang WQ, Yan ZZ, Zhang YH, Chen LZ, Lin GH. Mangroves: obligate or facultative halophytes? A review. Trees, 2011; **25**: 953-963.

[29] Tikumporn A, Vipak J, Kwanchai D. Some environmental factors relating growth of *Avicennia alba* Bl. and *Avicemmia marina* (Forsk.) Vierh. Proceeding of 48th Kasetsart University Annual Conference: Natural Resources and Environment, 2010; **48**: 132-139.

[30] Toma T. Ecological study on water quality and soil condition with development of rhizosphere of Mangrove ecosystem. PhD dissertation of United Graduate School of Agriculture of Ehime University, 1992; pp.125, (in Japanese with English abstract).

[31] Tomlinson PB. The Botany of Mangroves, 2nd ed. Cambridge University Press, 2016; pp.432.

[32] Wang M, Wang WQ, Lin GS. Sanya Mangrove forest, 2019; pp.202. Scientific Publisher, Beijing (in Chinese).

[33] Xu L, Wang M, Xin CP, Liu C, Wang WQ. Mangrove distribution in relation to seasonal water salinity and ion compartmentation: a field study along a freshwater dominated river. Hydrobiologia, 2019; https://doi.org/10.1007/s10750-019-04119-7

[34] Kitao M, Utsugia H, Kuramoto S, Tabuchi T, Fujimoto K, Lihpaid S. Light-dependent photosynthetic characteristics indicated by chlorophyll fluorescence in five mangrove species native to Pohnpei Island, Micronesia. Physiologia Plantarum, 2003; **117**: 376-382.

[35] Hagihara A. Mangrove stands as carbon storage. *In*: The 21st century COE of The Univ of Ryukyu (ed.) Biodiversity of coral reef and island ecosystems of the Ryukyu, 2006; 258-277, Tokai University Press (in Japanese).

[36] Hoque ATMR, Sharma S, Hagihara A. Above and belowground carbon acquisition of mangrove *Kandelia obovata* trees in Manko wetland, Okinawa, Japan. International Journal of Environment, 2011; **1**: 7-13.

[37] Suwa R. The Production Ecology of a subtropical mangrove in Manko Wetland on Okinawa Island, Tropical Ecology Letters, 2013; **90**: 2-9.

[38] Birkland TA, Herabat P, Little RG, Wallace WA. The impact of the December 2004 Indian Ocean tsunami on tourism in Thailand. Earthquake Spectra, 2006; **22**: 889-900.

[39] Rajendran CP. Historical accounts of sea disturbances from South India and their bearing on the penultimate predecessor of the 2004 Tsunami. Seismological Research Letters, 2019; **90**: 774-783.

[40] Zhao S, Hong HS, Zhang LP, Chen WQ. Emergy value of mangrove ecosystem services in China. Resources Science, 2007; **29**: 147-154 (in Chinese with English abstract).

[41] Lovelock CE, Cahoon DR, Friess DA, Guntenspergen GR, Krauss KW, Reef R, Rogers K, Saunders ML, Sidik F, Swales A, Saintilan N, Thuyen LX, Triet T, The vulnerability of Indo-Pacific mangrove forests to sea-level rise. 2015; Nature, **526**: 559-563.

[42] Pickering MD, Reef R, Vafeidis AT, Hinkel J, Nicholls RJ, Brown S. Future response of global coastal wetlands to sea-level rise. Nature, 2018; **561**: https://doi.org/ 10.1038/s41586-018-0476-5

[43] Soper FM, MacKenzie RA, Sharma S, Cole TG, Litton CM, Sparks JP. Non-native mangroves support carbon storage, sediment carbon burial, and accretion of coastal ecosystems, Global Change Biology, 2019; **25**: https://doi.org/10.1111/gcb.14813.

[44] Friess DA, Krauss KW, Horstman EM, Balk T, Bouma TJ, Galli D, Webb EL. Are all intertidal wetlands naturally created equal? Bottlenecks, thresholds and knowledge gaps to mangrove and saltmarsh ecosystems. Biological Reviews of the Cambridge Philosophical Society, 2012; **87**: 346-366.

[45] Saintilan N, Khan NS, Ashe E, Kelleway JJ, Rogers K, Woodroffe CD, Horton BP. Thresholds of mangrove survival under rapid sea level rise. Science, 2020; **368**: 1118-1121.

[46] Hogarth PJ. The Biology of Mangroves and Seagrasses, 3rd ed, 2015; Oxford University Press, 304pp.

[47] Wang WQ, Wang M. The mangroves of China, 2007; pp.186, Scientific Publisher, Beijing (in Chinese).

[48] Li RL, Chai MW, Qiu GY. Distribution, fraction, and ecological assessment of heavy metals in sediment-plant system in mangrove forest, South China Sea. Plos One, 2016; DOI:10.1371/journal.pone.0147308.

[49] Koike T, Wang YN. A report of the meeting on mangrove ecosystem organized by Ecological Society of China. Japanese Journal of International Forest and Forestry, 2020; **107**: 28-33, (in Japanese).

[50] Fu XM, Wang YN, Shao CL, Wang CY, Li GQ, Liu GX, Sun SC, Zeng XQ, Ye ZJ, Guan HS. Investigation on the status of mangrove resources and medicinal research in China. II. Resources status, protection and management. Periodical of Ocean University of China, 2009; **39**: 705-711.

[51] Shao CL, Fu XM, Wang CY, Han L, Liu X, Fang YC, Li GQ, Liu GX, Guan HS. Investigation on the status of mangrove resources and medicinal research in China. III. Status of folk medicinal usage and medicinal research. Periodical of Ocean University of China, 2009; **39**: 712-718.

[52] Fan HQ, Wang WQ, Some thematic issues for mangrove conservation in China. Journal of Xiamen University (Natural Science), 2017; **56**: 323-330.

[53] Chen LZ, Wang WQ, Zhang YH, Lin GH. Recent progresses in mangrove conservation, restoration and research in China. Journal of Plant Ecology, 2009; **2**: 45-54 (in Chinese with English abstract).

[54] Chinese Ecological Society (Mangrove ecology WG). Program of working meeting of Chinese mangrove research group. Graduate School of Resources and Environment of Xiamen University, 2019; **9**: 1-19, (in Chinese).

[55] Chen QX, Zheng J, Zhou WP, Wang JW, Yang S, Li XW, Lu X, Liu XA. New *Kandelia obovata* Cultivar 'Longgang', Acta Horticulturae Sinica, 2019; **46**: 2927-2928 (in Chinese with English abstract).

[56] Lu X, Liu X, Wang JW, Yang S, Zhang LN, Ji HJ. Study on over-wintering methods for introduced *Kandelia obovata* in Jiangsu Province. Forestry Science & Technology 2019; **44**: 15-17. (in Chinese with English abstract).

[57] Hayes MA, Shor AC, Jesse A, Miller C, Kennedy JP, Feller I. The role of glycine betaine in range expansions; protecting mangroves against extreme freeze events. Journal of Ecology, 2019; **108**: Doi.org/10.1111/1365-2745.13243.

[58] Shi C, Ding H, Zan QJ, Li RL. Spatial variation and ecological risk assessment of heavy metals in mangrove sediments across China. Marine Pollution Bulletin, 2019; **143**: 115-124.

[59] Abrahim GMS, Parker RJ, Nichol SL. Distribution and assessment of sediment toxicity in Tamaki Estuary, Auckland, New Zealand. Environmental Geology, 2007; **52**: 1315-1323.

[60] Horiguchi H, Teranishi H, Niiya K, Aoshima K, Katoh T, Sakuragawa N, Kasuya M. Hypoproduction of erythropoietin contributes to anemia in chronic cadmium intoxication: clinical study on Itai-itai disease in Japan. Archives of Toxicology, 1994; **68**: 632-636.

[61] Sundaramanickam A, Shanmugam N, Cholan S, Kumaresan S, Madeswaran P, Balasubramanian T. Spatial variability of heavy metals in estuarine, mangrove and coastal ecosystems along Parangipettai, Southeast coast of India. Environmental Pollution 2016; **218**: 186-195.

[62] Bernhoft RA. Cadmium Toxicity and Treatment, Hindawi Publishing Corporation. The Scientific World Journal, 2013; Article ID **394652**, https://doi.org/10.1155/2013/394652

[63] Xie HW, Wen B, Guo Y, Shi YZ, Wu YH. Community characteristics and distribution of metal elements in mangroves in Futian of Shenzhen, China. Guihaia, 2010; **30:** 64-69 (in Chinese with English abstract).

[64] Zhang ZH, Hu G, Liang SC. Mangrove resources and conservation in Guangxi. Marine Environmental Science, 2007; **26:** 275-279.

[65] Long ER, MacDonald DD, Smith SL, Calder FD. Incidence of adverse biological effects within ranges of chemical concentrations in marine and estuarine sediments. Environmental Management, 1995; **19**: 81-97.

[66] Sugiura K, Muwamura K, Wu DD, Wang J, Wangang D, Konohira Y, Nakayama Y. Reclamation and subsequent change of land use of mangrove forests in the northern part of Hainan Island, P. R. China. Journal of Forest Planning, 2007; **41**: 271-280. (in Japanese with English abstract).

[67] Li RY, Li RL, Chai MW, Shen XX, Xu HL, Qiu GY. Heavy metal contamination and ecological risk in Futian mangrove forest sediment in Shenzhen Bay, South China. Marine Pollution Bulletin, 2015; **101:** 448-456.

[68] Tam NFY, Li SH, Lan CY, Chen GZ, Li MS, Wong YS. Nutrients and heavy metal contamination of plants and sediments in Futian mangrove forest. Hydrobiologia, 1995; **295:** 149-158.

[69] Guillén MT, Delgado J, Albanese S, Nieto JM, Lima A, Vivo BD. Heavy metals fractionation and multivariate statistical techniques to evaluate the environmental risk soils of Huelva Township (SW Iberian Peninsula). Journal of Geochemistry Exploration, 2012; **119/120**: 32-43.

[70] Sterckeman T, Goderniaux M, Sirguey C, Cornu JY, Nguyen C. Do roots or shoots control cadmium accumulation in the hyperaccumulator *Noccaea caerulescens*? Plant and Soil, 2015; **392**: 87-99.

[71] Hu YN, Cheng HF, Tao S. The challenges and solutions for cadmium-contaminated rice in China: a critical review. Environment International, 2016; **92-93**: 515-513.

[72] Farooq MA, Ali S, Hameed A, Bharwana SA, Rizwan M, Ishaque W, Farid M, Mahmood K, Iqbal Z. Cadmium stress in cotton seedlings: physiological, photosynthesis and oxidative damages alleviated by glycinebetaine. South African Journal of Botany, 2016; **104:** 61-68.

[73] Lovelock CE, Ball MC, Martin KC, Feller IC. Nutrient enrichment increases mortality of mangroves. PLoS One, 2012; **4:** 5600.

[74] Chai MW, Li RY, Shen XX, Tam NFY, Zan QJ, Li RL. Does ammonium nitrogen affect accumulation, subcellular distribution and chemical forms of cadmium *in Kandelia obovata*? Ecotoxicology and Environmental Safety, 2018; **162:** 430-437.

[75] Qiu GY, Omasa K, Sase S. An infrared-based coefficient to screen plant environmental stress: concept, test and applications. Functional Plant Biology, 2009; **36**: 990-997.

[76] Shen XX, Li RL, Chai MW, Yu K, Zan QJ, Qiu GY. Assessing the effect of extra nitrogen on *Kandelia obovata* growth under cadmium stress using high-resolution thermal infrared remote sensing and the three-temperature

model. Functional Plant Biology, 2018; **45**: 1162-1171.

[77] Marchener P. Marschner's Mineral Nutrition of Higher Plants, 3rd ed. 2012; pp. 672, Academic Press, DOI. org/10.1016/C2009-0-63043-9

[78] Wu XY, Cobbina SJ, Mao GH, Xu H, Zhang Z, Yang LQ. A review of toxicity and mechanisms of individual and mixtures of heavy metals in the environment. Environmental Science and Pollution Research, 2016; **23**: 8244-8259.

[79] Shen XX, Li RL, Chai MW, Cheng SS, Niu ZY, Qiu GY. Interactive effects of single, binary and trinary trace metals (lead, zinc and copper) on the physiological responses of *Kandelia obovata* seedlings. Environmental Geochemistry and Health, 2019; **41**: 135-148.

[80] Choudhary M, Jetley UK, Khan MA, Zutshi S, Fatma T. Effect of heavy metal stress on proline, malondialdehyde, and superoxide dismutase activity in the cyanobacterium *Spirulina platensis*-S5. Ecotoxicology and Environmental Safety, 2007; **66:** 204-209.

[81] Lewis M, Pryor R, Wilking L. Fate and effects of anthropogenic chemicals in mangrove ecosystems: a review. Environmental Pollution, 2011; **159**: 2328-2346.

[82] Masuchi Y. Bacterial flora in mangroves and their biological role. Microbes and Environments, 1998; **13**: 203-215. (in Japanese).

[83] Zhu P, Wang YP, Shi TT, Zhang XL, Huang GQ, Gong J. Intertidal zonation affects diversity and functional potentials of bacteria in surface sediments: a case study of the Golden Bay mangrove, China. Applied Soil Ecology, 2018; **130:** 159-168.

[84] Chuang PC, Young MB, Dale AW, Miller LG, Herrera-Silveira JA, Paytan A. Methane and sulfate dynamics in sediments from mangrove-dominated tropical coastal lagoons, Yucatan, Mexico. Biogensciences, 2016; **13**: 2981-3001.

[85] Guo XP, Lu DP, Niu ZS, Feng JN, Chen YR, Tou FY, Liu M, Yang Y. Bacterial community structure in response to environmental impacts in the intertidal sediments along the Yangtze Estuary, China. Marine Pollution Bulletin, 2018; **126**:141-149.

[86] Muyzer G, Stams AJ. The ecology and biotechnology of sulphate-reducing bacteria. Nature Reviews Microbiology, 2008; **6**, 441-454.

[87] GomesNCM,ClearlyDFR,PiresACC, Almeida A, Mendonca-Hagler LCS, Smalla K. Assessing variation in bacterial, composition between the rhizosphere of two mangrove tree species. Estuarine Coastal and Shelf Science, 2014; **139**: 40-45.

[88] Basak P, Majumder NS, Nag S, Bhattacharyya A, Roy D, Charkraborty A, SenGupta S, Roy A, Mukherjee A, PattanayakR, GhoshA, ChattopadhyayD, Bhattcharyya M. Spatial temporal analysis of bacterial diversity in sediments of Sundarbans using parallel 16S rRNA gene tag sequencing. Microbial Ecology, 2015; **69:** 500-511.

[89] Vane CH, Harrison I, Kim AW, Moss-hayes V, Vickers BP, Hong K. Organic and metal contamination in surface mangrove sediments of South China. Marine Pollution Bulletin, 2009; **58**: 134-144.

[90] Ma B, Dai ZM, Wang HZ, Dsouza M, Liu XM, He Y, Wu JJ, Rodrigues JLM, Gilbert JA, Brookes PC, Xu JM. Distinct biogeographic patterns for archaea, bacteria, and fungi along the vegetation gradient at the continental scale in eastern China. Systems 2017; **2**. UNSP e00174-16.

[91] Tong TL, Li RL, Wu SJ, Xie SG. The distribution of sediment bacterial community in mangrove across China was governed by geographic location and eutrophication. Marine Pollution Bulletin, 2019; **140:** 198-203.

[92] Marcos MS, Barboza AD, Keijzer RM, Laanbroek HJ. Tide as steering factor in structuring archaeal and bacterial ammonia-oxidizing communities in mangrove forest soils dominated by *Avicennia germinans* and *Rhizophora mangle*. Microbial Ecology, 2018; **75:** 997-1008.

[93] Li RL, Tong TL, Wu SJ, Chai MW, Xie SG. Multiple factors govern the biogeographic distribution of archaeal community in mangroves across China. Estuarine Coastal and Shelf Science, 2019; **231**: 106414.

[94] Prato E, Biandolino F, Palapiano I, Giandomenico S, Denti G, Calò M, Spada L, Leo AD. Proximate, fatty acids and metals in edible marine bivalves from Italian market: beneficial and risk for consumers health. Science of the Total Environment, 2019; **648**: 153-163.

[95] Shenzhen Statistics Bureau. Shenzhen Statistical Yearbook. China Statistics Press, Beijing, 2017. (in Chinese).

[96] Gong Y, Chai MW, Ding H, Shi C, Wang Y, Li RL. Bioaccumulation and human health risk of shellfish contamination to heavy metals and As in most rapid urbanized Shenzhen, China. Environmental Science and Pollution Research, 2020; **27**: 2096-2106.

[97] Chai MW, Ding H, Shen XX, Li RL. Contamination and ecological risk of polybrominated diphenyl ethers (PBDEs) in surface sediments of mangrove wetlands: A nationwide study in China. Environmental Pollution, 2019a; **249**, 992-1001.

[98] Environmental Chemistry Society (ed.). An inconvenience chemical in global environment, Blue-backs of Kodansha-publisher, 2019; Tokyo, (in Japanese).

[99] Ohishi Y, Inoue M. Forest Education, Kaisei-sha Press, 2012; pp. 239. Kaisei-sha Publisher, Otsu. (in Japanese).

[100] Li RL, Yu LY, Chai MW, Wu HL, Zhu XS. The distribution, characteristics and ecological risks of microplastics in the mangroves of Southern China. Science of the Total Environment, 2020; **708:** 135025.

[101] Cetin B, Yurdakul S, Odabasi M. Spatio-temporal variations of atmospheric and soil polybrominated diphenyl ethers (PBDEs) in highly industrialized region of Dilovasi. Science of the Total Environment, 2019; **646**: 1164-1171.

[102] Chao HR, Shy CG, Wang SL, Chih-Cheng Cheng S, Koh TW, Chen FA, Chang-Chien GP, Tsou TC. Impact of non-occupational exposure to polybrominated diphenyl ethers on menstruation characteristics of reproductive-age females. Environment International, 2010; **36:** 728-735.

[103] Chai MW, Li RL, Shi C, Shen XX, Li RY, Zan QJ. Contamination of polybrominated diphenyl ethers (PBDEs) in urban mangroves of Southern China. Science of the Total Environment, 2019b; **646**: 390-399.

[104] Zhu HW, Wang Y, Wang XW, Luan TG, Tam NFY. Distribution and accumulation of polybrominated diphenyl ethers (PBDEs) in Hong Kong mangrove sediments. Science of the Total Environment, 2014; **468/469:** 130-139.

[105] Hu JC, Jin J, Wang Y, Ma ZH, Zheng WJ. Levels of polybrominated

diphenyl ethers and hexabromo-cyclododecane in the atmosphere and tree bark from Beijing, China. Chemosphere, 2011; **84:** 355-360.

[106] Sanders RA. Urban vegetation impacts on the hydrology of Dayton, Ohio. Urban Ecology, 1986; **9:** 361-376.

[107] Tu W, Hu ZW, Li LF, Cao JZ, Jiang JC, Li QP, Li QQ. Portraying urban functional zones by coupling remote sensing imagery and human sensing data. Remote Sensing, 2018; **10:** 141. DOI:10.3390/rs10010141

[108] Yang XP, Fang ZX, Yin L, Li JY, Zhou Y, Lu SW. Understanding the spatial structure of urban commuting using mobile phone location data: a case study of Shenzhen, China. Sustainability, 2018; **19**: 1435. URL: https://www.mdpi.com/2071-1050/10/5/1435/

[109] Tam NFY, Ke L, Wang XH, Wong YS. Contamination of polycyclic aromatic hydrocarbons in surface sediments of mangrove swamps. Environmental Pollution, 2001; **114:** 255-263.

[110] Li FL, Zeng XK, Yang JD, Zhou K, Zan QJ, Lei AP, Tam NFY. Contamination of polycyclic aromatic hydrocarbons (PAHs) in surface sediments and plants of mangrove swamps in Shenzhen, China. Marine Pollution Bulletin, 2014; **85:** 590-596.

[111] Zheng GJ, Lam MHW, Lam PKS, Richardson BJ, Man BKW, Li AMY. Concentrations of persistent organic pollutants in surface sediments of the mudflat and mangroves at Mai Po Marshes Nature Reserve, Hong Kong. Marine Pollution Bulletin, 2000; **40**: 1210-1214.

[112] Tam NFY, Yao MWY. Concentrations of PCBs in coastal mangrove sediments of Hong Kong. Marine Pollution Bulletin, 2002; **44**: 642-651.

[113] Vane CH, Harrison I, Kim AW, Moss-Hayes V, Vickers BP, Hong K. Organic and metal contamination in surface mangrove sediments of South China. Marine Pollution Bulletin, 2009; **58**: 134-144.

[114] Zhao B, Zhou YW, Chen GZ. The effect of mangrove reforestation on the accumulation of PCBs in sediment from different habitats in Guangdong, China. Marine Pollution Bulletin, 2012; **64**: 1614-1619.

[115] Zhang ZW, Li JL, Sui T, Cheng MM, Sun KF. Organic contamination in mangrove ecosystems of China. Ecological Science, 2017; **36:** 232-240 (in Chinese with English abstract).

[116] Tam NFY, Wong TWY, Wong YS. A case study on fuel oil contamination in a mangrove swamp in Hong Kong. Marine Pollution Bulletin, 2005; **51:** 1092-1100.

[117] Tang Y, Fang Z, Yu S, 2008. Heavy metals, polycyclic aromatic hydrocarbons and organochlorine pesticides in the surface sediments of mangrove swamps from coastal sites along the Leizhou Peninsula, South China. Acta Oceanologica Sinica, 2008, **27**: 42-53 (in Chinese with English abstract).

[118] Sharma S, MacKenzie RA, Tieng T, Soben K, Tulyasuwan N, Resanond A, Blate C, Litton GM. The impacts of degradation, deforestation and restoration on mangrove ecosystem carbon stocks across Cambodia. Science of the Total Environment, 2019; **706**: https://doi.org/10.1016/j.scitotenv.2019.135416

[119] Hwang GJ, Hung PH, Chen NS, Liu GZ. Mindtool-assisted in field learning (MAIL): an advanced ubiquitous learning project in Taiwan. Journal of Educational Technology & Society, 2014; **17**: 4-16.

[120] Li Y, Ru ZZ, Cheng HR, Wang ZL, Zan X. Study on management modes

of mangrove wetlands in Shenzhen, China. Guangdong Forestry Science and Technology, 2013; **29**: 31-37 (in Chinese with English abstract).

[121] Qin WH, Qiu QW, Zhang Y, Shen XX. Experiences of management and conservation in Hong Kong's Maipo Nature Reserve. Wetland Science & Management, 2010; **6**: 34-37 (in Chinese with English abstract).

[122] Sun YX, Luo XJ, Mo L, Zhang Q, Wu JP, Chen SJ, Zou FS, Mai BX. Brominated flame retardants in three terrestrial passerine birds from South China: geographic pattern and implication for potential sources. Environmental Pollution, 2012; **162**: 381-388.

[123] Komiyama A, Pounngparn S, Kato T. Common allometric equations for estimating the tree weight of mangroves. Journal of Tropical Ecology, 2005; **21**: 471-477.

[124] Fu HF, Tao YJ, Wang WQ. Some issues about the impacts of sea level rise on mangroves in China. Chinese Journal of Ecology, 2014; **33**: 2842-2848 (in Chinese with English abstract).

Chapter 5

The Commercial Value of Mangrove-Based Pigments as Natural Dye for Batik Textiles

Delianis Pringgenies, Ali Ridlo, Lutfianna Fatma Dewi and Ali Djunaedi

Abstract

Mangrove, or bakau as it is known in Indonesia, is one of the vegetations commonly found along the shallow coasts, estuaries, deltas and protected coastal areas and are still influenced by rising tides. After the Aceh tsunami disaster, mangrove restoration was intensively conducted in coastal areas all over Indonesia and was made into a special conservation program by the government. Mangrove is distinguishable by its big, wooden stilt roots, sharpening tip in the form of supporting leaves. The roots of the mangrove tree are morphologically distinguishable into heart root which grows into the ground and the stilt root which appear to grabs onto the surface of the ground. Mangrove forests serve several important ecological roles: they act as filters which turns saline water into fresh water, buffer from seawater intrusion, prevent erosion and abrasion, hold sediments to form new habitats, feeding ground, nursery ground, and spawning ground for a number of aquatic wildlife. Mangrove forest also possess economical functions such as as source of income, industrial ingredients for the locals and as source of new mangrove seedlings. Mangunhardjo Village, Urban Community of Mangunhardjo, Mangkang Area, Kecamatan of Tugu, Semarang City, Indonesia was an area dotted with brackish water pond. However, the area had been suffering from the effects of climate change, being inundated by overflow of river and seawater intrusion (rob). These disasters caused decline in the productivity of the ponds in the area. In an effort to combat the adverse effect of environmental change in the area, the locals of Mangunhardjo village decided to shift their livelihood by restoring the surrounding mangrove forest. Mangrove conservation at Mangunhardjo Village was conducted through activities of the program such as mangrove planting, mangrove-based food production, and mangrove waste management by applications of bioactivator bacteria for mangrove composting and production of mangrove-based natural dye for batik fabric. Mangrove-based natural dye for batik fabric from *Rhizopora mucronata* mangrove waste is a quite promising product and increases people's income.

Keywords: batik, mangrove, natural dye, *Rhizopora mucronata*, waste

1. Introduction

One of the most valuable and potential natural resources in coastal areas of Indonesia is the mangrove forest. With a coastal line exceeding 80,000 km,

Indonesia possesses 4.2 million ha total area of mangrove forest [1]. In 2014, the total coverage area of mangrove forest in Indonesia is 4,227,800 ha, which comprises approximately 25.79% global coverage area of mangrove forest [2]. However, mangrove forests in Indonesia continue to be threatened albeit relatively more protected as private sectors, NGOs, and environmental protection communities strive to preserve what is left.

There have been efforts to restore ecological, socio-cultural and socio-economical functions of mangrove forest, particularly in the northern coast of Java Island. One of most common restoration efforts is the replanting of mangrove trees.

Mangunhardjo Village, Urban Community of Mangunhardjo, Mangkang Area, Kecamatan of Tugu, Central Java Province, Indonesia was an area which mostly consisted of brackish water ponds. This area had been impacted by climate change, which caused yearly flood from the overflow of Beringin River and seawater intrusion (rob), reducing the production capacity of the brackish water ponds. The community in this area was affected by this condition, considering the brackish water ponds were the main source of income for many people in the area, as presented in **Figure 1**. To cope with this adversity, many people in the Urban Community of Mangunhardjo shifted their livelihood.

Most of the brackish water fish farmers became motorcycle taxi drivers, fragrant oil sellers, drinking water refill providers, fish-based food producers, mangrove tree seeders, etc. Today, what was once brackish water pond area has been turned into a mangrove forest restoration site. The existence of mangrove in the area has met the criteria of mangrove ecosystem. The ecosystem is protected from waves and currents as to support conservation of coastal areas in Semarang [3]. The shift in the socio-economic landscape of the community in Mangunhardjo Village came along with the progress of mangrove forest restoration. Once the restoration has shown viability, many in the community changed their focus on utilizing mangrove as a source of income. The restoration of the mangrove forest has had positive impact to the welfare of its surrounding community. In addition to enhancing the natural beauty and livability of the area and mitigating the negative impact of climate change, mangrove forest can provide a source of income to the surrounding community. Mangrove can provide economic benefit as the mangrove restoration project provide employment opportunity by sustainable planting and selling mangrove seedlings, producing and selling mangrove-based food, sourcing mangrove as basic materials for bioactivator in compost, and processing mangrove waste into natural textile dye.

Figure 1.
The site of Mangunhardjo Village, Urban Community of Mangunhardjo, Mangkang Area, Kecamatan of Tugu, Cental Java Province, Indonesia before mangrove forest restoration (a) and after mangrove forest restoration (b).

2. Mangrove tree planting

Due to the decrease in productivity of brackish water pond, the community of Mangunhardjo, Semarang, took part in mangrove forest restoration project. This activity involved almost everyone from all ages in the community. In addition to plating the mangrove used to alleviate ecological stress through enhancing biodiversity by conservation activities, the local community also nurse seedlings to be sold all over the country. Today, ecosystem of mangrove forest and its diverse plant and animal life strives in Mangunhardjo village. Healthy mangrove ecosystem provide shelter and feeding ground for various marine life, such as fish, prawn and crab, which in turn provides additional source of income for the surrounding community.

3. Mangrove-based food

Mangrove fruit can be processed into snacks and food, such as chips, syrup, brownies, klepon, sticks and other kind of snacks. The species used in making food are Lindur (*Bruquiera gymnorrhiza*), Api-api (*Avecennia sp*), and Pidada (*Sonneratia sp*), (*Rizhopora sp*) Lindur fruit is rich in carbohydrate, higher than that of rice. Mangrove fruit has tannin content, which gives it a bitter taste. To lower its tannin contents, the fruits are boiled or immersing overnight, before they are processed. Boiling or immersion has proven to reduce the tannin content of mangrove fruits by 40%. The fruits are also made into flour, to preserve its quality. Storing mangrove fruits as flours halts its metabolism and giving it a longer shelf life due to the lack of water, making it a viable ingredient for various food.

R Mangrove-based food made and marketed by women of the fishing community of Mangunhardjo village are chips, syrup, sticks, klepon, and cakes, as presented in **Figure 2**.

Figure 2.
Cookies made from Avecenia mangrove fruit.

4. Mangrove-based bacterial bioactivator for composting

Microorganisms associated with mangrove waste synthesize secondary metabolites similar to their host. These microbes are viable source of new compounds. Symbiont bacteria of mangrove waste are bacteria which thrive in association with mangrove waste. These associated bacteria contribute in the cycle of nutrition of its host and are decomposing agents for the waste. Compounds produced by symbiont bacteria has the potential to be used as precursors for the biosynthesis metabolism of immunity against pathogenic bacteria and other predators [4]. Microbes are the most numerous of all the organisms living on water, and as symbiotes of other organisms [5]. One way that bacteria contributes to its ecosystem is to act as a decomposer in breaking down organic materials such as dead leaves around mangrove plants. Due to bacterial activities, dead mangrove leaves are eventually broken down into nutrition. One of the processes in mangrove ecosystem which significantly contribute to the biodiversity in the water is decomposition, or specifically, the disintegration of mangrove leaves into nutrition. Disintegration is a step in the decomposition process, which in turn will produce important nutrients within the food chain, through the productivity of the surrounding waters [6]. Decomposition bacteria are groups of bacteria with the capability of decomposing other dead microorganisms into its basic building blocks, all of which will return to the environment. These decomposition bacteria are categorized into saprophytic organisms, due to its ability to break down organic compounds in nature. Saprophytic bacteria break down dead plants or animals and remains or waste of other organisms [7, 8]. Mangrove waste is a supply of organic materials to mangrove ecosystems, which maintains the carrying capacity of the surrounding area [9].

Microbes isolated from plants with bioactive compounds have been known to have similar compound to its host and, in some cases, even indicate greater activity than that of its host [10]. A study on symbiont microbes with bioactivator potential found four viable species, namely Pseudomonas sp., Flavobacterium sp., Acinetobacter sp., and Bacillus subtilis. The consortium of the 4 species can act as organic waste decomposer and restore the color and odor of fresh water [11]. The symbiont bacteria from mangrove waste have seen application in bioactivator products which has been used by the community in Urban Community of Tembalang, Semarang, Central Java to process organic waste into compost, as presented in **Figure 3** below.

Figure 3.
Liquid Bioactivator products from symbiont bacteria of mangrove waste, which are used in making compost.

5. Using mangrove waste as natural textile dye

The latest development in fashion industry sees a demand for breakthrough from designers and scholars to create textile materials and clothes that are creative, innovative and marketable. Batik, as one of the most sought after fashion products in Indonesia, are mostly made using synthetic dye. Synthetic dye has its advantages, namely its availability, range of colors and the practicality of its application. However, the use of synthetic dye pose health risk of consumers, and even greater threat to the environment. Due to its carcinogenic nature, the use of these dyes in fabric may trigger allergy reaction. The process by which these dyes is made also presents environmental hazards. Therefore, there is an opportunity to reintroduce natural dye as a safer, more environmentally friendly alternative. Batik clothes and fabric made using natural dye have high commercial value because of its artistry, unique colors and the sustainability by which they are made and sourced. The use of natural dye in batik also give impressions of ethnic-look and exclusiveness.

Rhizophora mucronata is one of the potential mangrove species to be used in the production of natural dye. Other than being an important species for the mangrove restoration project in Mangunhardjo village, Semarang, *R. mucronata* still sees limited utilization by the surrounding community. Yet, parts of mangrove have been known to be used as natural dye in several other areas in Indonesia such as Papua and Takisung. [12] mentioned that *R. mucronata* is a natural tanning agent commonly used in textile industry and can produce color variation depending on the mordant used. A number of studies also indicate the potency of *R. mucronata* as dye material. In Bontang, Borneo, fruit of *R. mucronata* is used as a material in dye production for the local industry [13]. One study also found that natural dye made from *R. mucronata* passed the quality test with a predicate of 'fine' [14]. Although color pigments of *R. mucronata* parts can be sourced as a material for natural dye, there are more color variation and and ways to retain colors that has yet to be tested.

Pigments of *R. mucronata* is a unique potential of this species of mangrove. The pigment content can become an asset through effective and efficient utilization, which can provide economic value to the community around mangrove ecosystem. Therefore, further studies on pigments of bark, propagules, and leaves of *R. mucronata* in relation to their application in batik as dye materials.

Batik has experience a rise in popularity among both the locals and foreigners in the last few decades. The increasing demand of batik products also creates increasing demand for and use of synthetic dye. This is due to the fact that synthetic dyes are marketed at a lower price point and have better color retention compared to natural dyes. However, as more and more consumers become more aware of environmental issues, fabrics with natural dyes becomes more popular in the market. Synthetic dye has been known to be carcinogenic, and the waste from its production poses danger for the environment. [15] mentioned that synthetic dye is mutagenic and non-degradable in nature. Orange II is one example of the most prevalent artificial dye in the industry. This artificial dye has been known to not easily broken down by natural means. The waste from production and use of synthetic dyes has also been known to contain high levels of heavy metals such as chromium, zinc, copper, etc. [15, 16] wrote that waste water from textile production activity was found to pose health hazard to the surrounding community in Palembang, due to its high content of corrosive chemicals, organic pollutants, and high levels of acidity in its waste. Pollutants from synthetic dye production and use contains high leves of heavy metals, and intermediate dyes which are mutagenic in nature [17]. Not only does this damage the environment, it also pose health hazard to the community.

Compared to stock of sythetic dyes, the availability of natural dyes is more limited since artificial dyes are mass produced and have better distribution chain whereas natural dye often see limited production and must be sourced directly from its native area. Yet, not all sythetic dyes in Indonesian market is produced within the country. *R. mucronata* with its application potential as a material for natural dye can be found all over Indonesia, yet there has been limited commercial exploitation for this use. Studies of *R. mucronata* parts to be used as dyes for batik fabric are expected to contribute to the novelty of *R. mucronata* as an alternative source for dye in batik textile industry.

6. The biology of *R. mucronata*

R. mucronata is a species prevalent in the Indo-Pacific region. In Indonesia, *R. mucronata* is known locally as "Bakau Hitam" (lit. Black Mangrove), "Bakau Korap", "Bakau Merah" (lit. Red Mangrove), "Angka Hitam", "Belukap", "Dongoh Korap", "Jankar", "Lenggayong", and "Lolaro". This mangrove often become the choice plant in mangrove restoration programs [18].

R. mucronata is classified into the genus Rhizophora. One distinguishing feature of this species is its broad leaves. There are two other species within the Rhizophora genus, namely *Rhizophora apiculata* andRhizophora stylosa along with two hybrids, namely Rhizophora lamarckii (a hybrid between *R. apiculata* and R. stylosa) and Rhizophora annamalayana (a hybrid between R.apiculata and *R. mucronata*) [19].

The tree of *R. mucronata* can reach a height of 30 meters. The trunk diameter can grow up to 70 cm with bark which is dark, mostly black and a horizontal crevice. Stilt roots and aerial roots grow on the lower branches of the tree. The stilt roots can be quite sizable and are woody. The stilt roots of *R. mucronata* are usually abortive, whereas the lateral roots can be quite numerous in one tree and extend from the tip of the branch as well as possessing numerous branches on itself, which are also known as stilt roots/hoop/pile-like which supports the tree. Aerial roots can sometimes be found in the lower branches. The trunk itself is enclosed cylindrical in form, with bark that are black or dark red, has a coarse, scaly texture, and with horizontal crevices formed around the bark [20].

The leaves of *R. mucronata* has layers with green stalks. The leaves can reach a length of 23 cm. They are typically elliptical with narrow tips. The propagules of *R. mucronata* have an egg-like shape. The color of the fruit varies from green to brownish. The base of the fruit has a coarse texture and typically monocots. When ripe, the cotyledon neck will turn yellow.

R. mucronata is a species with the highest tolerance for sandy environment, compared to other Rhizophora species. This species is commonly found in tidal area with sand substrates [21]. *R. mucronata* thrives in mud with fine, grainy soil and is believed to be one of the mangrove species capable of surviving during inundation by high tide [22].

7. Chemical composition of *Rhizophora mucronata*

Mangrove commonly contains compounds such as alkaloid, flavonoid, phenol, terpenoid, steroids, and saponins [23]. Proximate analysis of *R. mucronata* fruit by [24] found that there were 46.63% of water, 1.96% of fat, 0.41% of protein, 1.25% of ash and 22.29% of carbohydrate. [25] identified the phytochemical contents of *R. mucronata* bark, and found Positive results on phenolic compounds (including flavonoid and tannins), and believed that the tannins are drawn in the methanol extract with testing using FeCl3. In addition to phenolic group compounds, secondary metabolites such as terpenoids/steroids, alkaloid and saponins were also found in the bark of *R. mucronata*, only in this study the results were obtained by the use of multiple solvents (ethyl acetate and methanol) and varying reagents.

The leaves of *R. mucronata* was indicated as the most effective part to be used in extraction process [26]. It was found that the leaves of *R. mucronata* are rich in phenolic compounds, consisting of several flavonoid, phenolic acids, and tannin [27]. *R. mucronata* leaves contains dihydroflavonol with free 5-OH and 7-OH, with restored raffinose at 3-OH, caffeic acid, vanillic acid, p-hydroxybenzoate acid, and tannin, believed to be catechin tannin. [28] elaborated that the extract of *R. mucronata* leaves, both fresh and dried, and extracted using sterilized distilled water contains the following phytochemical constituents: alkaloid, carboxylic acid, coumarin, flavonoid, phenol, protein, amino acid, quinone, resin, saponin, steroids/phtyosterols, tannins, xanthoprotein.

8. The use of *Rhizophora mucronata*

In Madagascar, the wood of *R. mucronata* is extensively used in making boats and fishing nets for fish and shrimps, and is domestically used as a construction material for fences, housings and cooking fuel [29]. The indigenous people of Papua has also been using *R. mucronata* as materials for fence poles, walls and boats. In addition, fruits if *R. mucronata* has been used to treat diarrhea. Whereas in general, parts of the Rhizophora tree are brewed into alcoholic drinks in the Wondama Bay area [30]. In the field of biochemistry, bark from *R. mucronata* containing polysaccharides has been used as an in-vitro treatment for human immunodeficiency virus (HIV) [31].

[32] stated that *R. mucronata* is one of eight types of sources for natural dye used by the people of Papua. This species is used for several purposes such as a material for dye, food ingredients, and cosmetics [33]. *R. mucronata* bark from Takisung area has been used for dyeing batik cloth [34]. [35] wrote that *R. mucronata* bark, which has a natural brown pigment, is used as a textile dye because its tannin content reaches 30%. [36] successfully used the stem and leaf waste of *R. mucronata* as a natural dye for batik on cotton and silk fabrics.

9. Batik

Batik is a form of textile product that are generally used in the form of various crafts, tablecloths, sheets, and clothing. In 2009, UNESCO awarded Indonesian Batik as an Intangible Cultural Heritage of Humanity. The uniqueness of batik products are often found in their style, use, and design which are not only attractive to the local market, but also to the international market. In the period between January to June 2014, Batik became one of the commodity groups that had the highest export value in Central Java, compared to the other two commodities, namely textiles and textile goods. Textile and textile goods have the largest contribution of 36.84% of total export value or approximately US$189.01 million. This export value shows that textiles and textile products in Central Java have a huge potential as a contributor to the country's foreign exchange.

10. Textile dye

Color becomes visible to the eye when there is absorption of a portion of the color spectrum in the visible area by molecules. The molecular structure is responsible for the presence of compounds that absorb visible light, which will be interpreted as colors. Molecules in plants consist of chains of carbon, oxygen, and hydrogen as main compounds and a few additional heteroatoms such as nitrogen. Molecules that absorb visible light are filled by chains of alternating and single carbon bonds that are alternating or conjugated. The longer the double bond, the more vivid the colors will appear. This bond can absorb visible light in certain areas, which provides coloration to the compounds [37].

Textile dyes, based on the materials from which they are sourced, are classified into two namely Natural Dyes (ZPA) and Synthetic Dyes (ZPS). ZPA is a dye obtained from natural ingredients, which generally comes from the extracts of plants or animals. ZPS is artificial dyes or syntheses made by chemical reactions using the basic ingredients of charcoal, coal, or petroleum which are the result of aromatic hydrocarbon derivatives such as benzene, naphthalene, and anthracene [38].

The intensity of the color produced in natural dyes depends on the type of coloring matter. Coloring matter is the substance that determines the hue of natural dyes and is an organic compound. The classification of natural dyes based on coloring matter is divided into four groups namely mordant dyes, direct dyes, acid/base dyes, and laver dyes. Mordant (natural) dyes in the coloring process must be combined with a metal oxide complex to form an insoluble dye. Natural dyes in mordant dyes have good color resistance potential, for example Moridin dyes from Noni roots. Direct dyes are retained to the fabric fibers based on hydrogen bonds, making the color retention low, for example Curcumin from turmeric. Acid/base dyes consist of a combination of acid and base groups, such as flavonoid pigments. The last group is laver dye. These dyes must go through the process of reduction–oxidation (redox) in the fabric dyeing process. In addition, laver dyes are also known as the oldest dyes in the world because they have the best color retention among the three other classes of natural dyes. One example of laver dyes is Indigo from torn leaves [13]. Natural dyes that have been explored from plants and have been used in fabric coloring include sengon leaves (*Albizia falcataria*) as silk fabric dyes, mangosteen rind (*Garcinia mangostana*) as natural dyes on cotton fabrics, *Morinda citrifolia* bark on *Morinda citrifolia* cotton cloth, purple sweet potato (*Ipomea batatas*), etc. [39–42]. Extraction of natural dyes is mostly carried out using polar solvents such as distilled water, ethyl acetate, methanol, acetone and n-hexane [25, 37, 43].

11. Dye extract

In general, the results of extraction from leaves, bark, and propagules show brown color with different color density variations (**Figure 4**). The brown color indicates the presence of tannin [11]. Previous studies have found that high levels of tannin produce a dense color on tea leaves [44]. Several factors such as the extraction temperature below 100°C, the type of solvent (polar) used for extraction, particle size, and extraction time are things that need to be taken into account in producing quality tannins [45]. Tannins are found in the bark, fruit (propagules), and leaves of *R. mucronata* [46].

However, there is variation in tannin content in each part of the tree. The content of tannin *R. mucronata* has similarities with tannin derived from *Ceriops tagal* bark,

Figure 4.
Batik fabric coloring from mangrove waste extraction with lime, tunjung and alum fixations.

which is soluble in distilled water that has polar properties [47, 48]. The tannin content produced from extraction using distilled water did not differ significantly compared to other solvents with similar polarity.

12. FTIR analysis and UV-Vis spectrophotometry

UV Vis spectropometry to extracts of *R. mucronata* leaves, bark, and propagules extracted at 70°C shows the maximum absorbance located at a wavelength of 412 nm. This shows the existence of conjugated C=C and C=O bonds. The maximum absorbance value obtained at wavelengths between 300 and 550 nm indicates the presence of $\pi \rightarrow \pi^*$ denoting conjugated C=C and n transitions $\rightarrow \pi^*$ in the form of chromophore C=O [49]. Tannins are classified as natural polyphenol compounds which contain phenolic hydroxyl groups and carboxyl groups. In addition, there are also chromophore groups which generally give color to a compound. The C=C conjugated bonds and C=O are included in the chromophore group, thus supporting the notion that the brown color that arises from extraction is caused by the presence of tannin content.

Subsequent testing to see the absorption pattern using an infrared spectrophotometer. Test results on the three types of dye extracts on leaves, bark and propagules showed a similar absorption pattern. Absorption in the range of wave numbers 3500 to 3000 cm^{-1} and 2000 to 1500 cm^{-1} indicates the presence of C-H groups. The C-O group is also indicated although it must be further analyzed in the fingerprint area. The C-O group forms an aromatic compound, which is part of the tannin together with the O-H and -CH2 groups [50]. The solid-shaped extract of D_{70} shows a different absorption pattern. O-H, C-H, C=O ester, and C-O-C ether groups are indicated. The existence of these four types of functional groups shows that the flavonoid compound is indicated in D_{70} extract. This is supported by research conducted by [51] who found that the flavonoid compounds from the flavonone group had the OH functional group bound, aliphatic CH, C=O, C=C Aromatic, C-O and C-H aromatics. [52] revealed that flavonoids are building blocks of proanthocyanidin compounds which are condensed tannins. Flavan-3-ols polymer compounds consisting of (+)-catechin and (−)-epicathecin are the main constituents of the group of flavonoid compounds that fall into the category of condensed tannins. To support these findings, Total Phenol Content and Total Flavonoid Content were conducted.

13. Total phenol content and total flavonoid content

Determination of total phenol in D_{70} extract was carried out using the Folin–Ciocalteu reagent. This reagent is sensitive in reducing compounds such as polyphenols and in its reaction will show blue when measured by spectrophotometer [53]. The test results showed that the phenol content were 2.4950 mg GAE/g. The presence of phenol in the extract can be an indication of tannin. In general tannins are high molecular weight polyphenol compounds, which naturally form complexes with protein [54]. Testing of total flavonoids was also carried out and the results obtained were 0.6516 mg QE/g. Flavonoids are still included in polyphenol compounds and usually consist of flavones, flavonols, and condensed tannins, which are secondary metabolites of plants [53, 54].

14. Fabric with mangrove-based dye

Fabric dyeing using natural dyes of *R. mucronata* was done by immersing dry, white cotton fabric with the dye and then air-drying them. This step was repeated three times. When immersing the fabric into the dye solution, the fabric undergoes swelling so that the pores of the fabric fibers will open and the dye can be absorbed into the fiber together with the dye solution. Dyes that have been absorbed into the fiber will be bound by reactive groups on cellulose fibers in the form of hydroxyl groups (OH) and form hydrogen bonds. The finished dyed fabric was then aerated with protection from sun exposure. After the cloth dries, the cloth is then immersed in alum color-fixating agent ($KAl(SO4)2.12H_2O$). When dyeing, the dye is absorbed into the fabric fibers. But in general there are substances on the fabric surface that block the process, so fixation agents such as alum are needed to help the absorption of dyes on the fabric and increase color retention by binding the dye molecules to the fibers of the fabric. The reaction between the fabric which has been dyed and fixated by alum (KAl(SO4)2.12H2O) does not produce complex salts but compounds which are ionically bonded [55].

The colors produced by the three types of dyes from *R. mucronata* after fixation with alum through qualitative observations based on [56] were shades of tawny/tenné brown. The tawny brown digital code according to [57] is AE6938. Tawny brown can be described as a light brown hue with a combination of brown and orange The tawny brown color that was obtained after fixation with alum was not much different from the color before the addition of alum. This is in accordance with the nature of alum which gives out hue according to its original color [58].

15. Color retention test

The results of the color retention test of dyed fabrics from parts of the *R. mucronata* through extraction with temperature variations and fixation using alum showed permanent color properties. The value of Gray Scale and Staining Scale in the color retention test against fabric rubbing showed almost the same results, namely in the category 4 (fine) and 4–5 (fine). In the soap washing test, the average results showed 3–4 (adequate) to 4 (fine). Leaf extracts heated at 70°C (D_{70}) and bark extract heated at the same temperature (K_{70}) consistently showed a staining scale value of 4 showing the 'fine' category at three replications of the test. In general, all test results met the minimal SNI standards of 3.

Tannin commonly used as dyes are found in *Ceriops tagal* mangrove bark and can produce a brownish red color [48]. The leaves have 15% less tannin content.

Types of tannins in *Ceriops tagal* and *R. mucronata* are tannins condensed with procyanidin types. Tannin extraction from plants is strongly influenced by the composition of the solvent used [59]. The optimal solvent will be able to produce tannins in large quantities. In addition to the dyes obtained, the color retention of the fabric also depends on the fixation agent. Staining quality test results that showed the category of 'adequate' and 'fine due to the use of alum fixation, creating strong molecular bond which in turn contributes to good color retention. According to [59], the strength of the bond that occurs between fabric fibers and dyes determines the color retention during the washing process. Dyes strongly retained in the fabric fibers will create better, more vibrant colors after being washed.

16. Conclusions

Mangroves can be used as natural dyes on batik cloth, because of their high availability and positive impact on the household economy of the local community. *Rhizophora mucronata*, a mangrove species commonly found in the coastal areas of Semarang, is used in conservation efforts and beneficial in the field of fisheries. The existence of this species of mangrove forests can improve the catch and welfare of the local fishing community.

Author details

Delianis Pringgenies*, Ali Ridlo, Lutfianna Fatma Dewi and Ali Djunaedi
Department of Marine Science, Faculty of Fisheries and Marine Science, Diponegoro University, Semarang, Indonesia

*Address all correspondence to: delianispringgenies@lecturer.undip.ac.id

References

[1] Tarigan, M.S. 2008. Sebaran dan Luas Hutan Mangrove di Wilayah Pesisir Teluk Pising Utara Pulau Kabaena Provinsi Sulawesi Tenggara. Makara Sains, 12 (2): 108-112.

[2] Hamilton, S.E. dan D. Casey. 2016. Creation of a high spatio-temporal resolution global database of continuous mangrove forest cover for the 21st century (CGMFC-21). Journal of Macroecology, 25 (6): 729-738.

[3] Zaky, C. Suryono, and R. Pribadi, "Kajian Kondisi Lahan Mangrove di Desa Bedono, Kecamatan Sayung, Kabupaten Demak dan Kelurahan Mangunharjo, Kecamatan Tugu, Kota Semarang," Journal of Marine Research, vol. 1, no. 2, pp. 88-97, Feb. 2013.

[4] Taylor, W. J., dan R. Ford., 2007. Chickpea. In: Chittarajan, K., Genome Mapping and Molecular Breeding in Plants, 3 (6): 109-122

[5] Pelczar, M. J., dan E. C. S. Chan. 2008. Dasar-Dasar Mikrobiologi Jilid I. Jakarta: UI Press.

[6] Widhitama S, Purnomo PW, Suryanto A. 2016. Produksi dan Laju Dekomposisi Serasah Mangrove Berdasarkan Tingkat Kerapatannya di Delta Sungai Wulan, Demak, Jawa Tengah. Diponegoro. Journal of Maquares. 5: 311-319.

[7] Todar, K. 2008. Online Textbook of Bacteriology. http://www.text book of bacteriology.net/index.html

[8] Saraswati, R. dan Prihatini, T. 2004. Teknologi pupuk mikroba untuk meningkatkan efisiensi pemupukan dan keberlanjutan sistem produksi padi sawah. Dalam: Fahmuddin,A,. et al. Tanah sawah dan teknologi pengelolaannya. Pusat Penelitian dan Pengembangan Tanah dan Agroklimat. Bogor 40

[9] Lestari, T. 2015. Kumpulan teori untuk kajian pustaka penelitian kesehatan. Yogyakarta: Nuha Medika.

[10] Krinsky, N.I., dan E. J. Johnson. 2005. Carotenoid Actions and Their Relation to Health And Disease. Molecular Aspects of Medicine, 26: 459-516.

[11] Pringgenies, D., R. Widiyadmi., D. Ariyanto., R. Idris., dan A. Djunaedi. 2018. Bakteri Konsorsium dari Serasah Mangrove untuk Produksi Kompos. Jurnal Pengelolaan Perairan, 1(2): 19-26.

[12] Lacasse, K. dan W. Baumann. 2004. Textile Chemicals: Environmental Data and Facts. Springer Science & Business Media, Berlin, 1184 p.

[13] Sulistyo, I. dan B. Budi. 2013. Pemberdayaan Masyarakat melalui Program Corporate Social Responsibility (CSR) berupa Pengembangan Pewarna Alami dari Buah Mangrove Spesies Rhizophora mucronata untuk Batik Khas Bontang Kalimantan Timur. Jurnal Komunikasi Massa, 6 (2): 135-150.

[14] Kwartiningsih, E., Paryanto, W. Agung W., E. Mastuti, R. Sonia A., dan Y. Pipit N. 2014. Pemanfaatan Ekstrak Buah Mangrove (*Rhizophora mucronata*) untuk Pewarna Alami Batik. Seminar Nasional Tekstil, Bandung.

[15] Agustina, T.E., E. Nurisman, Prasetyowati, N. Haryani, L. Cundari, A. Novisa, dan O. Khristina. 2011. Pengolahan Air Limbah Pewarna Sintetis dengan Menggunakan Reagen Feton. Dalam: Prosiding Seminar Nasional AVoER ke-3 di Palembang Tanggal 26-27 Oktober 2011. Fakultas Teknik Universitas Sriwijaya, pp. 260-266.

[16] Indrianingsih, A.W. dan C. Darsih. 2013. Natural Dyes from Plants Extract and Its Applications in Indonesian Textile Small Medium Scale Enterprise.

Technical Implementation Unit for Chemical Engineering Processes, 11(1): 16-22.

[17] Tanziz, R. 2009. Laporan Teknis : Identifikasi Logam Berat dalam Zat Warna Tekstil, Balai Besar Tekstil, Bandung.

[18] Duke, N., K. Kathiresan, S.G. Salmo III, E.S. Fernando, J.R. Peras, S. Sukardjo, dan T. Miyagi. 2010. *Rhizophora mucronata*. The IUCN Red List of Threatened Species 2010: e.T178825A7618520. http://dx.doi.org/10.2305/IUCN.UK.2010-2.RLTS.T178825A7618520.en (11 April 2016).

[19] Setyawan, A.D., Y.I. Ulumuddin, dan P. Ragavan. 2014. Review: Mangrove hybrid of Rhizophora and its parental species in Indo-Malayan region. Nusantara Bioscience, 6 (1): 69-81.

[20] Orwa, C., A. Mutua, R. Kindt, R. Jamnadass, dan S. Anthony. 2009. *Rhizophora mucronata*.http://www.worldagroforestry.org/treedb/AFTPDFS/Rhizophora.

[21] Noor, R.Y., M. Khazali, dan I.N.N. Suryadiputra. 1999. Panduan Pengenalan Mangrove di Indonesia. PHKA/WI-IP, Bogor.

[22] Tan, R. 2001. Rhizophora mucronata. http://www.naturia.per.sg/buloh/plants/rhizophora%20mucronata.htm (12 Mei 2016).

[23] Kordi K, Guhufran H. 2012. Ekosistem Mangrove, Potensi, Fungsi, dan Pengelolaan. Jakarta : Rineka Cipta.

[24] Bunyapraphatsara, N., Srisukh, V., Jutiviboonsuk, A., Sornlek, P., Thongbainoi, W., Chuakul, W., Fong, H.H.S., Pezzuto, J.M. and Kosmeder, J. 2002. Vegetables from the mangrove areas. Thai Journal of Phytopharmacy 9(1): 1-12. (1) (PDF) Study of ripe Rhizophora mucronata fruit flour as functional food for antidiabetic.

[25] Pradana, D., D. Suryanto, dan Yunasfi. 2014. Uji Daya Hambat Ekstrak Kulit kayu Rhizophora mucronata terhadap Pertumbuhan Bakteri Aeromonas hydrophila, Streptococcus agalactiae, dan Jamur Saprolegnia sp. Secara In Vitro. Aquacoastmarine, 2(1): 78-92.

[26] Yasmon, A. 2000. Sensitifitas *Vibrio Parahaemolyticus* terhadap Ekstrak Mangrove Rhizopora Apiculata di Dalam Lumpur dan Air Laut. Skripsi Sarjana Fakultas Perikanan dan IlmuKelautan Universitas Riau. 37p.

[27] Sutjihati, R., S. Soetarno, dan S. Kusmardiyani. 1995. Pemeriksaan Senyawa Fenolik Daun *Rhizophora mucronata* lamk. (Rhizoporaceae), Suatu Tumbuhan Mangrove [Skripsi]. Sekolah Farmasi ITB. http://bahan-alam.fa.itb.ac.id (12 Mei 2016).

[28] Babuselvam, M., K. Kathiresan, S. Ravikumar, M. Uthiraselvam, dan E. Rajabudeen. 2012. Scientific evaluation of aqueous extracts of fresh and dried leaves from Rhizophora mucronata lamk (Rhizophoracea) in Rats. African Journal of Pharmacy and Pharmacology, 6 (11): 814-817.

[29] Rasolofo, M.V. 1997. Use of Mangroves by Traditional Fishermen in Madagascar. Mangroves and Salt Marshes Vol. 1(4): 243-253.

[30] Arobaya, Agustina Y.S. dan Freddy Pattiselanno. 2010. Potensi mangrove dan Manfaatnya bagi Kelompok Etnik di Papua. Biota, 5(3): 494-500.

[31] Premanathan, M., Arakaki, R, Izumi, Kandasamy, K. Nakano, Masatoshi, Yamamoto, N. Nakashima, H. (1999). Antiviral properties of a mangrove plant, *Rhizophora apiculata* Blume, against human immunodeficiency virus. Antiviral Research. 44. 113-122. 10.1016/S0166-3542(99)00058-3.

[32] Makabori, S. 1999. Teknik silvikultur jenis-jenis tanaman penghasil warna alam Irian Jaya. Seminar Menggali Potensi Warna Alam Irian Jaya. Departemen Kehutanan dan Perkebunan Propinsi Irian Jaya.

[33] Wibowo, A. 2003. Identifikasi Jenis-jenis Tumbuhan Penghasil Warna Alami dan Pemanfaatannya dalam Kehidupan Suku Hatam di Kampung Mbenti Distrik Anggi Kabupaten Manokwari. [Skripsi]. Manokwari: Fakultas Kehutanan Universitas Negeri Papua.

[34] Hamidah, S. 2006. Rendemen dan Kadar Tanin Kulit Kayu Bakau (Rhizophora mucronata Lamck) dari daerah Takisung. Jurnal Hutan Tropis Borneo (18): 15-23..

[35] Prabhu, K.H. dan A.S. Bhute. 2012. Plant based natural dyes and mordnats: A Review. J. Nat. Prod. Plant Resour., 2(6): 649-664.

[36] Pulungan, A.S.S. 2014. Pengaruh Fiksasi terhadap Ketuaan Warna dengan Menggunakan Pewarna Alami Batik dari Limbah Mangrove. Dalam: Prosiding Seminar Nasional Biologi dan Pembelajarannya di Medan tanggal 23 Agustus 2014. Medan, pp: 297 – 301.

[37] Budimarwanti, C. dan S. Handayani. 2010. Efektivitas Katalis Asam Basa pada Sintesis 2-hidroksikalkon, Senyawa yang Berpotensi sebagai Zat Warna. . Dalam: Prosiding Seminar Nasional Kimia dan Pendidikan Kimia Tanggal 30 Oktober 2010. Yogyakarta, pp.

[38] Isminingsih. 1978. Pengantar Kimia Zat Warna. Bandung: STTT Press.

[39] Kusriniati, D. 2008. Pemanfaatan Daun Sengon (Albizia falcataria) sebagai Pewarna Kain Sutera menggunakan Mordan Tawas dengan Konsentrasi yang Berbeda. Teknobuga, 1 (1): 7-14..

[40] Manurung, M. 2012. Aplikasi Kulit Buah Manggis (Garcinia mangostana L.) sebagai Pewarna Alami pada Kain Katun secara Pre-Mordanting. Jurnal Kimia, 6(2): 183-190.

[41] Thomas, M., M. Manurung, dan I.A.R.A. Asih. 2013. Pemanfaatan Zat Warna Alam dari Ekstrak Kulit Akar Mengkudu (Morinda citrifolia Linn) pada Kain Katun. Jurnal Kimia, 7(2): 119-126.

[42] Winarti, S., U. Sarofa, dan D. Anggrahini. 2008. Ekstraksi dan Stabilitas Ubi Jalar Ungu (Ipomoea batatas L.) sebagai Pewarna Alami. Jurnal Teknik Kimia, 3(1): 207-214.

[43] Artati, E.K. dan Fadilah. 2007. Pengaruh Kecepatan Putar Pengadukan dan Suhu Operasi pada Ekstraksi Tanin dari Jambu Mete dengan Pelarut Aseton. Ekuilibrium, 6(1): 33-38.

[44] Adriana, N., R. Batubara, dan E. Julianti. 2015. Nilai Kesukaan Konsumen terhadap The Daun Gaharu (Aquilaria malaccensis Lamk.) berdasarkan Letak Daun pada Batang. Peronema Forestry Science Journal, 4(4): 1-5.

[45] Hamidah, S. 2006. Rendemen dan Kadar Tanin Kulit Kayu Bakau (Rhizophora mucronata Lamck) dari daerah Takisung. Jurnal Hutan Tropis Borneo (18): 15-23.

[46] Musman, M. 2010. Tanin Rhizophora mucronata sebagai Moluskosida Keong Mas (Pomacea canaliculata). Bionatura, Jurnal Ilmu-ilmu Hayati dan Fisik, 12(3): 184-189.

[47] Handayani, P.A. dan I. Maulana. 2013. Pewarna Alami Batik dari Kulit Soga Tinggi (Ceriops tagal) dengan Metode Ekstraksi. Jurnal Bahan Alam Terbarukan, (2): 1-6.

[48] Jansen, P.C.M dan D. Cardon. 2005. Dyes and Tannins. Volume 3. Backhuys Publishers, Wageningen, 215 p.

[49] Sastrohamidjojo, H., 2001, Spektroskopi, 3-4, 11, Liberty Press, Yogyakarta.

[50] Hutauruk, S. 2004. Uji Aktivitas Tanin pada Daun Jati Belanda (Guazuma ulmifolia Lamk.) sebagai Tabir Surya. [Skripsi]. Fakultas Matematika dan Ilmu Pengetahuan Alam, Universitas Diponegoro, Semarang, 32 hlm.

[51] Parubak, Apriani Sulu. 2013. Senyawa Flavonoid yang Bersifat Antibakteri dari Akway (Drimys becarina Gibbs). Chemistry Progress, 6(1): 34-37.

[52] Dixon RA, Xie DY, and Sharma SB. 2005. Proanthocyanidins--a final frontier in flavonoid research?. New Phytol. 2005 Jan;165(1):9-28.

[53] Sultana, M., P.K. Verma, R. Raina, S. Prawez, dan M.A. Dar. 2012. Quantitative Analysis of Total Phenolic, Flavonoids, and Tannin Contents in Acetone and n-Hexane Extracts of Ageratum conyzoides. Journal of ChemTech Research, 4(3): 996-999.

[54] Baba, S.A. dan S.A. Malik. 2015. Determination of Total Phenolic and Flavonoid Content, Antimicrobial, and Antioxidant Activity of A Root Extract of Arisaema jacquemontii Blume. Journal of Taibah University for Science, 9: 449-454.

[55] Prayitno, R.E., S. Wijana, dan B.S. Diyah. 2014. Pengaruh Nahan FIksasi terhadap Ketahnan Luntur dan Intensitas Warna Kain Mori Batik Hasil Pewarnaan Daun Alpukat (*Persea Americana* Mill.). Jurnal Lulusan TIP FTP UB, 1-8.

[56] Maerz, A.J. dan M.R. Paul. 1930. A Dictionary of Color. http://people.csail.mit.edu/jaffer/Color/M.htm (19 Agustus 2016).

[57] Ridgway, R. 1912. Color Standards and Color Nomenclature. http://people.csail.mit.edu/jaffer/Color/R.htm.

[58] Widowati, T.B. dan G. Sutapa. 2012. Pemanfaatan Bagian Cabang dan Pucuk Cabang Dalbergia latifolia sebagai Pewarna Alami Kain Batik. Dalam: Prosiding Seminar Nasional Masyarakat Peneliti Kayu Indonesia (MAPEKI) XVI di Makassar Tanggal 6-7 November 2012.

[59] Yuliana, P. 2014. Ekstraksi Senyawa Tanin dan Saponin dari Tanaman serta Efeknya terhadap Fermentasi Rumen dan Metanogenesis In Vitro. [Tesis]. Sekolah Pascasarjana, Institut Pertanian Bogor, Bogor, p.37.

Chapter 6

Mangrove Ecosystem Restoration after Oil Spill: Bioremediation, Phytoremediation, Biofibers and Phycoremediation

Ícaro Thiago Andrade Moreira, Célia Karina Maia Cardoso, Evelin Daiane Serafim Santos Franco, Isadora Machado Marques, Gisele Mara Hadlich, Antônio Fernando de Souza Queiroz, Ana Katerine de Carvalho Lima Lobato and Olívia Maria Cordeiro de Oliveira

Abstract

Environmental accidents involving spills of oil and its derivatives in mangroves present themselves as difficult problems to be solved in the short term, as for example in the construction of emergency strategies to combat the arrival of oil stains and fragments. Petroleum its derivatives and the residues generated in this chain, have a complex mixture of hydrocarbons and are considered dangerous substances. This mixture is difficult to degrade and can cause multiple problems in the ecosystem. Our developed biofiber barrier removes oil more than five times in relation to its mass in a simple way and in a short time. However when the spilled oil reaches the mangroves, other biotechnologies were developed and applied such as phytoremediation (87% efficiency), the use of microalgae (94% efficiency) and the use of fungi and bacteria (70% efficiency). This chapter will present biotechnologies developed, patented and applied in cases of oil spills in tropical mangrove of Brazil. These generated biotechnologies have been applied together with civil society in tropical ecosystems that were hit by the Venezuelan oil spill in 2019. The use of advanced molecular biology (studies of genomics, transcriptome, proteomics and metabolomics) in the biotechnologies presented has shown a promising path to faster, viable economically and ecologically correct mangrove restoration.

Keywords: petroleum, bioeconomy, biodegradation, recovery of degraded areas, urban wastewater treatment

1. Introduction

The control of losses during the production, transport and disposal of oil is fundamental in preventing environmental accidents, in guaranteeing the safety of the community and in the preservation of natural resources [1]. Process losses, such as leaks and ruptures, generate a double cost for companies and environmental pollution [2]. First, the cost of repairing the image, as a damaged image can trigger the closing of contracts with other companies and a reduction in sales. The second cost is to repair the environmental damage caused to people and environmental resources. It is important to note that if the company does not meet the legal criteria, it may still have additional costs with the request for licenses not previously requested. Hence, the need to establish an action plan for the prevention and containment of environmental accidents, that is, to think of all strategies from the containment of the problem to the repair or recovery of the affected area.

The first action to be considered before the arrival of the spot is what is the measure or set of containment measures that will precede the arrival of the spot on the coast. Especially because at first the objective is to prevent the stain from spreading and reaching the coastline, and to carry out this action, the support of the government, fire department, NGOs, scientists, companies and other professionals is necessary, depending on the severity of the problem [1]. Recently, the coast of northeastern Brazil was the victim of a natural disaster, during this period (2019–2020) some scholars sought answers, such as the origin of oil, how to contain and treat it [3].

Biosorbents are efficient in combating environmental accidents, since the bioproducts generated can act as adsorbents and co-products of added value and can adsorb contaminants [4, 5]. They thus allow the reduction of the arrival of the oil slick in the mangroves, in addition to assisting in phytoremediation through the adsorption of hydrocarbons. Coconut and sisal fibers, thermally and chemically pretreated [6–8], have high adsorption efficiency of aromatic hydrocarbons, such as anthracene and methylene blue [9, 10].

In biotechnological processes, it is common to use microorganisms to generate bioproducts of interest in the industry, such as drugs, beverages and fertilizers, as well as in the treatment of contaminated sediments and soils [11]. Rhizospheric microorganisms, also present in plants, can be applied in the bioremediation of areas contaminated by oil spills [12]. It is important to note that plants have abilities such as phytoextraction, phytostabilization, photodegradation and phytostabilization, which assist in the action of rhizospheric microorganisms. In phytostimulation, for example, the plant releases enzymes, metabolites and nutrients, which influence rhizodegradation and plant growth, given that there are bacteria that promote plant growth [12].

Phytoremediation also has the advantage of acting in the control of carbon dioxide emissions, which is a product of the biodegradation of hydrocarbons generated in the bioprocess [13]. It is also important to highlight the application of microalgae (phytoremediation) in the biofixation of carbon dioxide generated during the photosynthesis process [14]. In this way, phytoremediation is aligned with the Paris Protocol, the 17 UN SDGs and also with the circular economy, since the biomass generated can be reused directly in industrial processes.

2. Biotechnologies developed and applied in the restoration of oil-impacted mangroves

2.1 Phytoremediation with mangrove plants

The mangrove is strongly impacted by environmental accidents, due to the richness of biodiversity and abundance of resources, which are used by riverside communities as a source of subsistence [15]. Nowadays, one of the main problems associated with the mangrove ecosystem is environmental accidents, resulting from failures in the process, during the transportation, production, storage and disposal of oil [16].

The presence of heavy metals in petroleum-contaminated areas also interferes in the rehabilitation of the mangrove ecosystem [17–20]. It is important to point out that certain metals, such as Al, Fe, Pb, Cr, Cu, Zn and Ni influence the biodegradation process of petroleum, which may make it impossible or reduce the efficiency of the implemented remediation strategy. [19] performed an integrated evaluation of the metals Al, Fe, Pb, Cr, Cu, Zn and Ni with the physical–chemical parameters (pH, temperature, dissolved oxygen and salinities), in order to assess how these factors contributed to the response variable, that is, in the percentage of removal of total hydrocarbons from oil, by phytoremediation and bioremediation. Pearson's correlation indicated the strength and direction of the linear relationship between the variables, and the principal component analysis (PCA) was applied to explain the events that were not fully clarified by Pearson's correlation.

The experiment was carried out for three months in a greenhouse near the mangrove, where samples of sediment and residual oil were collected. The tidal regime was simulated daily in each unit, in order to provide models for the remediation conditions closest to a marshy ecosystem. Altogether, 72 seedlings were selected and cultivated, 36 of which were *Avicennia schaueriana* and 36 were *Rizophora mangle* [18, 19].

Seedlings of mangrove plants were collected at low tide, taking into account their height (average of 3 months), defining a standard sample. [19] found after fifteen days of application, that intrinsic bioremediation showed greater efficiency in relation to phytoremediation, which required a longer period to reach maximum efficiency. As [20] this result was expected, since the plants have a longer response time to the pollutant, which varies according to the species and the conditions of the natural system studied or simulated in the laboratory. It is also important to highlight the peculiarity of the mangrove sediment, which has low or no oxygen availability, medium salinity and low diversity of microorganisms in relation to other environmental compartments. These factors can provide greater or lesser biodegradation of toxic compounds. However, the impact of petroleum on mangroves also depends on the types of pollutants, concentration, toxicity, distribution and the retention time [18]. In many cases, the sediment can behave as a reservoir of pollutants such as heavy metals from the marine or terrestrial environment and, thus, the toxicity of the contamination is greater and the degree of difficulty in removing the organic compounds derived from oil becomes more complex [15].

After three months, [19] proved that phytoremediation showed a removal of compounds in the sediments from 33.0 to 4.0 μg / g initially, while the intrinsic bioremediation decreased from 33.0 to 9.0 μg/g (**Figure 1**). Thus, phytoremediation was able to remove the sediment about 20% more than intrinsic bioremediation.

A higher efficiency of phytoremediation, in relation to the removal of the intrinsic bioremediation of petroleum hydrocarbons was observed. The presence of metals did not influence directly on Bioremediation, except for Cu, which may have moderately inhibited greater efficiency in the process (**Table 1**). However, Ni and Al seem to have been absorbed by mangrove plants, while they were removed from the hydrocarbons, which may have favored more the growth of microorganisms in the rhizosphere, besides the stimulation by the allelopathic compounds. Finally, it was emphasized that the implementation of the Phytoremediation model in areas impacted by oil activities can be very important, since it is an inexpensive, environmentally friendly and socially correct technique. Moreover, this process may also contribute to reduce global warming through carbon.

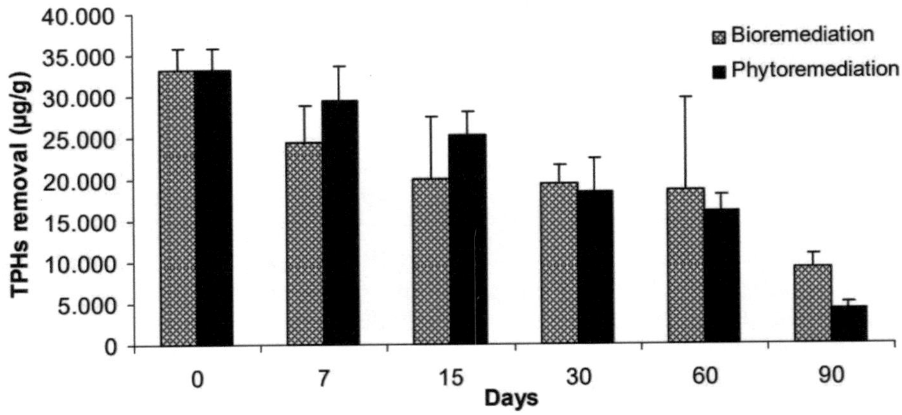

Figure 1.
Temporal removal of total petroleum hydrocarbons (TPHs) during the application of intrinsic bioremediation and phytoremediation in mangrove sediments.

2.2 Bioreactors

The mangrove is one of the coastal ecosystems impacted by socioeconomic and environmental activities, which directly or indirectly affect the distribution of hydrocarbons derived from oil [21]. Marine sediment is one of the main matrices for the deposition of organic contaminants in mangroves, as the biogeochemical characteristics contribute to the accumulation of these compounds and associated with environmental legislation limit the local application of biotechnologies [22]. Thus, studies in bioreactors are great strategies for studying bioprocesses for treating sediments contaminated by oil, without inflicting the requirements established by Brazilian environmental legislation [19–24].

To address this issue, [24] evaluated the efficiency of the bioreactor in the biodegradation kinetics of Total Petroleum Hydrocarbons (TPHs) in the Campos basin present in mangrove sediment. To this end, [24] monitored over 187 days the variation of the physical–chemical and biological characteristics in the sediment and in the circulation water in the bioreactor, used in the simulation of high and low tides.

The methodological steps were carried out in two phases. In the first phase, only bioremediation was tested in the bioreactor, in phase 2, bioremediation plus phytoremediation was tested. The kinetics of oil hydrocarbon biodegradation was evaluated using the [25] model (Eq. (1)), previously selected by [24] in a systematic review. The experimental design was composed of two treatments, the control (sediment without oil) and the contaminated (sediment with oil), both in triplicate, totaling six simulation units, with 6 mini reservoirs/simulation unit.

$$\text{Substrate consumption speed} = \mu\text{máx.S}\left(\text{S0} - \text{X0} - \text{S}\right) / \left(\text{Ks} + \text{S}\right) \qquad (1)$$

In Eq. (1), KS represents the concentration of half saturation, the S was the concentration in time t (final bioprocess time), µmáx was maximum growth rate and the S0 the concentration in time t0 (initial bioprocess time).

The method 3051A (extraction in microwaves) was used for the determination of HTP and HPA in sediment [26]. Subsequently, the extracts were injected in a gas-operated chromatograph (VARIAN brand, model CP3800), with DB 5 capillary column and flame ionization detector (CG / FID) for the determination of total petroleum hydrocarbons. HPAs were detected by gas chromatography (Mark

VARIAN) coupled to a mass spectrometer (GC / MS). In the water samples the extraction method applied was that of U.S. EPA 3510C (liquid–liquid extraction) [27]. Subsequently, the extract was concentrated with the aid of a rotovaporator, model R-215 and, finally, transferred to vials and sent for reading in the gaseous chromatographs. HTP reading was performed on the CP3800 gas chromatograph, with a DB 5 capillary column and flame ionization detector (CG/FID), and HPA determination on the gas chromatograph (VARIAN) coupled to a mass spectrometer (GC/MS). The colony forming units were counted through the micro droplet technique [28] and the culture medium used was Nutrient Agar.

The maximum and specific growth rates of microorganisms were similar in the simulation units in the contaminated treatment, but the speeds obtained were different between the units. In the contaminated treatment the value of the maximum rate in the simulation units ranged from 0.10 day −1 to 0.14 day −1, the temperature ranged from 25° C to 30° C and the pH ranged from 6.5 to 8,5 in the simulation units, a suitable range for biodegradation of petroleum hydrocarbons [29].

The mangrove ecosystem is influenced by the variation of the tides, and to simulate this process in the bioreactor, water from the São Paulo river estuary (class 1 saline water) was used [30].

The catabolism and anabolism reactions that occur during the biodegradation of compounds are influenced by abiotic and biotic factors that interact in biological systems, both natural and simulated. Understanding how these factors influence the response of microorganisms is of fundamental importance in phytoremediation and bioremediation studies.

The uses of predictive models bring satisfactory results for understanding interactions during phytoremediation and bioremediation processes. Therefore, to understand how these variables influence or contributed to the response of microorganisms, [24] generated a generalized linear model, which did not evaluate the contribution of independent variables in the variable response, and also did a Pearson correlation analysis to verify how variables correlated with each other.

In all contaminated treatment simulation units, estuary water significantly influenced the growth of microorganism cells in the sediment. Water pH and salinity had a positive contribution to the number of colony forming units, the number of microbial cells increased with increasing pH and salinity of water used in the bioreactor ($p < 0.05$). Total nitrogen, as well as total organic carbon and total hydrocarbon concentration of petroleum significantly decreased in the sediment with the increase of colony forming units (p value <0.05). We also observed a reduction in the number of colonies with increasing temperature (**Table 2**). Therefore, we believe that microbial cell growth is associated with total nitrogen consumption and total oil hydrocarbon catabolism. Probably, this total nitrogen reduction in the mangrove sediment is available to detect fixed bacteria as rhizospheric roots, such as those that were probably stimulated with the activation of enzymes and metabolites by plant roots.

Correlation analysis of the variables indicated that there was a negative correlation between the number of colony forming units and the dissolved oxygen concentration[1] (**Tables 1** and **3**). This correlation is expected since the mangrove sediment is anoxic, so we believe that the bacteria present in this sediment are facultative anaerobic) [32].

The temperature had a positive correlation with dissolved oxygen concentration and salinity (**Table 2**). Therefore, we believe that increasing the temperature in

[1] The correlation test evaluated the interaction of the independent variables with the response variable, and the pvalue was used to assess the degree of significance of this interaction, so the p <0.05 indicates that the null hypothesis was rejected, and the result was significant [31].

	pH	Eh	T	Sal	DO	Cu	Zn	Pb	Cr	Ni	Fe	Al	TOM	TOC	TN	P	TPH
pH	1																
Eh	−0.92	1															
T	-0.23	0.18	1														
Sal	-0.8	0.58	0.31	1													
DO	-0.31	-0.07	0.1	0.74	1												
Cu	-0.21	-0.09	0.04	0.73	0.9	1											
Zn	-0.04	0.03	0.93	0.27	0.09	0.17	1										
Pb	-0.39	0.28	0.76	0.71	0.46	0.57	0.85	1									
Cr	-0.17	0.23	0.88	0.32	-0.02	0.11	0.97	0.86	1								
Ni	-0.01	0.08	0.64	0.29	0.03	0.3	0.87	0.84	0.92	1							
Fe	-0.46	0.32	0.79	0.65	0.5	0.39	0.72	0.84	0.7	0.54	1						
Al	0.48	-0.34	0.43	-0.3	-0.22	-0.14	0.59	0.33	0.57	0.6	0.4	1					
TOM	-0.47	0.22	0.05	0.85	0.86	0.89	0.13	0.6	0.14	0.27	0.57	-0.05	1				
TOC	0.5	-0.23	-0.71	-0.82	-0.73	-0.68	-0.67	-0.87	-0.6	-0.5	0-0.82	0	-0.65	1			
TN	0.67	-0.38	-0.5	-0.91	-0.83	-0.68	-0.39	-0.72	-0.35	-0.23	-0.83	0.13	-0.79	0.92	1		
P	-0.52	0.24	0.43	0.91	0.86	0.87	0.46	0.83	0.44	0.46	0.76	-0.02	0.9	-0.91	-0.93	1	
TPH	0,52	-0.16	-0.24	-0.85	0.91	-0.83	-0.18	-0.54	-0.09	-0.07	-0.49	0.44	-0.74	0.84	0.87	0.86	1

Table 1.
Pearson correlation in intrinsic bioremediation.

Variables	Estimate	Std. Error	z value	Pr (>|z)
Intercept	72987.51	727.12	100.38	<2e-16
TOC (sediment)	-757.15	22.96	-32.97	<2e-16
TPH (sediment)	-0.01	0.00	-370.31	<2e-16
NT (sediment)	-107861.61	1138.39	-94.75	<2e-16
DO (Water)	6855.73	31.41	218.26	<2e-16
P (Water)	-31438.05	126.42	-248.69	<2e-16
P (sediment)	631.38	3.11	202.94	<2e-16
pH (water)	11954.19	63.08	189.50	<2e-16
Salinity (water)	440.45	2.58	170.48	<2e-16
Temperature (water)	-6374.67	30.03	-212.24	<2e-16

Table 2.
Contribution of water and sediment biogeochemical variables in the growth of colony forming units in simulation units.

bioreactors generates higher evaporation rate in the system, increased salinity and reduced number of colonies forming units. The reduction of colony forming units causes increased availability of dissolved oxygen in the water used to simulate tidal variations.

One of the factors that may indicate the degradation of petroleum hydrocarbons is the increase in the number of microorganism cells. [22] identified that a number of *Bacillus subtilis* bacteria cells were using hydrocarbon biodegradation from the petroleum bioremediation experiment. Analyzing a growth curve of unit 1, it was possible to observe the microorganisms reached in the stationary phase in the final stage of the experiment, from the 153 days of experiment.

In unit 2, the stationary phase of growth of microorganisms was not observed. However, just as in unit 1 there was variation of exponential death phase in the ranges from 42 to 62 days (death phase), 62 to 99 days (exponential), 99 to 118 days (death), 118 to 153 days (exponential) and 153 to 187 days (death). The same was observed for unit 3, which was also not detected in the stationary phase, but there was exponential phase variation in the 42 to 99 days (death), 99 to 118 days (exponential), and 118 to 153 (death) intervals. And 153 to 187 (exponential) (**Table 4**).

The unit 2 in the second phase showed an efficiency of 85.93% in the biodegradation of low molecular weight Aromatic Polycyclic Hydrocarbons compared to the first phase of the bioprocess experiment (49.18%) (**Table 4**) [11].

The three simulation units presented greater removal of hydrocarbon from the sediment in phase 2 of the bioprocess experiment. Thus, we believe that oil contaminated sediment remediation was efficient with bioprocess joining. And, In relation to biodegradation of total petroleum hydrocarbons, unit 2 stood out with the percentage of biodegradation of petroleum from 81,25% and GCR from 0,04 mg.kg-1.day-1 (**Table 5**).

It is important to note that in the first phase of the unit 3 of bioprocess experiment the concentration of low molecular weight hydrocarbons increased in the sediment (**Table 4**). Probably, the increase of low molecular weight HPAs was the result of catabolism of high molecular weight HPAs.

The substrate consumption velocity differed significantly between units, with a confidence level of 5%. There was no interaction of velocity with time and unit. However, at certain time intervals (153 until 187 days and 62 until 99 days) there was similarity in velocities.

Variables	Intercept (CFU)	TOC	TPH	NT	DO	P (water)	P (sediment)	pH	Salinity	Temperature
Intercept (CFU)	1.00									
TOC	0.57	1.00								
TPH	-0.42	-0.02	1.00							
NT	-0.07	-0.36	0.15	1.00						
OD	-0.54	-0.26	0.48	-0.12	1.00					
$P_{(water)}$	0.12	0.36	0.64	0.08	0.26	1.00				
P (sediment)	-0.18	-0.40	-0.49	-0.44	-0.02	-0.53	1.00			
pH	-0.65	-0.26	0.06	0.20	-0.06	-0.35	-0.15	1.00		
Salinity	0.24	0.03	-0.19	-0.31	0.06	0.07	0.39	-0.58	1.00	
Temperature	-0.06	-0.19	0.29	-0.24	0.54	0.33	0.40	-0.71	0.54	1.00

[1]*pvalor < 0.05.*

Table 3.
Correlation between the variables of the generalized linear model and the CFU response variable.

UNIT	% of biodegradation of PHA of low molecular weight	
	Step 1- Bioremediation	Step 2- Bioremediation + Phytoremediation
1	76,38	82,16
2	49,18	85,93
3	55,28 of increase	83,63

Table 4.
Percentage of low molecular weight aromatic polycyclic hydrocarbon (PHA) removal from mangrove sediment.

SIMULATION UNITS	GCR (ppb.day-1)	GCR (mg.kg-1. day-1)	% BIOD.
UNIT 1	10,82	0,01	46,05
NIT 2	40,14	0,04	81,25
UNIT 3	22,14	0,02	64,95
	24,37 ± 14,79	0,02 ± 0,01	64,08 ± 17,61

Table 5.
GCR in the simulation units.

2.3 Phycoremediation with marine microalgae

The produced water is one of the effluents generated after oil extraction. It is estimated that the production of water produced/oil can reach approximately 3: 1 (v/v) of barrels/day [31]. Organic compounds from produced water such as polycyclic aromatic hydrocarbons (PAHs) are classified as highly hazardous pollutants mangrove ecosystems because they are compounds recalcitrant, carcinogenic, teratogenic and mutagenic. When PAHs come into contact with plants and animals, toxic effects are alarming by oxidative stresses, genetic mutations and biomagnification.

Mangrove swamps are coastal ecosystems of great ecological importance to tropical countries. These environments return biomass and nutrients to the sea and act as ecologically nurseries of marine organisms. However, according to the Environmental Sensitivity Index for Coastal Areas published by NOAA, the mangrove habitat is classified as a tropical habitat sensitive to oil spills due to the difficulties of implementing a contingency plan [33].

Therefore, it is necessary to use biotechnologies that increase the efficiency in removing PAHs before being released into water bodies. The use of microalgae-based biotechnologies is characterized as an self-sustaining treatment, since these microscopic beings are photosynthesizes and can be mixotrophic, that is, they perform photosynthesis using sunlight as an energy source and biofixation of CO2 for their own development, but are also able to absorb organic carbons from polycyclic aromatic hydrocarbons as a potential source of energy [34, 35]. Thus, the objective of this research is to evaluate the potential of the removal of polycyclic aromatic hydrocarbons (PAH's) for the treatment of produced water using a photobioreactor system with the marine microalgae species *Nannochloropsis oculata*.

To assess the potential, the removal of polycyclic aromatic hydrocarbons (PAHs) by the photobioreactor system with marine microalgae an experiment was carried out with gradients of different concentrations, for 28 days in the laboratory. The samples produced water were provided by the company Petróleo Brasileiro S.A. (Petrobras). The saline water samples, for dilution of concentrations, were collected at the pier of Porto da Barra whose bounding coordinates were: 13°00′14.12"S and

38°32′01.81"O. The marine microalgae species *Nannochloropsis oculata* was acquired from the Culture Collection of Algae at The University of Texas at Austin (UTEX).

A set of photobioreactors with different gradients of produced water concentration diluted in saline water was assembled, establishing five gradients was called: Photobioreactor with 25%, 50%, 75% 100% (v/v) of produced water and photobioreactor with Conway medium (control). The monitoring of the removal of PAHs chosen in this research was planned in five intervals of different times: consider the 1st day, 7th day, 14th day, 21st day, 28th day. The evaluation of the removal of PHAs was made from the Liquid–liquid extraction was performed in the laboratory [36]. All samples extracted were transferred for injection into the gas chromatograph coupled with mass spectrometer (GC–MS).

The monitoring of microalgae growth was determined through cellular density (cell number ml^{-1}) from the correlation curve (Eq. (2)) between cell count using a Neubauer camera [37] and spectrophotometer absorbance (Agilent Cary 60 UV–Vis) using 680 nm wavelength. The experiment was carried out in triplicate and one-way analysis of variance (ANOVA) with a significance level of $p < 0.05$ was used for statistical analysis. The analyses were conducted using BioEstat 5.3 software.

$$n^{\circ}\frac{Cel}{ml} = (3E+07)\,x\,OD680nm + (8E+06)\;\; \mathrm{R}^2 = 0{,}9231 \tag{2}$$

The first five days corresponded to the microalgae acclimatization process to the new cultivation medium (**Figure 2**). After these days of experiment, the growth of microalgae *Nannochloropsis oculata* increased by 25% in all photobioreactors, presenting resistance to organic contaminants until reaching cell stabilization from the 15th day. The results show that its higher growth was presented on the 19th day and in Conway cultivation medium with a concentration of 6.50×10^7 cells ml^{-1}, followed by cultivation in water produced at 25% presenting final result 5.24×10^7 cells ml^{-1} and the cultivation in water produced at 50% with final concentration of 4.09×10^7 cells ml^{-1}.

Considering 75% produced water its highest growth was presented on the 17th day with about 2.77×10^7 cells ml^{-1} subsequently begins its decay. 100% produced water presented lower growth of the marine microalgae where it remained constant during the experiment, with about $1.17\ 10^7$ cells ml^{-1}. Constant growth may have been influenced by ions present in the desemulsifiers inserted in the produced water, where they are introduced directly before the water was collected in the well.

The ions of the demulsifying, when at high concentrations, have the ability to neutralize the cellular surface by inhibiting the multiplication of them, being reported the same results in crops above 50% of water produced in experiments developed by [34].

Although microalgae growth is reduced when the water concentrations produced in the medium increase, an increase in efficiency in the removal of total PAHs by microalgae was also increased when the water concentrations produced are also increased, there was a reduction of PAH in low molecular weight compounds and those of high molecular weight in all produced water concentrations, and the largest reduction was observed in 100% produced water, presenting about 3.016 µg L^{-1} reducing to 152.22 µg L^{-1} of light compounds. This reduction represents about 94% efficiently (**Table 6**).

Removal of PAHs in water using microalgae can occur through two pathways: by intracellular bioaccumulation or biodegradation [38]. Biodegradation of polycyclic aromatic compounds by marine microalgae occurs under mixotrophic conditions, with compounds oxidized by oxireductose enzymes, in addition, can

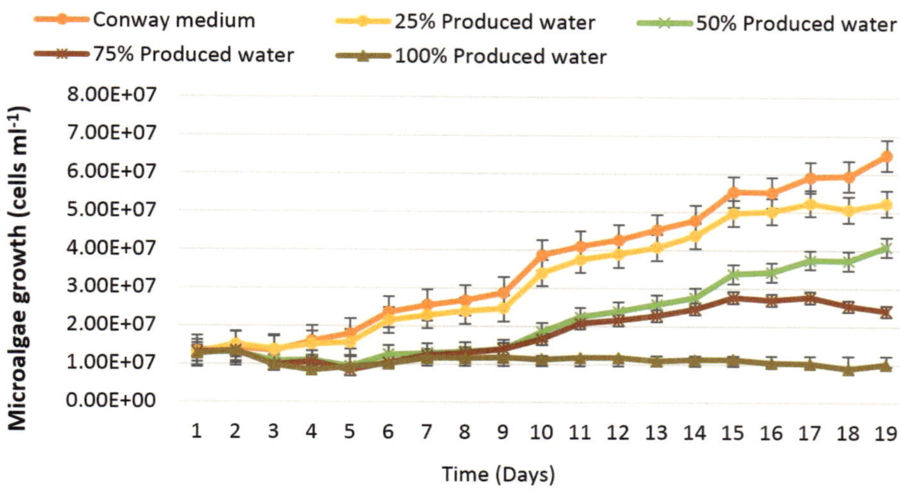

Figure 2.
Effect of different water concentrations produced in the growth of the Nannochloropsis oculata *microalgae measured in cells mL^{-1}. Concentrations: Conway medium, 25% produced water, 50% produced water, 75% produced water and 100% water produced. Data are presented mean value ± margin of error, n = 3.*

form hydroxylated intermediate fractions, justifying the increase in intermediate fractions in the medium [39].

Microalgae are presented as a profitable biotechnology for the efficient removal of PAHs, being exploited as a sustainable source in the treatment of different types of effluents and can be reused for the generation of biofuels, enabling its multi-restorative effects to prevent PAHs from reaching sensitive ecosystems such as mangroves.

2.4 Biofibers

Biotechnologies such as the use of biosorbents and phytoremediation can be applied in situ if there are specialists who know the technique and the ecosystems affected. Adsorbent containment barriers are alternative technology used as a response to spills to contain and recover oil, preventing stains from spreading affecting sensitive areas, as mangroves [40]. Currently the main sorbent materials used in the commercialized barriers are synthetic, however these materials are expensive, are not biodegradable, which makes it difficult to use [7]. In this scenario, some strategies are studied to make the use of sorbents economically viable, through the implementation of natural fibers. The main advantages of natural fibers are low cost, low abrasiveness, not toxic, low density, as well as ecological and social aspects, due to better recyclability and biodegradability [41, 42]. The coconut and sisal fibers, thermally and chemically pre-treated, have high efficiency of oil adsorption in marine and sweet waters, and can be applied from the construction of a containment bar [43–45]. These fibers can also be applied directly to coastal surfaces (beach sediments, mangrove sediments, exposed rocks) and with little contact time and in a simple way is able to remove oil. The use of biofibers should be encouraged, as it allows the reuse of oil adsorbed in the industry for various purposes. This biotechnology is directly aligned with the circular economy [40, 42].

The crushed and sieved coconut fibers were stirred in NaOH solution (5% w /v) at room temperature (approximately 25°C) for 1 hour for mercerization. Then, a wash was carried out with distilled water until a constant pH to ensure the removal of all residual solvent in the fiber [43]. This same procedure was used for acetylation

Concentration of polycyclic aromatic hydrocarbons (μg L^{-1})									
	Initial concentration			Final concentration			Total composition of PAHs		
	LMW (2 - 4 rings)	**IMW (4 rings)**	**HMW (5 - 6 rings**	**LMW (2 - 4 rings)**	**IMW (4 rings)**	**LMW (5 - 6 rings**	**Initial**	**Final**	**Removal efficiency**
Conway medium	102.8 ±4.92	8.26± 0.35	5.84± 0.31	94± 4.91	9.99± 0.77	1.34± 0.02	116.9±4.78	105.33± 2.98	10%
25% produced water	687.84 ±19.23	14.09± 1.13	8.3± 0.98	92.44± 4.90	13.84± 1.02	2.32± 0.13	710.23± 4.88	108.6± 2.57	85%
50% produced water	1537.25± 41.48	13.52± 0.47	5.38± 0.71	127.38± 13.84	9.08± 0.71	1.87± 0.01	1556.15± 4.61	138.33± 2.06	91%
75% produced water	1823.72±166.45	13.29± 0.57	5.02± 0.12	131.79± 8.74	14.56± 0.91	5.43± 0.01	1842.03± 4.86	151.78± 1.05	92%
100% produced water	3016.43± 93.75	13.59± 0.89	7.38± 0.49	152.22± 1.91	35.73± 4.05	6.89± 0.02	3037.4± 2.40	194.84± 4.56	94%

Low Molecular Weight (LMW), Intermediate Molecular Weight (IMW) and Hight Molecular Weight (HMW). Data shown as the mean ± SD, n = 3.

Table 6.
Removal efficiency of polycyclic aromatic hydrocarbons in relation to the molecular compounds present in the five photobioreactors with different gradients concentrations of water produced diluted in saline water.

and treatment with Protic Ionic Liquids (PIL). In acetylation, mercerized fibers were immersed in a solution of acetic anhydride and glacial acetic acid (1.5:1.0 by mass) with 12 drops of sulfuric acid at 80°C for 3 hours. For the treatment with PIL, a sample of coconut fiber was added in 2-hydroxyidoethylammonium acetate [2-HEA] [Ac] at 80°C for 2 hours [44]. These fibers (in natura, pretreated by mercerization followed by acetylation and with PIL) were characterized from the morphology by the Scanning Electron Microscope (SEM).

After the pre-treatment procedures, the fibers were weighed (0.5 g) and conditioned in mini barriers made from TNT (non-woven fabric) to continue the sorption and kinetics tests. The tests were performed in a thermostatic bath, with reciprocal movements of approximately 126 cycles/minute and temperature of 25°C (average temperature of the marine environment). The kinetic experiment was conducted in beakers with 95 mL of saline water and 5 mL of oil from the Campos Basin with the mini-barriers in contact with the oil slick for 120 minutes, in which samples were taken, in triplicate, in the time intervals of 5, 20, 40, 60, 90 and 120 min [7]. In the sorption equilibrium experiment, the oil concentration was varied for the construction of the isotherms [45]. After testing, the samples were cold dried in the lyophilize and weighed. The sorption capacity of the fibers was determined through Eq. (3), where S is the adsorption capacity (sorbate g/g of sorbent), S_0 (g) is the initial mass of the fiber and S_f (g) is the final mass of the fiber after adsorption [6, 7]. The tests were performed by the barriers with in natura and pre-treated fibers.

$$S = \left(S_f - S_o\right) / S_o \tag{3}$$

Through SEM analysis, it was possible to observe a large irregularity and pores on the surface of fibers in natura (**Figure 3a**). After the treatments, the mercerized/acetylated fiber increased the rough area of the cross section, in comparison with the fiber in natura (**Figure 3b**). The fiber with PIL, on the other hand, had a higher number of pores (**Figure 3c**), resulting from the cleaning by treatment with this organic solvent. Thus, chemically treated fibers have more space available for adsorption through the pores and the roughened surface compared to fibers in natura.

The kinetic results of the adsorption of the barriers with coconut fibers are shown in **Figure 4**. In all fibers studied (in natura, mercerized/acetylated and with PIL) the kinetic behavior was very similar. There was a marked sorption up to 5 minutes and then the sorption remained practically constant. This happens because the initial number of pores and available surface in the fibers are occupied over time, reducing the availability and consequently the sorption capacity [6, 7, 42, 45–47]. From these results, it can be concluded that the time of 5 minutes has more significant efficiency in adsorption, requiring a minimum contact time between the

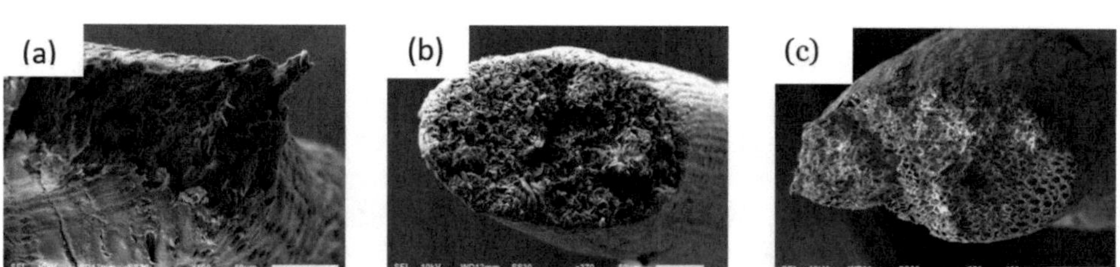

Figure 3.
SEM coconut fiber (a) in natura (b) treated with mercerization/acetylation (c) treated with PIL.

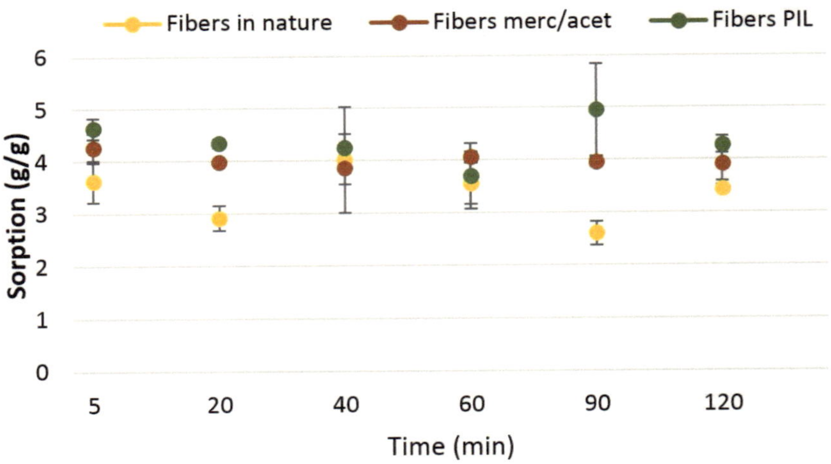

Figure 4.
Comparison of the kinetic behavior of sorption among all coconut fibers.

adsorbent material and the adsorbate to removal of crude oil in the marine environment, in addition to the contact technology operator time with toxic oil.

The equilibrium sorption was 4.00 g / g for fresh coconut fiber, 4.27 g / g for mercerized / acetylated fiber and 5.37 g/g for PIL fiber. Therefore, the fiber with PIL adsorbed 20.5% more than the mercerized/acetylated coconut fiber and 25.5% more than the fresh fiber. (b).

The result of the higher sorption of the treated coconut fibers can be explained by the chemical, structural and morphological modification presented in relation to the natural fibers through the characterizations. The greater quantity and density of pores resulting from the removal of chemical constituents, such as lignin and hemicellulose, waxes and impurities, made the pores of coconut fibers clear and consequently increased the surface area for interaction with oil.

3. Conclusions

Our study with phytoremediation in mangroves, showed that it is possible to accelerate the process of removing oil hydrocarbons in sediments when using the mechanisms of plants, their rhizosphere and the associated microorganisms. Phytoremediation is the most suitable technique for mangrove areas, since sediments have low oxygen solubility and have granulometric characteristics that increase the residence time of persistent organic pollutants such as PAHs. Based on the results found, it can be said that the barriers with chemically treated fibers are more efficient than in natura to be used in the containment and cleaning of oil spilled in marine environments so that it does not affect sensitive areas such as mangroves. The barrier composed of the fiber treated with PIL obtained greater oil sorption, followed by the fiber treated by mercerization-acetylation and finally the fiber in natura. These fiber barriers that were produced by our group can be used during emergency combat of oil stains in estuarine waters, preventing oil from reaching the sediment. They can also be used as sponges to clean oil already adhered to the surface of plants and mangrove sediments, preventing the infiltration of hydrocarbons. In this study it was identified that the *Nannochloropsis oculata* marine microalgae used for the removal of polycyclic aromatic hydrocarbons in produced water showed greater efficiency in the produced water with 94% removal, demonstrating that this marine microalgae is able to contribute to the degradation of organic pollutants and to prevent PAHs from reaching sensitive ecosystems such as mangroves. Microalgae photobioreactors can be used in the treatment of effluents from the oil industry that are released into the mangrove. In addition, the use of microalgae biorefineries has already been used to remedy river waters, and may be an option during the emergency combat of oil spills in mangrove estuarine waters. A sequential application of bioremediation and adequacy contributed positively to the biodegradation of petroleum hydrocarbons. There were improvements in the quality of the sediment, due to the variation of the physical–chemical characteristics provided by the action of rhizosmic microorganisms, stimulated by enzymes released by the plants, during the oil hydrocarbon metabolism. This process was also noticeable in the growth curve of microorganisms and in the variation in the speed of consumption of petroleum hydrocarbons. The studies that our group has been carrying out for more than 10 years show that there is not a single biotechnology that can restore oil-impacted mangroves. Each biotechnology presented here has its particular contribution in removing pollutants in the various environmental matrices of the ecosystem. Recently our group has advanced in the studies of advanced molecular biology (studies of genomics, transcriptome, proteomics and metabolomics) for the improvement of bioprocesses in a faster restoration and with economic viability (bioeconomy).

Acknowledgements

We thank CAPES and FAPESB for providing scholarships to the graduate students involved, we also thank CNPQ for providing scholarships to researchers involved in the studies in addition to financing the DEMBPETRO project.

Note

We want to honor with this pubication our eternal advisor Professor Ph.D. Jorge Alberto Trigüis (in memoriam) for all the teachings passed to the area of geochemistry in Brazil and in the world. We also want to thank our family members for supporting us in the career we have followed, especially in this pandemic moment where we have to organize ourselves with the different demands (professional and private). For this reason, our tribute to the future great professionals Yasmin M. P. Andrade Moreira and Rafaella M. P. Andrade Moreira.

Author details

Ícaro Thiago Andrade Moreira[1*], Célia Karina Maia Cardoso[1], Evelin Daiane Serafim Santos Franco[2], Isadora Machado Marques[3], Gisele Mara Hadlich[4], Antônio Fernando de Souza Queiroz[5], Ana Katerine de Carvalho Lima Lobato[6] and Olívia Maria Cordeiro de Oliveira[7]

1 Program in Chemical Engineering (PPEQ-UFBA), Federal University of Bahia, Salvador, Bahia, Brazil

2 Geochemistry: Petroleum and Environment (POSPETRO-UFBA), Federal University of Bahia, Salvador, Bahia, Brazil

3 Industrial Engineering (PEI-UFBA), Federal University of Bahia, Salvador, Bahia, Brazil

4 Department of Geography, Federal University of Bahia (UFBA), Salvador, Bahia, Brazil

5 Department of Oceanography, Federal University of Bahia (UFBA), Salvador, Bahia, Brazil

6 Program in Chemical Engineering (PPEQ), Salvador University (UNIFACS), Brazil

7 Department of Geophysics, Geosciences Institute, Federal University of Bahia (UFBA), Salvador, Bahia, Brazil

*Address all correspondence to: icarotam@ufba.br

References

[1] Almeida A. G., Vinnem J. E. Major accident prevention illustrated by hydrocarbon leak case studies: A comparison between Brazilian and Norwegian offshore functional petroleum safety regulatory approaches. Journals & Books. 2020: 121: 652-665.

[2] Yanting Z., Liyun X. Research on Risk Management of Petroleum Operations. Energy Procedia. 2011:5:2330-2334. DOI: https://doi.org/10.1016/j.egypro.2011.03.400.

[3] Oliveira O. M.C. et al. Environmental disaster in the northeast coast of Brazil: Forensic geochemistry in the identification of the source of the oily material. Marine Pollution Bulletin. 2020: 160: 1-7.

[4] Zeng C., Hu. 2018 petroleum & chemical industry development report. Chinese Journal of Chemical Engineering. 2019: 27: 2606-2614.

[5] Aghaalikhani A. et al. Poplar from phytoremediation as a renewable energy source: gasification properties and pollution analysis. Energy Procedia. 2017: 142: 924-931.

[6] Cardoso, C. K. M., Santana, R. S. G. de, Silva, V. L. da, Meirelles, A. C. L. E., Mattedi, S., Moreira, Ícaro T. A., & Lobato, A. K. de C. L. (2020). Kinetic and equilibrium study of petroleum adsorption using pre-treated coconut fibers. Research, Society and Development, 9(7), e523974413. DOI: https://doi.org/10.33448/rsd-v9i7.4413

[7] Annunciado, TR, Sydenstricker, THD, Amico, SC. Experimental investigation of various vegetable fibers as sorbent materials for oil spills. Marine Pollution Bulletin. 2005;50:1340-1346. DOI: https://doi.org/10.1016/j.marpolbul.2005.04.043

[8] CALDAS, A. S.; VIANA, Z. C. V.; SANTOS, V. L. C. S. Fibers of *Cocos nucifera* L. as sorbent of petroleum in marine environment. Acta Brasiliensis, [S.l.], v. 1, n. 1, p. 13-18, jan. 2017. DOI: https://doi.org/10.22571/Actabra1120179.

[9] Silva C.C. et al. Influence of plasma treatment on the physical and chemical properties of sisal fibers and environmental application in adsorption of methylene blue. Journals & Books. 2020: 23: 1-9.

[10] Lopes F. C. S. M. R. Ag/TiO2 photocatalyst immobilized onto modified natural fibers for photodegradation of anthracene. Chemical Engineering Science. 2020: 227: 115939.

[11] Streltsova E., Linton J. D. Biotechnology patents in BRICS countries: strategy and dynamics. Trends in Biotechnology. 2018: 36: 642-645.

[12] Santos J. J., Maranho L. T. Rhizospheric microorganisms as a solution for the recovery of soils contaminated by petroleum: A review. Journal of Environmental Management. 2018:20:104-113.

[13] Straathof A. J. J. et al. Major research challenges in sustainable industrial biotechnology .Biotechnology trends. 2019: 37: 1042-1050.

[14] Arbib Z. et al. Capability of different microalgae species for phytoremediation processes: Wastewater tertiary treatment, CO2 bio-fixation and low cost biofuels production. Water Research. 2014: 465-474.

[15] Duke N. C. Oil spill impacts on mangroves: Recommendations for operational planning and action based on a global review. Marine Pollution Bulletin. 2016: 109: 700-715.

[16] Mafi-Gholami D., Jaafari A., Zenner E. K., Kamari A. N., Bui D. T. Spatial modeling of exposure of mangrove ecosystems to multiple environmental hazards. Science of the Total Environment. 2020: 740: 1-11.

[17] Wua S., Lia R., Xieb S., Shi C. Depth-related change of sulfate-reducing bacteria community in mangrove sediments: The influence of heavy metal contamination. Marine Pollution Bulletin. 2019: 443-450.

[18] Moreira, I TA, Oliveira, OMC, Triguis, JÁ, Queiroz, AFS, Santos, AMP; Martins, CMS, Silva, CS, Jesus, RS. Phytoremediation using Rizophora mangle L. in mangrove sediments contaminated by persistent total petroleum hydrocarbons (TPH's). Microchemical Journal (Print). 2011:99:376-382. DOI: https://doi.org/10.1016/j.microc.2011.06.011

[19] Moreira, I TA, Oliveira, OMC, Triguis, JA, Queiroz, AFS, Barbosa, RM, Anjos, JASA, Reyes, CY, Silva, CS, Trindade, MCLF, Rios, MC. Evaluation of the effects of metals on biodegradation of total petroleum hydrocarbons. Microchemical Journal (Print). 2013:110:215-220. DOI: https://doi.org/10.1016/j.microc.2013.03.020.

[20] Matamoros V, Gutiérrez R, Ferrer I, García J, Bayona JM. Capability of microalgae-based wastewater treatment systems to remove emerging organic contaminants: A pilot-scale study, J. Hazard. Mater. 2015;288:34-42. DOI: https://doi.org/10.1016/j.jhazmat.2015.02.002.

[21] Araújo M. P., Hamancher C., Farias C. O., Martinho, P., Chaves F. O. C., Soares, M. L. C. Assessment of brazilian mangroves hydrocarbon contamination from a latitudinal perspective. Marine Pollution Bulletin. 150 (2020) 12p.

[22] Gravitol A. D., Marchetti A. A, Wetler –Tonini T. M. V., Gatts EN. C., Thompson F. L., Rezende C. E. Bacterial interactions and implications for oil biodegradation process in mangrove sediments. Marine Pollution Bulletin. 118 (2017) p. 221-228.

[23] Moreira, I TA, Oliveira, OMC, Triguis, JA, Queiroz, AFS, Ferreira, SLC, Martins, CMS, Silva, ACM, Falcão, BA. Phytoremediation in mangrove sediments impacted by persistent total petroleum hydrocarbons (TPH s) using Avicennia schaueriana. Marine Pollution Bulletin. 2013:67:130-136. DOI: https://doi.org/10.1016/j.marpolbul.2012.11.024

[24] Franco E. D. S. S.F. Microbial kinetic modeling applied to the degradation of Campos's oil present in mangrove sediment [dissertation]. Salvador: Federal university of Bahia; 2018.

[25] Simkin S., Alexander M. Models for mineralization kinetics with the variables of substrate concentration and population density. Aplied and Environmental Microbiology. 47 (1984) p. 1299-1306.

[26] Usepa. Method Usepa 3051A. Microwave Assisted Acid Digestion of Sediments, Sludges, Soils and Oils. Revision, fev. 2007. https://www.epa.gov/sites/production/files/2015-12/documents/3051a.pdf (acessado em 02 março 2017).

[27] Usepa. Method U.S EPA 3510C. Separatory funnel liquid-liquid extraction. 8p. 2015.https://www.epa.gov/sites/production/files/2015-12/documents/3510c.pdf (acessado em 03 março 2017).

[28] Romeiro R. S. Métodos em bacteriologia de plantas. Viçosa: UFV. 2001 279p.

[29] Colombo J.C. et al. Determination of hydrocarbons sources using n-alkane and polyaromatic hidrocarbon distribution indexes. Case study: Rio de La Plate estuary, Argentina.

Environmental Science Technology 23 (1989) p.888-894.

[30] Brasil, Resolução CONAMA n°357, de 17 de março de 2005. Classificação de águas, doces, salobras e salinas do Território Nacional. Publicado no D.O.U.

[31] Arriada, A.D.E.A. Cultivo da microalga marinha nannochloropsis oculata (eustigmatophyceae) em água de produção de petróleo : Viabilidade de crescimento, produção de lipídieo e açúcares. [Thesis] Rio Grande: Universidade Federal do Rio Grande; 2014.

[32] Moreira, I TA, Oliveira, OMC; Azwell, T, Queiroz, AFS; Nano, RMW, Souza, ES, Anjos, JASA, Assunção, RV, Guimarães, LM. Strategies of Bioremediation for the Degradation of Petroleum Hydrocarbons in the Presence of Metals in Mangrove Simulated. Clean (Weinheim. Print). 2016:44:631-637. DOI: https://doi.org/10.1002/clen.201300939

[33] Moreira ITA, Oliveira OMC, Triguis JA, Queiroz AFS, Ferreira, SLC, Martins, CMS, Silva ACM, Falcão BA. Phytoremediation in mangrove sediments impacted by persistent total petroleum hydrocarbons (TPH's) using Avicennia schaueriana. Marine Pollution Bulletin. 2013;67:130-136.

[34] Ammar SH, Khadim HJ, Mohamed AI. Cultivation of Nannochloropsis oculata and *Isochrysis galbana* microalgae in produced water for bioremediation and biomass production, Environ. Technol. Innov. 2018;10:132-142. DOI: https://doi.org/10.1016/j.eti.2018.02.002.

[35] Matamoros V, Gutiérrez R, Ferrer I, García J, Bayona JM. Capability of microalgae-based wastewater treatment systems to remove emerging organic contaminants: A pilot-scale study, J. Hazard. Mater. 2015;288:34-42. DOI: https://doi.org/10.1016/j.jhazmat.2015.02.002.

[36] Moreira ITA. Investigação dos possíveis impactos ecológicos do petróleo sobre comunidades biológicas estuarinas na Baía de Todos os Santos e no sul da Bahia: OSA como uma ferramenta norteadora. 2014. 207 f. [Thesis] Salvador: Universidade Federal da Bahia; 2014.

[37] Lourenço SO. Cultivo de Microalgas Marinhas: Princípios e aplicações. Editora RiMa, São Carlos, São Paulo, 2006. 588 p.

[38] Subashchandrabose S.R, Ramakrishnan B, Megharaj M, Venkateswarlu K, Naidu R. Mixotrophic cyanobacteria and microalgae as distinctive biological agents for organic pollutant degradation, Environ. Int. 2013;51:59-72. DOI: https://doi.org/10.1016/j.envint.2012.10.007.

[39] Deshmukh S, Kumar R, Bala K. Microalgae biodiesel: A review on oil extraction, fatty acid composition, properties and effect on engine performance and emissions. Fuel Processing Technology. 2019;191:232-247. DOI: https://doi.org/10.1016/j.fuproc.2019.03.013.

[40] Ossai, IC, Ahmed, A, Hassan, A, Hamid, FS. Remediation of soil and water contaminated with petroleum hydrocarbon: A review. Environmental Technology & Innovation. 2020;17:100526. DOI: https://doi.org/10.1016/j.eti.2019.100526

[41] Pessôa, TS, Ferreira, LEL, Silva, MP, Neto, LMP, Nascimento, BF, Fraga, TJM, Jaguaribe, EF, Cavalcante, JV, Sobrinho, MAM. Açaí waste beneficing by gasification process and its employment in the treatment of synthetic and raw textile wastewater. Journal of Cleaner Production. 2019;240:118047. DOI: https://doi.org/10.1016/j.jclepro.2019.118047

[42] Luis-Zarate, VH, Rodriguez-Hernandez, C, Alatriste-Mondragon, F, Chazaro-Ruiz, LF, Rangel-Mendez, JR. Coconut endocarp and mesocarp as both biosorbents of dissolved hydrocarbons in fuel spills and as power a source when exhausted. Journal of Environmental Managenment. 2018;211:103-111. DOI: https://doi.org/10.1016/j.jenvman.2018.01.041

[43] Cabral, MMS, Abud, AKS, Rocha, MSRS, Almeida, RMRG, Gomes, MA. Composição da fibra da casca de coco verde in natura e após pré-tratamentos químicos. ENGEVISTA. 2017;19:99-108. DOI: https://doi.org/10.22409/engevista.v19i1.802

[44] Rocha, EGA, Costa, AC, Aznar, M. Use of Protoct Ionic Liquids as Biomass Pretreatment for Lignocellulosic Ethanol Production. Chemical Engeneering Transactions. 2016;37:397-402. DOI: 10.3303/CET1437067

[45] Nwadiogbu, JO, Ajiwe, VIE, Okoye, PAC. Removal of crude oil from aqueous medium by sorption onhydrophobic corncobs: Equilibrium and kinetic studies. Journal of Taibah University for Science. 2016;10:56-63. DOI: https://doi.org/10.1016/j.jtusci.2015.03.014

[46] Abdelwahab, O, Nasr, SM, Thabet, WM. Palm fibers and modified palm fibers adsorbents for different oils. Alexandria Engineering Journal. 2017;56: 749-755. DOI: https://doi.org/10.1016/j.aej.2016.11.020

[47] Din, RU, Nikogeorgos, N, Jellesen, MS, Shabadi, R, Ambat, R. Influence of steam-based pre-treatment using acidic chemistries on the adhesion performance of powder coated aluminium alloy AA6060. International Journal of Adhesion and Adhesives. 2017;74:167-176. DOI: ttps://doi.org/10.1016/j.ijadhadh.2017.01.008

Section 3

Conservation and Management

Chapter 7

Stieglers Gorge Dam Construction: Potential Impacts and Possible Mangrove Restoration Options in the Rufiji Delta, Tanzania

Claude Gasper Mung'ong'o

Abstract

The varied importance of mangroves has long been recognized. And so have been the threats to their existence, leading to various actions taken locally by local communities, national governments, and through international agreements for the protection and integration of human livelihood needs in a manner that balances conservation goals and goals of socio-economic development. In Tanzania the mangrove conservation ethos began during the German colonization of Tanganyika and has been persistent in the age of high globalization. In an effort to deepen our understanding of the dynamics of global, national and local nature conservation, this chapter documents the various strategies and approaches used in mangrove restoration elsewhere in the world generally, and then specifically in the Rufiji Delta. The chapter further unpacks the contrasting socio-political interests behind the efforts to conserve mangroves worldwide and in Tanzania. It does so by looking at three competing narratives, i.e. the mainstream perspective, the neo-liberal perspective, and the local cultural perspective, acting at a number of nested scales from the local grassroots to the national and the global scales.

Keywords: ecosystem goods, ecosystem services, Africa, coastal community, conservation management, socioeconomic

1. Introduction

1.1 Framing the problem

Mangroves have provided critical services to humans and the ecosystems. They have done this well and their ecological, economic, cultural and esthetic importance values have long been recognized. It was not until recently in the Anthropocene when the threats to their existence have been magnified with losses of the habitat of more than 50% reported in some parts of the globe due to their commoditization [1–5]. It is thus that the mangrove conservation ethos has particularly been persistent now than ever before in human history [6].

In an attempt to deepen our understanding of the dynamics of nature conservation reinforced by dominant discourses of the 1990s, neo-liberalism and ecological modernization, this chapter documents the strategies and approaches used in mangrove restoration in Tanzania, in general, and specifically in the Rufiji Delta. The chapter further unpacks the contrasting interests behind the efforts to conserve mangroves worldwide. This is done by looking at three competing moral narratives operating at nested scales from the grassroots to achieve conservation with social justice through the national, regional to the global and back again [7–9]. The three narratives interrogated here are the mainstream ecological conservation narrative, the neo-liberal economic narrative and the local cultural narrative at the grassroots [10] as discussed later in this chapter.

1.2 Methodology

Data collection for this study involved a desk-based literature search during September–November, 2019. Published and gray literature and secondary sources were purposively sampled using key words in the Google search engine. Key words such as mangroves, restoration, commoditization, ecosystem goods and services, Rufiji Delta, et cetera, were used to create a document set for known mangrove countries in the world.

A qualitative content analysis approach was then used to analyze each of the relevant research categories. Through a directed approach each document was reviewed to identify and collate evidence for each of the research themes and the level at which they operated. Three levels were identified: namely, local factors that operate at the community, village and household level; national factors that operate at the state level; and international factors that operate beyond the level of the state.

Content analysis was then followed by discourse and narrative analyses that have long been a major part of political ecology. While discourse analysis was an epistemological exercise, narrative analysis was important for understanding how environmental knowledge of specific events was communicated [11]. These narratives were deemed important for policymakers as they would make arguments for controlling the actions of certain, often unknowing groups; to achieve desired environmental outcomes. The rest of this chapter is structured according to the results of these latter analyses.

2. Results and discussion

2.1 Defining the mangrove ecosystem

The word "mangrove" refers to trees, families of the plant, and the ecosystem that has adapted to flourish in tidal zones in tropical and sub-tropical regions. Mangroves have been defined variously by many people; but they all agreed that mangroves are salt-tolerant evergreen forests found at sheltered coastlines, shallow-water lagoons, estuaries, rivers or deltas in intertidal areas around the world [12, 13].

They comprise around 73 species covering an area of over 150,000 km^2 spread in 123 countries [1]. (See also **Figure 1**). Over two-thirds of the mangroves exist in just 12 countries, with Indonesia accounting for over 20% of the global mangrove area. With about 8% of the total mangrove estate, Brazil has the largest contiguous mangrove forest cover. In tropical estuaries of Indonesia and Brazil and deltas like the Rufiji, there grows some of the largest mangrove trees in the world, reaching heights of 30 m or more, with extensive roots penetrating into soft mud deposits. Mangrove trees growing in the sediments of a carbonated shoreline and in arid,

Figure 1.
Mangroves are common along tropical and subtropical coastlines around the world, and among the most biologically important systems on earth [12]. ***Source:*** *NASA earth observatory as described by Twilley and Rovai [14].*

very salty regions along the Red Sea are so much smaller that they look like stunted "ornamental trees in public parks" [15].

Mangrove forests in mainland Tanzania are categorized as State Forest Reserves by the Forest Act of 2002 [16]. They occur along almost the entire coastline in continuous or fragmented stands [17]. Recent estimates by the National Forest Resources Monitoring and Assessment (NAFORMA) indicate that mangroves cover approximately 158,100 ha, which is about 0.3% of the total forest area in the country [18].

Despite the commercial value of mangroves, "blue" carbon ecosystems are globally being lost twice as fast as tropical rainforests [19]. They are threatened by changing climate, natural impacts such as hurricanes, and human impacts such as deforestation and alterations in freshwater management regimes. Approximately 35% of mangrove trees were lost in this way during the last two decades of the 20th century [20].

Ngongolo *et al* [21] note, for example, that by 2000 the total estimate for mangroves was 137,760 km^2, representing a decline from 198,000 km^2 of mangroves in 1980, and 157,630 km^2 in 1990. These losses represent about 2.0% per year between 1980 and 1990, and 0.7% per year for the period 1990–2000. Hence, the anticipated task of mangrove restoration is immense.

2.2 Mangrove's ecosystem services

Ecosystem services are the benefits that people get from ecosystems such as mangroves [21, 22]. Forests moderate the amount and type of water we get from a river. It also decreases both the erosion and run-off of a place. They also provide food security as far as the variety of biodiversity they contain and general economic development. The goods and services they provide have the capacity to bring resilience among smallholder farmers, through diversification of incomes and livelihoods such as fishing.

Ecosystem services are necessary for people's livelihoods and wellbeing. They include provisioning, regulating, and cultural services that directly affect people, and supporting services needed to maintain other services (**Figure 2**). This is from the provisioning of food and water, to disease regulation and maintenance of general conditions of the natural environment. Biodiversity is behind every ecosystem services.

The concept of ecosystem services has been used in diverse ways by different interests to justify different kinds of interventions that at times might be totally opposed. The concept has historically been depicted as a one-way flow of services from ecosystems to people. Jeffers *et al.* [21] argue, however, that this conceptualization is inaccurate. It neglects the reality that humans have often also contributed to the maintenance and enhancement of ecosystems, as evidenced in many traditional and indigenous societies.

Secondly, the ecosystem services idea has mainly been used to justify forest conservation in ways open to critique for its neo-liberalization of nature [25, 26] or disempowerment of communities in some developing countries such as Madagascar. On the other hand, the discourse of ecosystem services has also served the liberating agendas of traditional populations and family farm lobbies in places like the Brazilian Amazon, where the ecosystem services concept has been mobilized by diverse actor interests in real-life situations that have led "to complex, regionally particular and fundamentally political outcomes" [27].

Irrespective of the range of species and forest types, the manifold ecological role of mangrove ecosystems is, economically and socially, highly significant. The international discourse on mangroves hypothesizes that they play an important part in shaping the physical coastline by trapping sediments and stabilizing the coast.

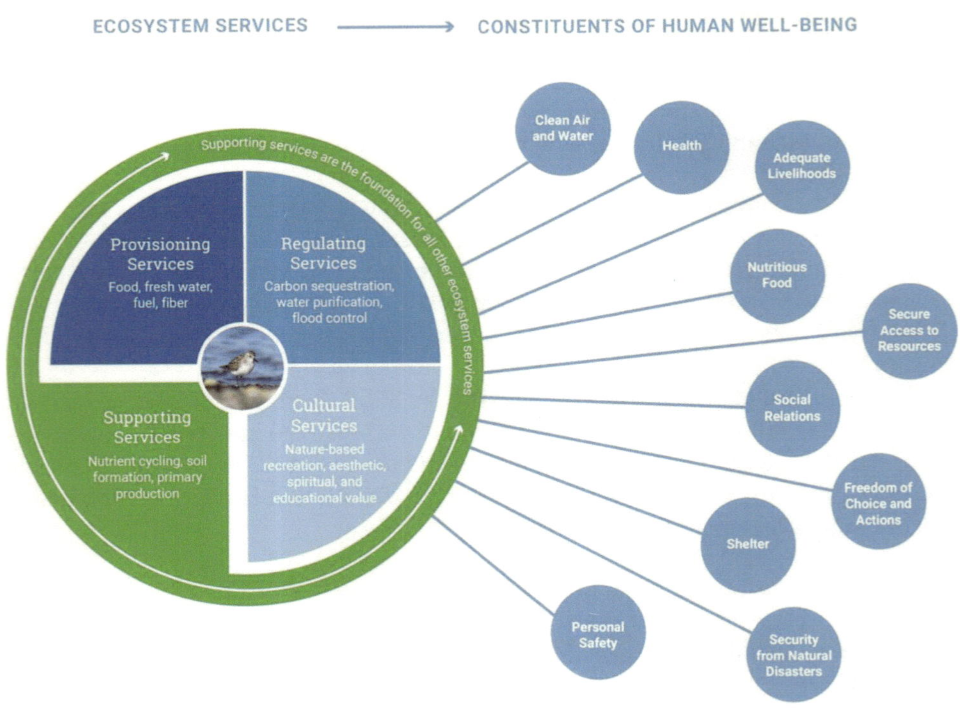

Figure 2.
Healthy ecosystems benefit human well-being. Humans have often also contributed to the maintenance and enhancement of ecosystems in a reciprocal fashion [23]. ***Source:*** *Adopted and modified from Chapin* et al. *[24].*

Moreover, the mangroves are highly productive ecosystem that can store carbon in sediment for long period; therefore, they have the potential of providing an efficient CO2 sink [11].

Mangrove plantation provides the possibility of forestation for the Clean Development Mechanism (CDM) project worldwide. Recent studies revealed that mangrove restoration can continue to combat climate change after 25–30 years [28]. Anthropogenic interventions decrease the capacity of ecosystems to provide goods and services as discussed above.

2.3 The mangrove estate of the Rufiji Delta

Located between latitudes 7°50′ and 8°03′ S and longitudes 39°15′ and 32°17′E, the Rufiji Delta is about 200 km south of Dar-es-Salaam (**Figure 3**). The lower Rufiji valley starts upstream from Stiegler's Gorge, some 180 km from the Indian Ocean, and into the Selous Game Reserve. Below the gorge the river fans out into an outer plain with numerous lakes before entering its lower floodplain. This part of the floodplain gradually widens until the river branches out and forms the 23 km wide and 65 km long cobra like hood of the delta [30].

Before reaching the Mafia Channel in the East, the river passes through 20 islands and 31 villages, and supports the largest contiguous block of mangrove

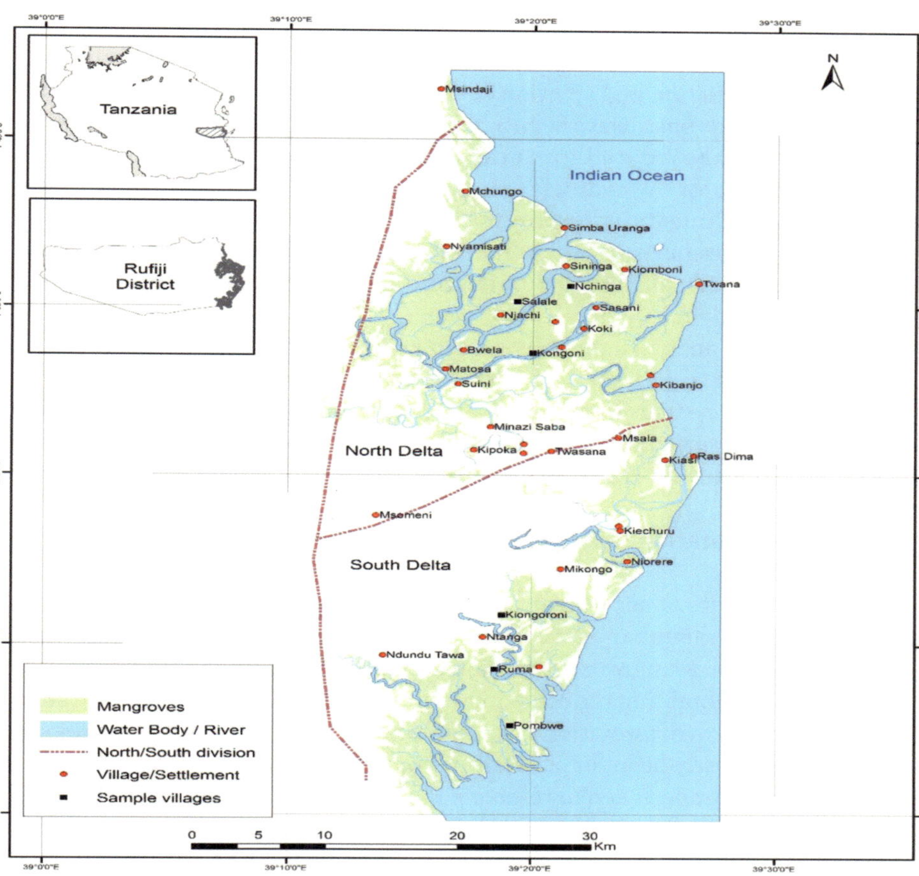

Figure 3.
*Location of the delta area in Rufiji District, Tanzania. The crescent shaped Rufiji Delta has extensive, estuarine mangrove forests that constitute almost half of Tanzania's mangrove forests (**Source:** Adopted from Mwansasu [29]).*

forest (53,255 ha) in East Africa. Eight mangrove species are reported to occur and are well represented in the delta, i.e. *Avicennia marina, Sonneratia alba, Ceriops tagal, Lumnitzera racemosa, Bruguiera gymnorrhiza, Rhizophora mucronata, Xylocarpus granatum* and *Heritiera littoralis* [31]. Two particular species, *Xylocarpus molluccensis* and *Pemphis acidula,* are missing in the delta. These are characteristically rare in the region due to a limited geomorphological niche [17].

In the delta mangroves are cleared mostly for rice farming and timber to feed urban Tanzania, including Zanzibar, and some parts of the United Arab Emirates. Areas dominated by *H. littoralis* are more favored for rice farming while *C. tagal, R. mucronata* and *B. gymnorrhiza* are heavily cut for poles. *X. granatum,* and more recently *S. alba,* are logged for timber [32].

Rufiji River is also endowed with the greatest fish potential along Tanzania's coastline, supporting about 80% of all prawn fisheries in the country [33, 34]. With a mean annual flow of approximately 800 m3/, Rufiji is one of the largest rivers in Africa and drains 20% of mainland Tanzania through three major tributaries, the Great Ruaha, the Kilombero and the Luwegu. Together, the rivers provide approximately 18%, 62% and 15% of Rufiji's annual flow, respectively [35].

The basin is targeted for major socio-economic development over the next two decades as part of the Southern Agricultural Growth Corridor of Tanzania (SAGCOT). Its water resources are central to the development plans. If the plans go on as arranged targets are met, irrigation water demand will increase by 7 billion m^3/per year. 2.4 giga watts of new hydropower are expected to be produced from the controversial Stigler's Gorge. Many of these developments will be in the most valued landscapes and ecosystems of Kilombero and Lower Rufiji sub-basins [36].

Coastal sedimentation and siltation of coastal waters due to agricultural expansion on the highlands has always been a blessing rather than a curse. Of primary threat will probably be reduced stream and peak water flows due to the proposed river impoundment for the hydroelectricity power (HEP) station, with the associated trapping of sediments in the proposed dam [37]. This is further expected to have significant consequences with respect to increased salt water intrusion and diminished nutrient availability for agriculture and altered natural vegetation in the floodplain and the delta.

The delta is also an internationally recognized wetland protected under the Ramsar Convention and a system of UNESCO's World Heritage Sites upstream [38]. These aquatic systems have historically provided valuable ecosystem services, including the *mlau* agriculture performed by the Warufiji as discussed elsewhere by Ochieng [30], Duvail and Hamerlynck [35], and others.

2.4 Mangrove restoration initiatives/options

The characteristics of ecosystems, such as species composition, tree cover or growth conditions, modulate the type and magnitude of ecosystem services that can flow to societies. Mangrove restoration is an important strategy for reversing plant decline and rebuilding the ecosystem services lost due to deforestation and degradation. Mangrove restoration has usually been in the form of replants of single species, and has mostly been for silvicultural purposes. More recently replants have, however, also been undertaken to re-create the lost ecosystem functions [17]. The Indian Ocean region, for example, saw the rapid expansion of government and NGO-funded mangrove replants after the 2004 Indian Ocean tsunami to maximize the coastal protection function provided by mangroves.

Large-scale planting projects have, nevertheless, had mixed success. Many causes have contributed to the low success of plantation interventions, including:

i) biological causes such as pest infestations, ii) unsuitable physical locations, and iii) the socio-economic aspects, as elaborated in Section 3.3 below.

In addition, knowledge of the ecological processes has been added in some projects to increase restoration success. One such approach has been the Ecological Mangrove Restoration (EMR), a community based restoration practice that used several ecological principles to support natural decolonization [39]. This approach shifts the emphasis from seedling planting to prior physical site preparation. For example, the hydrology and topography of a site can be restored to allow natural regeneration of selected mangrove species [40].

2.5 Mangroves in the mainstream development narrative

Nested within mainstream values and definitions of development in Tanzania, several assumptions tend to have guided a policy narrative that has supported mangrove forest conservation in the country:

1. assumptions of inefficiency in local resource use and management regimes [28];

2. inadequate perceptions about the socio-economic benefits of local aquaculture production system, and

3. the central government may be seeing mangrove restoration from the angle of ecological services, while local people view mangrove forests as part of their culture and source of livelihoods.

For example, local aquaculture production systems are considered by regional and district officials to be economically inefficient and incongruent with broader national economic development goals for wetlands [21]. Such assumptions and perceptions are even captured in formal policy documents [41].

In contrast, traditional use of common property mangrove resources is not accorded the same economic value in mainstream discourse despite the numerous benefits provided to local communities [42, 43]. Such conceptions of local inefficiencies and the economic productivity of external aquaculture and carbon sequestration for REDD+ situate well within the broader value sets and definitions of "development" and encourage neo-liberal narratives in developing countries like Tanzania [44, 45].

Moreover, in the case of Rufiji study area, values supportive of common property ecosystems held by indigenous groups have been usurped by an influx of different interest groups, including international environmental NGOs with their political agendas and economic modes of production [44, 46, 47].

The practice of mangrove restoration is based on restoration ecology, which aims to help the recovery of resilience and capacity of ecosystems to adapt to degradation and other damage. Since environmental impacts are ongoing, successful restoration of an ecosystem implies not merely to recreate its former condition, but to strengthen its capacity to adapt to change over time [48]. One of these capacities is the management aspects of mangrove conservation.

After the acknowledgement that strict protection of mangroves did not work in Tanzania, experiments have recently been unfolding in the Rufiji Delta and elsewhere in the country. Three different models of community engagement have been tried in the delta with varying degrees of success [49] as discussed below.

3. Managing mangroves by restoration and reserves

Declines in the extent of mangrove forest cover have a long history in the Rufiji Delta [29, 32]. The Rufiji mangrove forest was the first to be declared a forest reserve in Tanzania during the German colonial period in 1898 [45]. The delicate socio-ecological balance was, however, upset during the course of successive German, British, and the national governments.

The British colonial government adopted and expanded a strict protection approach in the 1920s and 1930s [49]. The Forest Ordinance of 1957 allowed for the creation of forest reserves by government decree after considering "any objections" by interested parties to this de jure transfer of rights from local communities to the state. The independent state expanded mangrove forest reserves in the 1960s and has repeatedly used its authority over mangrove forests to exert control over the Rufiji Delta communities and resources. In 1987, for instance, the current Tanzania Forest Service (formerly Forestry and Beekeeping Division) declared a ban on the cutting of all mangroves in the northern Rufiji Delta, with forest officers posted to the area to enforce this ban [12].

By creating 'forest reserves' for the exclusive use of the government, local communities were effectively excluded from using these socially, culturally, and economically important resources. Meanwhile, various large-scale extractive projects were proposed for the delta, including commercial shrimp harvesting [12]. Elsewhere in the Coast Region and the country a devolution of resource management to local government, in combination with improved road access and the opening-up of the Tanzanian economy, led to increasingly unsustainable use of other forests in general.

Evidence from literature shows that traditional government agencies have not been effective in protecting forest ecosystems, including mangrove forests. Constraints of capacity and the economic position of many district agencies have resulted in few controls on the scale and intensity of mangrove conservation in the delta, despite an increasingly well-established legal framework for forest conservation [29, 32].

Tanzania was the first country in Africa to develop a mangrove management plan [49]. Although it was never implemented, the 1991 National Mangrove Management Plan crafted by the TFS was the first attempt at halting mangrove conversion alongside monitoring and regulating the use of the resource. Strict mangrove protection entailed actively excluding people living in and around mangroves from accessing and using mangroves for their survival, while the government controlled the harvest and export of mangrove products, particularly timber and building poles [13].

Tanzania's protectionist policies generally achieved limited and short term success in some locations around the country, with general failure in most mangrove areas of the country. At the local level, to the mid-1990s local residents had, from the colonial period, actively resisted state-led protection of forests, including mangroves [49].

As a result, efforts by the government to continue with strict protection approaches, such as establishment of new marine parks in the late 1990s and mangrove forest reserves continued to face serious implementation challenges, including resistance from coastal residents who complained that these forest reserves and marine parks marginalized them from their main livelihoods [46].

Lack of an appropriate institutional framework for the allocation of management rights and responsibilities between the local government and the national state, as well as weak government enforcement capacity at the national level [33, 51] have been the main culprits of this failure. It was because of these weaknesses, for example, that Wang *et al.* [43] and Mwansasu [29] noted the ambiguous features of the forest reserve in the Delta - that there were legally established village settlements within it which relied on mangroves and the associated marine environment for a range of ecosystem goods and services. Population estimates by then indicated that over 49,000 people lived in and around the delta, directly engaging in rice farming, mangrove cutting for poles and timber, and fishing [32].

This was possible partly because of political influence at the national and local levels. Mshale *et al.* [49] point out that politicians at various levels in the Rufiji Delta had been issuing statements that encouraged unsustainable use of mangrove forests and mangrove clearance for paddy rice farming to gain residents' political support, particularly during election times. Such statements sent confusing messages to the populace. While elective politicians often promoted mangrove clearance when this was pertinent to increase votes, the civil service maintained a strict protectionist approach. Often, the rural poor who depended on mangroves for their subsistence continued to be characterized by the state institutions as culprits of the degradation of the resource [17, 52].

The 1990s can be dubbed the age of policy shifts. Many natural resource management policies were changed during this period. Since then, the approach to mangrove protection changed, with new efforts being more and more directed toward collaborative management of the trees with local communities. This paradigm shift later led to the emergence of what I call "community appeasement forest management" that was seen as an appropriate alternative to state control with its ambiguous institutional arrangement for ensuring management of forest resources, including mangroves. The turn toward community appeasement forest management was motivated by a number of reasons, including conflicts over forest use between local residents and outsiders seeking to use the forests, and political interests at the national and sub-national levels as discussed above [49].

I would argue in this context that mangrove restoration, including replanting, has in fact also been an attempt at community appeasement rather than a forest management practice adopted to regenerate areas that have been seriously degraded. Mangrove restoration has thus been conducted in various parts of the country by the Mangrove Management Programme, as well as in Tanga and Mbweni by local coastal management efforts. In the Rufiji Delta WWF's Rufiji-Mafia-Kilwa Seascape (RUMAKI) programme has invested substantial donor resources since 2005 to help communities in the delta, as well as in Mafia and Kilwa Districts, to secure long-term mangrove co-management rights [8, 53, 54].

The philosophy behind this supposedly "new approach" has arguably been to show local stakeholders that the restoration projects and the protected areas were in fact theirs, providing a range of benefits such as access to carbon financing schemes, eco-tourism revenue and sustainable sale of commercially valuable timber and non-timber products. It has been the carrot side of the Equation [44, 45]. Three different models of community engagement have, therefore, been tried – with varying degrees of success as discussed below.

4. Three models of community engagement

4.1 Individual taungya farming with permits

This is a TFS system whereby individuals are given permits to farm forest plots with the aim of clearing them of weeds. Farmers apply for renewable one-year licenses allowing them to continue farming rice in exchange for facilitating the natural regeneration of mangrove trees on their plots. Once the trees reach a certain height, their shade renders rice paddies less productive, and farmers must move elsewhere to repeat the process.

This scheme has not fared well though. Farmers have found it one-sided – imposing a lot of conservation responsibility on the farmers in exchange for meager returns. It has also been creating insecurity. People know that once the mangroves have re-grown farmers will be kicked out, so there is a perverse incentive for farmers to intentionally prevent mangrove recovery.

The written contracts have also been problematic. Many people in the delta are illiterate, and they fear anything that is written and requires to be signed. People feel like they are getting tricked. As one respondent in a focus group discussion remarked: "*perhaps there is something written there that we don't understand...?*" During the introduction phase of one of these projects many communities refused to sign these contracts due to such apprehensions [12]. After so many years of mistrust and harsh policies, people (especially pastoralists and "squatters" around protected areas) do not always trust government's intentions [49].

4.2 Group rehabilitation

This is another rehabilitation strategy that has been tried for the mangroves in the Rufiji Delta, with the support of the UNDP and UNEP. Local collectives of 15–30 men and women were assigned an area of mangrove forest to rehabilitate, and were paid for each day they used replanting or weeding the young trees. Under this arrangement, small-scale mangrove replanting was undertaken between 2009 and 2010 as a community project by 100–200 community members from 10 villages over about 70 ha of former mangrove habitat, of which around 45 ha were abandoned rice farming plots [12].

Communities initially embraced the project, but as one project official confessed sometime later, some villagers complained about favoritism, saying they felt excluded from the scheme (Mshale, pers. comm.) Even though TFS were emphatic that the project would be expanded to ensure benefits were shared by as many people as possible, the program could not manage to give people a sense of ownership over the forest. As the official later noted:

> *"These people are providing casual labour, but they don't have any other rights over the areas that they are replanting. So the moment you stop paying them, they won't be able to come and work for you."* (Mshale, pers. comm.)

Hence, the future of the program itself was uncertain because it relied heavily upon donor support. Once the funds dried up, the system could not be sustained. And because these schemes also failed to confer long-term management rights or responsibilities on community members, those involved were not incentivized to ensure the trees' survival and in many cases people returned to farming the areas once the scheme failed.

4.3 Community co-management of mangrove resources, including Joint Forest Management

Community participation as a neo-liberal approach to natural resources management in Tanzania has become the most important approach within the forestry sector following its inclusion in the National Forest Policy of 1998, Land Act of 1999 and the Forest Act of 2002. Nevertheless, the efficacy of the move to designate responsibility of forest management to local communities remains unclear [51, 55].

Building interest in poverty reduction to enhance sustainable resource management and strengthen the rights of communities to access and manage mangrove forests to improve livelihoods and resource conditions is one thing. But, whether this has translated into actual poverty reduction in practice is another thing altogether. So, the dilemma persists in poor countries like Tanzania on account of building a strong synergy of community participation and poverty reduction, *at least in theory* (emphasis added) [56].

The most promising approach, according to CIFOR research, has been the Joint Forest Management (JFM) scheme being tried in the Rufiji Delta as part of the Participatory Forest Management program [49]. By the time of this study TFS had negotiated with individual communities in four (4) Rufiji villages to draw up plans for sharing the costs and benefits of managing the mangrove forest. Though the state retains ultimate ownership of the mangroves, the scheme transfers some decision-making power to the local people.

By the time of the Mshale *et al* [49] study, the communities had embraced the scheme. The JFM system appeared to provide broader rights and benefits than the other two mechanisms. It meant that community members' actions such as harvesting timber, poles, charcoal, firewood and other products were no longer criminalized. What is needed is proper management and making sure that the benefits and costs are equitably and fairly distributed among community members.

In particular, it needs ensuring women's meaningful participation in decision-making in a culture where women are traditionally meant to stay silent during group meetings. That could need providing a separate space for women to debate ideas among themselves, before bringing their concerns to the wider community. However, implementation efforts and tangible evaluation of progress continues to lag behind conceptual development [57].

The proposed mechanism for REDD+ offers significant potential for conserving forests to reduce negative impacts of climate change. Tanzania has been one of nine pilot countries for the United Nations REDD Program, receiving significant funding from the Norwegian, Finnish and German governments. The country is also a participant in the World Bank's Forest Carbon Partnership Facility. In combination, these interventions aim to mitigate GHG emissions, provide an income to rural communities and conserve biodiversity for the market [58].

As already noted above the mangrove plantation is expected to be one of the options of afforestation in the Clean Development Mechanism (CDM) project through its ability to accumulate sequestered carbon below-ground [58]. However, it is the economics of the initiative that is of interest to us here. The establishment of the UN-REDD Program in Tanzania, for example, illustrates the challenges that face many developing countries. As noted by Burgess *et al.* [59], the challenges have included inadequate baseline forestry data sets needed to calculate reference emission levels.

It has also involved inadequate government capacity and insufficient experience of implementing REDD+ type measures at operational levels. In addition, for REDD+ to succeed, current users of forest resources must adopt new practices, including the equitable sharing of benefits that accrue from REDD+ implementation. This challenge is compounded by failure of conservation (as a form of land use) to compete effectively with alternative land uses [60].

For example, it is reported that the annual global economic value of ecosystem services is estimated to be between US$200,000 - US $900,000 per hectare [12]. The value of commercial mangrove timber products and poles in the Rufiji mangrove delta and flood plain in Tanzania is around $771,789 per year, fuel wood extraction $156,000 per year and honey extraction $9000 per year [19]. Most of the latter benefits go straight into the communities' household economy but stand to be foregone under global conservation.

Büscher *et al* [59] provide an interesting thesis on this phenomenon. They note that in the spirit of "ecological modernization", modern environmental problems and related crises are in fact themselves increasingly becoming conceptualized as opportunities for capitalist expansion rather than vehicles of poverty reduction. In a study that measured the impact of a national community-based conservation and poverty reduction initiative in Tanzania, it was found that from 2007 to 2015, the impacts of Wildlife Management Areas (WMAs) on wealth of the local communities were small and variable, with no clear evidence of widespread poverty reduction [61].

On the other hand, another study using five (5) years of photographic data capture-recapture found greater densities of livestock and lower densities of wildlife inside a WMA. After the management changes, the study documented significantly higher densities of wild ungulate species and lower densities of domestic ungulates in the WMA [62]. Giraffes' survival and population growth rate were both found to have increased in response to the management changes, indicating that the WMAs were effectively providing habitat and protection for wild ungulates while generally excluding domestic livestock [62].

4.4 Mangroves in the neo-liberal perspective

Neo-liberal conservation takes many forms, but more significant it reframes conservation in terms of market mechanisms [63]. Neo-liberal conservation is being defined as the decentralization of environmental governance, or a shift in responsibility for formal resource management from state to local institutions and new forms of commoditization and commercialization of nature that emerge in these contexts in order to fund conservation efforts. Advocates of market-based conservation argue that such markets will increase conservation funding and increase environmental-friendly businesses. People say it will promote participatory conservation and protect native property rights. Others say it promotes environmental consciousness, thus enhancing more effective and efficient conservation [64, 65].

Ever since the first Earth Summit in 1992 which was perceived to be a potential regulatory constraint on the operation of business, corporate interests have struggled to draw the sting out of the regulatory measures that the Summit recommended. States are said to have out-maneuvered NGOs around the Convention on Biological Diversity (CBD) to produce a convention which prioritized generating profit from genetic resources over protecting the environment [63]. Business interests have thus enjoyed access to prime slots during Conferences to the Parties of the CBD. Crucial to all these developments has been the formation of a transnational capitalist class fostering alliances and giving business interests the space and support they require [66].

It has been observed that much as conservation NGOs need the capital and legitimacy businesses provide, it has been the corporate interest which has reached out to conservation groups [63]. Corporate interests are seeking to make money out of new opportunities in mangrove restoration and conservation. They are looking for new profits in ecological modernization rather than biodiversity conservation. Although the gains for biodiversity are less clear within conservation organizations, the result of their embrace by commercial interests is that there has been a "near universal conflation of nature and capital [which] has established itself as a dominant view" (*ibid* [67]. See also [25].

Costanza *et al.* [68] assessed the economic value of 17 ecosystem services for 16 different habitats. They found out that the value of the whole biosphere was approximately US$16–54 trillion per year or US$33 trillion per year on average. They note that those figures may probably be higher today. Hence, they insist that drastic measures need to be taken if we are to prevent further widespread and irreversible loss of these ecosystems and *sustain their ability to generate new profit avenues worldwide* [64] (emphasis added). See also [69, 70].

However, market-based conservation has also been observed to lead to primitive accumulation, accumulation by dispossession and green grabbing [64]. All these processes have been proved to have negative impacts on local people's access to natural resources, food security, human rights, and the environment. Mariki [64] reports, for example, that some WMAs in Tanzania have disappointed the local people as people were persuaded to demarcate their land for conservation, take management responsibility and benefit from the resources through tourism. Instead in WMAs like Enduimet the central government has retained the power of strategic decision making over the WMA and only minimal benefits have been realized by the local communities.

4.5 Mangroves from a local cultural perspective

> *With a grim look on her face, Maimuna (Maimuna Ramadhani, 39-year old lady, Mchungu village) laments over the dwindling number of fish in the area for which she blames the destruction of mangroves. "The trees are harvested without replanting and now even when there is a high tide, it floods our homes," she says. "Why don't you do something about it?" we ask. "Because it is not my place to do so," she says.* (Mshale, pers. comm.)

Formal sources specifically linking people and mangroves, in particular their management and tenure regimes are lacking. However, basing on knowledge on linguistic analyses of some of the ancestors' myths and other oral traditions, coupled with the written narratives by European explorers, mangroves seem to the colonizers to have long been wastelands [13]. Nevertheless, to a majority of the local people, the mangrove landscapes were communal territories, inhabited, managed into multiple use systems, governed by access and use rights and controlled by local customs [64].

A second point that can also be surmised from these histories is the diversity of the human establishments in the mangroves and the highly varied forms of customary tenure with their "simple" knowledge of resources, the presence of spirits and supernatural creatures, with which the populations had to negotiate [13].

South of the Rufiji Delta, the island of Kilwa establishes a remarkable evidence of an urban and commercial civilization built in the mangroves dating from 9th to the 16th centuries. The historian Sheriff [71] reports of palaces and mosques, testifying of a glorious past of the sultanate, until the arrival of the Portuguese, who destroyed the estate in 1505 and monopolized the trade of gold, textile, spices, ivory and slaves.

The colonial institutions, then the independent State competed to destroy the former order in the form of traditional institutions [13]. They also tried to marginalize the traditional users. The tragedy of the commons observed elsewhere by Hardin [72] became their explanatory theory. But observers such as McCay and Acheson [73] see a new struggle brewing up by the present generation that is trying to grapple with in the Rufiji Delta and elsewhere in the country.

In general, however, the main image of mangrove swamps, the one that emerges from the narratives of European voyagers and missionaries in the 17th century, echoed throughout the 18th and 19th centuries in the writings of colonial agents, is one of a hostile and impenetrable environment [13]. For hygienic and productivity aims, mangrove swamps were reclaimed under the control of colonizers, who finally became their landlords.

Thus the first enclosures of mangroves were both public and colonial and aimed at converting mangrove swamps into rice fields. The following more recent enclosures (especially from the years 1970s) joins what certain authors qualified as "green imperialism" and aim at making it world heritages through conservation [13]. Nevertheless, the effectiveness of such protection has been highly variable, with several protected areas failing to halt mangrove decline because they were poorly designed or lacked enforcement.

In Tanzania there were greater mangrove gains between 2009 and 2015 than in other years [40]. This is probably due to conservation projects initiated in the delta in late 1990s and implemented by WWF-Tanzania, the TFS Agency and the Rufiji District Authority aimed to restore the deforested areas [17]. The projects are now promoting Payment for Ecosystem Services (PES) as the most rational approach to environmental management. PES uses the language of economics to convince potentially resistant policymakers, corporate actors and domestic populations, particularly in developing countries, to farther ecological goals such as biodiversity conservation.

Thus, environmental nongovernmental organizations (ENGOs), academics and international organizations alike spend considerable effort trying to "translate" the worthiness of the environment into the mutually intelligible language of neoliberal economics, in order to convince policymakers and economic actors of the validity of the conservation actions [9, 74].

Nevertheless, a study on the outcomes of CDM projects in Argentina a few years ago did not give such a hopeful picture at all. The study demonstrated that under the current Argentina's energy policy framework, the income by the selling of CERs covered less than 6% of the incremental costs for renewable energy projects [75]. A sensitivity analysis to evaluate the impact of CDM in the coverage of incremental costs for renewable energy based on the prices of both the energy in the local market and the CER demonstrated that the best conditions would only cover 15% of those costs. The contribution of CDM to technology transfer in Argentina was minor considering that 45% of the projects qualified as type III where technological learning and capacity building were limited at the level of operation and maintenance of a foreign technology [75].

In another research on REDD+ implementation in two case study villages in Tanzania, Scheba [65] argued that the emergence and nature of market-based conservation are complex and more shaped by structural challenges than is commonly acknowledged. The research identified three important challenges:

1. the politics surrounding the establishment of community-based forest management;

2. the mismatch between formal governance institutions and actual practices on the ground; and

3. the fickleness of income from carbon sales and alternative livelihood opportunities [65].

The challenges were conceptualized not only as teething problems. The results questioned the very fundamental assumptions of market-based conservation. Moreover, adopting neo-liberal approaches like PES may cause problems for the effective management of nature at the local level, and by extension, the implementation of the environmental regimes themselves. According to Scheba [65], market-based instruments share a common objective of re-framing landscapes as providers of "ecosystem services" and rural communities as latent eco-entrepreneurs, who can cultivate and sell the services as commodities for profit. "Selling nature to save it" is promoted as the best philosophy of achieving sustainable rural livelihoods in the face of mounting environmental crises and persistent poverty.

In fact, some authors (e.g. [67]) even think the commoditization of nature is part of broader project of neo-liberal globalization and encompasses a number of scientific and political techniques meant to bring non-market and non-economic materials, processes and things, including those that are considered part of natural ecosystems and are objects of traditional conservation, into the logics of economics and markets [65, 76–78].

Many indigenous peoples are concerned about the ways in which carbon markets commodify nature [79]. A market-based view prioritizes cost-effective strategies and the commoditization of ecological services, thereby utilizing the same economic tools and logic of capitalism that is also the underlying cause of the climate change problem. It is thus that some observers emphasize that market-based conservation is more complex, contested in practice, with mixed outcomes than is implied in the mainstream narrative [65]. It produces both benefits and risks. It produces benefits in the form of nature protection, political inclusion, and economic opportunities, while risks include physical displacement, loss of livelihoods, increased human–wildlife conflicts, and unequal distribution of benefits to some local community groups.

5. Concluding remarks

Since the late 2010 mangroves have become an important focus of market-based carbon-oriented nature conservation. A lot of work by different conservation organizations has been focused on framing or branding mangroves as particularly charismatic and valuable, but vulnerable, ecosystems. Beyond carbon, conserved mangroves and other coastal ecosystems are framed as untapped resources for ecosystem services, including coastal protection, fisheries, water purification, and conservation of marine and coastal biodiversity.

However, what the neo-liberal approach describes as "effective management" has not always been compatible with effective management of resources. For practical and ethical reasons, practitioners need to be critical of any assumption that neo-liberal economics is always an appropriate framework upon which to base local environmental management. Initiatives like REDD+ just represent the latest in a long line of efforts to tap global markets for conservation finance.

It is thus that current difficulties in the REDD+ mechanism are essentially symptomatic of inherent deficiencies in the market-based conservation in general. The fundamental problem is that conservation markets are intended to counter the conventional extractive markets, which generate profit by externalizing environmental costs.

Conservation markets seek to reverse this by internalizing these costs within the payments they provide to forest managers. Yet to function as market mechanisms, *payments must provide at least as much revenue as the extractive markets they replace* (emphasis added), covering not only opportunity costs of extraction but also the social and environmental costs that this extraction externalizes.

Author details

Claude Gasper Mung'ong'o
Institute of Resource Assessment, University of Dar es Salaam, Dar es Salaam, Tanzania

*Address all correspondence to: claude_gasper@yahoo.com

References

[1] Kairo, JG, Dahdouh-Guebas, F, Bosire, J. and Koedam, N. (2001). Restoration and management of mangrove systems - a lesson for and from the East African region. *South African Journal of Botany*, 67: 383-389.

[2] Iftekhar, MS (2008). Functions and development of reforested mangrove areas: A review. *International Journal of Biodiversity Science and Management* **4:** 1-14.

[3] Friess, DA (2016). Ecosystem services and disservices of mangrove forests: Insights from historical colonial observations, *Forests* 2016, *7*: 183; doi:10.3390/f7090183

[4] Romañach, SS, DeAngelis, DL, Koh, HL, Li, Y, *et al* (2018). Conservation and restoration of mangroves: Global status, perspectives, and prognosis. *Ocean and Coastal Management* 154: 72-82.)

[5] Serrano, D; Duarte, C; Lovelock, C; Lavery, P; Atwood, TB (2019). Australia's hidden opportunity to cut carbon emissions, and make money in the process. *The Conversation,* November.

[6] Duke, NC; Meynecke, JO; Dittmann, S; Ellison, AM; *et al* (2007). A world without mangroves? *Science.* **317** (5834): 41-42. doi:10.1126/science.317.5834.41b. PMID 17615322.

[7] Blaikie, P; Brookfield, H (1987). Land degradation and society. London: Methuen.

[8] Nindi, SJ, Machano, H, Rubens, J (2014). Remote sensing study of Rufiji Mangroves, Tanzania from 1990 to 2010. WWF Tanzania Country Office, Dar es Salaam.

[9] Armstrong, C (2016). Fairness, free-riding and rainforest protection. *Political Theory,* 44(1): 106-130.

[10] Havnevik, K (1993). Tanzania - The limits to development from above. Nordic Africa Institute and Mkuki na Nyota Publishers, Dar es Salaam.

[11] Mcleod, Elizabeth, Gail L. Chmura, Steven Bouillon, Rodney Salm, Mats Björk, Carlos M. Duarte, Catherine E. Lovelock, William H. Schlesinger, and Brian R. Silliman (2011). "A blueprint for blue carbon: toward an improved understanding of the role of vegetated coastal habitats in sequestering CO2." *Frontiers in Ecology and the Environment* 9 (10): 552-560.

[12] Alongi, DM (2002). Present state and future of the world's mangrove forests. *Environmental Conservation,* 29 (3): 331-349. doi:10.1017/s0376892902000231. ISSN 0376-8929.

[13] Cormier-Salem, M-C (2017). Mangrove grabbing: an exploration of changes in mangrove tenure from a political ecology perspective. H. Artaud & A. Surallés (eds). The sea within: maritime tenure and cosmopolitical debates, IWGIA,, pp.143-162. ffird-01824202ff

[14] Twilley, R; Rovai, A (2019). Mapping the world's 'blue carbon' hot spots in coastal mangrove forests. *The Conversation,* January 11.

[15] UNEP (2014). The Importance of mangroves to people: A call to action, edited by van Bochove, J., Sullivan, E., Nakamura, T. United Nations Environment Programme World Conservation Monitoring Centre, Cambridge. 128 pp.

[16] URT (2002). The Forest Act of 2002. Dar es Salaam: Govt Printer

[17] Monga, E; Mangora, MM; Mayunga, JS (2018). Mangrove cover change detection in the Rufiji Delta in Tanzania. *WIO Journal of Marine Science,* 17 (2): 1-10.

[18] Dahdouh-Guebas, F; Hettiarachchi, S; Lo Seen, D; Batelaan, O, *et al* (2005). Transitions in ancient inland freshwater resource management in Sri Lanka affect biota and human populations in and around coastal lagoons. *Current Biology* 15: 579-586.

[19] Turpie, JK (2000). The use and value of natural resources of the Rufji Floodplain and Delta, Tanzania. Technical Report No. 17. 98

[20] MEA [Millennium Ecosystem Assessment]. 2005. Ecosystems and human well-being: Synthesis. Washington, DC: Island Press.

[21] Jeffers, ES; Nogué, S; Willis, KJ (2015). The role of palaeoecological records in assessing ecosystem services. *Quat. Sci. Rev.* 112, 17-32.

[22] Samuelson, L, Bengtsson, K, Celander, T, Johansson, O, Jägrud, L, *et al.* (2015). Water, forests, people – building resilient landscapes. Report Nr. 36. SIWI, Stockholm.

[23] Comberti, C, Thornton, TF, Wyllie de Echeverria, V, Patterson, T (2015). Ecosystem services or services to ecosystems? Valuing cultivation and reciprocal relationships between humans and ecosystems. *Global Environmental Change,* 34 247-262.

[24] Chapin, FS III, Kofinas, GP; Folke, C (eds.) (2009). Principles of ecosystem stewardship: Resilience based natural resource management in a changing world. New York: Springer.

[25] Bakker, K (2005). Neo-liberalizing nature? Market environmentalism in water supply in England and Wales. *Annals of the Association of American Geographers,* 95(3): 542-565.

[26] Ngongolo, K, Mtoka, S, and Mahulu, A (2015). Challenges and opportunities for restoring the threatened mangroves, *Journal of Scientific Research & Reports* 5(5): 352-360; Article no.JSRR.2015.102.

[27] Kull CA, Arnauld de Sartre, X, Castro-Larrañaga, M (2015). The political ecology of ecosystem services. *Geoforum* 61: 122-134 (DOI: 10.1016/j.geoforum.2015.03.004)

[28] Mangora, MM (2011) Poverty and institutional management stand-off: a restoration and conservation dilemma for mangrove forests in Tanzania. *Wetlands Ecology and Management* 19 (6): 533-541.

[29] Mwansasu, S (2016). Causes and perceptions of environmental change in the mangroves of Rufiji Delta, Tanzania. Implications for sustainable livelihood and conservation. PhD Thesis, Stockholm University.

[30] Ochieng, CA (2002). Research Master Plan for the Rufiji floodplain and delta, 2003-2013. Rufiji Environment Management Project, Technical Report 28.

[31] Wagner, GM; Sallema-Mtui, R (2016). The Rufiji estuary: Climate change, anthropogenic pressures, vulnerability assessment and adaptive management strategies. In: S Diop, P Scheren, and JF Machiwa, (Eds.), Estuaries: A lifeline of ecosystem services in the Western Indian Ocean, Estuaries of the World. doi 10.1007/978-3-319-25370-1_12

[32] Peter, L (2013). Assessment of the status of mangrove vegetation and their degradation in Rufiji Delta in Tanzania. MSc Thesis, University of Ghana.

[33] Semesi AK (1992). Developing management plans for the mangrove forest reserves of mainland Tanzania. *Hydrobiologia* 247: 1-10.

[34] Rweyemamu, C (1998). Advocacy on mangroves restoration: Tanzania case study. Paper presented at the

Eastern Africa Regional Workshop on Mangroves and Shrimp Aquaculture, in Mombasa, Kenya from February 3-6, 1998.

[35] Duvail, S; Hamerlynck, O (2007). The Rufiji River flood: plague or blessing? *International Journal of Biometeorology* 52:33-42. DOI 10.1007/s00484-007-0105-8

[36] Pickaver, A; Sadacharan, D. (Eds) (2007).The benefits of inter-linking coastal and river management: Twenty case studies world-wide indicate opportunities and constraints. *Coastline Reports* 10. EUCC - The Coastal Union, Leiden.

[37] Smith, CDM (2016). Environmental flows in Rufiji River Basin assessed from the perspective of planned development in the Kilombero and Lower Rufiji Sub-Basins. Report to the United States Agency for International Development. Available for download at: https://dec.usaid.gov/dec/content/search.aspx .

[38] World Bank (2019). Tanzania: Country environmental analysis – Environmental trends and threats, and pathways to improved sustainability. Washington, DC: World Bank.

[39] Lewis, RR (2005). Ecological engineering for successful management and restoration of mangrove forests. *Ecological Engineering,* **24** (4): 403-418. doi:10.1016/j.ecoleng.2004.10.003.

[40] Bosire, JO *et al*. (2008). Functionality of restored mangroves: A review. *Aquatic Botany* **89**, 251-259.

[41] URT (1998). National Forest Policy of 1998. Dar es Salaam: Govt Printer.

[42] Lowe, C (2000). Global markets, local injustice in Southeast Asian Seas: the live fish trade and local fishers in the Togean Islands of Sulawesi. In: Zerner, C. (Ed.), People, Plants, and Justice: the Politics of Nature Conservation. Columbia University Press, New York, pp. 234-258.

[43] Wang, Y, Bonynge, G, Nugranad, J, Traber, M, *et al* (2003). Remote sensing of mangrove change along the Tanzania coast. *Marine Geodesy* 26: 35-48.

[44] Muganda, A. (2004). Tanzania's economic reforms and lessons learned. In Proceedings of the Scaling up Poverty Reduction: A Global Learning Process and Conference, Shanghai, China, 25-27 May.

[45] Thompson, BS (2018). The political ecology of mangrove forest restoration in Thailand: Institutional arrangements and power dynamics. *Land Use Policy*, 78:503-514. https://doi.org/10.1016/j.landusepol.2018.07.016

[46] Beymer-Farris, BA; Bassett, TJ (2012). The REDD menace: Resurgent protectionism in Tanzania's mangrove forests. *Global Environmental Change* 22: 332-341.

[47] Beymer-Farris, BA, Bryceson, I; Peter, CM (2019). The human rights dimensions of conservation and climate change initiatives in Coastal Tanzania: Examples of villagers' successful struggles for their rights, in: Yanda, PZ, I Bryceson, H Mwevura and CG Mung'ong'o (Eds). Climate change and coastal resources in Tanzania. Studies in socio-ecological systems' vulnerability, resilience and governance. London: Springer Climate; pp. 169-203; ISBN 978-3-030-04896-9.

[48] Temmerman, S; Meire, P; Bouma, TJ; Herman, PMJ; *et al* (2013). Ecosystem-based coastal defence in the face of global change. *Nature*, 504: 79-83.

[49] Mshale, B; Senga, M; Mwangi, E (2017). Governing mangroves: Unique challenges for managing Tanzania's coastal forests. Bogor, Indonesia: CIFOR; Washington, DC: USAID

Tenure and Global Climate Change Program.

[50] Sunseri, T. (2007). Fueling the city: Dar es Salaam and the evolution of colonial forestry, 1892-1960. In: Brennan, JR, Burton, A, Lawi, Y (eds). History from an emerging African Metropolis. Dar es Salaam. Mkuki na Nyota, pp 13-75.

[51] Kulindwa, K, Sosovele, H; Mgaya, Y (2001). Socio-economic dimensions of biodiversity loss in Tanzania. Dar es Salaam: Dar es Salaam University Press.

[52] Kihiyo, VBM (1998). Forest policy changes in Tanzania: towards community participation in forest management. The World Bank/WBI s CBNRM Initiative, February

[53] Moshy, VH; Bryceson, I; Mwaipopo, R (2015). Social ecological changes, livelihoods and resilience among fishing communities in Mafia Island Marine Park, Tanzania. *Forum for Development Studies*, *42*, 529-553. doi:10.1080/080394 10.2015.1065906

[54] Burgess, ND, Mwakalila, S, Munishi, P *et al.* (2010). Getting ready for REDD+ in Tanzania: a case study of progress and challenges. *Oryx,* 44(3) 339-351.

[55] Blomley, T; Iddi, S (2009). Participatory forest management in Tanzania: 1993-2009. Lessons learned and experiences to date. Division of Forestry and Beekeeping, Ministry of Natural Resources and Tourism, United Republic of Tanzania, September.

[56] Adhikari B, Baig SP, Iftikhar UA (2010). The use and management of mangrove ecosystems in Pakistan. *Journal of Environmental Development,* 19(4):446-467.

[57] Katikiro R, Macusi E, Ashoka Deepananda KHM (2013) Changes in fisheries and social dynamics in Tanzanian coastal fshing communities. *West Indian Ocean J Mar Sci* 12(2):95-110

[58] Patil V, Singh A, Naik N, Seema U and Sawant B (2012). Carbon sequestration in mangroves ecosystems. *Journal of Environmental Research and Development* 7(1A), July–September.

[59] Büscher, B; Sullivan, S; Neves, K; Igoe, J; Brockington, D (2014). Towards a synthesized critique of neoliberal biodiversity conservation. *Capitalism Nature Socialism* 23 (2), pp. 4-30. ISSN 15483-290.

[60] Kideghesho JR; Rija AA; Mwamende KA; Selemani IS (2013) Emerging issues and challenges in conservation of biodiversity in the rangelands of Tanzania. *Nature Conservation* 6: 1-29. doi: 10.3897/natureconservation.6.5407

[61] Keane, A; Lund, JF; Bluwstein, J; Burgess, ND *et al* (2019). Impact of Tanzania's Wildlife Management Areas on household wealth. *Nature Sustainability*.

[62] Lee, DE (2018) **Evaluating conservation effectiveness in a Tanzanian community wildlife management area.** *The Journal of Wildlife Management*; DOI: 10.1002/jwmg.21549

[63] Brockington, D; Duffy, R (2010). Capitalism and conservation: The production and reproduction of biodiversity conservation. *Antipode* 42(3):469-484; May DOI: 10.1111/j.1467-8330.2010.00760.x

[64] Mariki, SB (2016) Commercialization of nature: Can market-based mechanisms deliver positive conservation and development outcomes? *Open Journal of Social Sciences*, **4**, 61-69.http://dx.doi.org/10.4236/jss.2016.46007

[65] Scheba, A (2018) Market-based conservation for better livelihoods? The promises and fallacies of REDD+ in Tanzania. *Land,* 7, 119; doi:10.3390/land7040119

[66] Sklair, L (2001). The transnational capitalist class. Oxford: Blackwells.

[67] Spalding, M; Kainuma, M; Collins, L (2010). World atlas of mangroves. London, UK: Washington, DC: Earthscan.

[68] Costanza R, de Groot R, Sutton P, van der Ploeg S, *et al.* (2014). Changes in the global value of ecosystem services. *Global Environmental Change* 26: 152-158.

[69] Young, TR (2006). The legal framework for MPAs - successes and failures in their incorporation into national legislation. FAO Fisheries Report No 825.

[70] Kuboja, BN (2013). Policies and legal frameworks for Marine Protected Areas governance in Tanzania mainland: their potential and limitations for achieving conservation and livelihood goals. United Nations-Nippon Foundation of Japan Fellowship Programme, December.

[71] Sheriff, A (1987). Slaves, spices & ivory in Zanzibar, Eastern African studies. James Currey (Oxford), EAEP (Nairobi), Mkuki na Nyota (Dar Es Salaam). Ohio University Press, Atenas.

[72] Hardin, G. (1968). The tragedy of the commons. *Science* 162: 1243-1247.

[73] McCay, B; Acheson, JM (eds) (1987). The question of the commons: the culture and ecology of communal resources. University of Arizona Press, Tucson, Arizona.

[74] Fuentes-George, K 2013. Diversity: How problematizing the commoditization of nature affects regime effectiveness, *Global Environmental Politics* November. DOI: 10.1162/GLEP_a_00202

[75] Blanco, G; Córdoba, V; Baldi, R; Fernández, M; Santalla, E (2016) Outcomes of the clean development mechanism in Argentina. *American Journal of Climate Change,* 5, 431-447. http://dx.doi.org/10.4236/ajcc.2016.54032

[76] Igoe, J; Brockington, D (2007) Neo-liberal conservation: A brief introduction. *Conservation and Society*, 5, 432-449.

[77] Schuppert, F (2016). Carbon sink conservation and global justice: Benefiting, free riding and non-compliance. *Res Publica*, *22*, 99-116. https://doi.org/10.1007/s11158-015-9314-x

[78] Huff, A; Tonui, C (2017). Making 'mangroves together': Carbon, conservation and co-management in Gazi Bay, Kenya, STEPS Working Paper 95, Brighton: STEPS Centre.

[79] Osborne, T, Bellante, L, vonHedemann, N (2014). Indigenous peoples and REDD+: A critical perspective. Indigenous People's Biocultural Climate Change Assessment Initiative. IPCCA and Public Political Ecology Lab, University of Arizona.

Chapter 8

Conservation Management of Planted Mangroves through Evaluating Ecosystem Services in Baros Village Bantul Regency, Indonesia

Djumanto

Abstract

Planting mangrove trees on sandy land in Baros village into forest conservation has many ecological, economic, social, and tourism benefits for the surrounding community. The mangrove conservation in Baros village is artificial conservation managed by the men and women Baros youth. The coastal area of Baros village is often affected by tidal flooding, which causes losses to agriculture, fisheries, and livestock. In the early 2000s, an NGO assisted at research sites in mangrove restoration in a mangrove restoration area in the lagoon of Baros village to prevent abrasion and sea intrusion and protect agricultural areas. Restored mangroves can grow well to bring ecological, biological, economic, and social benefits. The local government of Bantul has designated the Baros mangrove forest as a reserve of a coastal park conservation area. The existence of the tree is beneficial ecological, biological, economic, and social. The Baros village youth group made various efforts to increase mangrove trees' area so that their benefits were sustainable. Managers and the village government and tour guides are expected to accommodate the existence of culture and local wisdom. Also, increasing community participation, fisheries, agriculture, and animal husbandry activities can provide socio-economic benefits for the community and the wider community's welfare.

Keywords: community, diversity, ecotourism, youth group, Yogyakarta

1. Introduction

Yogyakarta is one of the fourth most considerable cities in Indonesia by population, located in the southern part of the Java island. It became the capital of the Republic of Indonesia in January 1946–August 1950. It became a Special Province of Yogyakarta because of its enormous role in the founding of the Republic of Indonesia. It is a province that has many nicknames, namely as a student city, cultural city, tourist city, culinary city, artist city, batik city, palace-city, traditional market city, warm city, bakpia city, and so on.

Yogyakarta has a variety of tourist and cultural attractions that are very interesting to be visited and enjoyed by local and foreign tourists. Tourist attractions in Yogyakarta are quite complete, ranging from ecotourism, natural attractions such as beaches and mountains, historical tourism, religious tourism, shopping tourism, culinary tourism, and cultural tourism. Therefore, Yogyakarta ranks second as a tourist destination after Bali Island. Besides having full tourism potential, it also has several beautiful and challenging natural environments. One of them is the Merapi-Kaliurang lava tour, the off-road jeep Merapi lava tour, which is also an attraction for tourists to visit and try it. Some of these attractions include an excellent warm air climate, stunning views of the sandbanks, a culture that is still profoundly engraved with its people, and the friendliness of its citizens [1].

Bantul is a regency city located on the southern side of the Special Province of Yogyakarta. It has a 16.8 km long beach, which is very interesting as a tourist spot and various other activities. One of the tourist attractions is the sandbank or dune along the south coast with a width of 1–2 km. Residents partly use the dune area as a sultan ground for agriculture, animal husbandry, fisheries, tourist areas, and conservation areas. Local people also plant with evergreen shrimp trees (*Casuarina equisetifolia*) for a protector of crops and shade for visitors or tourists. Also, some areas of the southern coast of Bantul serve as turtle landing areas for laying eggs. Species of turtles that often land are hawksbill turtles (*Eretmochelys imbricata*), curved turtles (*Lepidochelys olivacea*), green turtles (*Chelonia mydas*) and leatherback turtles (*Dermochelys coriacea*). They use Pelangi Beach Depok, Samas Beach, Goa Cemara Beach, and Baru Pandansimo Beach as landing points for laying eggs. Furthermore, some people who are members of nongovernmental organizations carry out turtle conservation by saving and hatching to be released into the sea after they are old enough [2].

Bantul beach also has a natural attraction with the presence of mangrove forest conservation areas in the village of Baros. Administratively, the mangrove forest located in Baros village, Kretek sub-district, Bantul regency. The Baros Mangrove Conservation Area situated between Depok beach and Samas Beach in Yogyakarta. The mangrove forest is a conservation area that was initiated by the family of Baros young women (KP2B) since 2003. The Baros coastal area is planted with several types of mangroves, *namely Avicennia lanata, Avicennia officinalis, Nypa fruticans, Rhizophora mucronata, Rhizophora stylosa*, and *Sonneratia caseolaris*. The plants are used for area conservation due to natural disasters, such as coastal abrasion, crop failure, drought, intrusion, tsunami threats, and high salt levels that make crops wither. Natural hazards that routinely hit Baros beach have made people aware of cheap and durable natural protection. The position of Baros beach, which is at the mouth of the Opak river, causes various types of waste to be stranded on the beach, for example, wood debris. The community processes wood debris waste into high-value decoration. Besides, the Baros youth family collaborated with NGOs of Relung initiatives to carried out mangrove planting and various other coastal plants. With the existence of environmental conservation activities in Baros Beach, Baros's youth family has increasingly gained the community's trust in managing the environment of the Baros village sustainably [3].

Baros mangrove conservation ecosystem is a dynamic area that has biotic and abiotic resources. Biodiversity consists of birds, land animals, terrestrial plants, mangroves, aquatic biota, while abiotic natural source includes lagoons, sandy beaches, river mouths. The area behind the mangrove is for agriculture, animal husbandry, and fisheries. Various components interact and are interdependent with each other, which has great potential in supporting the community's economy, so it must be managed optimally and sustainably [3]. The Baros mangrove conservation area can have a positive impact on the surrounding community, including the following: (a) improve the economy of the surrounding community. (b) They are protecting rice fields from abrasion. (c) Protecting crops from the brunt of sea salt. (d) Opening job opportunities and new business opportunities. (e) Increasing public awareness to protect and preserve the environment, (f) Providing social, economic, and physical comfort. (g) Foster a sense of ownership of mangroves for community members. And (h) open networks with government, NGOs, universities, communities, and stakeholders to manage mangroves.

The conservation mangrove area of Baros is an artificial natural tourism area that has very diverse natural resource potentials, but its management is not yet optimal. It is necessary to study their resource potential, the obstacles faced by managers, and opportunities for their development. This paper aims to identify the benefits of the existence of mangroves, and opportunities to increase their profits and sustainable management strategies while still considering their natural protective and coastal functions. Besides, it is as a reference for policymakers. It can add insight into the importance of mangrove forests as a natural tourist attraction as a place of learning to foster public awareness of the sustainability of mangroves.

This paper was written based on field observations, interviews with the KP2B management, and Baros village officials, research results, and based on literature studies from various sources. Data or information from resource persons is analyzed descriptively to get an understanding of the level of mangrove utilization for the Baros village community and the role of the society in maintaining the sustainability of coastal ecosystems, as well as providing input for Baros mangrove conservation managers.

2. The benefits of Baros mangrove ecosystem

The mangrove forest ecosystem in Baros is an essential natural resource located on the southern coast of the Baros village and the west side of the Opak river estuary. The ecosystem has four main functions, namely, physical, biological, economic, and social services. The physical service is as a windbreak, filtering pollutants coming from upstream rivers and the sea, anchoring waves and rising tides, flood protection, waste repellent, and preventing seawater intrusion to land. The biological function is as a place for the parent and larvae of aquatic biota that is to spawn (spawning ground), nursery ground, and as a place to find food (feeding ground) for fish and other marine biotas. The direct economic function is as a producer of wood for building materials and industrial raw materials, food and medicines, animal feed sources, a place for grazing and raising poultry. The indirect economic function is for tourist attractions, places of education, research, its existence can protect crops. Also, mangroves become primary producers, forming stable micro-ecosystems between marine and terrestrial ecosystems. The social function is the existence of mangroves capable of bridging the formation of social groups of Baros youth families, breeder groups, farmer groups, marketing women's groups [4]. The existence of mangroves on the Baros coast can provide ecosystem services directly and indirectly. For example, direct services filter out dust and salt vapors, provide oxygen, provide shade for visitors, and protect beaches. For example, indirect ecosystem services preserve agricultural, livestock, and fisheries areas, and prevent seawater intrusion.

The existence of mangroves is a characteristic of the coastal area, however the Opak estuary and the Bantul coast until 2003 were in the form of a very dry and unproductive dune. Furthermore, starting in 2003, some university students in Yogyakarta, NGOs, and the local community planted several species of mangrove trees. Since then, the mangrove area has slowly increased. Mangrove tree seedlings transported from the city of Cilacap, which was 170 km from Bantul. The Species of mangrove seedlings planted were *Rhizophora* sp., *Bruguera* sp., *Nypa* sp., *Soneratia* sp.

The Regent of Bantul in 2014 issued Decree No. 284 of 2014 concerning the Reserve of Conservation Areas for Coastal Parks in Bantul Regency [5]. The Baros mangrove forest area is designated as a conservation area with a total area of 132 hectares divided into three zones, namely the core zone (10 ha), transition zone (94 ha), and buffer zone (28 ha). The Baros mangrove forest's current area is still less than 10 hectares, so a gradual expansion is needed. The development of the mangrove forest area is carried out by planting mangrove seedlings that are propagated in locations or transported from other cities (**Figure 1**).

2.1 The physical existence benefits

The existence of mangroves on the Baros beach will provide many benefits to the surrounding ecosystem directly or indirectly. The direct benefits of the mangrove

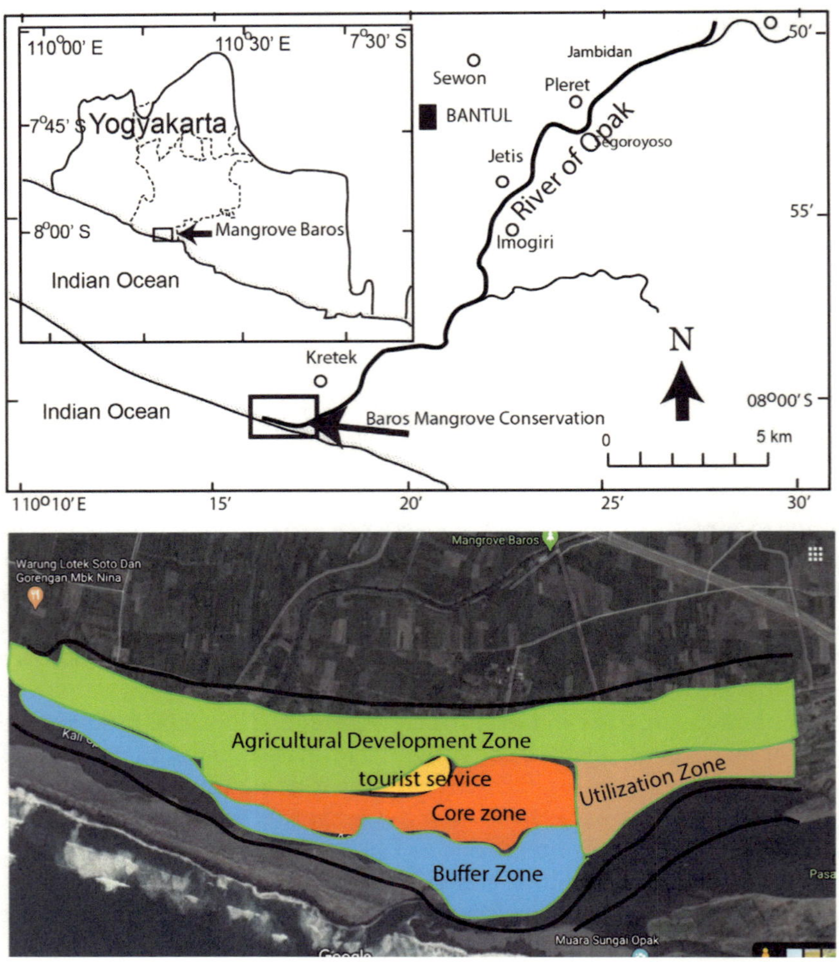

Figure 1.
The map shows the reserving area of coastal garden conservation area in Bantul regency. The zone consists of a 10-ha core zone, a 28-ha buffer zone, and transition 94-ha zone.

ecosystem as a habitat, as a protection, and source of raw materials. Indirect benefits provide a fresh ecosystem, provide food, absorb carbon, and provide nutrients for aquatic organisms.

2.1.1 Mangrove as a habitat

Baros mangrove ecosystem becomes a habitat for various species of native or migratory birds, living temporarily or permanently. Mangroves provide space for birds to breed and maintain their chicks. Species of birds found before mangroves are only a few species, for example, Blekog Sawah (*Ardeola speciosa*), rice field finches (*Lonchura punctulata*), finches (*Pycnonotus aurigaster*), etc. Blekog sawah comes during the growing season, while rice field finches come just before the rice harvest. After the mangrove area planted, various types of birds stopped by permanently or in-migration. Every rice planting season, several species of birds will migrate alternately [5].

The presence of birds in the Baros mangrove area can be an indicator of functional space. The number of bird species increases due to the availability of material to make nests, and there is no disruption to birds. Water birds are the dominant bird species in the mangrove area because mangroves provide habitat for feeding, breeding, and growing chicks. The species of birds that found are Java bondol (*Lonchura leucogastroides*), Peking bondol (*Lonchura punctulata*), and beach trinil (*Trunga hypoleucos*). The bird is active in mangrove trees of *Avicennia alba* and *Rhizophora apiculata*. The number of bird species in the Baros mangrove found as many as 21 families consisting of 48 species [5].

2.1.2 Mangroves as coastal protection

The existence of mangroves can protect the coastline, crops, and groundwater. (1) A robust mangrove root system helps to form a natural barrier against storm surges and floods [6], trapping river and terrestrial sediments [7], thus protecting coastline areas and slowing erosion. (2) The presence of mangrove plants can reduce and affect airflow. The sea breeze that blows hard will decrease its speed after passing mangrove plants. Strong winds that blow directly can break the branches and leaves of vegetable crops. Also, high salt levels contained in the sea breeze will be absorbed by mangrove leaves so that the salt content also reduced [8]. Mangrove plants are resistant to salt, while crops are very vulnerable. Sea breeze with high salt content can kill crops. Salt carried by the sea breeze will stick to vegetables. Salt is absorbing water so that salt attached to plants will absorb plant water, which can cause plants to wither. (3) The existence of mangroves can withstand the flow of seawater through a healthy root system so that saltwater does not intrude on agricultural land. The farm area behind the mangrove is protected and can be used for farming. Farmers can plant crops only once per year before there are mangroves, but now thanks to mangroves, agriculture can be done 2–3 times per year [4].

2.1.3 Mangrove as source of raw materials

Coastal communities use plants in mangrove ecosystems for various purposes, namely as medicinal materials, firewood, animal feed, building materials, decoration materials. (1) Daruju or Mountain Thistle (*Acanthus montanus*) is a bush that grows in the Baros mangrove ecosystem (**Figure 2**). This plant has a unique shape, large-toothed leaf edges, and sharp edges. The leaves are shiny green and slightly yellowish. People use daruju seeds for stomach worming and cough

Figure 2.
(A) The community uses the leaves and seeds of daruju (Acanthus montanus) as a traditional and herbal medicine for several types of diseases. (B) The daruju that grows on the edge of the lagoon to form thickets as a hiding place for fish larvae.

medicines. Also, shrimp farmers use daruju leaves to treat white disease in their shrimp. (2) Baros residents easily find a lot of dry wood scattered on the beach. The wood comes from the headwaters or the sea carried by the waves. Timber from the headwaters of the river is for firewood, while wood from the sea becomes driftwood. Driftwood is wood that has long been adrift in the sea. Driftwood has a definite, dry, lightweight, and unique shape, so it is suitable for decoration raw materials. They separate wood into driftwood and firewood. Good quality driftwood is used as raw material for handicrafts, while low quality is used as firewood. Conservation area managers forbid anyone from collecting mangroves for firewood. Firewood and driftwood are sold to collectors, making it a household income. (3) Mangrove leaves and grass contain a lot of protein so that it can be used as animal feed ingredients. *Avicennia* sp. can be used for animal feed. In addition, grass that grows around mangroves is also collected as animal feed. Cattle farmers often have difficulty in getting forage feed, especially in the dry season. The presence of grass in the Baros mangrove ecosystem can be an alternative source of animal feed. In Baros, there is a group of breeders with around 35 cows, some of which have green grasslands in the form of grass from the Baros mangrove area. (4) Mangrove wood has a waterproof and robust character, so it is suitable for building materials or ship materials. Large mangrove stems can be used as household building materials. Mangroves that are already high enough can be used as blocks for housing construction [9].

2.1.4 Providing fresh air

Mangrove in Baros is in excellent condition with the color of fresh green leaves, and no damaged mangroves are found. The distance between trees is less than two meters, which is classified as dense and has a high enough biomass potential. A healthy tree has healthy chlorophyll so that during photosynthesis, it can produce a lot of oxygen. Mangrove forests can produce oxygen (O_2), which is essential for life. The existence of mangrove forests in Baros becomes so important and attracts people to visit [9]. Every month there are around 700–1000 people visit the Baros mangrove forest to feel healthy ecosystems and fresh air.

2.1.5 Providing food

Mangrove ecosystems can support food security by providing food that is an aquatic biota that lives in the mangrove ecosystem, and food that is processed using mangrove plant parts. Mangroves can produce substantial amounts of organic material as food. The use of mangroves as a food source can improve family nutrition.

The existence of mangroves becomes very important in the life cycle of various types of fish, shrimp, shellfish, mollusks, and other aquatic organisms. Many kinds of shrimps and crabs that have economic value utilize mangroves as nurseries. Many species live in mangroves or visit temporarily in young stadia using mangroves to find food. The existence of good mangrove forests will support the life and production of marine biota that can be used as a source of food for the community [5].

2.1.6 Absorption carbon

During photosynthesis, mangroves absorb CO_2 gas from the atmosphere and convert it into organic carbon, then stored as carbon biomass in roots, stems, and leaves. The amount of CO_2 used by mangroves has a positive relationship with the total amount of biomass. CO_2 gas that is absorbed is getting more significant when the mangrove plants are getting denser. The results of photosynthesis are used for horizontal and vertical growth. The larger the diameter of the tree, the higher the CO_2 gas absorbed by the tree to be converted into organic material. Mangrove forests store more carbon than most tropical rain forests [10].

2.1.7 Providing nutrients

Mangrove provides organic material and nutrients through the production of litter. Litter production is an important part of the transfer of organic matter from vegetation to the soil. The nutrients produced from litter are essential in the growth of mangroves and as a source of detritus for estuary ecosystems in supporting the lives of various aquatic organisms. Mangrove litter productivity is a source for fishery's productivity in estuaries and contributors of nutrients to the surrounding waters. It makes mangroves play an important role as a nutrient cycle chain for aquatic organisms. Mangrove plants are a potential food source, in various forms, for all the biota that live in the mangrove ecosystem. The basic component of the food chain in the mangrove ecosystem comes from mangrove plants that fall into the litter. Mangrove leaves provide many benefits to the surrounding organisms. Mangrove leaves contribute 81% of the total litter production, while the rest comes from twigs and plant reproductive organs [11]. As much as 5% of the total leaf production is consumed directly by herbivorous organisms, while 95% enters the aquatic environment as detritus. Mangrove dominated by Rhizophora sp. with a high density can produce an average of a litter of 4.05 $g/m^2/day$. Nutrient potential production from litter produced is $C = 0.35$, $N = 0.009$ and $P = 0.0008$ $g/m^2/day$ [12].

2.2 Biological function

One of the essential functions of mangrove forests is biological functions, namely, their ability to support life around them. The biological function that is carried out by the Baros mangrove forest is as a spawning ground, nursery ground, and feeding ground for aquatic biota, as a natural habitat and to provide food. Some fish species that migrate or settle in the mangrove ecosystem may spawn, care for their offspring, take refuge, or find prey [13].

Species of fish inhabitants of the Baros mangrove ecosystem based on their ability to adapt to salinity are grouped into three, namely the original inhabitants of freshwater, brackish water, and marine species. The number of fish species found in the mangrove ecosystem is 36 species consisting of 3 brackish water species, 11 freshwater species, and 22 marine species [12].

2.2.1 Spawning ground

The Baros mangrove ecosystem experiences changes in salinity that occur every day. Extreme changes in physical parameters affect the number of fish species that can adapt to brackish water. Brackish water species are species that all stages of life exist in brackish waters. Brackish water species are the least amount compared to seawater or freshwater species. This species uses brackish or mangrove water ecosystems to breed, care for, and raise their offspring [3]. This species is most vulnerable to extreme physical and chemical changes in brackish waters. Efforts to improve mangrove ecosystems can save the survival of brackish water species. Brackish water species found in the Baros ecosystem are *Acentrogobius caninus*, *Ambassis interrupta* and *Glossogobius giuris*.

2.2.2 Nursery ground

Freshwater and marine fish species found in the mangrove ecosystem are mostly in the larval stage. Eggs that hatch into larvae spawned in the sea or upstream of the river are carried by currents to reach the mangrove ecosystem by utilizing tidal or river flow. The waters of the mangrove ecosystem provide habitat for larvae protection. Besides, the mangrove ecosystem is very fertile so that it can provide food for aquatic biota larvae stage. The proportion of larvae of marine species is more than that of freshwater species. The Baros mangrove ecosystem is essential as a nursery ground for most economically important marine fish, such as snapper, mugil, and sardine [3].

2.2.3 Feeding ground

The number of species and individual fish in the juvenile stage is higher than in the young fish stage. Omnivorous and detritivores groups dominate fish in the immature stage. Mangroves, as first-rate producers, can produce large amounts of detritus from leaves and twigs. Mangrove ecosystems can provide organic nutrients that fertilize brackish water, the more mangrove trees, the higher leaf litter production. The number of leaves that fall will produce more detritus. The more leaf waste is produced, the more food is available. Mangrove ecosystems and health conditions affect the capacity of young fish to feed in mangroves. The healthier mangrove ecosystems, the more able to accommodate fish biomass [3].

Season and water conditions affect the abundance of juveniles in mangrove ecosystems. The number of biomass and young fish species is found more in the rainy season. The amount of available nutrients is more abundant in the rainy season. Likewise, most fish species spawn in the rainy season. The abundance of juvenile fish is also affected by the lunar cycle. In the full moon and new moon, the juvenile number is less.

2.3 Economic function

Baros mangrove ecosystem has an economic function or economic value. The commercial service is the role value of mangrove forests in creating employment for the community, so they have income. Economic value shows the role of conservation of mangrove resources in producing activities that can be valued. Economic value includes the amount of utilization and nonvalue utilization of resources. Utilization value consists of the direct amount of use, the indirect cost of use, and the value of choice. Nonutilization value includes the value of existence.

Direct use value is the value obtained from the production of goods that is resulting from the direct utilization of resources. The direct use value includes the use of wood and the capture of fishery commodities in the mangrove area. Indirect use-value is the value assigned to the benefit of resources in the environment. Indirect use values, for example, the cost of mangrove areas as a barrier to seawater intrusion, and the amount of a wall as waves, or other environmental benefits felt by the local community. The choice value is the benefit value of a resource that is stored or maintained for future use. The choice value is the value of the biodiversity of mangrove flora and fauna. The existence value is the value given by the community to resources for the various benefits [4].

The total economic benefit value of the Baros mangrove area is $ 14,846 US ha/year. The direct economic benefits value of $ 1738 US ha/year, and the indirect economic benefits value of $ 11,615 US ha/year. The optional benefits value about $ 15 ha/year. The existence value for $ 1478 US ha/year. The highest percentage of economic benefit value is the indirect benefit of the mangrove area at 78.2% [4].

2.4 Social function

Social service is the role of mangrove forests to improve the welfare of life or improve the social welfare problems of local communities. The social function of the Baros mangrove is closely related to the formation of the family of Baros young women (KP2B) in the 80s. KP2B members are all Baros villagers who are in first grade of high school. KP2B, in collaboration with NGO Relung in 2003, initiated the planting of mangrove trees in Baros Village. The purpose of planting mangrove trees is to protect the village from the threat of abrasion, intrusion, and tsunami and to protect plants from blowing sea breezes high in salt, which causes crops to wither.

The family of Baros young women empowers citizens to play an active role in improving their welfare by forming five working groups. Workgroup activities are related to mangrove conservation. The working groups formed are Avicennia working group, Mino Tirtohargo, Andini Lestari-Karya Manunggal, Mangrove Farmer Group Association—Women Farmer Group, and Processing and Marketing Group.

Each working group conducts a business that supports mangrove sustainability. They are namely (1) Avicennia working group does business in the field of processing marine driftwood waste. (2) The Mino Tirtohargo working group conducts fishing activities. (3) Andini Lestari-Karya Manunggal working group conducts business in cattle and duck farming. (4) A joint working group of farmers—women farmer groups carrying out agricultural activities. (5) The processing and marketing working group carries out learning activities and consumption services in the form of traditional food. The actions of each working group are as follows.

Avicennia's working group activity is an effort to process driftwood waste into handicrafts with artistic value. This effort is motivated by concerns about the amount of garbage in the Baros mangrove area. Garbage often covers or breaks young mangroves, thus disrupting mangrove conservation efforts. Driftwood is wood waste that has long been oscillated in the sea and then stranded in mangroves or beaches. Processing driftwood waste is an effort to reduce garbage and to increase income. Driftwood waste processed into handicrafts with high artistic value. The final product processed driftwood is a type of craft items in the form of wall displays, wall clocks, decorative lamps, tissue boxes, miniature Christmas trees.

The Mino Tirtohargo working group carried restocking of mangrove crab and fishing. The crabs find food, grow and breed in the mangrove area. Mangrove crabs can grow well and reproduce in areas with mangrove areas. Stocking crabs can increase the stock and catch of fishermen. Before doing restocking, fishermen get small crabs in the dry season only. After restocking and mangrove have grown well, fishermen can get crab catch every day. Restocking is also carried out by universities to increase fish stocks and fishermen's catches. The fish stocked is milkfish (*Chanos chanos*) in the year of 2014 at the amount of 20,000 individuals. Milkfish restocking aims to increase fish stocks as fish consumption catch and fishermen income. Milkfish restocking is done by considering that the Baros mangrove area is a suitable habitat for milkfish. Milkfish restocking can balance the population structure of the mangrove ecosystem.

The Andini Lestari working group conducts a cattle breeding business using group cages. The location of cattle farms is in an integrated livestock utilization zone adjacent to the agricultural land and the Baros mangrove area. Andini Lestari's working group has 35 units of cages with three types of cows, namely Simetal, Lemusin, and Java, which are managed in an integrated manner. Cattle are fed grass obtained in the Baros mangrove conservation area, to increase household income.

Manunggal working group works in the field of raising ducks using a cage. It is located in the ranch area to the north of the Baros mangrove forest. Breeders herd ducks during the day in the mangroves, so the ducks get natural food and nutrients. Ducks are herded into cages in the afternoon, so ducks lay eggs at night in pet cages. The eggs are then processed into high-quality salted eggs.

Working groups of Farmers and Women Farmers are active in agriculture. They grow agricultural commodities, which include vegetables, shallots, and rice. Each type of agrarian product has a different treatment and planting period, for example, vegetables 40–50 days, onions 50–60 days, and rice 90–110 days. Rice farming activities are carried out in groups such as plowing soil, planting rice seeds, cleaning weeds, and harvesting rice. The location of agricultural land is in an integrated agrarian zone. This location is behind the mangrove, so it is protected from the sea breeze [3]. The existence of mangroves can protect integrated agricultural areas from exposure to sea breezes and tidal floods.

The Processing and Marketing working group consists of mothers who provide lessons on traditional cooking and guest consumption services. The types of traditional cuisine taught are typical village dishes, namely Cucur, Adrem, Rempeyek Udang, Nasi Wiwit. Baros traditional food is served to support mangrove ecotourism activities. The process of making traditional food is part of the mangrove ecotourism material. Traditional cuisine guides are mothers of traditional cake craft craftsmen.

3. Problems encountered

Baros mangrove forest managers face several obstacles to maintain environmental and forest health. The management of the KP2B organization often faces many obstacles in managing conservation areas, but the most prominent is the management of waste and the quality of human resources. Constraints or problems in managing mangrove conservation areas are mostly from outside, and a small portion is from local locations. Some limitations and solutions are explained below.

3.1 Garbage disturbance

3.1.1 Rubbish characteristics

The Baros Mangrove Conservation Area is in the lagoon of the Opak river and south of the village of Baros. The headwaters of the Opak river are located on the slopes of Mount Merapi, stretching from north to south. This river has a flow length of about 65 km and an area of flow of about ±1398.18 km^2. The river flows through the city with a population density of around 1194 people/km^2, carrying garbage from settlements and agriculture, consisting of various types and shapes. The volume of waste that has accumulated in the Opak river estuary is increasing, making it a scourge for mangrove conservation area managers [14].

Garbage piles at the Opak river mouth come from residents along the riverbank and residents in the village of Baros. Garbage contribution from residents along the riverbanks is estimated at 1800 kg/day or 180 m^3/day, whereas the contribution of rubbish from Baros villagers is estimated at 1000 kg/day or 100 m^3/day. Total garbage accumulation in the river mouth is expected to reach 2800 kg/day or 280 m^3/day. Types of waste consist of plastic, rubber, Styrofoam, cans, wood, glass, cloth. The most dominant type of waste is plastic, then timber, and the least is cloth [15].

Garbage scattered along the coast of Baros originates from upstream rivers and debris carried by seawater. The volume of waste increases when river water overflows during high rainfall. Every rainy season arrives, household rubbish that settles at the bottom of the river then flows with the river flows toward the beach. Waste that flows directly into the sea will eventually be pushed by the waves back to the beach so that, in the end, the beach is full of rubbish. Every high rainfall and abundant river flow, the waste carried from the river to the sea will increase.

3.1.2 Garbage management

The main problem in dealing with garbage is to change the paradigm, behavior, and public awareness. The community views waste as something that is useless and has no value. Society considers waste as everything that is thrown away, rejected, ignored, unwanted material, or worthless. This mindset must be straightened out so that it considers garbage as a valuable object. Those who are involved in the work of utilizing waste are slogan "in your opinion, the things you have disposed of are rubbish, but for us, these items are a blessing for life." We must see waste as raw material or material that has economic value. Abundant waste can be processed into objects that have artistic value or sale value. Communities must be able to manage waste by recycling, reusing, purifying, or purifying [16].

Garbage that scattered along the coast of Baros consisting of organic and inorganic materials. Types of organic waste, such as used wood for household furniture, can be recycled into valuable items. It is improving by changing used materials into useful new articles and of economic value. Recycling can reduce the volume of waste, minimize the use of new materials, reduce the cost of producing goods, increase household income, and create new jobs. Recycling begins with the activity of sorting waste, grouping similar goods, processing into new products. The rest of the recycled material can be used for firewood so that all organic waste is used up. Production goods can be sold to tourists as souvenirs, furniture stores, or exported.

The most inorganic waste is plastic used food envelopes, beverages, and household needs. Plastic has many advantages compared to other materials, for example, cheap, durable, lightweight, resilient, strong, so that dependence on plastic is very high. Each person, on average, discards plastic as waste as much as 0.45 kg/day, while waste production is as much as 0.9 kg/day. Plastic waste reaches 20–25% of all types of inorganic waste. Plastic waste mostly causes the death of young mangroves by covering leaves or breaking stems. Also, seedling cannot grow if the space is occupied by plastic or other rubbish. Plastic waste management is done by reusing, recycling, and making other products. Reusing plastic waste, for example, used plastic bottles for sowing mangrove seeds, gallon bottles for vertical cultivation of vegetable crops. Plastic waste that is completely unused, it can be melted down to make new products, such as flower pots or other products [9].

3.2 Human resources quality

Human resources play an essential role in the successful management of the Baros mangrove conservation area. Conservation area managers need someone who has the determination and love for the environment, innovative and creative. Love of a healthy environment will make someone persistent in saving plants as the lungs of the world. The population of Baros village who completed education to graduate from high school was 42%, while those who graduated from college were 8.8% [4]. The learning average of Baros residents is junior high school graduates, so they are classified as common knowledge. Although the average learning level is moderate, they have high innovation and creativity. Skilled human resources are needed to realize the creation of an independent and sustainable mangrove conservation area. Creative human resources are required to face various challenges in managing mangrove areas.

A manager is a group of people consisting of the head of the Baros youth group, village heads, hamlet heads, and elders to manage the Baros mangrove conservation area. Compact managers can take advantage of mangrove forests optimally and sustainably. Mangrove managers need skilled human resources capable of creating healthy mangrove conservation areas. Improving the quality of HR education is very influential in the successful management of natural resources. Some ways to improve the quality of human resources can be through formal and informal learning. Also, human resource development can be achieved through skills training, certification, or competency testing. Residents can certify as managers of marine and fisheries conservation areas. National standardization professional bodies carry out certification and competency tests for those who need them for free.

4. Mangrove management strategies

The Bantul Regency Government has designated the Baros mangrove forest area as a coastal park conservation area. Conservation areas with reserve status indicate a minimum level of management. The government has not allocated funds to carry out activities and has not formed an organizational management unit following the provisions. Bantul has a mangrove forest conservation area of 132 ha, which is managed by the zoning system. The area of each zone is a 10-ha core zone, a restricted use zone of 28 ha, and the other zone is 94 ha. The next step is to make a management plan for the Coastal Park-Conservation Area in Bantul Regency [17]. The area management plan must be implemented to accommodate the interests of many parties while maintaining the minimum negative impact.

The development of conservation areas is directed to ecotourism activities based on mangrove conservation. Mangrove conservation is managed by continuing to plant mangroves until the plant covers all core zones. However, mangrove planting must consider the representation of mangrove plant species to produce high biodiversity of mangroves. High diversity and plant density can increase the heterogeneity of animals and aquatic biotics. At present, there are 25 species of mangrove plants, 48 species of birds, and 36 species of fish [4]. Mangroves that need to be propagated are red pedada (*Sonneratia caseolaris*) and perepat (*Sonneratia alba*). These plants quick increase, local plants, so they are adaptable, have many biological and ecological benefits. However, the local community has not been able to seed, and the success rate of plants is low. Conservation area managers as well researchers must collaborate with the government and other parties to increase the diversity of mangrove species via restoration.

The next step is to empower all levels of society to be actively involved in supporting the success of ecotourism. Farmers cultivate various types of plants under local conditions as an attraction. Farmers do livestock activities according to the rules, so it is interesting to be visited and carried out by visitors. Production of traditional food preparations is carried out following health standards, and produces attractive products, and is characteristic of the village of Baros [16].

The next step regulates the zonation of mangrove areas so that the available land use is following its purpose. The goal is that there is harmony between the mangrove area and other designation areas, such as agricultural, livestock, and fishery areas. The existence of mangroves can protect agrarian crops from tidal flooding, seawater intrusion, and a strong sea breeze. Also, mangroves are home to various species of birds that can help farmers cope with pest attacks. Conversely, agricultural activities can support the lives of birds by providing prey hunting grounds [18]. Managers need to invite experts to provide education to local communities about the life of birdlife, fish biota, and mangrove ecology, to create harmony in the mangrove ecosystem with its environment.

Sustainable mangrove management strategies by carrying out ecotourism activities that integrated with other businesses under their designation. Ecological tourism activities that combine various attractions by involving different sectors, namely forestry, agriculture, fisheries, and animal husbandry. The forestry sector presents a scene of observing the diversity of mangroves, the ecological role of mangroves as a buffer for the environment, the variety of bird species. Various attractions can be carried out in the core zone, for example, observing bird behavior, planting mangroves, arranging mangroves, and researching the diversity of aquatic biota. In the field of agriculture, it can present the attraction of farming activities in the utilization zone as well as the observation and introduction of crops. The fisheries sector can show the attractiveness of raising fish in ponds, spawning, breeding, harvesting fish. Fisheries activities in the utilization zone can provide fish restocking, crab restocking, and fishing activities. Animal husbandry activities can present attractions offering livestock activities, for example, feeding, harvesting livestock products [2].

Managers need to do promotions so that the number of tours increases. However, an increase in tourists must be followed by an increase in services. It is necessary to conduct a study of the capacity of tourist visits to achieve maximum capacity and satisfaction. Mangrove managers can collaborate with a competent institution to survey tourist carrying capacity. The results of the study can provide information on the time and number of visits, set the types of attractions, and information on facilities for tourists and tourism service managers. The results of the carrying capacity study were disseminated to stakeholders to get responses.

One of the critical jobs that must be done for conservation area managers is to collaborate with stakeholders that can have a positive impact. Besides, good governance can have a positive effect on management performance and increase revenue and generate economic activity in conservation areas and people's purchasing power. Guidance to the community so that an increase in income will have an impact on raising public awareness in supporting the preservation of regional resources.

5. Conclusions

Mangrove forests can directly act as coastal protection, storm barriers, strengthen coastal areas for sea wave disturbance, and control coastal erosion. Mangrove restoration measures can build resilience in coastal communities, reduce storm disruption, tidal flooding, and high wave disturbances, and protect surrounding agricultural, livestock, and fishery areas. Mangrove conservation areas can provide welfare and provide new jobs for the surrounding community, as a source of food, food, medicinal materials, and habitat for terrestrial and aquatic animals. Healthy mangrove areas can provide ecosystem services that can generate economic activities for the community and stakeholders.

The management strategy of Mangrove Baros is by implementing ecotourism, integrated with other companies that are mutually beneficial, arrangement of the tourist area so that it can attract visitors by providing tracks to explore the mangrove area, provision of complete public facilities, transportation, health, and safety facilities, and the addition of various attractions that are interactive with visitors. The attractions are adjusted to the interests of visitors.

Conservation managers need to accommodate research and development interests related to the area to support management activities. The use of conservation area resources considers all benefits and does not harm the inhabitants. Agricultural activities, fisheries, and livestock in conservation areas are managed to the maximum to provide socioeconomic benefits to the community. Collaboration with partners is crucial to create economic activities and increase stakeholder income.

Acknowledgements

The author would like to thank Mr. Dwi Ratmanto for providing information about the management of conservation areas in the Village of Baros. This paper is part of research activity on fish communities in the mangrove conservation area of Baros Village, Yogyakarta Special Province.

Conflict of interest

The authors declare no conflict of interest.

Author details

Djumanto
Aquatic Resources Management Laboratory, Department of Fisheries, Faculty of Agriculture, Gadjah Mada University, Indonesia

*Address all correspondence to: lely4192@yahoo.com

References

[1] Suparwoko W. Tourism development in Indonesia. In: Towards Responsible Tourism: A Community-Based Approach to Tourism Development in Indonesia. Saarbrücken, Jerman: Lambert Academic Publishing (LAP); 2012. pp. 62-68

[2] Budiantoro A. The sustainable ecotourism potential development with special reference to Oliveridley Sea turtle (*Lepidochelys olivacea*) along Bantul beaches, Indonesia. IOP Conference Series: Earth and Environmental Science. 2019;**391**:1-11. DOI: 10.1088/1755-1315/391/1/0120692

[3] Purwaningrum H. Development of mangrove forest ecotourism Baros village, Titihargo village, Kretek district, Bantul regency. Journal of Tourism and Economic. 2020;**3**(1):31-40

[4] Trialfhianty TI, Suadi D. Economic valuation of mangrove resource in baros coast Tirtohargo village sub-district of Kretek. Kawistara. 2014;**4**(2):111-224

[5] Qiptiyah M, Broto BW, Setiawan H. Bird's diversity in mangrove area of Rawa Aopa Watumohai National Park. Wallacea Forestry Research Journal. 2013;**2**(1):41-50

[6] Zhang K, Liu H, Li Y, Xu H, Shen J, Rhome J, et al. The role of mangroves in attenuating storm surges. Estuarine, Coastal and Shelf Science. 2012;**102-103**:11-23

[7] Victor S, Golbuu Y, Wolanski E, et al. Fine sediment trapping in two mangrove-fringed estuaries exposed to contrasting land-use intensity, Palau, Micronesia. Wetlands Ecology and Management. 2004;**12**:277-283. DOI: https://doi.org/10.1007/s11273-005-8319-1

[8] Takle ES, Chen TC, Wu X. Protection from wind and salt spray. In: Coastal Protection in the Aftermath of the Indian Ocean Tsunami: What Role for Forests and Trees? Proceedings of the Regional Technical Workshop; 28-31 August 2006; Khao Lak, Thailand. 2006. pp. 65-80

[9] Wahyuni I. Analysis of production and potentials of mangrove little hara elements in natural reserve dua Serang, Banten. Biodidaktika. 2016;**11**(2):66-76

[10] Donato D, Kauffman J, Murdiyarso D, et al. Mangroves among the most carbon-rich forests in the tropics. Nature Geoscience. 2011;**4**: 293-297. DOI: 10.1038/ngeo1123

[11] Rafael A, Calumpong HP. Comparison of litter production between natural and reforested mangrove areas in Central Philippines. AACL Bioflux. 2018;**11**(4):1399-1414

[12] Suryaperdana Y, Soewardi K, Mashar A. Relationship of mangrove environment to the shrimps and milkfish production in silvofishery area of Blanakan, Subang, West Java. Bonorowo Wetlands. 2012;**2**(2):74-85

[13] Mahyudin RP. Sustainable waste management strategy. Enviroscienteae. 2014;**10**(1):33-40

[14] Yusfi RN, Damanhuri TP. Study of solid waste characteristics and recycle potential in Cikapundung riverbanks. Journal of Environmental Engineering. 2012;**18**(2):155-166

[15] Kusumasari B, Alam Q. Bridging the gaps: The role of local government capability and the management of a natural disaster in Bantul, Indonesia. Natural Hazards. 2012;**60**:761-779. DOI: 10.1007/s11069-011-0016-1

[16] Jackson B, Sparks JLD, Brown C, Boyd DS. Understanding the co-occurrence of tree loss and modern slavery to improve efficacy

of conservation actions and policies. Conservation Science and Practice. 2020;**183**:1-13. DOI: 10.1111/csp2.183

[17] Anonymous. 2014. Decree of the Regent of Bantul, Number 284 in 2014, Concerning Reserving Coastal Area Conservation Area in Bantul District.

[18] Ferichani M, Prasetya DA. System of rice intensification increases rice productivity on saline soil. Paddy and Water Environment. 2017;**15**:649-657. DOI: 10.1007/s10333-017-0584-0